INTRODUCTORY FUNCTIONAL ANALYSIS WITH APPLICATIONS

Erwin Kreyszig
University of Windsor

JOHN WILEY & SONS
New York • Chichester • Brisbane • Toronto • Singapore

***Library of Congress Cataloging in Publication Data*:**

Kreyszig, Erwin.
Introductory functional analysis with applications.

Bibliography: p.
1. Functional analysis. I. Title.
QA320.K74 515'.7 77-2560

Printed in the Republic of Singapore

10 9

PREFACE

Purpose of the book. Functional analysis plays an increasing role in the applied sciences as well as in mathematics itself. Consequently, it becomes more and more desirable to introduce the student to the field at an early stage of study. This book is intended to familiarize the reader with the basic concepts, principles and methods of functional analysis and its applications.

Since a textbook should be written for the student, I have sought to bring basic parts of the field and related practical problems within the comfortable grasp of senior undergraduate students or beginning graduate students of mathematics and physics. I hope that graduate engineering students may also profit from the presentation.

Prerequisites. The book is elementary. A background in undergraduate mathematics, in particular, linear algebra and ordinary calculus, is sufficient as a prerequisite. Measure theory is neither assumed nor discussed. No knowledge in topology is required; the few considerations involving compactness are self-contained. Complex analysis is not needed, except in one of the later sections (Sec. 7.5), which is optional, so that it can easily be omitted. Further help is given in Appendix 1, which contains simple material for review and reference.

The book should therefore be accessible to a wide spectrum of students and may also facilitate the transition between linear algebra and advanced functional analysis.

Courses. The book is suitable for a one-semester course meeting five hours per week or for a two-semester course meeting three hours per week.

The book can also be utilized for shorter courses. In fact, chapters can be omitted without destroying the continuity or making the rest of the book a torso (for details see below). For instance:

Chapters 1 to 4 or 5 makes a very short course.

Chapters 1 to 4 and 7 is a course that includes spectral theory and other topics.

Content and arrangement. Figure 1 shows that the material has been organized into five major blocks.

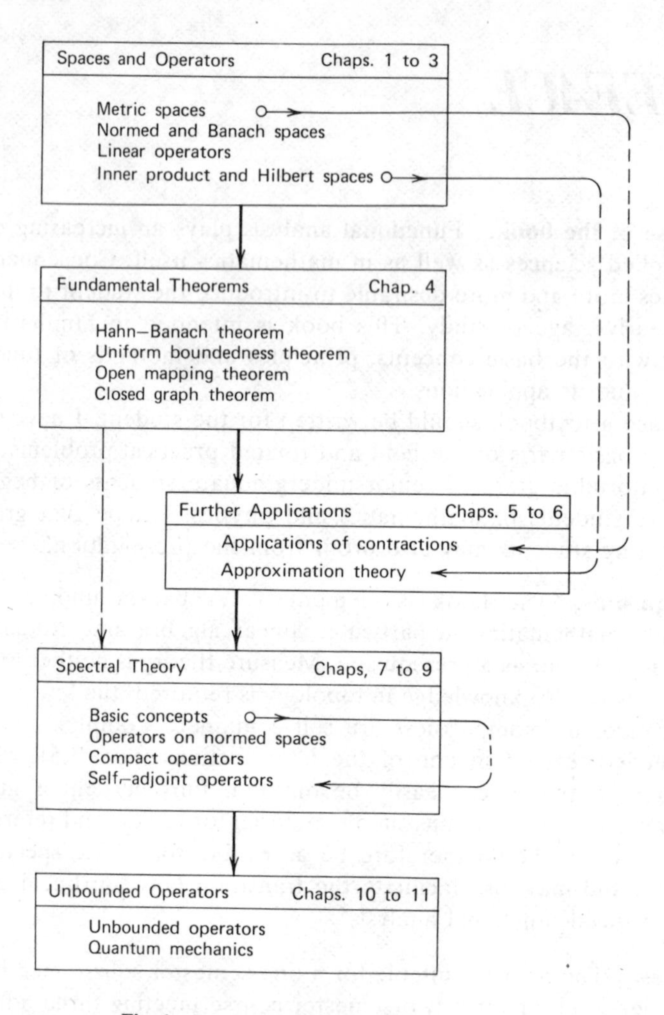

Fig. 1. Content and arrangement of material

Hilbert space theory (Chap. 3) precedes the basic theorems on normed and Banach spaces (Chap. 4) because it is simpler, contributes additional examples in Chap. 4 and, more important, gives the student a better feeling for the difficulties encountered in the transition from Hilbert spaces to general Banach spaces.

Chapters 5 and 6 can be omitted. Hence after Chap. 4 one can proceed directly to the remaining chapters (7 to 11).

Spectral theory is included in Chaps. 7 to 11. Here one has great flexibility. One may only consider Chap. 7 or Chaps. 7 and 8. Or one may focus on the basic concepts from Chap. 7 (Secs. 7.2. and 7.3) and then immediately move to Chap. 9, which deals with the spectral theory of bounded self-adjoint operators.

Applications are given at various places in the text. Chapters 5 and 6 are separate chapters on applications. They can be considered in sequence, or earlier if so desired (see Fig. 1):

Chapter 5 may be taken up immediately after Chap. 1.

Chapter 6 may be taken up immediately after Chap. 3.

Chapters 5 and 6 are optional since they are not used as a prerequisite in other chapters.

Chapter 11 is another separate chapter on applications; it deals with unbounded operators (in quantum physics), but is kept practically independent of Chap. 10.

Presentation. The material in this book has formed the basis of lecture courses and seminars for undergraduate and graduate students of mathematics, physics and engineering in this country, in Canada and in Europe. The presentation is detailed, particularly in the earlier chapters, in order to ease the way for the beginner. Less demanding proofs are often preferred over slightly shorter but more advanced ones.

In a book in which the concepts and methods are necessarily abstract, great attention should be paid to motivations. I tried to do so in the general discussion, also in carefully selecting a large number of suitable examples, which include many simple ones. I hope that this will help the student to realize that abstract concepts, ideas and techniques were often suggested by more concrete matter. The student should see that practical problems may serve as concrete models for illustrating the abstract theory, as objects for which the theory can yield concrete results and, moreover, as valuable sources of new ideas and methods in the further development of the theory.

Problems and solutions. The book contains more than 900 carefully selected problems. These are intended to help the reader in better understanding the text and developing skill and intuition in functional analysis and its applications. Some problems are very simple, to encourage the beginner. Answers to odd-numbered problems are given in Appendix 2. Actually, for many problems, Appendix 2 contains complete solutions.

The text of the book is self-contained, that is, proofs of theorems and lemmas in the text are given in the text, not in the problem set. Hence the development of the material does not depend on the problems and omission of some or all of them does not destroy the continuity of the presentation.

Reference material is included in Appendix 1, which contains some elementary facts about sets, mappings, families, etc.

References to literature consisting of books and papers are collected in Appendix 3, to help the reader in further study of the text material and some related topics. All the papers and most of the books are quoted in the text. A quotation consists of a name and a year. Here are two examples. "There are separable Banach spaces without Schauder bases; cf. P. Enflo (1973)." The reader will then find a corresponding paper listed in Appendix 3 under Enflo, P. (1973). "The theorem was generalized to complex vector spaces by H. F. Bohnenblust and A. Sobczyk (1938)." This indicates that Appendix 3 lists a paper by these authors which appeared in 1938.

Notations are explained in a list included after the table of contents.

Acknowledgments. I want to thank Professors Howard Anton (Drexel University), Helmut Florian (Technical University of Graz, Austria), Gordon E. Latta (University of Virginia), Hwang-Wen Pu (Texas A and M University), Paul V. Reichelderfer (Ohio University), Hanno Rund (University of Arizona), Donald Sherbert (University of Illinois) and Tim E. Traynor (University of Windsor) as well as many of my former and present students for helpful comments and constructive criticism.

I thank also John Wiley and Sons for their effective cooperation and great care in preparing this edition of the book.

ERWIN KREYSZIG

CONTENTS

NOTATIONS

In each line we give the number of the page on which the symbol is explained.

A^{C}	Complement of a set A 18, 609
A^{T}	Transpose of a matrix A 113
$B[a, b]$	Space of bounded functions 228
$B(A)$	Space of bounded functions 11
$BV[a, b]$	Space of functions of bounded variation 226
$B(X, Y)$	Space of bounded linear operators 118
$B(x; r)$	Open ball 18
$\tilde{B}(x; r)$	Closed ball 18
c	A sequence space 34
c_0	A sequence space 70
$\mathbf{C}$	Complex plane or the field of complex numbers 6, 51
$\mathbf{C}^n$	Unitary n-space 6
$C[a, b]$	Space of continuous functions 7, 61
$C'[a, b]$	Space of continuously differentiable functions 110
$C(X, Y)$	Space of compact linear operators 411
$\mathfrak{D}(T)$	Domain of an operator T 83
$d(x, y)$	Distance from x to y 3
$\dim X$	Dimension of a space X 54
δ_{jk}	Kronecker delta 114
$\mathscr{E} = (E_\lambda)$	Spectral family 494
$\|f\|$	Norm of a bounded linear functional f 104
$\mathscr{G}(T)$	Graph of an operator T 292
I	Identity operator 84
inf	Infimum (greatest lower bound) 619
$L^p[a, b]$	A function space 62
l^p	A sequence space 11
l^∞	A sequence space 6
$L(X, Y)$	A space of linear operators 118
$M^\perp$	Annihilator of a set M 148
$\mathcal{N}(T)$	Null space of an operator T 83
0	Zero operator 84
$\varnothing$	Empty set 609

$\mathbf{R}$	Real line or the field of real numbers 5, 51
$\mathbf{R}^n$	Euclidean n-space 6
$\mathscr{R}(T)$	Range of an operator T 83
$R_\lambda(T)$	Resolvent of an operator T 370
$r_\sigma(T)$	Spectral radius of an operator T 378
$\rho(T)$	Resolvent set of an operator T 371
s	A sequence space 9
$\sigma(T)$	Spectrum of an operator T 371
$\sigma_c(T)$	Continuous spectrum of T 371
$\sigma_p(T)$	Point spectrum of T 371
$\sigma_r(T)$	Residual spectrum of T 371
span M	Span of a set M 53
sup	Supremum (least upper bound) 619
$\\|T\\|$	Norm of a bounded linear operator T 92
T^*	Hilbert-adjoint operator of T 196
$T^\times$	Adjoint operator of T 232
T^+, T^-	Positive and negative parts of T 498
T_λ^+, T_λ^-	Positive and negative parts of $T_\lambda = T - \lambda I$ 500
$T^{1/2}$	Positive square root of T 476
Var(w)	Total variation of w 225
$\xrightarrow{w}$	Weak convergence 257
X^*	Algebraic dual space of a vector space X 106
X'	Dual space of a normed space X 120
$\\|x\\|$	Norm of x 59
$\langle x, y\rangle$	Inner product of x and y 128
$x \perp y$	x is orthogonal to y 131
$Y^\perp$	Orthogonal complement of a closed subspace Y 146

INTRODUCTORY FUNCTIONAL ANALYSIS WITH APPLICATIONS

CHAPTER 1
METRIC SPACES

Functional analysis is an abstract branch of mathematics that originated from classical analysis. Its development started about eighty years ago, and nowadays functional analytic methods and results are important in various fields of mathematics and its applications. The impetus came from linear algebra, linear ordinary and partial differential equations, calculus of variations, approximation theory and, in particular, linear integral equations, whose theory had the greatest effect on the development and promotion of the modern ideas. Mathematicians observed that problems from different fields often enjoy related features and properties. This fact was used for an effective unifying approach towards such problems, the unification being obtained by the omission of unessential details. Hence the advantage of such an *abstract approach* is that it concentrates on the essential facts, so that these facts become clearly visible since the investigator's attention is not disturbed by unimportant details. In this respect the abstract method is the simplest and most economical method for treating mathematical systems. Since any such abstract system will, in general, have various concrete realizations (concrete *models*), we see that the abstract method is quite versatile in its application to concrete situations. It helps to free the problem from isolation and creates relations and transitions between fields which have at first no contact with one another.

In the abstract approach, one usually starts from a set of elements satisfying certain axioms. The nature of the elements is left unspecified. This is done on purpose. The theory then consists of logical consequences which result from the axioms and are derived as theorems once and for all. This means that in this axiomatic fashion one obtains a mathematical structure whose theory is developed in an abstract way. Those general theorems can then later be applied to various special sets satisfying those axioms.

For example, in algebra this approach is used in connection with fields, rings and groups. In functional analysis we use it in connection with *abstract spaces;* these are of basic importance, and we shall consider some of them (Banach spaces, Hilbert spaces) in great detail. We shall see that in this connection the concept of a "space" is used in a very wide and surprisingly general sense. An *abstract space* will be a set of (unspecified) elements satisfying certain axioms. And by choosing different sets of axioms we shall obtain different types of abstract spaces.

The idea of using abstract spaces in a systematic fashion goes back to M. Fréchet (1906)[1] and is justified by its great success.

In this chapter we consider metric spaces. These are fundamental in functional analysis because they play a role similar to that of the real line **R** in calculus. In fact, they generalize **R** and have been created in order to provide a basis for a unified treatment of important problems from various branches of analysis.

We first define metric spaces and related concepts and illustrate them with typical examples. Special spaces of practical importance are discussed in detail. Much attention is paid to the concept of completeness, a property which a metric space may or may not have. Completeness will play a key role throughout the book.

Important concepts, brief orientation about main content

A *metric space* (cf. 1.1-1) is a set X with a *metric* on it. The metric associates with any pair of elements (*points*) of X a *distance*. The metric is defined axiomatically, the axioms being suggested by certain simple properties of the familiar distance between points on the real line **R** and the complex plane **C**. Basic examples (1.1-2 to 1.2-3) show that the concept of a metric space is remarkably general. A very important additional property which a metric space may have is *completeness* (cf. 1.4-3), which is discussed in detail in Secs. 1.5 and 1.6. Another concept of theoretical and practical interest is *separability* of a metric space (cf. 1.3-5). Separable metric spaces are simpler than nonseparable ones.

1.1 Metric Space

In calculus we study functions defined on the real line **R**. A little reflection shows that in limit processes and many other considerations we use the fact that on **R** we have available a distance function, call it d, which associates a *distance* $d(x, y) = |x - y|$ with every pair of points

[1] References are given in Appendix 3, and we shall refer to books and papers listed in Appendix 3 as is shown here.

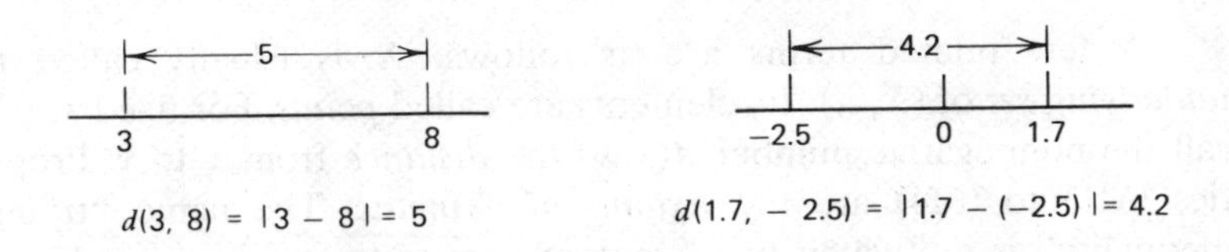

Fig. 2. Distance on **R**

$x, y \in \mathbf{R}$. Figure 2 illustrates the notation. In the plane and in "ordinary" three-dimensional space the situation is similar.

In functional analysis we shall study more general "spaces" and "functions" defined on them. We arrive at a sufficiently general and flexible concept of a "space" as follows. We replace the set of real numbers underlying **R** by an *abstract* set X (set of elements whose nature is left unspecified) and introduce on X a "distance function" which has only a few of the most fundamental properties of the distance function on **R**. But what do we mean by "most fundamental"? This question is far from being trivial. In fact, the choice and formulation of axioms in a definition always needs experience, familiarity with practical problems and a clear idea of the goal to be reached. In the present case, a development of over sixty years has led to the following concept which is basic and very useful in functional analysis and its applications.

1.1-1 Definition (Metric space, metric). A *metric space* is a pair (X, d), where X is a set and d is a *metric on* X (or *distance function on* X), that is, a function defined[2] on $X \times X$ such that for all $x, y, z \in X$ we have:

(M1) d is real-valued, finite and nonnegative.

(M2) $d(x, y) = 0$ if and only if $x = y$.

(M3) $d(x, y) = d(y, x)$ (*Symmetry*).

(M4) $d(x, y) \leq d(x, z) + d(z, y)$ **(Triangle inequality).** ∎

[2] The symbol $\times$ denotes the *Cartesian product* of sets: $A \times B$ is the set of all ordered pairs (a, b), where $a \in A$ and $b \in B$. Hence $X \times X$ is the set of all ordered pairs of elements of X.

A few related terms are as follows. X is usually called the *underlying set* of (X, d). Its elements are called *points.* For fixed x, y we call the nonnegative number $d(x, y)$ the *distance* from x to y. Properties (M1) to (M4) are the *axioms of a metric.* The name "triangle inequality" is motivated by elementary geometry as shown in Fig. 3.

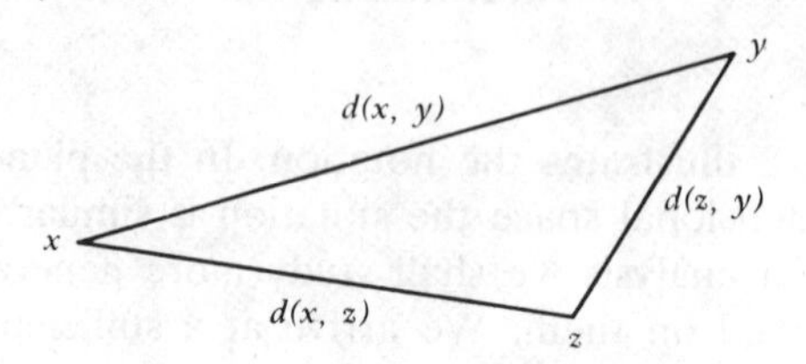

Fig. 3. Triangle inequality in the plane

From (M4) we obtain by induction the *generalized triangle inequality*

$$d(x_1, x_n) \leqq d(x_1, x_2) + d(x_2, x_3) + \cdots + d(x_{n-1}, x_n). \tag{1}$$

Instead of (X, d) we may simply write X if there is no danger of confusion.

A **subspace** $(Y, \tilde{d})$ of (X, d) is obtained if we take a subset $Y \subset X$ and restrict d to $Y \times Y$; thus the metric on Y is the restriction[3]

$$\tilde{d} = d|_{Y \times Y}.$$

$\tilde{d}$ is called the metric **induced** on Y by d.

We shall now list examples of metric spaces, some of which are already familiar to the reader. To prove that these are metric spaces, we must verify in each case that the axioms (M1) to (M4) are satisfied. Ordinarily, for (M4) this requires more work than for (M1) to (M3). However, in our present examples this will not be difficult, so that we can leave it to the reader (cf. the problem set). More sophisticated

[3] Appendix 1 contains a review on mappings which also includes the concept of a restriction.

metric spaces for which (M4) is not so easily verified are included in the next section.

Examples

1.1-2 Real line R. This is the set of all real numbers, taken with the usual metric defined by

$$d(x, y) = |x - y|. \tag{2}$$

1.1-3 Euclidean plane $\mathbf{R}^2$. The metric space $\mathbf{R}^2$, called the *Euclidean plane*, is obtained if we take the set of ordered pairs of real numbers, written[4] $x = (\xi_1, \xi_2)$, $y = (\eta_1, \eta_2)$, etc., and the *Euclidean metric* defined by

$$d(x, y) = \sqrt{(\xi_1 - \eta_1)^2 + (\xi_2 - \eta_2)^2} \qquad (\geqq 0). \tag{3}$$

See Fig. 4.

Another metric space is obtained if we choose the same set as before but another metric d_1 defined by

$$d_1(x, y) = |\xi_1 - \eta_1| + |\xi_2 - \eta_2|. \tag{4}$$

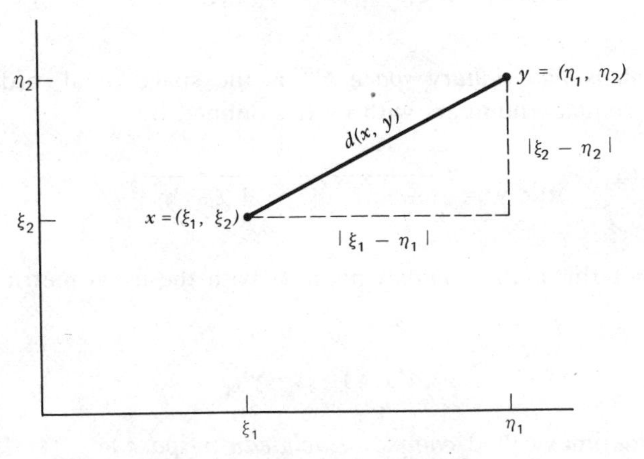

Fig. 4. Euclidean metric on the plane

[4] We do not write $x = (x_1, x_2)$ since $x_1, x_2, \cdots$ are needed later in connection with sequences (starting in Sec. 1.4).

This illustrates the important fact that from a given set (having more than one element) we can obtain various metric spaces by choosing different metrics. (The metric space with metric d_1 does not have a standard name. d_1 is sometimes called the *taxicab metric.* Why? $\mathbf{R}^2$ is sometimes denoted by E^2.)

1.1-4 Three-dimensional Euclidean space $\mathbf{R}^3$. This metric space consists of the set of ordered triples of real numbers $x=(\xi_1, \xi_2, \xi_3)$, $y=(\eta_1, \eta_2, \eta_3)$, etc., and the *Euclidean metric* defined by

$$d(x, y)=\sqrt{(\xi_1-\eta_1)^2+(\xi_2-\eta_2)^2+(\xi_3-\eta_3)^2} \quad (\geqq 0). \tag{5}$$

1.1-5 Euclidean space $\mathbf{R}^n$, unitary space $\mathbf{C}^n$, complex plane $\mathbf{C}$. The previous examples are special cases of *n-dimensional Euclidean space* $\mathbf{R}^n$. This space is obtained if we take the set of all ordered n-tuples of real numbers, written

$$x=(\xi_1, \cdots, \xi_n), \qquad y=(\eta_1, \cdots, \eta_n)$$

etc., and the *Euclidean metric* defined by

$$d(x, y)=\sqrt{(\xi_1-\eta_1)^2+\cdots+(\xi_n-\eta_n)^2} \quad (\geqq 0). \tag{6}$$

n-dimensional unitary space $\mathbf{C}^n$ is the space of all ordered n-tuples of *complex* numbers with metric defined by

$$d(x, y)=\sqrt{|\xi_1-\eta_1|^2+\cdots+|\xi_n-\eta_n|^2} \quad (\geqq 0). \tag{7}$$

When $n=1$ this is the *complex plane* $\mathbf{C}$ with the usual metric defined by

$$d(x, y)=|x-y|. \tag{8}$$

($\mathbf{C}^n$ is sometimes called *complex Euclidean n-space.*)

1.1-6 Sequence space l^∞. This example and the next one give a first impression of how surprisingly general the concept of a metric space is.

As a set X we take the set of all bounded sequences of complex numbers; that is, every element of X is a complex sequence

$$x=(\xi_1, \xi_2, \cdots) \qquad \text{briefly} \qquad x=(\xi_j)$$

such that for all $j=1, 2, \cdots$ we have

$$|\xi_j| \leqq c_x$$

where c_x is a real number which may depend on x, but does not depend on j. We choose the metric defined by

$$d(x, y)=\sup_{j\in\mathbf{N}} |\xi_j-\eta_j| \tag{9}$$

where $y=(\eta_j)\in X$ and $\mathbf{N}=\{1, 2, \cdots\}$, and sup denotes the supremum (least upper bound).[5] The metric space thus obtained is generally denoted by l^∞. (This somewhat strange notation will be motivated by 1.2-3 in the next section.) l^∞ is a *sequence space* because each element of X (each point of X) is a sequence.

1.1-7 Function space $C[a, b]$. As a set X we take the set of all real-valued functions $x, y, \cdots$ which are functions of an independent real variable t and are defined and continuous on a given closed interval $J=[a, b]$. Choosing the metric defined by

$$d(x, y)=\max_{t\in J} |x(t)-y(t)|, \tag{10}$$

where max denotes the maximum, we obtain a metric space which is denoted by $C[a, b]$. (The letter C suggests "continuous.") This is a *function space* because every point of $C[a, b]$ is a function.

The reader should realize the great difference between calculus, where one ordinarily considers a single function or a few functions at a time, and the present approach where a function becomes merely a single point in a large space.

[5] The reader may wish to look at the review of sup and inf given in A1.6; cf. Appendix 1.

1.1-8 Discrete metric space. We take any set X and on it the so-called *discrete metric* for X, defined by

$$d(x, x) = 0, \qquad d(x, y) = 1 \qquad (x \neq y).$$

This space (X, d) is called a *discrete metric space.* It rarely occurs in applications. However, we shall use it in examples for illustrating certain concepts (and traps for the unwary). ∎

From 1.1-1 we see that a metric is defined in terms of axioms, and we want to mention that axiomatic definitions are nowadays used in many branches of mathematics. Their usefulness was generally recognized after the publication of Hilbert's work about the foundations of geometry, and it is interesting to note that an investigation of one of the *oldest* and simplest parts of mathematics had one of the most important impacts on *modern* mathematics.

Problems

1. Show that the real line is a metric space.

2. Does $d(x, y) = (x - y)^2$ define a metric on the set of all real numbers?

3. Show that $d(x, y) = \sqrt{|x - y|}$ defines a metric on the set of all real numbers.

4. Find all metrics on a set X consisting of two points. Consisting of one point.

5. Let d be a metric on X. Determine all constants k such that (*i*) kd, (*ii*) $d + k$ is a metric on X.

6. Show that d in 1.1-6 satisfies the triangle inequality.

7. If A is the subspace of l^∞ consisting of all sequences of zeros and ones, what is the induced metric on A?

8. Show that another metric $\tilde{d}$ on the set X in 1.1-7 is defined by

$$\tilde{d}(x, y) = \int_a^b |x(t) - y(t)| \, dt.$$

9. Show that d in 1.1-8 is a metric.

10. (Hamming distance) Let X be the set of all ordered triples of zeros and ones. Show that X consists of eight elements and a metric d on X is defined by $d(x, y)$ = number of places where x and y have different entries. (This space and similar spaces of n-tuples play a role in switching and automata theory and coding. $d(x, y)$ is called the *Hamming distance* between x and y; cf. the paper by R. W. Hamming (1950) listed in Appendix 3.)

11. Prove (1).

12. (Triangle inequality) The triangle inequality has several useful consequences. For instance, using (1), show that

$$|d(x, y)-d(z, w)| \leqq d(x, z)+d(y, w).$$

13. Using the triangle inequality, show that

$$|d(x, z)-d(y, z)| \leqq d(x, y).$$

14. (Axioms of a metric) (M1) to (M4) could be replaced by other axioms (without changing the definition). For instance, show that (M3) and (M4) could be obtained from (M2) and

$$d(x, y) \leqq d(z, x)+d(z, y).$$

15. Show that nonnegativity of a metric follows from (M2) to (M4).

1.2 Further Examples of Metric Spaces

To illustrate the concept of a metric space and the process of verifying the axioms of a metric, in particular the triangle inequality (M4), we give three more examples. The last example (space l^p) is the most important one of them in applications.

1.2-1 Sequence space s. This space consists of the set of all (bounded or unbounded) sequences of complex numbers and the metric d

defined by

$$d(x, y) = \sum_{j=1}^{\infty} \frac{1}{2^j} \frac{|\xi_j - \eta_j|}{1 + |\xi_j - \eta_j|}$$

where $x = (\xi_j)$ and $y = (\eta_j)$. Note that the metric in Example 1.1-6 would not be suitable in the present case. (Why?)

Axioms (M1) to (M3) are satisfied, as we readily see. Let us verify (M4). For this purpose we use the auxiliary function f defined on $\mathbf{R}$ by

$$f(t) = \frac{t}{1+t}.$$

Differentiation gives $f'(t) = 1/(1+t)^2$, which is positive. Hence f is monotone increasing. Consequently,

$$|a+b| \leqq |a| + |b|$$

implies

$$f(|a+b|) \leqq f(|a| + |b|).$$

Writing this out and applying the triangle inequality for numbers, we have

$$\begin{aligned} \frac{|a+b|}{1+|a+b|} &\leqq \frac{|a|+|b|}{1+|a|+|b|} \\ &= \frac{|a|}{1+|a|+|b|} + \frac{|b|}{1+|a|+|b|} \\ &\leqq \frac{|a|}{1+|a|} + \frac{|b|}{1+|b|}. \end{aligned}$$

In this inequality we let $a = \xi_j - \zeta_j$ and $b = \zeta_j - \eta_j$, where $z = (\zeta_j)$. Then $a + b = \xi_j - \eta_j$ and we have

$$\frac{|\xi_j - \eta_j|}{1+|\xi_j - \eta_j|} \leqq \frac{|\xi_j - \zeta_j|}{1+|\xi_j - \zeta_j|} + \frac{|\zeta_j - \eta_j|}{1+|\zeta_j - \eta_j|}.$$

If we multiply both sides by $1/2^j$ and sum over j from 1 to ∞, we obtain $d(x, y)$ on the left and the sum of $d(x, z)$ and $d(z, y)$ on the right:

$$d(x, y) \leqq d(x, z) + d(z, y).$$

This establishes (M4) and proves that s is a metric space.

1.2-2 Space $B(A)$ of bounded functions. By definition, each element $x \in B(A)$ is a function defined and bounded on a given set A, and the metric is defined by

$$d(x, y) = \sup_{t \in A} |x(t) - y(t)|,$$

where sup denotes the supremum (cf. the footnote in 1.1-6). We write $B[a, b]$ for $B(A)$ in the case of an interval $A = [a, b] \subset \mathbf{R}$.

Let us show that $B(A)$ is a metric space. Clearly, (M1) and (M3) hold. Also, $d(x, x) = 0$ is obvious. Conversely, $d(x, y) = 0$ implies $x(t) - y(t) = 0$ for all $t \in A$, so that $x = y$. This gives (M2). Furthermore, for every $t \in A$ we have

$$\begin{aligned} |x(t) - y(t)| &\leqq |x(t) - z(t)| + |z(t) - y(t)| \\ &\leqq \sup_{t \in A} |x(t) - z(t)| + \sup_{t \in A} |z(t) - y(t)|. \end{aligned}$$

This shows that $x - y$ is bounded on A. Since the bound given by the expression in the second line does not depend on t, we may take the supremum on the left and obtain (M4).

1.2-3 Space l^p, Hilbert sequence space l^2, Hölder and Minkowski inequalities for sums. Let $p \geqq 1$ be a fixed real number. By definition, each element in the space l^p is a sequence $x = (\xi_j) = (\xi_1, \xi_2, \cdots)$ of numbers such that $|\xi_1|^p + |\xi_2|^p + \cdots$ converges; thus

$$\sum_{j=1}^{\infty} |\xi_j|^p < \infty \qquad (p \geqq 1, \text{ fixed}) \tag{1}$$

and the metric is defined by

$$d(x, y) = \left(\sum_{j=1}^{\infty} |\xi_j - \eta_j|^p \right)^{1/p} \tag{2}$$

where $y=(\eta_j)$ and $\sum|\eta_j|^p<\infty$. If we take only real sequences [satisfying (1)], we get the *real space* l^p, and if we take complex sequences [satisfying (1)], we get the *complex space* l^p. (Whenever the distinction is essential, we can indicate it by a subscript $\mathbf{R}$ or $\mathbf{C}$, respectively.)

In the case $p=2$ we have the famous *Hilbert sequence space* l^2 with metric defined by

$$d(x, y)=\sqrt{\sum_{j=1}^{\infty}|\xi_j-\eta_j|^2}. \tag{3}$$

This space was introduced and studied by D. Hilbert (1912) in connection with integral equations and is the earliest example of what is now called a *Hilbert space.* (We shall consider Hilbert spaces in great detail, starting in Chap. 3.)

We prove that l^p is a metric space. Clearly, (2) satisfies (M1) to (M3) provided the series on the right converges. We shall prove that it does converge and that (M4) is satisfied. Proceeding stepwise, we shall derive

(a) an auxiliary inequality,
(b) the Hölder inequality from (a),
(c) the Minkowski inequality from (b),
(d) the triangle inequality (M4) from (c).

The details are as follows.

(a) Let $p>1$ and define q by

$$\frac{1}{p}+\frac{1}{q}=1. \tag{4}$$

p and q are then called **conjugate exponents.** This is a standard term. From (4) we have

$$1=\frac{p+q}{pq}, \qquad pq=p+q, \qquad (p-1)(q-1)=1. \tag{5}$$

Hence $1/(p-1)=q-1$, so that

$$u=t^{p-1} \qquad \text{implies} \qquad t=u^{q-1}.$$

Let α and β be any positive numbers. Since $\alpha\beta$ is the area of the rectangle in Fig. 5, we thus obtain by integration the inequality

$$\alpha\beta \leqq \int_0^\alpha t^{p-1}\,dt + \int_0^\beta u^{q-1}\,du = \frac{\alpha^p}{p} + \frac{\beta^q}{q}. \tag{6}$$

Note that this inequality is trivially true if $\alpha = 0$ or $\beta = 0$.

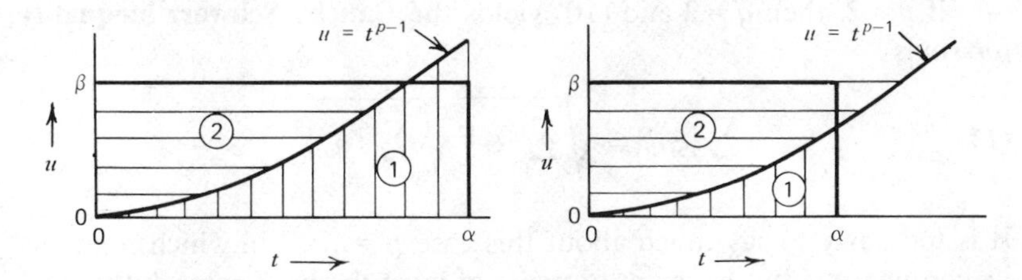

Fig. 5. Inequality (6), where ① corresponds to the first integral in (6) and ② to the second

(b) Let $(\tilde{\xi}_j)$ and $(\tilde{\eta}_j)$ be such that

$$\sum |\tilde{\xi}_j|^p = 1, \qquad \sum |\tilde{\eta}_j|^q = 1. \tag{7}$$

Setting $\alpha = |\tilde{\xi}_j|$ and $\beta = |\tilde{\eta}_j|$, we have from (6) the inequality

$$|\tilde{\xi}_j \tilde{\eta}_j| \leqq \frac{1}{p} |\tilde{\xi}_j|^p + \frac{1}{q} |\tilde{\eta}_j|^q.$$

If we sum over j and use (7) and (4), we obtain

$$\sum |\tilde{\xi}_j \tilde{\eta}_j| \leqq \frac{1}{p} + \frac{1}{q} = 1. \tag{8}$$

We now take any nonzero $x = (\xi_j) \in l^p$ and $y = (\eta_j) \in l^q$ and set

$$\tilde{\xi}_j = \frac{\xi_j}{\left(\sum |\xi_k|^p\right)^{1/p}}, \qquad \tilde{\eta}_j = \frac{\eta_j}{\left(\sum |\eta_m|^q\right)^{1/q}}. \tag{9}$$

Then (7) is satisfied, so that we may apply (8). Substituting (9) into (8) and multiplying the resulting inequality by the product of the denominators in (9), we arrive at the **Hölder inequality** *for sums*

$$\textbf{(10)} \qquad \sum_{j=1}^{\infty} |\xi_j \eta_j| \leqq \left(\sum_{k=1}^{\infty} |\xi_k|^p \right)^{1/p} \left(\sum_{m=1}^{\infty} |\eta_m|^q \right)^{1/q}$$

where $p > 1$ and $1/p + 1/q = 1$. This inequality was given by O. Hölder (1889).

If $p = 2$, then $q = 2$ and (10) yields the **Cauchy-Schwarz inequality** *for sums*

$$\textbf{(11)} \qquad \sum_{j=1}^{\infty} |\xi_j \eta_j| \leqq \sqrt{\sum_{k=1}^{\infty} |\xi_k|^2} \sqrt{\sum_{m=1}^{\infty} |\eta_m|^2}.$$

It is too early to say much about this case $p = q = 2$ in which p equals its conjugate q, but we want to make at least the brief remark that this case will play a particular role in some of our later chapters and lead to a space (a Hilbert space) which is "nicer" than spaces with $p \neq 2$.

(c) We now prove the **Minkowski inequality** *for sums*

$$\textbf{(12)} \qquad \left(\sum_{j=1}^{\infty} |\xi_j + \eta_j|^p \right)^{1/p} \leqq \left(\sum_{k=1}^{\infty} |\xi_k|^p \right)^{1/p} + \left(\sum_{m=1}^{\infty} |\eta_m|^p \right)^{1/p}$$

where $x = (\xi_j) \in l^p$ and $y = (\eta_j) \in l^p$, and $p \geqq 1$. For finite sums this inequality was given by H. Minkowski (1896).

For $p = 1$ the inequality follows readily from the triangle inequality for numbers. Let $p > 1$. To simplify the formulas we shall write $\xi_j + \eta_j = \omega_j$. The triangle inequality for numbers gives

$$\begin{aligned} |\omega_j|^p &= |\xi_j + \eta_j| \, |\omega_j|^{p-1} \\ &\leqq (|\xi_j| + |\eta_j|) |\omega_j|^{p-1}. \end{aligned}$$

Summing over j from 1 to any fixed n, we obtain

$$(13) \qquad \sum |\omega_j|^p \leqq \sum |\xi_j| \, |\omega_j|^{p-1} + \sum |\eta_j| \, |\omega_j|^{p-1}.$$

To the first sum on the right we apply the Hölder inequality, finding

$$\sum |\xi_j| \, |\omega_j|^{p-1} \leqq \left[\sum |\xi_k|^p \right]^{1/p} \left[\sum (|\omega_m|^{p-1})^q \right]^{1/q}.$$

On the right we simply have

$$(p-1)q=p$$

because $pq=p+q$; see (5). Treating the last sum in (13) in a similar fashion, we obtain

$$\sum |\eta_j||\omega_j|^{p-1} \leqq \left[\sum |\eta_k|^p\right]^{1/p} \left[\sum |\omega_m|^p\right]^{1/q}.$$

Together,

$$\sum |\omega_j|^p \leqq \left\{\left[\sum |\xi_k|^p\right]^{1/p} + \left[\sum |\eta_k|^p\right]^{1/p}\right\}\left(\sum |\omega_m|^p\right)^{1/q}.$$

Dividing by the last factor on the right and noting that $1-1/q=1/p$, we obtain (12) with n instead of ∞. We now let $n \longrightarrow \infty$. On the right this yields two series which converge because $x, y \in l^p$. Hence the series on the left also converges, and (12) is proved.

(d) From (12) it follows that for x and y in l^p the series in (2) converges. (12) also yields the triangle inequality. In fact, taking any $x, y, z \in l^p$, writing $z=(\zeta_j)$ and using the triangle inequality for numbers and then (12), we obtain

$$\begin{aligned} d(x, y) &= \left(\sum |\xi_j - \eta_j|^p\right)^{1/p} \\ &\leqq \left(\sum [|\xi_j - \zeta_j| + |\zeta_j - \eta_j|]^p\right)^{1/p} \\ &\leqq \left(\sum |\xi_j - \zeta_j|^p\right)^{1/p} + \left(\sum |\zeta_j - \eta_j|^p\right)^{1/p} \\ &= d(x, z) + d(z, y). \end{aligned}$$

This completes the proof that l^p is a metric space. ∎

The inequalities (10) to (12) obtained in this proof are of general importance as indispensable tools in various theoretical and practical problems, and we shall apply them a number of times in our further work.

Problems

1. Show that in 1.2-1 we can obtain another metric by replacing $1/2^j$ with $\mu_j > 0$ such that $\sum \mu_j$ converges.

2. Using (6), show that the geometric mean of two positive numbers does not exceed the arithmetic mean.

3. Show that the Cauchy-Schwarz inequality (11) implies

$$(|\xi_1| + \cdots + |\xi_n|)^2 \leqq n(|\xi_1|^2 + \cdots + |\xi_n|^2).$$

4. (Space l^p) Find a sequence which converges to 0, but is not in any space l^p, where $1 \leqq p < +\infty$.

5. Find a sequence x which is in l^p with $p > 1$ but $x \notin l^1$.

6. (Diameter, bounded set) The *diameter* $\delta(A)$ of a nonempty set A in a metric space (X, d) is defined to be

$$\delta(A) = \sup_{x,y \in A} d(x, y).$$

A is said to be *bounded* if $\delta(A) < \infty$. Show that $A \subset B$ implies $\delta(A) \leqq \delta(B)$.

7. Show that $\delta(A) = 0$ (cf. Prob. 6) if and only if A consists of a single point.

8. (Distance between sets) The *distance* $D(A, B)$ between two nonempty subsets A and B of a metric space (X, d) is defined to be

$$D(A, B) = \inf_{\substack{a \in A \\ b \in B}} d(a, b).$$

Show that D does *not* define a metric on the power set of X. (For this reason we use another symbol, D, but one that still reminds us of d.)

9. If $A \cap B \neq \phi$, show that $D(A, B) = 0$ in Prob. 8. What about the converse?

10. The *distance* $D(x, B)$ from a point x to a non-empty subset B of (X, d) is defined to be

$$D(x, B) = \inf_{b \in B} d(x, b),$$

in agreement with Prob. 8. Show that for any $x, y \in X$,

$$|D(x, B) - D(y, B)| \leq d(x, y).$$

11. If (X, d) is any metric space, show that another metric on X is defined by

$$\tilde{d}(x, y) = \frac{d(x, y)}{1 + d(x, y)}$$

and X is bounded in the metric $\tilde{d}$.

12. Show that the union of two bounded sets A and B in a metric space is a bounded set. (Definition in Prob. 6.)

13. (Product of metric spaces) The Cartesian product $X = X_1 \times X_2$ of two metric spaces (X_1, d_1) and (X_2, d_2) can be made into a metric space (X, d) in many ways. For instance, show that a metric d is defined by

$$d(x, y) = d_1(x_1, y_1) + d_2(x_2, y_2),$$

where $x = (x_1, x_2)$, $y = (y_1, y_2)$.

14. Show that another metric on X in Prob. 13 is defined by

$$\tilde{d}(x, y) = \sqrt{d_1(x_1, y_1)^2 + d_2(x_2, y_2)^2}.$$

15. Show that a third metric on X in Prob. 13 is defined by

$$\tilde{\tilde{d}}(x, y) = \max\,[d_1(x_1, y_1), d_2(x_2, y_2)].$$

(The metrics in Probs. 13 to 15 are of practical importance, and other metrics on X are possible.)

1.3 Open Set, Closed Set, Neighborhood

There is a considerable number of auxiliary concepts which play a role in connection with metric spaces. Those which we shall need are included in this section. Hence the section contains many concepts (more than any other section of the book), but the reader will notice that several of them become quite familiar when applied to Euclidean space. Of course this is a great convenience and shows the advantage of the terminology which is inspired by classical geometry.

We first consider important types of subsets of a given metric space $X=(X, d)$.

1.3-1 Definition (Ball and sphere). Given a point $x_0 \in X$ and a real number $r>0$, we define[6] three types of sets:

$$
\begin{aligned}
&(a) \quad B(x_0; r)=\{x \in X \mid d(x, x_0)<r\} \quad &\textbf{(Open ball)}\\
(1)\ &(b) \quad \tilde{B}(x_0; r)=\{x \in X \mid d(x, x_0) \leqq r\} \quad &\textbf{(Closed ball)}\\
&(c) \quad S(x_0; r)=\{x \in X \mid d(x, x_0)=r\} \quad &\textbf{(Sphere)}
\end{aligned}
$$

In all three cases, x_0 is called the *center* and r the *radius*. ∎

We see that an open ball of radius r is the set of all points in X whose distance from the center of the ball is less than r. Furthermore, the definition immediately implies that

$$
(2) \qquad S(x_0; r)=\tilde{B}(x_0; r)-B(x_0; r).
$$

Warning. In working with metric spaces, it is a great advantage that we use a terminology which is analogous to that of Euclidean geometry. However, we should beware of a danger, namely, of assuming that balls and spheres in an arbitrary metric space enjoy the same properties as balls and spheres in $\mathbf{R}^3$. This is not so. An unusual property is that a sphere can be empty. For example, in a discrete metric space 1.1-8 we have $S(x_0; r)=\varnothing$ if $r \neq 1$. (What about spheres of radius 1 in this case?) Another unusual property will be mentioned later.

Let us proceed to the next two concepts, which are related.

1.3-2 Definition (Open set, closed set). A subset M of a metric space X is said to be *open* if it contains a ball about each of its points. A subset K of X is said to be *closed* if its complement (in X) is open, that is, $K^C=X-K$ is open. ∎

The reader will easily see from this definition that an open ball is an open set and a closed ball is a closed set.

[6] Some familiarity with the usual set-theoretic notations is assumed, but a review is included in Appendix 1.

An open ball $B(x_0; \varepsilon)$ of radius ε is often called an ε-*neighborhood* of x_0. (Here, $\varepsilon > 0$, by Def. 1.3-1.) By a **neighborhood**[7] of x_0 we mean any subset of X which contains an ε-neighborhood of x_0.

We see directly from the definition that every neighborhood of x_0 contains x_0; in other words, x_0 is a point of each of its neighborhoods. And if N is a neighborhood of x_0 and $N \subset M$, then M is also a neighborhood of x_0.

We call x_0 an **interior point** of a set $M \subset X$ if M is a neighborhood of x_0. The **interior** of M is the set of all interior points of M and may be denoted by M^0 or Int (M), but there is no generally accepted notation. Int (M) is open and is the largest open set contained in M.

It is not difficult to show that the collection of all open subsets of X, call it $\mathcal{T}$, has the following properties:

(T1) $\varnothing \in \mathcal{T}$, $X \in \mathcal{T}$.

(T2) The union of any members of $\mathcal{T}$ is a member of $\mathcal{T}$.

(T3) The intersection of finitely many members of $\mathcal{T}$ is a member of $\mathcal{T}$.

Proof. (T1) follows by noting that $\varnothing$ is open since $\varnothing$ has no elements and, obviously, X is open. We prove (T2). Any point x of the union U of open sets belongs to (at least) one of these sets, call it M, and M contains a ball B about x since M is open. Then $B \subset U$, by the definition of a union. This proves (T2). Finally, if y is any point of the intersection of open sets $M_1, \cdots, M_n$, then each M_j contains a ball about y and a smallest of these balls is contained in that intersection. This proves (T3). ∎

We mention that the properties (T1) to (T3) are so fundamental that one wants to retain them in a more general setting. Accordingly, one defines a **topological space** $(X, \mathcal{T})$ to be a set X and a collection $\mathcal{T}$ of subsets of X such that $\mathcal{T}$ satisfies the *axioms* (T1) to (T3). The set $\mathcal{T}$ is called *a topology for* X. From this definition we have:

A metric space is a topological space.

[7] In the older literature, neighborhoods used to be open sets, but this requirement has been dropped from the definition.

Open sets also play a role in connection with continuous mappings, where continuity is a natural generalization of the continuity known from calculus and is defined as follows.

1.3-3 Definition (Continuous mapping). Let $X=(X, d)$ and $Y=(Y, \tilde{d})$ be metric spaces. A mapping $T\colon X \longrightarrow Y$ is said to be *continuous at a point* $x_0 \in X$ if for every $\varepsilon > 0$ there is a $\delta > 0$ such that[8] (see Fig. 6)

$$\tilde{d}(Tx, Tx_0) < \varepsilon \qquad \text{for all } x \text{ satisfying} \qquad d(x, x_0) < \delta.$$

T is said to be *continuous* if it is continuous at every point of X. ∎

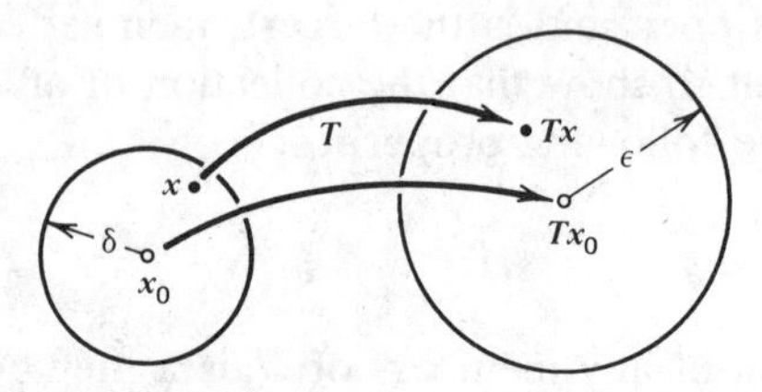

Fig. 6. Inequalities in Def. 1.3-3 illustrated in the case of Euclidean planes $X = \mathbf{R}^2$ and $Y = \mathbf{R}^2$

It is important and interesting that continuous mappings can be characterized in terms of open sets as follows.

1.3-4 Theorem (Continuous mapping). *A mapping T of a metric space X into a metric space Y is continuous if and only if the inverse image of any open subset of Y is an open subset of X.*

Proof. (*a*) Suppose that T is continuous. Let $S \subset Y$ be open and S_0 the inverse image of S. If $S_0 = \varnothing$, it is open. Let $S_0 \neq \varnothing$. For any $x_0 \in S_0$ let $y_0 = Tx_0$. Since S is open, it contains an ε-neighborhood N of y_0; see Fig. 7. Since T is continuous, x_0 has a δ-neighborhood N_0 which is mapped into N. Since $N \subset S$, we have $N_0 \subset S_0$, so that S_0 is open because $x_0 \in S_0$ was arbitrary.

(*b*) Conversely, assume that the inverse image of every open set in Y is an open set in X. Then for every $x_0 \in X$ and any

[8] In calculus we usually write $y = f(x)$. A corresponding notation for the image of x under T would be $T(x)$. However, to simplify formulas in functional analysis, it is customary to omit the parentheses and write Tx. A review of the definition of a mapping is included in A1.2; cf. Appendix 1.

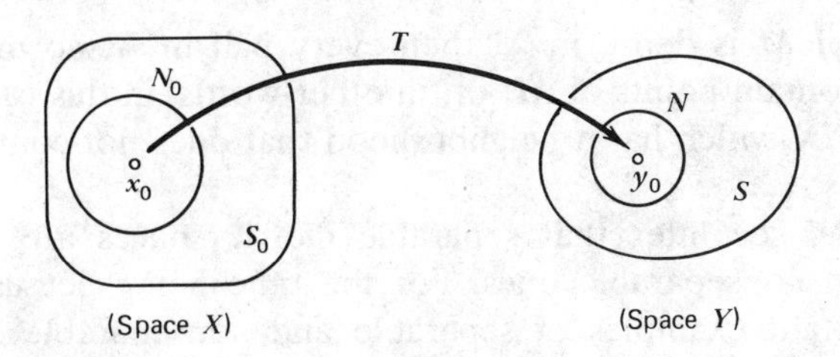

Fig. 7. Notation in part (*a*) of the proof of Theorem 1.3-4

ε-neighborhood N of Tx_0, the inverse image N_0 of N is open, since N is open, and N_0 contains x_0. Hence N_0 also contains a δ-neighborhood of x_0, which is mapped into N because N_0 is mapped into N. Consequently, by the definition, T is continuous at x_0. Since $x_0 \in X$ was arbitrary, T is continuous. ∎

We shall now introduce two more concepts, which are related. Let M be a subset of a metric space X. Then a point x_0 of X (which may or may not be a point of M) is called an **accumulation point** *of* M (or *limit point of* M) if every neighborhood of x_0 contains at least one point $y \in M$ distinct from x_0. The set consisting of the points of M and the accumulation points of M is called the **closure** *of* M and is denoted by

$$\bar{M}.$$

It is the smallest closed set containing M.

Before we go on, we mention another unusual property of balls in a metric space. Whereas in $\mathbf{R}^3$ the closure $\overline{B(x_0; r)}$ of an open ball $B(x_0; r)$ is the closed ball $\tilde{B}(x_0; r)$, this may not hold in a general metric space. We invite the reader to illustrate this with an example.

Using the concept of the closure, let us give a definition which will be of particular importance in our further work:

1.3-5 Definition (Dense set, separable space). A subset M of a metric space X is said to be *dense in* X if

$$\bar{M} = X.$$

X is said to be *separable* if it has a countable subset which is dense in X. (For the definition of a countable set, see A1.1 in Appendix 1 if necessary.) ∎

Hence if M is dense in X, then every ball in X, no matter how small, will contain points of M; or, in other words, in this case there is no point $x \in X$ which has a neighborhood that does not contain points of M.

We shall see later that separable metric spaces are somewhat simpler than nonseparable ones. For the time being, let us consider some important examples of separable and nonseparable spaces, so that we may become familiar with these basic concepts.

Examples

1.3-6 Real line R. *The real line* **R** *is separable.*

Proof. The set **Q** of all rational numbers is countable and is dense in **R**.

1.3-7 Complex plane C. *The complex plane* **C** *is separable.*

Proof. A countable dense subset of **C** is the set of all complex numbers whose real and imaginary parts are both rational.

1.3-8 Discrete metric space. *A discrete metric space X is separable if and only if X is countable.* (Cf. 1.1-8.)

Proof. The kind of metric implies that no proper subset of X can be dense in X. Hence the only dense set in X is X itself, and the statement follows.

1.3-9 Space l^∞. *The space l^∞ is not separable.* (Cf. 1.1-6.)

Proof. Let $y = (\eta_1, \eta_2, \eta_3, \cdots)$ be a sequence of zeros and ones. Then $y \in l^\infty$. With y we associate the real number $\hat{y}$ whose binary representation is

$$\frac{\eta_1}{2^1} + \frac{\eta_2}{2^2} + \frac{\eta_3}{2^3} + \cdots.$$

We now use the facts that the set of points in the interval $[0, 1]$ is uncountable, each $\hat{y} \in [0, 1]$ has a binary representation, and different $\hat{y}$'s have different binary representations. Hence there are uncountably many sequences of zeros and ones. The metric on l^∞ shows that any two of them which are not equal must be of distance 1 apart. If we let each of these sequences be the center of a small ball, say, of radius 1/3, these balls do not intersect and we have uncountably many of them. If M is any dense set in l^∞, each of these nonintersecting balls must contain an element of M. Hence M cannot be countable. Since M was an arbitrary dense set, this shows that l^∞ cannot have dense subsets which are countable. Consequently, l^∞ is not separable.

1.3-10 Space l^p. *The space l^p with $1 \leqq p < +\infty$ is separable.* (Cf. 1.2-3.)

Proof. Let M be the set of all sequences y of the form

$$y = (\eta_1, \eta_2, \cdots, \eta_n, 0, 0, \cdots)$$

where n is any positive integer and the η_j's are rational. M is countable. We show that M is dense in l^p. Let $x = (\xi_j) \in l^p$ be arbitrary. Then for every $\varepsilon > 0$ there is an n (depending on ε) such that

$$\sum_{j=n+1}^{\infty} |\xi_j|^p < \frac{\varepsilon^p}{2}$$

because on the left we have the remainder of a converging series. Since the rationals are dense in $\mathbf{R}$, for each ξ_j there is a rational η_j close to it. Hence we can find a $y \in M$ satisfying

$$\sum_{j=1}^{n} |\xi_j - \eta_j|^p < \frac{\varepsilon^p}{2}.$$

It follows that

$$[d(x, y)]^p = \sum_{j=1}^{n} |\xi_j - \eta_j|^p + \sum_{j=n+1}^{\infty} |\xi_j|^p < \varepsilon^p.$$

We thus have $d(x, y) < \varepsilon$ and see that M is dense in l^p.

Problems

1. Justify the terms "open ball" and "closed ball" by proving that (*a*) any open ball is an open set, (*b*) any closed ball is a closed set.

2. What is an open ball $B(x_0; 1)$ on $\mathbf{R}$? In $\mathbf{C}$? (Cf. 1.1-5.) In $C[a, b]$? (Cf. 1.1-7.) Explain Fig. 8.

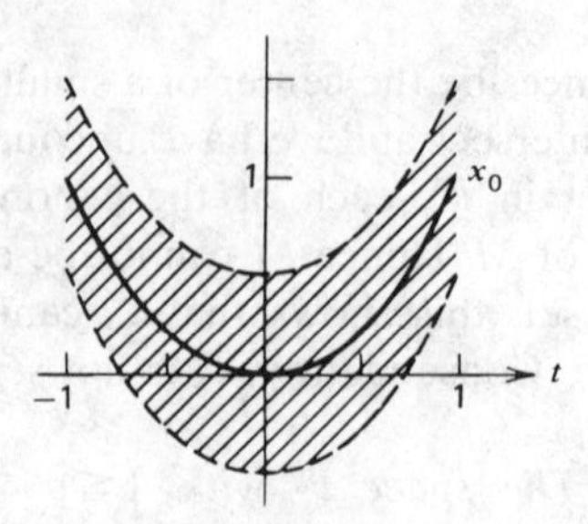

Fig. 8. Region containing the graphs of all $x \in C[-1, 1]$ which constitute the ε-neighborhood, with $\varepsilon = 1/2$, of $x_0 \in C[-1, 1]$ given by $x_0(t) = t^2$

3. Consider $C[0, 2\pi]$ and determine the smallest r such that $y \in \tilde{B}(x; r)$, where $x(t) = \sin t$ and $y(t) = \cos t$.

4. Show that any nonempty set $A \subset (X, d)$ is open if and only if it is a union of open balls.

5. It is important to realize that certain sets may be open and closed at the same time. (*a*) Show that this is always the case for X and $\varnothing$. (*b*) Show that in a discrete metric space X (cf. 1.1-8), every subset is open and closed.

6. If x_0 is an accumulation point of a set $A \subset (X, d)$, show that any neighborhood of x_0 contains infinitely many points of A.

7. Describe the closure of each of the following subsets. (*a*) The integers on $\mathbf{R}$, (*b*) the rational numbers on $\mathbf{R}$, (*c*) the complex numbers with rational real and imaginary parts in $\mathbf{C}$, (*d*) the disk $\{z \mid |z| < 1\} \subset \mathbf{C}$.

8. Show that the closure $\overline{B(x_0; r)}$ of an open ball $B(x_0; r)$ in a metric space can differ from the closed ball $\tilde{B}(x_0; r)$.

9. Show that $A \subset \bar{A}$, $\bar{\bar{A}} = \bar{A}$, $\overline{A \cup B} = \bar{A} \cup \bar{B}$, $\overline{A \cap B} \subset \bar{A} \cap \bar{B}$.

10. A point x not belonging to a *closed* set $M \subset (X, d)$ always has a nonzero distance from M. To prove this, show that $x \in \bar{A}$ if and only if $D(x, A) = 0$ (cf. Prob. 10, Sec. 1.2); here A is any nonempty subset of X.

11. **(Boundary)** A *boundary point* x of a set $A \subset (X, d)$ is a point of X (which may or may not belong to A) such that every neighborhood of x contains points of A as well as points not belonging to A; and the *boundary* (or *frontier*) of A is the set of all boundary points of A. Describe the boundary of (*a*) the intervals $(-1, 1)$, $[-1, 1)$, $[-1, 1]$ on

R; (b) the set of all rational numbers on **R**; (c) the disks $\{z \mid |z|<1\} \subset \mathbf{C}$ and $\{z \mid |z| \leqq 1\} \subset \mathbf{C}$.

12. (Space $B[a, b]$) Show that $B[a, b]$, $a<b$, is not separable. (Cf. 1.2-2.)

13. Show that a metric space X is separable if and only if X has a countable subset Y with the following property. For every $\varepsilon>0$ and every $x \in X$ there is a $y \in Y$ such that $d(x, y)<\varepsilon$.

14. (Continuous mapping) Show that a mapping $T: X \longrightarrow Y$ is continuous if and only if the inverse image of any closed set $M \subset Y$ is a closed set in X.

15. Show that the image of an open set under a continuous mapping need not be open.

1.4 Convergence, Cauchy Sequence, Completeness

We know that sequences of real numbers play an important role in calculus, and it is the metric on **R** which enables us to define the basic concept of convergence of such a sequence. The same holds for sequences of complex numbers; in this case we have to use the metric on the complex plane. In an arbitrary metric space $X=(X, d)$ the situation is quite similar, that is, we may consider a sequence (x_n) of elements $x_1, x_2, \cdots$ of X and use the metric d to define convergence in a fashion analogous to that in calculus:

1.4-1 Definition (Convergence of a sequence, limit). A sequence (x_n) in a metric space $X=(X, d)$ is said to *converge* or to *be convergent* if there is an $x \in X$ such that

$$\lim_{n \to \infty} d(x_n, x)=0.$$

x is called the *limit* of (x_n) and we write

$$\lim_{n \to \infty} x_n = x$$

or, simply,

$$x_n \longrightarrow x.$$

We say that (x_n) *converges to* x or *has the limit* x. If (x_n) is not convergent, it is said to be *divergent*. ∎

How is the metric d being used in this definition? We see that d yields the sequence of real numbers $a_n = d(x_n, x)$ whose convergence defines that of (x_n). Hence if $x_n \longrightarrow x$, an $\varepsilon > 0$ being given, there is an $N = N(\varepsilon)$ such that all x_n with $n > N$ lie in the ε-neighborhood $B(x; \varepsilon)$ of x.

To avoid trivial misunderstandings, we note that the limit of a convergent sequence must be a point of the space X in 1.4-1. For instance, let X be the open interval $(0, 1)$ on $\mathbf{R}$ with the usual metric defined by $d(x, y) = |x - y|$. Then the sequence $(\frac{1}{2}, \frac{1}{3}, \frac{1}{4}, \cdots)$ is not convergent since 0, the point to which the sequence "wants to converge," is not in X. We shall return to this and similar situations later in the present section.

Let us first show that two familiar properties of a convergent sequence (uniqueness of the limit and boundedness) carry over from calculus to our present much more general setting.

We call a nonempty subset $M \subset X$ a *bounded set* if its *diameter*

$$\delta(M) = \sup_{x,y \in M} d(x, y)$$

is finite. And we call a sequence (x_n) in X a **bounded sequence** if the corresponding point set is a bounded subset of X.

Obviously, if M is bounded, then $M \subset B(x_0; r)$, where $x_0 \in X$ is any point and r is a (sufficiently large) real number, and conversely.

Our assertion is now as follows.

1.4-2 Lemma (Boundedness, limit). *Let $X = (X, d)$ be metric space. Then:*

(a) *A convergent sequence in X is bounded and its limit is unique.*

(b) *If $x_n \longrightarrow x$ and $y_n \longrightarrow y$ in X, then $d(x_n, y_n) \longrightarrow d(x, y)$.*

Proof. **(a)** Suppose that $x_n \longrightarrow x$. Then, taking $\varepsilon = 1$, we can find an N such that $d(x_n, x) < 1$ for all $n > N$. Hence by the triangle inequality (M4), Sec. 1.1, for all n we have $d(x_n, x) < 1 + a$ where

$$a = \max\{d(x_1, x), \cdots, d(x_N, x)\}.$$

This shows that (x_n) is bounded. Assuming that $x_n \longrightarrow x$ and $x_n \longrightarrow z$, we obtain from (M4)

$$0 \leqq d(x, z) \leqq d(x, x_n) + d(x_n, z) \longrightarrow 0+0$$

and the uniqueness $x = z$ of the limit follows from (M2).

(b) By (1), Sec. 1.1, we have

$$d(x_n, y_n) \leqq d(x_n, x) + d(x, y) + d(y, y_n).$$

Hence we obtain

$$d(x_n, y_n) - d(x, y) \leqq d(x_n, x) + d(y_n, y)$$

and a similar inequality by interchanging x_n and x as well as y_n and y and multiplying by -1. Together,

$$|d(x_n, y_n) - d(x, y)| \leqq d(x_n, x) + d(y_n, y) \longrightarrow 0$$

as $n \longrightarrow \infty$. ∎

We shall now define the concept of completeness of a metric space, which will be basic in our further work. We shall see that completeness does *not* follow from (M1) to (M4) in Sec. 1.1, since there are *incomplete* (not complete) metric spaces. In other words, completeness is an additional property which a metric space may or may not have. It has various consequences which make complete metric spaces "much nicer and simpler" than incomplete ones—what this means will become clearer and clearer as we proceed.

Let us first remember from calculus that a sequence (x_n) of real or complex numbers converges on the real line **R** or in the complex plane **C**, respectively, if and only if it satisfies the *Cauchy convergence criterion*, that is, if and only if for every given $\varepsilon > 0$ there is an $N = N(\varepsilon)$ such that

$$|x_m - x_n| < \varepsilon \qquad \text{for all } m, n > N.$$

(A proof is included in A1.7; cf. Appendix 1.) Here $|x_m - x_n|$ is the distance $d(x_m, x_n)$ from x_m to x_n on the real line **R** or in the complex plane **C**. Hence we can write the inequality of the Cauchy criterion in the form

$$d(x_m, x_n) < \varepsilon \qquad (m, n > N).$$

And if a sequence (x_n) satisfies the condition of the Cauchy criterion, we may call it a *Cauchy sequence.* Then the Cauchy criterion simply says that a sequence of real or complex numbers converges on **R** or in **C** if and only if it is a Cauchy sequence. This refers to the situation in **R** or **C**. Unfortunately, in more general spaces the situation may be more complicated, and there may be Cauchy sequences which do not converge. Such a space is then lacking a property which is so important that it deserves a name, namely, completeness. This consideration motivates the following definition, which was first given by M. Fréchet (1906).

1.4-3 Definition (Cauchy sequence, completeness). A sequence (x_n) in a metric space $X = (X, d)$ is said to be *Cauchy* (or *fundamental*) if for every $\varepsilon > 0$ there is an $N = N(\varepsilon)$ such that

$$d(x_m, x_n) < \varepsilon \qquad \text{for every } m, n > N. \tag{1}$$

The space X is said to be *complete* if every Cauchy sequence in X converges (that is, has a limit which is an element of X). ∎

Expressed in terms of completeness, the Cauchy convergence criterion implies the following.

1.4-4 Theorem (Real line, complex plane). *The real line and the complex plane are complete metric spaces.*

More generally, we now see directly from the definition that complete metric spaces are precisely those in which the Cauchy condition (1) continues to be necessary and sufficient for convergence.

Complete and incomplete metric spaces that are important in applications will be considered in the next section in a systematic fashion.

For the time being let us mention a few simple incomplete spaces which we can readily obtain. Omission of a point a from the real line yields the incomplete space $\mathbf{R} - \{a\}$. More drastically, by the omission of all irrational numbers we have the *rational line* **Q**, which is incomplete. An open interval (a, b) with the metric induced from **R** is another incomplete metric space, and so on.

It is clear from the definition that in an arbitrary metric space, condition (1) may no longer be sufficient for convergence since the space may be incomplete. A good understanding of the whole situation is important; so let us consider a simple example. We take $X=(0, 1]$, with the usual metric defined by $d(x, y)=|x-y|$, and the sequence (x_n), where $x_n=1/n$ and $n=1, 2, \cdots$. This is a Cauchy sequence, but it does not converge, because the point 0 (to which it "wants to converge") is not a point of X. This also illustrates that the concept of convergence is not an intrinsic property of the sequence itself but also depends on the space in which the sequence lies. In other words, a convergent sequence is not convergent "on its own" but it must converge to some point in the space.

Although condition (1) is no longer sufficient for convergence, it is worth noting that it continues to be necessary for convergence. In fact, we readily obtain the following result.

1.4-5 Theorem (Convergent sequence). *Every convergent sequence in a metric space is a Cauchy sequence.*

Proof. If $x_n \longrightarrow x$, then for every $\varepsilon>0$ there is an $N=N(\varepsilon)$ such that

$$d(x_n, x)<\frac{\varepsilon}{2} \qquad \text{for all } n>N.$$

Hence by the triangle inequality we obtain for $m, n>N$

$$d(x_m, x_n)\leqq d(x_m, x)+d(x, x_n)<\frac{\varepsilon}{2}+\frac{\varepsilon}{2}=\varepsilon.$$

This shows that (x_n) is Cauchy. ∎

We shall see that quite a number of basic results, for instance in the theory of linear operators, will depend on the completeness of the corresponding spaces. Completeness of the real line **R** is also the main reason why in calculus we use **R** rather than the *rational line* **Q** (the set of all rational numbers with the metric induced from **R**).

Let us continue and finish this section with three theorems that are related to convergence and completeness and will be needed later.

1.4-6 Theorem (Closure, closed set). *Let M be a nonempty subset of a metric space (X, d) and $\bar{M}$ its closure as defined in the previous section. Then:*

(a) *$x \in \bar{M}$ if and only if there is a sequence (x_n) in M such that $x_n \longrightarrow x$.*

(b) *M is closed if and only if the situation $x_n \in M$, $x_n \longrightarrow x$ implies that $x \in M$.*

Proof. **(a)** Let $x \in \bar{M}$. If $x \in M$, a sequence of that type is $(x, x, \cdots)$. If $x \notin M$, it is a point of accumulation of M. Hence for each $n = 1, 2, \cdots$ the ball $B(x; 1/n)$ contains an $x_n \in M$, and $x_n \longrightarrow x$ because $1/n \longrightarrow 0$ as $n \longrightarrow \infty$.

Conversely, if (x_n) is in M and $x_n \longrightarrow x$, then $x \in M$ or every neighborhood of x contains points $x_n \neq x$, so that x is a point of accumulation of M. Hence $x \in \bar{M}$, by the definition of the closure.

(b) M is closed if and only if $M = \bar{M}$, so that (b) follows readily from (a). ∎

1.4-7 Theorem (Complete subspace). *A subspace M of a complete metric space X is itself complete if and only if the set M is closed in X.*

Proof. Let M be complete. By 1.4-6(a), for every $x \in \bar{M}$ there is a sequence (x_n) in M which converges to x. Since (x_n) is Cauchy by 1.4-5 and M is complete, (x_n) converges in M, the limit being unique by 1.4-2. Hence $x \in M$. This proves that M is closed because $x \in \bar{M}$ was arbitrary.

Conversely, let M be closed and (x_n) Cauchy in M. Then $x_n \longrightarrow x \in X$, which implies $x \in \bar{M}$ by 1.4-6(a), and $x \in M$ since $M = \bar{M}$ by assumption. Hence the arbitrary Cauchy sequence (x_n) converges in M, which proves completeness of M.

This theorem is very useful, and we shall need it quite often. Example 1.5-3 in the next section includes the first application, which is typical.

The last of our present three theorems shows the importance of convergence of sequences in connection with the continuity of a mapping.

1.4-8 Theorem (Continuous mapping). *A mapping $T: X \longrightarrow Y$ of a metric space (X, d) into a metric space $(Y, \tilde{d})$ is continuous at a point*

$x_0 \in X$ *if and only if*

$$x_n \longrightarrow x_0 \qquad \text{implies} \qquad Tx_n \longrightarrow Tx_0.$$

Proof. Assume T to be continuous at x_0; cf. Def. 1.3-3. Then for a given $\varepsilon > 0$ there is a $\delta > 0$ such that

$$d(x, x_0) < \delta \qquad \text{implies} \qquad \tilde{d}(Tx, Tx_0) < \varepsilon.$$

Let $x_n \longrightarrow x_0$. Then there is an N such that for all $n > N$ we have

$$d(x_n, x_0) < \delta.$$

Hence for all $n > N$,

$$\tilde{d}(Tx_n, Tx_0) < \varepsilon.$$

By definition this means that $Tx_n \longrightarrow Tx_0$.

Conversely, we assume that

$$x_n \longrightarrow x_0 \qquad \text{implies} \qquad Tx_n \longrightarrow Tx_0$$

and prove that then T is continuous at x_0. Suppose this is false. Then there is an $\varepsilon > 0$ such that for every $\delta > 0$ there is an $x \neq x_0$ satisfying

$$d(x, x_0) < \delta \qquad \text{but} \qquad \tilde{d}(Tx, Tx_0) \geqq \varepsilon.$$

In particular, for $\delta = 1/n$ there is an x_n satisfying

$$d(x_n, x_0) < \frac{1}{n} \qquad \text{but} \qquad \tilde{d}(Tx_n, Tx_0) \geqq \varepsilon.$$

Clearly $x_n \longrightarrow x_0$ but (Tx_n) does not converge to Tx_0. This contradicts $Tx_n \longrightarrow Tx_0$ and proves the theorem. ∎

Problems

1. **(Subsequence)** If a sequence (x_n) in a metric space X is convergent and has limit x, show that every subsequence (x_{n_k}) of (x_n) is convergent and has the same limit x.

2. If (x_n) is Cauchy and has a convergent subsequence, say, $x_{n_k} \longrightarrow x$, show that (x_n) is convergent with the limit x.

3. Show that $x_n \longrightarrow x$ if and only if for every neighborhood V of x there is an integer n_0 such that $x_n \in V$ for all $n > n_0$.

4. (Boundedness) Show that a Cauchy sequence is bounded.

5. Is boundedness of a sequence in a metric space sufficient for the sequence to be Cauchy? Convergent?

6. If (x_n) and (y_n) are Cauchy sequences in a metric space (X, d), show that (a_n), where $a_n = d(x_n, y_n)$, converges. Give illustrative examples.

7. Give an indirect proof of Lemma 1.4-2(b).

8. If d_1 and d_2 are metrics on the same set X and there are positive numbers a and b such that for all $x, y \in X$,

$$ad_1(x, y) \leqq d_2(x, y) \leqq bd_1(x, y),$$

show that the Cauchy sequences in (X, d_1) and (X, d_2) are the same.

9. Using Prob. 8, show that the metric spaces in Probs. 13 to 15, Sec. 1.2, have the same Cauchy sequences.

10. Using the completeness of **R**, prove completeness of **C**.

1.5 Examples. Completeness Proofs

In various applications a set X is given (for instance, a set of sequences or a set of functions), and X is made into a metric space. This we do by choosing a metric d on X. The remaining task is then to find out whether (X, d) has the desirable property of being complete. To prove completeness, we take an arbitrary Cauchy sequence (x_n) in X and show that it converges in X. For different spaces, such proofs may vary in complexity, but they have approximately the same general pattern:

(i) Construct an element x (to be used as a limit).
(ii) Prove that x is in the space considered.
(iii) Prove convergence $x_n \longrightarrow x$ (in the sense of the metric).

We shall present completeness proofs for some metric spaces which occur quite frequently in theoretical and practical investigations.

The reader will notice that in these cases (Examples 1.5-1 to 1.5-5) we get help from the completeness of the real line or the complex plane (Theorem 1.4-4). This is typical.

Examples

1.5-1 Completeness of $\mathbf{R}^n$ and $\mathbf{C}^n$. *Euclidean space* $\mathbf{R}^n$ *and unitary space* $\mathbf{C}^n$ *are complete.* (Cf. 1.1-5.)

Proof. We first consider $\mathbf{R}^n$. We remember that the metric on $\mathbf{R}^n$ (the Euclidean metric) is defined by

$$d(x, y) = \left(\sum_{j=1}^{n} (\xi_j - \eta_j)^2 \right)^{1/2}$$

where $x = (\xi_j)$ and $y = (\eta_j)$; cf. (6) in Sec. 1.1. We consider any Cauchy sequence (x_m) in $\mathbf{R}^n$, writing $x_m = (\xi_1^{(m)}, \cdots, \xi_n^{(m)})$. Since (x_m) is Cauchy, for every $\varepsilon > 0$ there is an N such that

$$\text{(1)} \qquad d(x_m, x_r) = \left(\sum_{j=1}^{n} (\xi_j^{(m)} - \xi_j^{(r)})^2 \right)^{1/2} < \varepsilon \qquad (m, r > N).$$

Squaring, we have for $m, r > N$ and $j = 1, \cdots, n$

$$(\xi_j^{(m)} - \xi_j^{(r)})^2 < \varepsilon^2 \qquad \text{and} \qquad |\xi_j^{(m)} - \xi_j^{(r)}| < \varepsilon.$$

This shows that for each fixed j, $(1 \leqq j \leqq n)$, the sequence $(\xi_j^{(1)}, \xi_j^{(2)}, \cdots)$ is a Cauchy sequence of real numbers. It converges by Theorem 1.4-4, say, $\xi_j^{(m)} \longrightarrow \xi_j$ as $m \longrightarrow \infty$. Using these n limits, we define $x = (\xi_1, \cdots, \xi_n)$. Clearly, $x \in \mathbf{R}^n$. From (1), with $r \longrightarrow \infty$,

$$d(x_m, x) \leqq \varepsilon \qquad (m > N).$$

This shows that x is the limit of (x_m) and proves completeness of $\mathbf{R}^n$ because (x_m) was an arbitrary Cauchy sequence. Completeness of $\mathbf{C}^n$ follows from Theorem 1.4-4 by the same method of proof.

1.5-2 Completeness of l^∞. *The space* l^∞ *is complete.* (Cf. 1.1-6.)

Proof. Let (x_m) be any Cauchy sequence in the space l^∞, where $x_m = (\xi_1^{(m)}, \xi_2^{(m)}, \cdots)$. Since the metric on l^∞ is given by

$$d(x, y) = \sup_j |\xi_j - \eta_j|$$

[where $x = (\xi_j)$ and $y = (\eta_j)$] and (x_m) is Cauchy, for any $\varepsilon > 0$ there is an N such that for all $m, n > N$,

$$d(x_m, x_n) = \sup_j |\xi_j^{(m)} - \xi_j^{(n)}| < \varepsilon.$$

A fortiori, for every fixed j,

$$|\xi_j^{(m)} - \xi_j^{(n)}| < \varepsilon \qquad (m, n > N). \tag{2}$$

Hence for every fixed j, the sequence $(\xi_j^{(1)}, \xi_j^{(2)}, \cdots)$ is a Cauchy sequence of numbers. It converges by Theorem 1.4-4, say, $\xi_j^{(m)} \longrightarrow \xi_j$ as $m \longrightarrow \infty$. Using these infinitely many limits $\xi_1, \xi_2, \cdots$, we define $x = (\xi_1, \xi_2, \cdots)$ and show that $x \in l^\infty$ and $x_m \longrightarrow x$. From (2) with $n \longrightarrow \infty$ we have

$$|\xi_j^{(m)} - \xi_j| \leqq \varepsilon \qquad (m > N). \tag{2*}$$

Since $x_m = (\xi_j^{(m)}) \in l^\infty$, there is a real number k_m such that $|\xi_j^{(m)}| \leqq k_m$ for all j. Hence by the triangle inequality

$$|\xi_j| \leqq |\xi_j - \xi_j^{(m)}| + |\xi_j^{(m)}| \leqq \varepsilon + k_m \qquad (m > N).$$

This inequality holds for every j, and the right-hand side does not involve j. Hence (ξ_j) is a bounded sequence of numbers. This implies that $x = (\xi_j) \in l^\infty$. Also, from (2*) we obtain

$$d(x_m, x) = \sup_j |\xi_j^{(m)} - \xi_j| \leqq \varepsilon \qquad (m > N).$$

This shows that $x_m \longrightarrow x$. Since (x_m) was an arbitrary Cauchy sequence, l^∞ is complete.

1.5-3 Completeness of *c*. The space c consists of all convergent sequences $x = (\xi_j)$ of complex numbers, with the metric induced from the space l^∞.

The space c is complete.

Proof. c is a subspace of l^∞ and we show that c is closed in l^∞, so that completeness then follows from Theorem 1.4-7.

We consider any $x = (\xi_j) \in \bar{c}$, the closure of c. By 1.4-6(*a*) there are $x_n = (\xi_j^{(n)}) \in c$ such that $x_n \longrightarrow x$. Hence, given any $\varepsilon > 0$, there is an N such that for $n \geqq N$ and all j we have

$$|\xi_j^{(n)} - \xi_j| \leqq d(x_n, x) < \frac{\varepsilon}{3},$$

in particular, for $n = N$ and all j. Since $x_N \in c$, its terms $\xi_j^{(N)}$ form a convergent sequence. Such a sequence is Cauchy. Hence there is an N_1 such that

$$|\xi_j^{(N)} - \xi_k^{(N)}| < \frac{\varepsilon}{3} \qquad (j, k \geqq N_1).$$

The triangle inequality now yields for all j, $k \geqq N_1$ the following inequality:

$$|\xi_j - \xi_k| \leqq |\xi_j - \xi_j^{(N)}| + |\xi_j^{(N)} - \xi_k^{(N)}| + |\xi_k^{(N)} - \xi_k| < \varepsilon.$$

This shows that the sequence $x = (\xi_j)$ is convergent. Hence $x \in c$. Since $x \in \bar{c}$ was arbitrary, this proves closedness of c in l^∞, and completeness of c follows from 1.4-7. ∎

1.5-4 Completeness of l^p. *The space l^p is complete; here p is fixed and* $1 \leqq p < +\infty$. (Cf. 1.2-3.)

Proof. Let (x_n) be any Cauchy sequence in the space l^p, where $x_m = (\xi_1^{(m)}, \xi_2^{(m)}, \cdots)$. Then for every $\varepsilon > 0$ there is an N such that for all $m, n > N$,

$$d(x_m, x_n) = \left(\sum_{j=1}^{\infty} |\xi_j^{(m)} - \xi_j^{(n)}|^p \right)^{1/p} < \varepsilon. \tag{3}$$

It follows that for every $j = 1, 2, \cdots$ we have

$$|\xi_j^{(m)} - \xi_j^{(n)}| < \varepsilon \qquad (m, n > N). \tag{4}$$

We choose a fixed j. From (4) we see that $(\xi_j^{(1)}, \xi_j^{(2)}, \cdots)$ is a Cauchy sequence of numbers. It converges since **R** and **C** are complete (cf.

1.4-4), say, $\xi_j^{(m)} \longrightarrow \xi_j$ as $m \longrightarrow \infty$. Using these limits, we define $x = (\xi_1, \xi_2, \cdots)$ and show that $x \in l^p$ and $x_m \longrightarrow x$.

From (3) we have for all $m, n > N$

$$\sum_{j=1}^{k} |\xi_j^{(m)} - \xi_j^{(n)}|^p < \varepsilon^p \qquad (k = 1, 2, \cdots).$$

Letting $n \longrightarrow \infty$, we obtain for $m > N$

$$\sum_{j=1}^{k} |\xi_j^{(m)} - \xi_j|^p \leqq \varepsilon^p \qquad (k = 1, 2, \cdots).$$

We may now let $k \longrightarrow \infty$; then for $m > N$

$$\sum_{j=1}^{\infty} |\xi_j^{(m)} - \xi_j|^p \leqq \varepsilon^p. \tag{5}$$

This shows that $x_m - x = (\xi_j^{(m)} - \xi_j) \in l^p$. Since $x_m \in l^p$, it follows by means of the Minkowski inequality (12), Sec. 1.2, that

$$x = x_m + (x - x_m) \in l^p.$$

Furthermore, the series in (5) represents $[d(x_m, x)]^p$, so that (5) implies that $x_m \longrightarrow x$. Since (x_m) was an arbitrary Cauchy sequence in l^p, this proves completeness of l^p, where $1 \leqq p < +\infty$. ∎

1.5-5 Completeness of $C[a, b]$. *The function space $C[a, b]$ is complete; here $[a, b]$ is any given closed interval on* $\mathbf{R}$. (Cf. 1.1-7.)

Proof. Let (x_m) be any Cauchy sequence in $C[a, b]$. Then, given any $\varepsilon > 0$, there is an N such that for all $m, n > N$ we have

$$d(x_m, x_n) = \max_{t \in J} |x_m(t) - x_n(t)| < \varepsilon \tag{6}$$

where $J = [a, b]$. Hence for any fixed $t = t_0 \in J$,

$$|x_m(t_0) - x_n(t_0)| < \varepsilon \qquad (m, n > N).$$

This shows that $(x_1(t_0), x_2(t_0), \cdots)$ is a Cauchy sequence of real numbers. Since $\mathbf{R}$ is complete (cf. 1.4-4), the sequence converges, say,

$x_m(t_0) \longrightarrow x(t_0)$ as $m \longrightarrow \infty$. In this way we can associate with each $t \in J$ a unique real number $x(t)$. This defines (pointwise) a function x on J, and we show that $x \in C[a, b]$ and $x_m \longrightarrow x$.

From (6) with $n \longrightarrow \infty$ we have

$$\max_{t \in J} |x_m(t) - x(t)| \leqq \varepsilon \qquad (m > N).$$

Hence for every $t \in J$,

$$|x_m(t) - x(t)| \leqq \varepsilon \qquad (m > N).$$

This shows that $(x_m(t))$ converges to $x(t)$ uniformly on J. Since the x_m's are continuous on J and the convergence is uniform, the limit function x is continuous on J, as is well known from calculus (cf. also Prob. 9). Hence $x \in C[a, b]$. Also $x_m \longrightarrow x$. This proves completeness of $C[a, b]$. ∎

In 1.1-7 as well as here we assumed the functions x to be real-valued, for simplicity. We may call this space the *real* $C[a, b]$. Similarly, we obtain the *complex* $C[a, b]$ if we take complex-valued continuous functions defined on $[a, b] \subset \mathbf{R}$. This space is complete, too. The proof is almost the same as before.

Furthermore, that proof also shows the following fact.

1.5-6 Theorem (Uniform convergence). *Convergence $x_m \longrightarrow x$ in the space $C[a, b]$ is uniform convergence, that is, (x_m) converges uniformly on $[a, b]$ to x.*

Hence the metric on $C[a, b]$ describes uniform convergence on $[a, b]$ and, for this reason, is sometimes called the *uniform metric.*

To gain a good understanding of completeness and related concepts, let us finally look at some

Examples of Incomplete Metric Spaces

1.5-7 Space Q. This is the set of all rational numbers with the usual metric given by $d(x, y) = |x - y|$, where $x, y \in \mathbf{Q}$, and is called the *rational line.* $\mathbf{Q}$ is not complete. (Proof?)

1.5-8 Polynomials. Let X be the set of all polynomials considered as functions of t on some finite closed interval $J = [a, b]$ and define a metric d on X by

$$d(x, y) = \max_{t \in J} |x(t) - y(t)|.$$

This metric space (X, d) is not complete. In fact, an example of a Cauchy sequence without limit in X is given by any sequence of polynomials which converges uniformly on J to a continuous function, not a polynomial.

1.5-9 Continuous functions. Let X be the set of all continuous real-valued functions on $J = [0, 1]$, and let

$$d(x, y) = \int_0^1 |x(t) - y(t)| \, dt.$$

This metric space (X, d) is not complete.

Proof. The functions x_m in Fig. 9 form a Cauchy sequence because $d(x_m, x_n)$ is the area of the triangle in Fig. 10, and for every given $\varepsilon > 0$,

$$d(x_m, x_n) < \varepsilon \qquad \text{when} \qquad m, n > 1/\varepsilon.$$

Let us show that this Cauchy sequence does not converge. We have

$$x_m(t) = 0 \text{ if } t \in [0, \tfrac{1}{2}], \qquad x_m(t) = 1 \text{ if } t \in [a_m, 1]$$

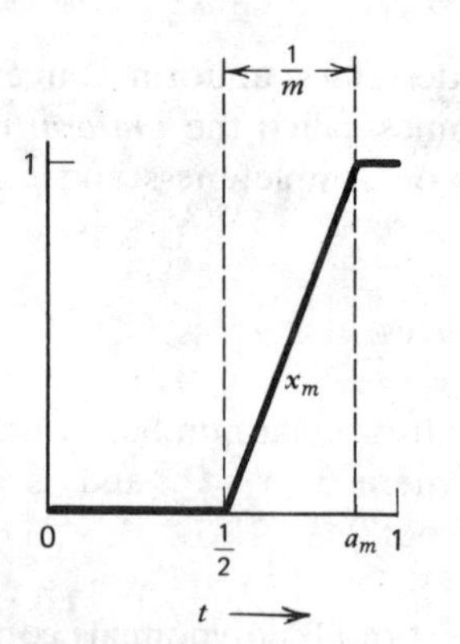

Fig. 9. Example 1.5-9

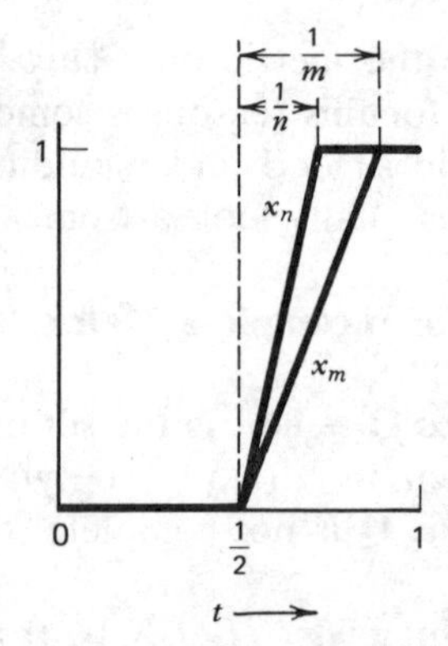

Fig. 10. Example 1.5-9

where $a_m = 1/2 + 1/m$. Hence for every $x \in X$,

$$d(x_m, x) = \int_0^1 |x_m(t) - x(t)|\, dt$$

$$= \int_0^{1/2} |x(t)|\, dt + \int_{1/2}^{a_m} |x_m(t) - x(t)|\, dt + \int_{a_m}^1 |1 - x(t)|\, dt.$$

Since the integrands are nonnegative, so is each integral on the right. Hence $d(x_m, x) \longrightarrow 0$ would imply that each integral approaches zero and, since x is continuous, we should have

$$x(t) = 0 \text{ if } t \in [0, \tfrac{1}{2}), \qquad x(t) = 1 \text{ if } t \in (\tfrac{1}{2}, 1].$$

But this is impossible for a continuous function. Hence (x_m) does not converge, that is, does not have a limit in X. This proves that X is not complete. ∎

Problems

1. Let $a, b \in \mathbf{R}$ and $a < b$. Show that the open interval (a, b) is an incomplete subspace of $\mathbf{R}$, whereas the closed interval $[a, b]$ is complete.

2. Let X be the space of all ordered n-tuples $x = (\xi_1, \cdots, \xi_n)$ of real numbers and

$$d(x, y) = \max_j |\xi_j - \eta_j|$$

where $y = (\eta_j)$. Show that (X, d) is complete.

3. Let $M \subset l^\infty$ be the subspace consisting of all sequences $x = (\xi_j)$ with at most finitely many nonzero terms. Find a Cauchy sequence in M which does not converge in M, so that M is not complete.

4. Show that M in Prob. 3 is not complete by applying Theorem 1.4-7.

5. Show that the set X of all integers with metric d defined by $d(m, n) = |m - n|$ is a complete metric space.

6. Show that the set of all real numbers constitutes an incomplete metric space if we choose

$$d(x, y) = |\arctan x - \arctan y|.$$

7. Let X be the set of all positive integers and $d(m, n) = |m^{-1} - n^{-1}|$. Show that (X, d) is not complete.

8. (Space $C[a, b]$) Show that the subspace $Y \subset C[a, b]$ consisting of all $x \in C[a, b]$ such that $x(a) = x(b)$ is complete.

9. In 1.5-5 we referred to the following theorem of calculus. If a sequence (x_m) of continuous functions on $[a, b]$ converges on $[a, b]$ and the convergence is uniform on $[a, b]$, then the limit function x is continuous on $[a, b]$. Prove this theorem.

10. (Discrete metric) Show that a discrete metric space (cf. 1.1-8) is complete.

11. (Space s) Show that in the space s (cf. 1.2-1) we have $x_n \longrightarrow x$ if and only if $\xi_j^{(n)} \longrightarrow \xi_j$ for all $j = 1, 2, \cdots$, where $x_n = (\xi_j^{(n)})$ and $x = (\xi_j)$.

12. Using Prob. 11, show that the sequence space s in 1.2-1 is complete.

13. Show that in 1.5-9, another Cauchy sequence is (x_n), where

$$x_n(t) = n \quad \text{if } 0 \leqq t \leqq n^{-2}, \qquad x_n(t) = t^{-\frac{1}{2}} \quad \text{if } n^{-2} \leqq t \leqq 1.$$

14. Show that the Cauchy sequence in Prob. 13 does not converge.

15. Let X be the metric space of all real sequences $x = (\xi_j)$ each of which has only finitely many nonzero terms, and $d(x, y) = \sum |\xi_j - \eta_j|$, where $y = (\eta_j)$. Note that this is a finite sum but the number of terms depends on x and y. Show that (x_n) with $x_n = (\xi_j^{(n)})$,

$$\xi_j^{(n)} = j^{-2} \quad \text{for } j = 1, \cdots, n \qquad \text{and} \qquad \xi_j^{(n)} = 0 \quad \text{for } j > n$$

is Cauchy but does not converge.

1.6 Completion of Metric Spaces

We know that the rational line **Q** is not complete (cf. 1.5-7) but can be "enlarged" to the real line **R** which is complete. And this "completion" **R** of **Q** is such that **Q** is dense (cf. 1.3-5) in **R**. It is quite important that an arbitrary incomplete metric space can be "completed" in a similar fashion, as we shall see. For a convenient precise formulation we use the following two related concepts, which also have various other applications.

1.6-1 Definition (Isometric mapping, isometric spaces). Let $X=(X, d)$ and $\tilde{X}=(\tilde{X}, \tilde{d})$ be metric spaces. Then:

(a) A mapping T of X into $\tilde{X}$ is said to be *isometric* or an *isometry* if T preserves distances, that is, if for all $x, y \in X$,

$$\tilde{d}(Tx, Ty) = d(x, y),$$

where Tx and Ty are the images of x and y, respectively.

(b) The space X is said to be *isometric* with the space $\tilde{X}$ if there exists a bijective[9] isometry of X onto $\tilde{X}$. The spaces X and $\tilde{X}$ are then called *isometric spaces.* ∎

Hence isometric spaces may differ at most by the nature of their points but are indistinguishable from the viewpoint of metric. And in any study in which the nature of the points does not matter, we may regard the two spaces as identical—as two copies of the same "abstract" space.

We can now state and prove the theorem that every metric space can be completed. The space $\hat{X}$ occurring in this theorem is called the **completion** of the given space X.

1.6-2 Theorem (Completion). *For a metric space $X=(X, d)$ there exists a complete metric space $\hat{X}=(\hat{X}, \hat{d})$ which has a subspace W that is isometric with X and is dense in $\hat{X}$. This space $\hat{X}$ is unique except for isometries, that is, if $\tilde{X}$ is any complete metric space having a dense subspace $\tilde{W}$ isometric with X, then $\tilde{X}$ and $\hat{X}$ are isometric.*

[9] One-to-one and onto. For a review of some elementary concepts related to mappings, see A1.2 in Appendix 1. Note that an isometric mapping is always injective. (Why?)

Proof. The proof is somewhat lengthy but straightforward. We subdivide it into four steps (a) to (d). We construct:

(a) $\hat{X} = (\hat{X}, \hat{d})$

(b) an isometry T of X onto W, where $\bar{W} = \hat{X}$.

Then we prove:

(c) completeness of $\hat{X}$,

(d) uniqueness of $\hat{X}$, except for isometries.

Roughly speaking, our task will be the assignment of suitable limits to Cauchy sequences in X that do not converge. However, we should not introduce "too many" limits, but take into account that certain sequences "may want to converge with the same limit" since the terms of those sequences "ultimately come arbitrarily close to each other." This intuitive idea can be expressed mathematically in terms of a suitable equivalence relation [see (1), below]. This is not artificial but is suggested by the process of completion of the rational line mentioned at the beginning of the section. The details of the proof are as follows.

(a) *Construction of* $\hat{X} = (\hat{X}, \hat{d})$. Let (x_n) and (x_n') be Cauchy sequences in X. Define (x_n) to be *equivalent*[10] to (x_n'), written $(x_n) \sim (x_n')$, if

$$\lim_{n\to\infty} d(x_n, x_n') = 0. \tag{1}$$

Let $\hat{X}$ be the set of all equivalence classes $\hat{x}, \hat{y}, \cdots$ of Cauchy sequences thus obtained. We write $(x_n) \in \hat{x}$ to mean that (x_n) is a member of $\hat{x}$ (a *representative* of the class $\hat{x}$). We now set

$$\hat{d}(\hat{x}, \hat{y}) = \lim_{n\to\infty} d(x_n, y_n) \tag{2}$$

where $(x_n) \in \hat{x}$ and $(y_n) \in \hat{y}$. We show that this limit exists. We have

$$d(x_n, y_n) \leqq d(x_n, x_m) + d(x_m, y_m) + d(y_m, y_n);$$

hence we obtain

$$d(x_n, y_n) - d(x_m, y_m) \leqq d(x_n, x_m) + d(y_m, y_n)$$

and a similar inequality with m and n interchanged. Together,

$$|d(x_n, y_n) - d(x_m, y_m)| \leqq d(x_n, x_m) + d(y_m, y_n). \tag{3}$$

[10] For a review of the concept of equivalence, see A1.4 in Appendix 1.

Since (x_n) and (y_n) are Cauchy, we can make the right side as small as we please. This implies that the limit in (2) exists because **R** is complete.

We must also show that the limit in (2) is independent of the particular choice of representatives. In fact, if $(x_n) \sim (x_n')$ and $(y_n) \sim (y_n')$, then by (1),

$$|d(x_n, y_n) - d(x_n', y_n')| \leqq d(x_n, x_n') + d(y_n, y_n') \longrightarrow 0$$

as $n \longrightarrow \infty$, which implies the assertion

$$\lim_{n\to\infty} d(x_n, y_n) = \lim_{n\to\infty} d(x_n', y_n').$$

We prove that $\hat{d}$ in (2) is a metric on $\hat{X}$. Obviously, $\hat{d}$ satisfies (M1) in Sec. 1.1 as well as $\hat{d}(\hat{x}, \hat{x}) = 0$ and (M3). Furthermore,

$$\hat{d}(\hat{x}, \hat{y}) = 0 \quad \Longrightarrow \quad (x_n) \sim (y_n) \quad \Longrightarrow \quad \hat{x} = \hat{y}$$

gives (M2), and (M4) for $\hat{d}$ follows from

$$d(x_n, y_n) \leqq d(x_n, z_n) + d(z_n, y_n)$$

by letting $n \longrightarrow \infty$.

(b) *Construction of an isometry* $T\colon X \longrightarrow W \subset \hat{X}$. With each $b \in X$ we associate the class $\hat{b} \in \hat{X}$ which contains the constant Cauchy sequence $(b, b, \cdots)$. This defines a mapping $T\colon X \longrightarrow W$ onto the subspace $W = T(X) \subset \hat{X}$. The mapping T is given by $b \longmapsto \hat{b} = Tb$, where $(b, b, \cdots) \in \hat{b}$. We see that T is an isometry since (2) becomes simply

$$\hat{d}(\hat{b}, \hat{c}) = d(b, c);$$

here $\hat{c}$ is the class of (y_n) where $y_n = c$ for all n. Any isometry is injective, and $T\colon X \longrightarrow W$ is surjective since $T(X) = W$. Hence W and X are isometric; cf. Def. 1.6-1(*b*).

We show that W is dense in $\hat{X}$. We consider any $\hat{x} \in \hat{X}$. Let $(x_n) \in \hat{x}$. For every $\varepsilon > 0$ there is an N such that

$$d(x_n, x_N) < \frac{\varepsilon}{2} \qquad (n > N).$$

Let $(x_N, x_N, \cdots) \in \hat{x}_N$. Then $\hat{x}_N \in W$. By (2),

$$\hat{d}(\hat{x}, \hat{x}_N) = \lim_{n\to\infty} d(x_n, x_N) \leqq \frac{\varepsilon}{2} < \varepsilon.$$

This shows that every ε-neighborhood of the arbitrary $\hat{x} \in \hat{X}$ contains an element of W. Hence W is dense in $\hat{X}$.

(c) *Completeness of* $\hat{X}$. Let $(\hat{x}_n)$ be any Cauchy sequence in $\hat{X}$. Since W is dense in $\hat{X}$, for every $\hat{x}_n$ there is a $\hat{z}_n \in W$ such that

$$\hat{d}(\hat{x}_n, \hat{z}_n) < \frac{1}{n}. \tag{4}$$

Hence by the triangle inequality,

$$\begin{aligned}\hat{d}(\hat{z}_m, \hat{z}_n) &\leqq \hat{d}(\hat{z}_m, \hat{x}_m) + \hat{d}(\hat{x}_m, \hat{x}_n) + \hat{d}(\hat{x}_n, \hat{z}_n)\\ &< \frac{1}{m} + \hat{d}(\hat{x}_m, \hat{x}_n) + \frac{1}{n}\end{aligned}$$

and this is less than any given $\varepsilon > 0$ for sufficiently large m and n because $(\hat{x}_m)$ is Cauchy. Hence $(\hat{z}_m)$ is Cauchy. Since $T\colon X \longrightarrow W$ is isometric and $\hat{z}_m \in W$, the sequence (z_m), where $z_m = T^{-1}\hat{z}_m$, is Cauchy in X. Let $\hat{x} \in \hat{X}$ be the class to which (z_m) belongs. We show that $\hat{x}$ is the limit of $(\hat{x}_n)$. By (4),

$$\begin{aligned}\hat{d}(\hat{x}_n, \hat{x}) &\leqq \hat{d}(\hat{x}_n, \hat{z}_n) + \hat{d}(\hat{z}_n, \hat{x})\\ &< \frac{1}{n} + \hat{d}(\hat{z}_n, \hat{x}).\end{aligned} \tag{5}$$

Since $(z_m) \in \hat{x}$ (see right before) and $\hat{z}_n \in W$, so that $(z_n, z_n, z_n, \cdots) \in \hat{z}_n$, the inequality (5) becomes

$$\hat{d}(\hat{x}_n, \hat{x}) < \frac{1}{n} + \lim_{m\to\infty} d(z_n, z_m)$$

and the right side is smaller than any given $\varepsilon > 0$ for sufficiently large n. Hence the arbitrary Cauchy sequence $(\hat{x}_n)$ in $\hat{X}$ has the limit $\hat{x} \in \hat{X}$, and $\hat{X}$ is complete.

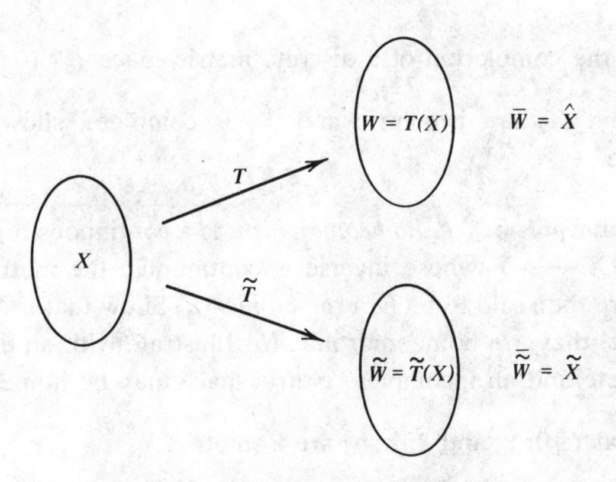

Fig. 11. Notations in part (*d*) of the proof of Theorem 1.6-2

(d) *Uniqueness of $\hat{X}$ except for isometries.* If $(\tilde{X}, \tilde{d})$ is another complete metric space with a subspace $\tilde{W}$ dense in $\tilde{X}$ and isometric with X, then for any $\tilde{x}, \tilde{y} \in \tilde{X}$ we have sequences $(\tilde{x}_n)$, $(\tilde{y}_n)$ in $\tilde{W}$ such that $\tilde{x}_n \longrightarrow \tilde{x}$ and $\tilde{y}_n \longrightarrow \tilde{y}$; hence

$$\tilde{d}(\tilde{x}, \tilde{y}) = \lim_{n \to \infty} \tilde{d}(\tilde{x}_n, \tilde{y}_n)$$

follows from

$$|\tilde{d}(\tilde{x}, \tilde{y}) - \tilde{d}(\tilde{x}_n, \tilde{y}_n)| \leq \tilde{d}(\tilde{x}, \tilde{x}_n) + \tilde{d}(\tilde{y}, \tilde{y}_n) \qquad \longrightarrow \qquad 0$$

[the inequality being similar to (3)]. Since $\tilde{W}$ is isometric with $W \subset \hat{X}$ and $\bar{W} = \hat{X}$, the distances on $\tilde{X}$ and $\hat{X}$ must be the same. Hence $\tilde{X}$ and $\hat{X}$ are isometric. ∎

We shall see in the next two chapters (in particular in 2.3-2, 3.1-5 and 3.2-3) that this theorem has basic applications to individual incomplete spaces as well as to whole classes of such spaces.

Problems

1. Show that if a subspace Y of a metric space consists of finitely many points, then Y is complete.

2. What is the completion of (X, d), where X is the set of all rational numbers and $d(x, y) = |x - y|$?

3. What is the completion of a discrete metric space X? (Cf. 1.1-8.)

4. If X_1 and X_2 are isometric and X_1 is complete, show that X_2 is complete.

5. (Homeomorphism) A *homeomorphism* is a continuous bijective mapping $T: X \longrightarrow Y$ whose inverse is continuous; the metric spaces X and Y are then said to be *homeomorphic.* (a) Show that if X and Y are isometric, they are homeomorphic. (b) Illustrate with an example that a complete and an incomplete metric space may be homeomorphic.

6. Show that $C[0, 1]$ and $C[a, b]$ are isometric.

7. If (X, d) is complete, show that $(X, \tilde{d})$, where $\tilde{d} = d/(1+d)$, is complete.

8. Show that in Prob. 7, completeness of $(X, \tilde{d})$ implies completeness of (X, d).

9. If (x_n) and (x_n') in (X, d) are such that (1) holds and $x_n \longrightarrow l$, show that (x_n') converges and has the limit l.

10. If (x_n) and (x_n') are convergent sequences in a metric space (X, d) and have the same limit l, show that they satisfy (1).

11. Show that (1) defines an equivalence relation on the set of all Cauchy sequences of elements of X.

12. If (x_n) is Cauchy in (X, d) and (x_n') in X satisfies (1), show that (x_n') is Cauchy in X.

13. (Pseudometric) A *finite pseudometric* on a set X is a function $d: X \times X \longrightarrow \mathbf{R}$ satisfying (M1), (M3), (M4), Sec. 1.1, and

$$\text{(M2*)} \qquad d(x, x) = 0.$$

What is the difference between a metric and a pseudometric? Show that $d(x, y) = |\xi_1 - \eta_1|$ defines a pseudometric on the set of all ordered pairs of real numbers, where $x = (\xi_1, \xi_2)$, $y = (\eta_1, \eta_2)$. (We mention that some authors use the term *semimetric* instead of *pseudometric.*)

14. Does

$$d(x, y) = \int_a^b |x(t) - y(t)| \, dt$$

define a metric or pseudometric on X if X is (*i*) the set of all real-valued continuous functions on $[a, b]$, (*ii*) the set of all real-valued Riemann integrable functions on $[a, b]$?

15. If (X, d) is a pseudometric space, we call a set

$$B(x_0; r) = \{x \in X \mid d(x, x_0) < r\} \qquad (r > 0)$$

an *open ball* in X with *center* x_0 and *radius* r. (Note that this is analogous to 1.3-1.) What are open balls of radius 1 in Prob. 13?

CHAPTER 2

NORMED SPACES. BANACH SPACES

Particularly useful and important metric spaces are obtained if we take a vector space and define on it a metric by means of a *norm.* The resulting space is called a *normed space.* If it is a complete metric space, it is called a *Banach space.* The theory of normed spaces, in particular Banach spaces, and the theory of linear operators defined on them are the most highly developed parts of functional analysis. The present chapter is devoted to the basic ideas of those theories.

Important concepts, brief orientation about main content

A *normed space* (cf. 2.2-1) is a *vector space* (cf. 2.1-1) with a metric defined by a *norm* (cf. 2.2-1); the latter generalizes the length of a vector in the plane or in three-dimensional space. A *Banach space* (cf. 2.2-1) is a normed space which is a complete metric space. A normed space has a completion which is a Banach space (cf. 2.3-2). In a normed space we can also define and use infinite series (cf. Sec. 2.3).

A mapping from a normed space X into a normed space Y is called an *operator.* A mapping from X into the scalar field $\mathbf{R}$ or $\mathbf{C}$ is called a *functional.* Of particular importance are so-called *bounded linear operators* (cf. 2.7-1) and *bounded linear functionals* (cf. 2.8-2) since they are continuous and take advantage of the vector space structure. In fact, Theorem 2.7-9 states that a *linear* operator is continuous if and only if it is bounded. This is a fundamental result. And vector spaces are of importance here mainly because of the linear operators and functionals they carry.

It is basic that the set of all bounded linear operators from a given normed space X into a given normed space Y can be made into a normed space (cf. 2.10–1), which is denoted by $B(X, Y)$. Similarly, the set of all bounded linear functionals on X becomes a normed space, which is called the *dual space* X' of X (cf. 2.10-3).

In analysis, infinite dimensional normed spaces are more important than finite dimensional ones. The latter are simpler (cf. Secs. 2.4, 2.5), and operators on them can be represented by matrices (cf. Sec. 2.9).

Remark on notation

We denote spaces by X and Y, operators by capital letters (preferably T), the image of an x under T by Tx (without parentheses), functionals by lowercase letters (preferably f) and the value of f at an x by $f(x)$ (with parentheses). This is a widely used practice.

2.1 Vector Space

Vector spaces play a role in many branches of mathematics and its applications. In fact, in various practical (and theoretical) problems we have a set X whose elements may be vectors in three-dimensional space, or sequences of numbers, or functions, and these elements can be added and multiplied by constants (numbers) in a natural way, the result being again an element of X. Such concrete situations suggest the concept of a vector space as defined below. The definition will involve a general field K, but in functional analysis, K will be **R** or **C**. The elements of K are called *scalars;* hence in our case they will be real or complex numbers.

2.1-1 Definition (Vector space). A *vector space* (or *linear space*) *over a field* K is a nonempty set X of elements $x, y, \cdots$ (called *vectors*) together with two algebraic operations. These operations are called *vector addition* and *multiplication of vectors by scalars*, that is, by elements of K.

Vector addition associates with every ordered pair (x, y) of vectors a vector $x+y$, called the *sum* of x *and* y, in such a way that the following properties hold.[1] Vector addition is commutative and associative, that is, for all vectors we have

$$x+y=y+x$$

$$x+(y+z)=(x+y)+z;$$

furthermore, there exists a vector 0, called the *zero vector*, and for every vector x there exists a vector $-x$, such that for all vectors we

[1] Readers familiar with groups will notice that we can summarize the defining properties of vector addition by saying that X is an additive abelian group.

have

$$x+0=x$$

$$x+(-x)=0.$$

Multiplication by scalars associates with every vector x and scalar α a vector αx (also written $x\alpha$), called the *product* of α and x, in such a way that for all vectors x, y and scalars α, β we have

$$\alpha(\beta x)=(\alpha\beta)x$$

$$1x=x$$

and the distributive laws

$$\alpha(x+y)=\alpha x+\alpha y$$

$$(\alpha+\beta)x=\alpha x+\beta x. \qquad \blacksquare$$

From the definition we see that vector addition is a mapping $X\times X\longrightarrow X$, whereas multiplication by scalars is a mapping $K\times X\longrightarrow X$.

K is called the **scalar field** (or *coefficient field*) of the vector space X, and X is called a **real vector space** if $K=\mathbf{R}$ (the field of real numbers), and a **complex vector space** if $K=\mathbf{C}$ (the field of complex numbers[2]).

The use of 0 for the scalar 0 as well as for the zero vector should cause no confusion, in general. If desirable for clarity, we can denote the zero vector by θ.

The reader may prove that for all vectors and scalars,

$$\text{(1a)} \qquad 0x=\theta$$

$$\text{(1b)} \qquad \alpha\theta=\theta$$

and

$$\text{(2)} \qquad (-1)x=-x.$$

[2] Remember that **R** and **C** also denote the real line and the complex plane, respectively (cf. 1.1-2 and 1.1-5), but we need not use other letters here since there is little danger of confusion.

Examples

2.1-2 Space $\mathbf{R}^n$. This is the Euclidean space introduced in 1.1-5, the underlying set being the set of all n-tuples of real numbers, written $x=(\xi_1, \cdots, \xi_n)$, $y=(\eta_1, \cdots, \eta_n)$, etc., and we now see that this is a real vector space with the two algebraic operations defined in the usual fashion

$$x+y=(\xi_1+\eta_1, \cdots, \xi_n+\eta_n)$$

$$\alpha x=(\alpha\xi_1, \cdots, \alpha\xi_n) \qquad (\alpha \in \mathbf{R}).$$

The next examples are of a similar nature because in each of them we shall recognize a previously defined space as a vector space.

2.1-3 Space $\mathbf{C}^n$. This space was defined in 1.1-5. It consists of all ordered n-tuples of complex numbers $x=(\xi_1, \cdots, \xi_n)$, $y=(\eta_1, \cdots, \eta_n)$, etc., and is a complex vector space with the algebraic operations defined as in the previous example, where now $\alpha \in \mathbf{C}$.

2.1-4 Space $C[a, b]$. This space was defined in 1.1-7. Each point of this space is a continuous real-valued function on $[a, b]$. The set of all these functions forms a real vector space with the algebraic operations defined in the usual way:

$$(x+y)(t)=x(t)+y(t)$$

$$(\alpha x)(t)=\alpha x(t) \qquad (\alpha \in \mathbf{R}).$$

In fact, $x+y$ and αx are continuous real-valued functions defined on $[a, b]$ if x and y are such functions and α is real.

Other important vector spaces of functions are (a) the vector space $B(A)$ in 1.2-2, (b) the vector space of all differentiable functions on $\mathbf{R}$, and (c) the vector space of all real-valued functions on $[a, b]$ which are integrable in some sense.

2.1-5 Space l^2. This space was introduced in 1.2-3. It is a vector space with the algebraic operations defined as usual in connection with sequences, that is,

$$(\xi_1, \xi_2, \cdots)+(\eta_1, \eta_2, \cdots)=(\xi_1+\eta_1, \xi_2+\eta_2, \cdots)$$

$$\alpha(\xi_1, \xi_2, \cdots)=(\alpha\xi_1, \alpha\xi_2, \cdots).$$

In fact, $x=(\xi_j)\in l^2$ and $y=(\eta_j)\in l^2$ implies $x+y\in l^2$, as follows readily from the Minkowski inequality (12) in Sec. 1.2; also $\alpha x\in l^2$.

Other vector spaces whose points are sequences are l^∞ in 1.1-6, l^p in 1.2-3, where $1\leqq p<+\infty$, and s in 1.2-1. ∎

A **subspace** of a vector space X is a nonempty subset Y of X such that for all $y_1, y_2\in Y$ and all scalars α, β we have $\alpha y_1+\beta y_2\in Y$. Hence Y is itself a vector space, the two algebraic operations being those induced from X.

A special subspace of X is the *improper subspace* $Y=X$. Every other subspace of X ($\neq\{0\}$) is called *proper.*

Another special subspace of any vector space X is $Y=\{0\}$.

A **linear combination** of vectors $x_1, \cdots, x_m$ of a vector space X is an expression of the form

$$\alpha_1x_1+\alpha_2x_2+\cdots+\alpha_mx_m$$

where the coefficients $\alpha_1, \cdots, \alpha_m$ are any scalars.

For any nonempty subset $M\subset X$ the set of all linear combinations of vectors of M is called the **span** of M, written

$$\operatorname{span} M.$$

Obviously, this is a subspace Y of X, and we say that Y is **spanned** or **generated** by M.

We shall now introduce two important related concepts which will be used over and over again.

2.1-6 Definition (Linear independence, linear dependence). Linear independence and dependence of a given set M of vectors $x_1, \cdots, x_r$ $(r\geqq 1)$ in a vector space X are defined by means of the equation

$$\alpha_1x_1+\alpha_2x_2+\cdots+\alpha_rx_r=0, \tag{3}$$

where $\alpha_1, \cdots, \alpha_r$ are scalars. Clearly, equation (3) holds for $\alpha_1=\alpha_2=\cdots=\alpha_r=0$. If this is the only r-tuple of scalars for which (3) holds, the set M is said to be *linearly independent.* M is said to be *linearly dependent* if M is not linearly independent, that is, if (3) also holds for some r-tuple of scalars, not all zero.

An arbitrary subset M of X is said to be *linearly independent* if every nonempty finite subset of M is linearly independent. M is said to be *linearly dependent* if M is not linearly independent. ∎

A motivation for this terminology results from the fact that if $M=\{x_1, \cdots, x_r\}$ is linearly dependent, at least one vector of M can be written as a linear combination of the others; for instance, if (3) holds with an $\alpha_r \neq 0$, then M is linearly dependent and we may solve (3) for x_r to get

$$x_r = \beta_1 x_1 + \cdots + \beta_{r-1} x_{r-1} \qquad (\beta_j = -\alpha_j/\alpha_r).$$

We can use the concepts of linear dependence and independence to define the dimension of a vector space, starting as follows.

2.1-7 Definition (Finite and infinite dimensional vector spaces). A vector space X is said to be *finite dimensional* if there is a positive integer n such that X contains a linearly independent set of n vectors whereas any set of $n+1$ or more vectors of X is linearly dependent. n is called the *dimension* of X, written $n = \dim X$. By definition, $X = \{0\}$ is finite dimensional and $\dim X = 0$. If X is not finite dimensional, it is said to be *infinite dimensional.* ∎

In analysis, infinite dimensional vector spaces are of greater interest than finite dimensional ones. For instance, $C[a, b]$ and l^2 are infinite dimensional, whereas $\mathbf{R}^n$ and $\mathbf{C}^n$ are n-dimensional.

If $\dim X = n$, a linearly independent n-tuple of vectors of X is called a **basis** *for* X (or a *basis in* X). If $\{e_1, \cdots, e_n\}$ is a basis for X, every $x \in X$ has a unique representation as a linear combination of the basis vectors:

$$x = \alpha_1 e_1 + \cdots + \alpha_n e_n.$$

For instance, a basis for $\mathbf{R}^n$ is

$$e_1 = (1, 0, 0, \cdots, 0),$$

$$e_2 = (0, 1, 0, \cdots, 0),$$

$$\cdots\cdots\cdots\cdots\cdots\cdots$$

$$\cdots\cdots\cdots\cdots\cdots\cdots$$

$$e_n = (0, 0, 0, \cdots, 1).$$

This is sometimes called the *canonical basis* for $\mathbf{R}^n$.

More generally, if X is any vector space, not necessarily finite dimensional, and B is a linearly independent subset of X which spans

X, then B is called a **basis** (or **Hamel basis**) for X. Hence if B is a basis for X, then every nonzero $x \in X$ has a unique representation as a linear combination of (finitely many!) elements of B with nonzero scalars as coefficients.

Every vector space $X \neq \{0\}$ has a basis.

In the finite dimensional case this is clear. For arbitrary infinite dimensional vector spaces an existence proof will be given by the use of Zorn's lemma. This lemma involves several concepts whose explanation would take us some time and, since at present a number of other things are more important to us, we do not pause but postpone that existence proof to Sec. 4.1, where we must introduce Zorn's lemma for another purpose.

We mention that all bases for a given (finite or infinite dimensional) vector space X have the same cardinal number. (A proof would require somewhat more advanced tools from set theory; cf. M. M. Day (1973), p. 3.) This number is called the **dimension** of X. Note that this includes and extends Def. 2.1-7.

Later we shall need the following simple

2.1-8 Theorem (Dimension of a subspace). *Let X be an n-dimensional vector space. Then any proper subspace Y of X has dimension less than n.*

Proof. If $n=0$, then $X=\{0\}$ and has no proper subspace. If $\dim Y=0$, then $Y=\{0\}$, and $X \neq Y$ implies $\dim X \geqq 1$. Clearly, $\dim Y \leqq \dim X = n$. If $\dim Y$ were n, then Y would have a basis of n elements, which would also be a basis for X since $\dim X = n$, so that $X = Y$. This shows that any linearly independent set of vectors in Y must have fewer than n elements, and $\dim Y < n$. ∎

Problems

1. Show that the set of all real numbers, with the usual addition and multiplication, constitutes a one-dimensional real vector space, and the set of all complex numbers constitutes a one-dimensional complex vector space.

2. Prove (1) and (2).

3. Describe the span of $M=\{(1,1,1),(0,0,2)\}$ in $\mathbf{R}^3$.

4. Which of the following subsets of $\mathbf{R}^3$ constitute a subspace of $\mathbf{R}^3$? [Here, $x=(\xi_1, \xi_2, \xi_3)$.]

(*a*) All x with $\xi_1=\xi_2$ and $\xi_3=0$.

(*b*) All x with $\xi_1=\xi_2+1$.

(*c*) All x with positive ξ_1, ξ_2, ξ_3.

(*d*) All x with $\xi_1-\xi_2+\xi_3=k=const.$

5. Show that $\{x_1, \cdots, x_n\}$, where $x_j(t)=t^j$, is a linearly independent set in the space $C[a, b]$.

6. Show that in an n-dimensional vector space X, the representation of any x as a linear combination of given basis vectors $e_1, \cdots, e_n$ is unique.

7. Let $\{e_1, \cdots, e_n\}$ be a basis for a complex vector space X. Find a basis for X regarded as a real vector space. What is the dimension of X in either case?

8. If M is a linearly dependent set in a complex vector space X, is M linearly dependent in X, regarded as a real vector space?

9. On a fixed interval $[a, b]\subset \mathbf{R}$, consider the set X consisting of all polynomials with real coefficients and of degree not exceeding a given n, and the polynomial $x=0$ (for which a degree is not defined in the usual discussion of degree). Show that X, with the usual addition and the usual multiplication by real numbers, is a real vector space of dimension $n+1$. Find a basis for X. Show that we can obtain a complex vector space $\tilde{X}$ in a similar fashion if we let those coefficients be complex. Is X a subspace of $\tilde{X}$?

10. If Y and Z are subspaces of a vector space X, show that $Y\cap Z$ is a subspace of X, but $Y\cup Z$ need not be one. Give examples.

11. If $M\neq\varnothing$ is any subset of a vector space X, show that span M is a subspace of X.

12. Show that the set of all real two-rowed square matrices forms a vector space X. What is the zero vector in X? Determine dim X. Find a basis for X. Give examples of subspaces of X. Do the symmetric matrices $x\in X$ form a subspace? The singular matrices?

13. (Product) Show that the Cartesian product $X=X_1\times X_2$ of two vector spaces over the same field becomes a vector space if we define the two algebraic operations by

$$(x_1, x_2)+(y_1, y_2)=(x_1+y_1, x_2+y_2),$$
$$\alpha(x_1, x_2)=(\alpha x_1, \alpha x_2).$$

14. (Quotient space, codimension) Let Y be a subspace of a vector space X. The *coset* of an element $x \in X$ with respect to Y is denoted by $x+Y$ and is defined to be the set (see Fig. 12)

$$x+Y=\{v \mid v=x+y, y \in Y\}.$$

Show that the distinct cosets form a partition of X. Show that under algebraic operations defined by (see Figs. 13, 14)

$$(w+Y)+(x+Y)=(w+x)+Y$$
$$\alpha(x+Y)=\alpha x+Y$$

these cosets constitute the elements of a vector space. This space is called the *quotient space* (or sometimes *factor space*) *of X by Y* (or *modulo* Y) and is denoted by X/Y. Its dimension is called the *codimension* of Y and is denoted by codim Y, that is,

$$\operatorname{codim} Y=\dim(X/Y).$$

15. Let $X=\mathbf{R}^3$ and $Y=\{(\xi_1, 0, 0) \mid \xi_1 \in \mathbf{R}\}$. Find X/Y, X/X, $X/\{0\}$.

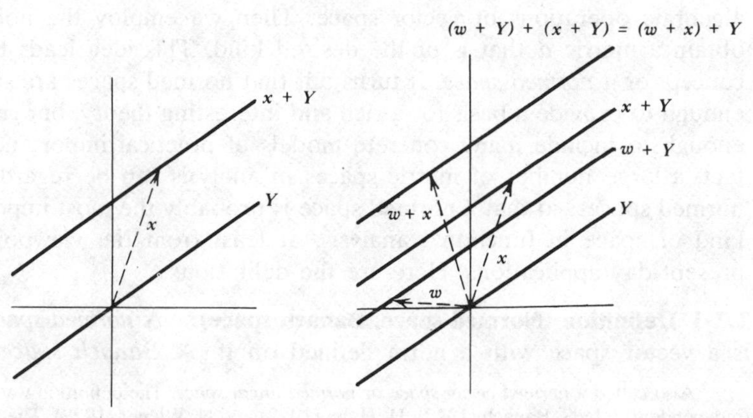

Fig. 12. Illustration of the notation $x+Y$ in Prob. 14

Fig. 13. Illustration of vector addition in a quotient space (cf. Prob. 14)

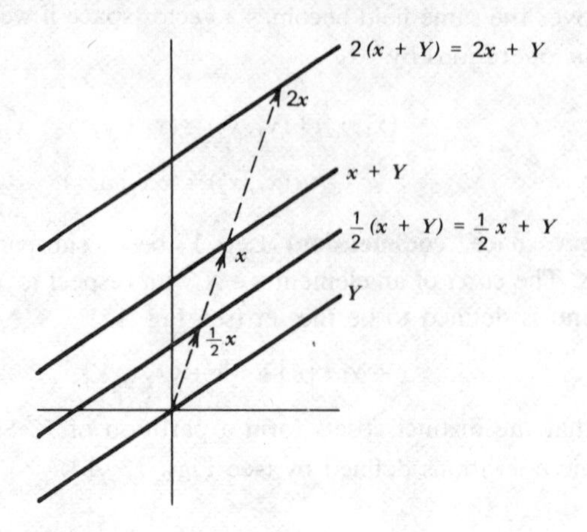

Fig. 14. Illustration of multiplication by scalars in a quotient space (cf. Prob. 14)

2.2 Normed Space. Banach Space

The examples in the last section illustrate that in many cases a vector space X may at the same time be a metric space because a metric d is defined on X. However, if there is no relation between the algebraic structure and the metric, we cannot expect a useful and applicable theory that combines algebraic and metric concepts. To guarantee such a relation between "algebraic" and "geometric" properties of X we define on X a metric d in a special way as follows. We first introduce an auxiliary concept, the *norm* (definition below), which uses the algebraic operations of vector space. Then we employ the norm to obtain a metric d that is of the desired kind. This idea leads to the concept of a *normed space.* It turns out that normed spaces are special enough to provide a basis for a rich and interesting theory, but general enough to include many concrete models of practical importance. In fact, a large number of metric spaces in analysis can be regarded as normed spaces, so that a normed space is probably the most important kind of space in functional analysis, at least from the viewpoint of present-day applications. Here are the definitions:

2.2-1 Definition (Normed space, Banach space). A *normed space*[3] X is a vector space with a norm defined on it. A *Banach space* is a

[3] Also called a *normed vector space* or *normed linear space.* The definition was given (independently) by S. Banach (1922), H. Hahn (1922) and N. Wiener (1922). The theory developed rapidly, as can be seen from the treatise by S. Banach (1932) published only ten years later.

complete normed space (complete in the metric defined by the norm; see (1), below). Here a **norm** on a (real or complex) vector space X is a real-valued function on X whose value at an $x \in X$ is denoted by

$$\|x\| \qquad \text{(read "norm of } x\text{")}$$

and which has the properties

(N1) $$\|x\| \geqq 0$$

(N2) $$\|x\| = 0 \quad \Longleftrightarrow \quad x = 0$$

(N3) $$\|\alpha x\| = |\alpha| \, \|x\|$$

(N4) $$\|x + y\| \leqq \|x\| + \|y\| \qquad (\textit{Triangle inequality});$$

here x and y are arbitrary vectors in X and α is any scalar.

A norm on X defines a metric d on X which is given by

$$\text{(1)} \qquad d(x, y) = \|x - y\| \qquad (x, y \in X)$$

and is called the *metric induced by the norm.* The normed space just defined is denoted by $(X, \|\cdot\|)$ or simply by X. ∎

The defining properties (N1) to (N4) of a norm are suggested and motivated by the length $|x|$ of a vector x in elementary vector algebra, so that in this case we can write $\|x\| = |x|$. In fact, (N1) and (N2) state that all vectors have positive lengths except the zero vector which has length zero. (N3) means that when a vector is multiplied by a scalar, its length is multiplied by the absolute value of the scalar. (N4) is illustrated in Fig. 15. It means that the length of one side of a triangle cannot exceed the sum of the lengths of the two other sides.

It is not difficult to conclude from (N1) to (N4) that (1) does define a metric. Hence normed spaces and Banach spaces are metric spaces.

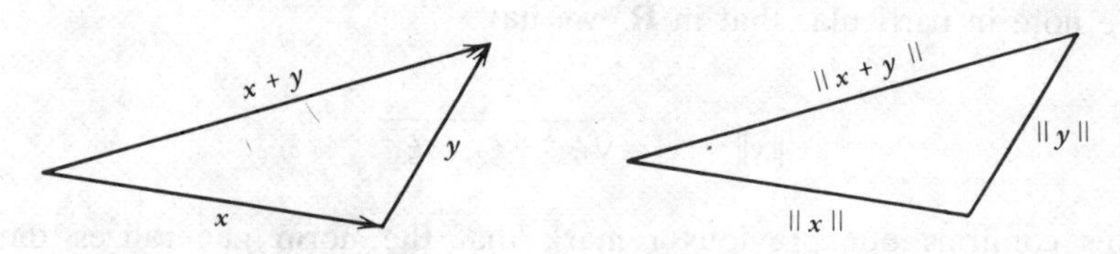

Fig. 15. Illustration of the triangle inequality (N4)

Banach spaces are important because they enjoy certain properties (to be discussed in Chap 4) which are not shared by incomplete normed spaces.

For later use we notice that (N4) implies

$$\big| \|y\| - \|x\| \big| \leqq \|y - x\|, \tag{2}$$

as the reader may readily prove (cf. Prob. 3). Formula (2) implies an important property of the norm:

The norm is continuous, that is, $x \longmapsto \|x\|$ *is a continuous mapping of* $(X, \|\cdot\|)$ *into* $\mathbf{R}$. (Cf. 1.3-3.)

Prototypes of normed spaces are the familiar spaces of all vectors in the plane and in three dimensional space. Further examples result from Secs. 1.1 and 1.2 since some of the metric spaces in those sections can be made into normed spaces in a natural way. However, we shall see later in this section that not every metric on a vector space can be obtained from a norm.

Examples

2.2-2 Euclidean space $\mathbf{R}^n$ and unitary space $\mathbf{C}^n$. These spaces were defined in 1.1-5. They are Banach spaces with norm defined by

$$\|x\| = \left(\sum_{j=1}^{n} |\xi_j|^2 \right)^{1/2} = \sqrt{|\xi_1|^2 + \cdots + |\xi_n|^2}. \tag{3}$$

In fact, $\mathbf{R}^n$ and $\mathbf{C}^n$ are complete (cf. 1.5-1), and (3) yields the metric (7) in Sec. 1.1:

$$d(x, y) = \|x - y\| = \sqrt{|\xi_1 - \eta_1|^2 + \cdots + |\xi_n - \eta_n|^2}.$$

We note in particular that in $\mathbf{R}^3$ we have

$$\|x\| = |x| = \sqrt{\xi_1^{\ 2} + \xi_2^{\ 2} + \xi_3^{\ 2}}.$$

This confirms our previous remark that the norm generalizes the elementary notion of the length $|x|$ of a vector.

2.2-3 Space l^p. This space was defined in 1.2-3. It is a Banach space with norm given by

$$\|x\| = \left(\sum_{j=1}^{\infty} |\xi_j|^p \right)^{1/p}. \tag{4}$$

In fact, this norm induces the metric in 1.2-3:

$$d(x, y) = \|x - y\| = \left(\sum_{j=1}^{\infty} |\xi_j - \eta_j|^p \right)^{1/p}.$$

Completeness was shown in 1.5-4.

2.2-4 Space l^∞. This space was defined in 1.1-6 and is a Banach space since its metric is obtained from the norm defined by

$$\|x\| = \sup_j |\xi_j|$$

and completeness was shown in 1.5-2.

2.2-5 Space $C[a, b]$. This space was defined in 1.1-7 and is a Banach space with norm given by

$$\|x\| = \max_{t \in J} |x(t)| \tag{5}$$

where $J = [a, b]$. Completeness was shown in 1.5-5.

2.2-6 Incomplete normed spaces. From the incomplete metric spaces in 1.5-7, 1.5-8 and 1.5-9 we may readily obtain incomplete normed spaces. For instance, the metric in 1.5-9 is induced by the norm defined by

$$\|x\| = \int_0^1 |x(t)|\, dt. \tag{6}$$

Can every incomplete normed space be completed? As a metric space certainly by 1.6-2. But what about extending the operations of a vector space and the norm to the completion? We shall see in the next section that the extension is indeed possible.

2.2-7 An incomplete normed space and its completion $L^2[a, b]$. The vector space of all continuous real-valued functions on $[a, b]$ forms a normed space X with norm defined by

$$\|x\| = \left(\int_a^b x(t)^2 \, dt \right)^{1/2}. \tag{7}$$

This space is not complete. For instance, if $[a, b] = [0, 1]$, the sequence in 1.5-9 is also Cauchy in the present space X; this is almost obvious from Fig. 10, Sec. 1.5, and results formally by integration because for $n > m$ we obtain

$$\|x_n - x_m\|^2 = \int_0^1 [x_n(t) - x_m(t)]^2 \, dt = \frac{(n-m)^2}{3mn^2} < \frac{1}{3m} - \frac{1}{3n}.$$

This Cauchy sequence does not converge. The proof is the same as in 1.5-9, with the metric in 1.5-9 replaced by the present metric. For a general interval $[a, b]$ we can construct a similar Cauchy sequence which does not converge in X.

The space X can be completed by Theorem 1.6-2. The completion is denoted by $L^2[a, b]$. This is a Banach space. In fact, the norm on X and the operations of vector space can be extended to the completion of X, as we shall see from Theorem 2.3-2 in the next section.

More generally, for any fixed real number $p \geqq 1$, the Banach space

$$L^p[a, b]$$

is the completion of the normed space which consists of all continuous real-valued functions on $[a, b]$, as before, and the norm defined by

$$\|x\|_p = \left(\int_a^b |x(t)|^p \, dt \right)^{1/p}. \tag{8}$$

The subscript p is supposed to remind us that this norm depends on the choice of p, which is kept fixed. Note that for $p = 2$ this equals (7).

For readers familiar with the Lebesgue integral we want to mention that the space $L^p[a, b]$ can also be obtained in a direct way by the use of the Lebesgue integral and Lebesgue measurable functions x on $[a, b]$ such that the Lebesgue integral of $|x|^p$ over $[a, b]$ exists and is finite. The elements of $L^p[a, b]$ are equivalence classes of those functions, where x is equivalent to y if the Lebesgue integral of $|x - y|^p$

over $[a, b]$ is zero. [Note that this guarantees the validity of axiom (N2).]

Readers without that background should not be disturbed. In fact, this example is not essential to the later development. At any rate, the example illustrates that completion may lead to a new kind of elements and one may have to find out what their nature is.

2.2-8 Space s. Can every metric on a vector space be obtained from a norm? The answer is no. A counterexample is the space s in 1.2-1. In fact, s is a vector space, but its metric d defined by

$$d(x, y) = \sum_{j=1}^{\infty} \frac{1}{2^j} \frac{|\xi_j - \eta_j|}{1 + |\xi_j - \eta_j|}$$

cannot be obtained from a norm. This may immediately be seen from the following lemma which states two basic properties of a metric d obtained from a norm. The first property, as expressed by (9a), is called the *translation invariance* of d.

2.2-9 Lemma (Translation invariance). *A metric d induced by a norm on a normed space X satisfies*

$$\text{(9)} \quad \begin{aligned} &\text{(a)} \qquad d(x+a, y+a) = d(x, y) \\ &\text{(b)} \qquad d(\alpha x, \alpha y) = |\alpha| \, d(x, y) \end{aligned}$$

for all $x, y, a \in X$ and every scalar α.

Proof. We have

$$d(x+a, y+a) = \|x + a - (y+a)\| = \|x - y\| = d(x, y)$$

and

$$d(\alpha x, \alpha y) = \|\alpha x - \alpha y\| = |\alpha| \, \|x - y\| = |\alpha| \, d(x, y). \qquad \blacksquare$$

Problems

1. Show that the norm $\|x\|$ of x is the distance from x to 0.

2. Verify that the usual length of a vector in the plane or in three dimensional space has the properties (N1) to (N4) of a norm.

3. Prove (2).

4. Show that we may replace (N2) by

$$\|x\|=0 \qquad \Longrightarrow \qquad x=0$$

without altering the concept of a norm. Show that nonnegativity of a norm also follows from (N3) and (N4).

5. Show that (3) defines a norm.

6. Let X be the vector space of all ordered pairs $x=(\xi_1, \xi_2)$, $y=(\eta_1, \eta_2), \cdots$ of real numbers. Show that norms on X are defined by

$$\|x\|_1=|\xi_1|+|\xi_2|$$

$$\|x\|_2=(\xi_1{}^2+\xi_2{}^2)^{1/2}$$

$$\|x\|_\infty=\max\{|\xi_1|, |\xi_2|\}.$$

7. Verify that (4) satisfies (N1) to (N4).

8. There are several norms of practical importance on the vector space of ordered n-tuples of numbers (cf. 2.2-2), notably those defined by

$$\|x\|_1=|\xi_1|+|\xi_2|+\cdots+|\xi_n|$$

$$\|x\|_p=(|\xi_1|^p+|\xi_2|^p+\cdots+|\xi_n|^p)^{1/p} \qquad (1<p<+\infty)$$

$$\|x\|_\infty=\max\{|\xi_1|, \cdots, |\xi_n|\}.$$

In each case, verify that (N1) to (N4) are satisfied.

9. Verify that (5) defines a norm.

10. (Unit sphere) The sphere

$$S(0; 1)=\{x \in X \mid \|x\|=1\}$$

in a normed space X is called the *unit sphere*. Show that for the norms in Prob. 6 and for the norm defined by

$$\|x\|_4=(\xi_1{}^4+\xi_2{}^4)^{1/4}$$

the ι nit spheres look as shown in Fig. 16.

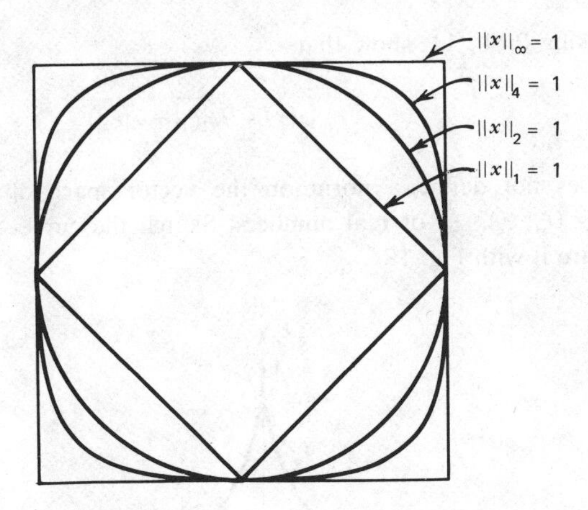

Fig. 16. Unit spheres in Prob. 10

11. (Convex set, segment) A subset A of a vector space X is said to be *convex* if $x, y \in A$ implies

$$M = \{z \in X \mid z = \alpha x + (1-\alpha)y, \quad 0 \leqq \alpha \leqq 1\} \subset A.$$

M is called a *closed segment* with *boundary points* x and y; any other $z \in M$ is called an *interior point* of M. Show that the *closed unit ball*

$$\tilde{B}(0; 1) = \{x \in X \mid \|x\| \leqq 1\}$$

in a normed space X is convex.

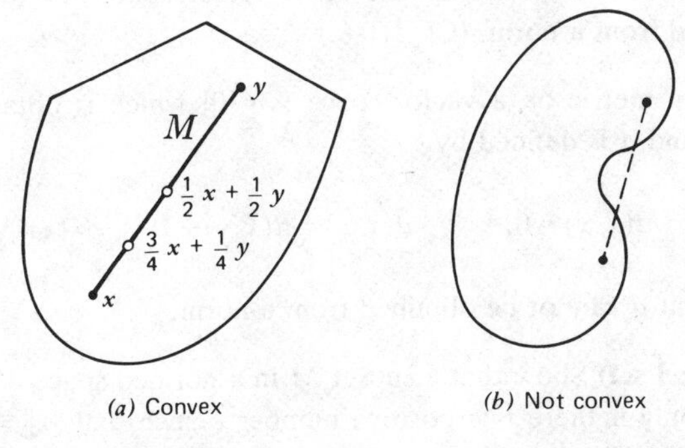

Fig. 17. Illustrative examples of convex and nonconvex sets (cf. Prob. 11)

12. Using Prob. 11, show that

$$\varphi(x)=(\sqrt{|\xi_1|}+\sqrt{|\xi_2|})^2$$

does not define a norm on the vector space of all ordered pairs $x=(\xi_1, \xi_2), \cdots$ of real numbers. Sketch the curve $\varphi(x)=1$ and compare it with Fig. 18.

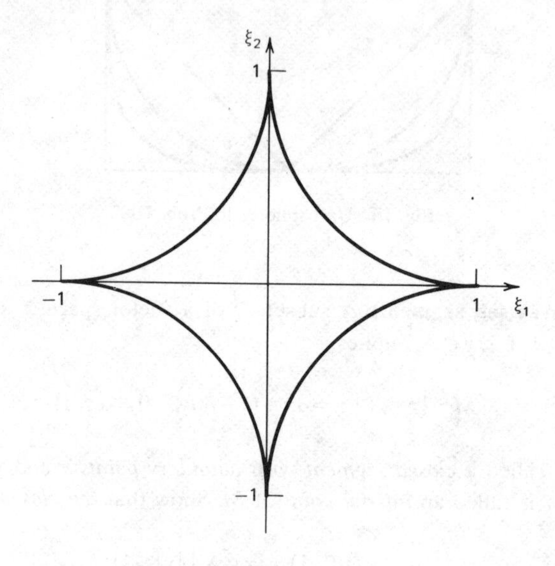

Fig. 18. Curve $\varphi(x)=1$ in Prob. 12

13. Show that the discrete metric on a vector space $X\neq\{0\}$ cannot be obtained from a norm. (Cf. 1.1-8.)

14. If d is a metric on a vector space $X\neq\{0\}$ which is obtained from a norm, and $\tilde{d}$ is defined by

$$\tilde{d}(x, x)=0, \qquad \tilde{d}(x, y)=d(x, y)+1 \qquad (x\neq y),$$

show that $\tilde{d}$ cannot be obtained from a norm.

15. (Bounded set) Show that a subset M in a normed space X is bounded if and only if there is a positive number c such that $\|x\|\leqq c$ for every $x\in M$. (For the definition, see Prob. 6 in Sec. 1.2.)

2.3 Further Properties of Normed Spaces

By definition, a **subspace** Y of a normed space X is a subspace of X considered as a vector space, with the norm obtained by restricting the norm on X to the subset Y. This norm on Y is said to be *induced* by the norm on X. If Y is closed in X, then Y is called a **closed subspace** of X.

By definition, a **subspace** Y of a Banach space X is a subspace of X considered as a normed space. Hence we do *not* require Y to be complete. (Some writers do, so be careful when comparing books.)

In this connection, Theorem 1.4-7 is useful since it yields immediately the following

2.3-1 Theorem (Subspace of a Banach space). *A subspace Y of a Banach space X is complete if and only if the set Y is closed in X.*

Convergence of sequences and related concepts in normed spaces follow readily from the corresponding definitions 1.4-1 and 1.4-3 for metric spaces and the fact that now $d(x, y) = \|x - y\|$:

(*i*) A sequence (x_n) in a normed space X is *convergent* if X contains an x such that

$$\lim_{n \to \infty} \|x_n - x\| = 0.$$

Then we write $x_n \longrightarrow x$ and call x the *limit* of (x_n).

(*ii*) A sequence (x_n) in a normed space X is *Cauchy* if for every $\varepsilon > 0$ there is an N such that

$$\|x_m - x_n\| < \varepsilon \qquad \text{for all } m, n > N. \tag{1}$$

Sequences were available to us even in a general metric space. In a normed space we may go an important step further and use series as follows.

Infinite series can now be defined in a way similar to that in calculus. In fact, if (x_k) is a sequence in a normed space X, we can associate with (x_k) the sequence (s_n) of *partial sums*

$$s_n = x_1 + x_2 + \cdots + x_n$$

where $n = 1, 2, \cdots$. If (s_n) is convergent, say,

$$s_n \longrightarrow s, \qquad \text{that is,} \qquad \|s_n - s\| \longrightarrow 0,$$

then the *infinite series* or, briefly, *series*

$$\sum_{k=1}^{\infty} x_k = x_1 + x_2 + \cdots \tag{2}$$

is said to *converge* or to *be* **convergent**, s is called the *sum* of the series and we write

$$s = \sum_{k=1}^{\infty} x_k = x_1 + x_2 + \cdots.$$

If $\|x_1\| + \|x_2\| + \cdots$ converges, the series (2) is said to be **absolutely convergent**. However, we warn the reader that in a normed space X, absolute convergence implies convergence if and only if X is complete (cf. Probs. 7 to 9).

The concept of convergence of a series can be used to define a "basis" as follows. If a normed space X contains a sequence (e_n) with the property that for every $x \in X$ there is a unique sequence of scalars (α_n) such that

$$\|x - (\alpha_1 e_1 + \cdots + \alpha_n e_n)\| \longrightarrow 0 \qquad (\text{as } n \longrightarrow \infty) \tag{3}$$

then (e_n) is called a **Schauder basis** (or *basis*) for X. The series

$$\sum_{k=1}^{\infty} \alpha_k e_k$$

which has the sum x is then called the *expansion* of x with respect to (e_n), and we write

$$x = \sum_{k=1}^{\infty} \alpha_k e_k.$$

For example, l^p in 2.2-3 has a Schauder basis, namely (e_n), where $e_n = (\delta_{nj})$, that is, e_n is the sequence whose nth term is 1 and all other terms are zero; thus

$$e_1 = (1, 0, 0, 0, \cdots)$$
$$e_2 = (0, 1, 0, 0, \cdots) \tag{4}$$
$$e_3 = (0, 0, 1, 0, \cdots)$$

etc.

If a normed space X has a Schauder basis, then X is separable (cf. Def. 1.3-5). The proof is simple, so that we can leave it to the reader (Prob. 10). Conversely, does *every* separable Banach space have a Schauder basis? This is a famous question raised by Banach himself about forty years ago. Almost all known separable Banach spaces had been shown to possess a Schauder basis. Nevertheless, the surprising answer to the question is no. It was given only quite recently, by P. Enflo (1973) who was able to construct a separable Banach space which has no Schauder basis.

Let us finally turn to the problem of completing a normed space, which was briefly mentioned in the last section.

2.3-2 Theorem (Completion). *Let $X = (X, \|\cdot\|)$ be a normed space. Then there is a Banach space $\hat{X}$ and an isometry A from X onto a subspace W of $\hat{X}$ which is dense in $\hat{X}$. The space $\hat{X}$ is unique, except for isometries.*

Proof. Theorem 1.6-2 implies the existence of a complete metric space $\hat{X} = (\hat{X}, \hat{d})$ and an isometry $A\colon X \longrightarrow W = A(X)$, where W is dense in $\hat{X}$ and $\hat{X}$ is unique, except for isometries. (We write A, not T as in 1.6-2, to free the letter T for later applications of the theorem in Sec. 8.2) Consequently, to prove the present theorem, we must make $\hat{X}$ into a vector space and then introduce on $\hat{X}$ a suitable norm.

To define on $\hat{X}$ the two algebraic operations of a vector space, we consider any $\hat{x}, \hat{y} \in \hat{X}$ and any representatives $(x_n) \in \hat{x}$ and $(y_n) \in \hat{y}$. Remember that $\hat{x}$ and $\hat{y}$ are equivalence classes of Cauchy sequences in X. We set $z_n = x_n + y_n$. Then (z_n) is Cauchy in X since

$$\|z_n - z_m\| = \|x_n + y_n - (x_m + y_m)\| \leqq \|x_n - x_m\| + \|y_n - y_m\|.$$

We define the sum $\hat{z} = \hat{x} + \hat{y}$ of $\hat{x}$ and $\hat{y}$ to be the equivalence class for which (z_n) is a representative; thus $(z_n) \in \hat{z}$. This definition is independent of the particular choice of Cauchy sequences belonging to $\hat{x}$ and $\hat{y}$. In fact, (1) in Sec. 1.6 shows that if $(x_n) \sim (x_n')$ and $(y_n) \sim (y_n')$, then

$(x_n+y_n) \sim (x_n'+y_n')$ because

$$\|x_n+y_n-(x_n'+y_n')\| \leqq \|x_n-x_n'\|+\|y_n-y_n'\|.$$

Similarly we define the product $\alpha\hat{x} \in \hat{X}$ of a scalar α and $\hat{x}$ to be the equivalence class for which (αx_n) is a representative. Again, this definition is independent of the particular choice of a representative of $\hat{x}$. The zero element of $\hat{X}$ is the equivalence class containing all Cauchy sequences which converge to zero. It is not difficult to see that those two algebraic operations have all the properties required by the definition, so that $\hat{X}$ is a vector space. From the definition it follows that on W the operations of vector space induced from $\hat{X}$ agree with those induced from X by means of A.

Furthermore, A induces on W a norm $\|\cdot\|_1$, whose value at every $\hat{y} = Ax \in W$ is $\|\hat{y}\|_1 = \|x\|$. The corresponding metric on W is the restriction of $\hat{d}$ to W since A is isometric. We can extend the norm $\|\cdot\|_1$ to $\hat{X}$ by setting $\|\hat{x}\|_2 = \hat{d}(\hat{0}, \hat{x})$ for every $\hat{x} \in \hat{X}$. In fact, it is obvious that $\|\cdot\|_2$ satisfies (N1) and (N2) in Sec. 2.2, and the other two axioms (N3) and (N4) follow from those for $\|\cdot\|_1$ by a limit process. ∎

Problems

1. Show that $c \subset l^\infty$ is a vector subspace of l^∞ (cf. 1.5-3) and so is c_0, the space of all sequences of scalars converging to zero.

2. Show that c_0 in Prob. 1 is a *closed* subspace of l^∞, so that c_0 is complete by 1.5-2 and 1.4-7.

3. In l^∞, let Y be the subset of all sequences with only finitely many nonzero terms. Show that Y is a subspace of l^∞ but not a closed subspace.

4. (Continuity of vector space operations) Show that in a normed space X, vector addition and multiplication by scalars are continuous operations with respect to the norm; that is, the mappings defined by $(x, y) \longmapsto x+y$ and $(\alpha, x) \longmapsto \alpha x$ are continuous.

5. Show that $x_n \longrightarrow x$ and $y_n \longrightarrow y$ implies $x_n+y_n \longrightarrow x+y$. Show that $\alpha_n \longrightarrow \alpha$ and $x_n \longrightarrow x$ implies $\alpha_n x_n \longrightarrow \alpha x$.

6. Show that the closure $\bar{Y}$ of a subspace Y of a normed space X is again a vector subspace.

7. (Absolute convergence) Show that convergence of $\|y_1\|+\|y_2\|+\|y_3\|+\cdots$ may not imply convergence of $y_1+y_2+y_3+\cdots$. *Hint.* Consider Y in Prob. 3 and (y_n), where $y_n=(\eta_j^{(n)})$, $\eta_n^{(n)}=1/n^2$, $\eta_j^{(n)}=0$ for all $j\neq n$.

8. If in a normed space X, absolute convergence of any series always implies convergence of that series, show that X is complete.

9. Show that in a Banach space, an absolutely convergent series is convergent.

10. (Schauder basis) Show that if a normed space has a Schauder basis, it is separable.

11. Show that (e_n), where $e_n=(\delta_{nj})$, is a Schauder basis for l^p, where $1\leqq p<+\infty$.

12. (Seminorm) A *seminorm* on a vector space X is a mapping $p\colon X\longrightarrow \mathbf{R}$ satisfying (N1), (N3), (N4) in Sec. 2.2. (Some authors call this a *pseudonorm.*) Show that

$$p(0)=0,$$

$$|p(y)-p(x)|\leqq p(y-x).$$

(Hence if $p(x)=0$ implies $x=0$, then p is a norm.)

13. Show that in Prob. 12, the elements $x\in X$ such that $p(x)=0$ form a subspace N of X and a norm on X/N (cf. Prob. 14, Sec. 2.1) is defined by $\|\hat{x}\|_0=p(x)$, where $x\in\hat{x}$ and $\hat{x}\in X/N$.

14. (Quotient space) Let Y be a closed subspace of a normed space $(X,\|\cdot\|)$. Show that a norm $\|\cdot\|_0$ on X/Y (cf. Prob. 14, Sec. 2.1) is defined by

$$\|\hat{x}\|_0=\inf_{x\in\hat{x}}\|x\|$$

where $\hat{x}\in X/Y$, that is, $\hat{x}$ is any coset of Y.

15. (Product of normed spaces) If $(X_1,\|\cdot\|_1)$ and $(X_2,\|\cdot\|_2)$ are normed spaces, show that the product vector space $X=X_1\times X_2$ (cf. Prob. 13, Sec. 2.1) becomes a normed space if we define

$$\|x\|=\max\left(\|x_1\|_1,\|x_2\|_2\right) \qquad [x=(x_1,x_2)].$$

2.4 Finite Dimensional Normed Spaces and Subspaces

Are finite dimensional normed spaces simpler than infinite dimensional ones? In what respect? These questions are rather natural. They are important since finite dimensional spaces and subspaces play a role in various considerations (for instance, in approximation theory and spectral theory). Quite a number of interesting things can be said in this connection. Hence it is worthwhile to collect some relevant facts, for their own sake and as tools for our further work. This is our program in this section and the next one.

A source for results of the desired type is the following lemma. Very roughly speaking it states that in the case of linear independence of vectors we cannot find a linear combination that involves large scalars but represents a small vector.

2.4-1 Lemma (Linear combinations). *Let $\{x_1, \cdots, x_n\}$ be a linearly independent set of vectors in a normed space X (of any dimension). Then there is a number $c>0$ such that for every choice of scalars $\alpha_1, \cdots, \alpha_n$ we have*

$$\|\alpha_1 x_1 + \cdots + \alpha_n x_n\| \geqq c(|\alpha_1| + \cdots + |\alpha_n|) \qquad (c>0). \tag{1}$$

Proof. We write $s = |\alpha_1| + \cdots + |\alpha_n|$. If $s=0$, all α_j are zero, so that (1) holds for any c. Let $s>0$. Then (1) is equivalent to the inequality which we obtain from (1) by dividing by s and writing $\beta_j = \alpha_j/s$, that is,

$$\|\beta_1 x_1 + \cdots + \beta_n x_n\| \geqq c \qquad \left(\sum_{j=1}^{n} |\beta_j| = 1\right). \tag{2}$$

Hence it suffices to prove the existence of a $c>0$ such that (2) holds for every n-tuple of scalars $\beta_1, \cdots, \beta_n$ with $\sum |\beta_j| = 1$.

Suppose that this is false. Then there exists a sequence (y_m) of vectors

$$y_m = \beta_1^{(m)} x_1 + \cdots + \beta_n^{(m)} x_n \qquad \left(\sum_{j=1}^{n} |\beta_j^{(m)}| = 1\right)$$

such that

$$\|y_m\| \longrightarrow 0 \qquad \text{as } m \longrightarrow \infty.$$

Now we reason as follows. Since $\sum |\beta_j^{(m)}| = 1$, we have $|\beta_j^{(m)}| \leqq 1$. Hence for each fixed j the sequence

$$(\beta_j^{(m)}) = (\beta_j^{(1)}, \beta_j^{(2)}, \cdots)$$

is bounded. Consequently, by the Bolzano-Weierstrass theorem, $(\beta_1^{(m)})$ has a convergent subsequence. Let β_1 denote the limit of that subsequence, and let $(y_{1,m})$ denote the corresponding subsequence of (y_m). By the same argument, $(y_{1,m})$ has a subsequence $(y_{2,m})$ for which the corresponding subsequence of scalars $\beta_2^{(m)}$ converges; let β_2 denote the limit. Continuing in this way, after n steps we obtain a subsequence $(y_{n,m}) = (y_{n,1}, y_{n,2}, \cdots)$ of (y_m) whose terms are of the form

$$y_{n,m} = \sum_{j=1}^{n} \gamma_j^{(m)} x_j \qquad \left(\sum_{j=1}^{n} |\gamma_j^{(m)}| = 1\right)$$

with scalars $\gamma_j^{(m)}$ satisfying $\gamma_j^{(m)} \longrightarrow \beta_j$ as $m \longrightarrow \infty$. Hence, as $m \longrightarrow \infty$,

$$y_{n,m} \quad \longrightarrow \quad y = \sum_{j=1}^{n} \beta_j x_j$$

where $\sum |\beta_j| = 1$, so that not all β_j can be zero. Since $\{x_1, \cdots, x_n\}$ is a linearly independent set, we thus have $y \neq 0$. On the other hand, $y_{n,m} \longrightarrow y$ implies $\|y_{n,m}\| \longrightarrow \|y\|$, by the continuity of the norm. Since $\|y_m\| \longrightarrow 0$ by assumption and $(y_{n,m})$ is a subsequence of (y_m), we must have $\|y_{n,m}\| \longrightarrow 0$. Hence $\|y\| = 0$, so that $y = 0$ by (N2) in Sec. 2.2. This contradicts $y \neq 0$, and the lemma is proved. ∎

As a first application of this lemma, let us prove the basic

2.4-2 Theorem (Completeness). *Every finite dimensional subspace Y of a normed space X is complete. In particular, every finite dimensional normed space is complete.*

Proof. We consider an arbitrary Cauchy sequence (y_m) in Y and show that it is convergent in Y; the limit will be denoted by y. Let $\dim Y = n$ and $\{e_1, \cdots, e_n\}$ any basis for Y. Then each y_m has a unique representation of the form

$$y_m = \alpha_1^{(m)} e_1 + \cdots + \alpha_n^{(m)} e_n.$$

Since (y_m) is a Cauchy sequence, for every $\varepsilon > 0$ there is an N such that $\|y_m - y_r\| < \varepsilon$ when $m, r > N$. From this and Lemma 2.4-1 we have for some $c > 0$

$$\varepsilon > \|y_m - y_r\| = \left\| \sum_{j=1}^{n} (\alpha_j^{(m)} - \alpha_j^{(r)}) e_j \right\| \geqq c \sum_{j=1}^{n} |\alpha_j^{(m)} - \alpha_j^{(r)}|,$$

where $m, r > N$. Division by $c > 0$ gives

$$|\alpha_j^{(m)} - \alpha_j^{(r)}| \leqq \sum_{j=1}^{n} |\alpha_j^{(m)} - \alpha_j^{(r)}| < \frac{\varepsilon}{c} \qquad (m, r > N).$$

This shows that each of the n sequences

$$(\alpha_j^{(m)}) = (\alpha_j^{(1)}, \alpha_j^{(2)}, \cdots) \qquad j = 1, \cdots, n$$

is Cauchy in **R** or **C**. Hence it converges; let α_j denote the limit. Using these n limits $\alpha_1, \cdots, \alpha_n$, we define

$$y = \alpha_1 e_1 + \cdots + \alpha_n e_n.$$

Clearly, $y \in Y$. Furthermore,

$$\|y_m - y\| = \left\| \sum_{j=1}^{n} (\alpha_j^{(m)} - \alpha_j) e_j \right\| \leqq \sum_{j=1}^{n} |\alpha_j^{(m)} - \alpha_j| \, \|e_j\|.$$

On the right, $\alpha_j^{(m)} \longrightarrow \alpha_j$. Hence $\|y_m - y\| \longrightarrow 0$, that is, $y_m \longrightarrow y$. This shows that (y_m) is convergent in Y. Since (y_m) was an arbitrary Cauchy sequence in Y, this proves that Y is complete. ∎

From this theorem and Theorem 1.4-7 we have

2.4-3 Theorem (Closedness). *Every finite dimensional subspace Y of a normed space X is closed in X.*

We shall need this theorem at several occasions in our further work.

Note that infinite dimensional subspaces need not be closed. *Example.* Let $X = C[0, 1]$ and $Y = \operatorname{span}(x_0, x_1, \cdots)$, where $x_j(t) = t^j$, so that Y is the set of all polynomials. Y is not closed in X. (Why?)

Another interesting property of a finite dimensional vector space X is that all norms on X lead to the same topology for X (cf. Sec. 1.3), that is, the open subsets of X are the same, regardless of the particular choice of a norm on X. The details are as follows.

2.4-4 Definition (Equivalent norms). A norm $\|\cdot\|$ on a vector space X is said to be *equivalent* to a norm $\|\cdot\|_0$ on X if there are positive numbers a and b such that for all $x \in X$ we have

$$a\|x\|_0 \leqq \|x\| \leqq b\|x\|_0. \tag{3}$$ ∎

This concept is motivated by the following fact.

Equivalent norms on X define the same topology for X.

Indeed, this follows from (3) and the fact that every nonempty open set is a union of open balls (cf. Prob. 4, Sec. 1.3). We leave the details of a formal proof to the reader (Prob. 4), who may also show that the Cauchy sequences in $(X, \|\cdot\|)$ and $(X, \|\cdot\|_0)$ are the same (Prob. 5).

Using Lemma 2.4-1, we can now prove the following theorem (which does *not* hold for infinite dimensional spaces).

2.4-5 Theorem (Equivalent norms). *On a finite dimensional vector space X, any norm $\|\cdot\|$ is equivalent to any other norm $\|\cdot\|_0$.*

Proof. Let $\dim X = n$ and $\{e_1, \cdots, e_n\}$ any basis for X. Then every $x \in X$ has a unique representation

$$x = \alpha_1 e_1 + \cdots + \alpha_n e_n.$$

By Lemma 2.4-1 there is a positive constant c such that

$$\|x\| \geqq c(|\alpha_1| + \cdots + |\alpha_n|).$$

On the other hand the triangle inequality gives

$$\|x\|_0 \leqq \sum_{j=1}^{n} |\alpha_j| \|e_j\|_0 \leqq k \sum_{j=1}^{n} |\alpha_j|, \qquad k = \max_j \|e_j\|_0.$$

Together, $a\|x\|_0 \leqq \|x\|$ where $a = c/k > 0$. The other inequality in (3) is now obtained by an interchange of the roles of $\|\cdot\|$ and $\|\cdot\|_0$ in the preceding argument. ∎

This theorem is of considerable practical importance. For instance, it implies that convergence or divergence of a sequence in a finite dimensional vector space does not depend on the particular choice of a norm on that space.

Problems

1. Give examples of subspaces of l^∞ and l^2 which are not closed.

2. What is the largest possible c in (1) if $X = \mathbf{R}^2$ and $x_1 = (1, 0)$, $x_2 = (0, 1)$? If $X = \mathbf{R}^3$ and $x_1 = (1, 0, 0)$, $x_2 = (0, 1, 0)$, $x_3 = (0, 0, 1)$?

3. Show that in Def. 2.4-4 the axioms of an equivalence relation hold (cf. A1.4 in Appendix 1).

4. Show that equivalent norms on a vector space X induce the same topology for X.

5. If $\|\cdot\|$ and $\|\cdot\|_0$ are equivalent norms on X, show that the Cauchy sequences in $(X, \|\cdot\|)$ and $(X, \|\cdot\|_0)$ are the same.

6. Theorem 2.4-5 implies that $\|\cdot\|_2$ and $\|\cdot\|_\infty$ in Prob. 8, Sec. 2.2, are equivalent. Give a direct proof of this fact.

7. Let $\|\cdot\|_2$ be as in Prob. 8, Sec. 2.2, and let $\|\cdot\|$ be any norm on that vector space, call it X. Show directly (without using 2.4-5) that there is a $b > 0$ such that $\|x\| \leqq b\, \|x\|_2$ for all x.

8. Show that the norms $\|\cdot\|_1$ and $\|\cdot\|_2$ in Prob. 8, Sec. 2.2, satisfy

$$\frac{1}{\sqrt{n}}\|x\|_1 \leqq \|x\|_2 \leqq \|x\|_1.$$

9. If two norms $\|\cdot\|$ and $\|\cdot\|_0$ on a vector space X are equivalent, show that (*i*) $\|x_n - x\| \longrightarrow 0$ implies (*ii*) $\|x_n - x\|_0 \longrightarrow 0$ (and vice versa, of course).

10. Show that all complex $m \times n$ matrices $A = (\alpha_{jk})$ with fixed m and n constitute an mn-dimensional vector space Z. Show that all norms on Z are equivalent. What would be the analogues of $\|\cdot\|_1$, $\|\cdot\|_2$ and $\|\cdot\|_\infty$ in Prob. 8, Sec. 2.2, for the present space Z?

2.5 Compactness and Finite Dimension

A few other basic properties of finite dimensional normed spaces and subspaces are related to the concept of compactness. The latter is defined as follows.

2.5-1 Definition (Compactness). A metric space X is said to be *compact*[4] if every sequence in X has a convergent subsequence. A subset M of X is said to be *compact* if M is compact considered as a subspace of X, that is, if every sequence in M has a convergent subsequence whose limit is an element of M. ∎

A general property of compact sets is expressed in

2.5-2 Lemma (Compactness). *A compact subset M of a metric space is closed and bounded.*

Proof. For every $x \in \bar{M}$ there is a sequence (x_n) in M such that $x_n \longrightarrow x$; cf. 1.4-6(a). Since M is compact, $x \in M$. Hence M is closed because $x \in \bar{M}$ was arbitrary. We prove that M is bounded. If M were unbounded, it would contain an unbounded sequence (y_n) such that $d(y_n, b) > n$, where b is any fixed element. This sequence could not have a convergent subsequence since a convergent subsequence must be bounded, by Lemma 1.4-2. ∎

The converse of this lemma is in general false.

Proof. To prove this important fact, we consider the sequence (e_n) in l^2, where $e_n = (\delta_{nj})$ has the nth term 1 and all other terms 0; cf. (4), Sec. 2.3. This sequence is bounded since $\|e_n\| = 1$. Its terms constitute a point set which is closed because it has no point of accumulation. For the same reason, that point set is not compact. ∎

However, for a finite dimensional normed space we have

2.5-3 Theorem (Compactness). *In a finite dimensional normed space X, any subset $M \subset X$ is compact if and only if M is closed and bounded.*

[4] More precisely, *sequentially compact;* this is the most important kind of compactness in analysis. We mention that there are two other kinds of compactness, but for metric spaces the three concepts become identical, so that the distinction does not matter in our work. (The interested reader will find some further remarks in A1.5. Appendix 1.)

Proof. Compactness implies closedness and boundedness by Lemma 2.5-2, and we prove the converse. Let M be closed and bounded. Let $\dim X = n$ and $\{e_1, \cdots, e_n\}$ a basis for X. We consider any sequence (x_m) in M. Each x_m has a representation

$$x_m = \xi_1^{(m)} e_1 + \cdots + \xi_n^{(m)} e_n.$$

Since M is bounded, so is (x_m), say, $\|x_m\| \leqq k$ for all m. By Lemma 2.4-1,

$$k \geqq \|x_m\| = \left\| \sum_{j=1}^{n} \xi_j^{(m)} e_j \right\| \geqq c \sum_{j=1}^{n} |\xi_j^{(m)}|$$

where $c > 0$. Hence the sequence of numbers $(\xi_j^{(m)})$ (j fixed) is bounded and, by the Bolzano-Weierstrass theorem, has a point of accumulation ξ_j; here $1 \leqq j \leqq n$. As in the proof of Lemma 2.4-1 we conclude that (x_m) has a subsequence (z_m) which converges to $z = \sum \xi_j e_j$. Since M is closed, $z \in M$. This shows that the arbitrary sequence (x_m) in M has a subsequence which converges in M. Hence M is compact. ∎

Our discussion shows the following. In $\mathbf{R}^n$ (or in any other finite dimensional normed space) the compact subsets are precisely the closed and bounded subsets, so that this property (closedness and boundedness) can be used for *defining* compactness. However, this can no longer be done in the case of an infinite dimensional normed space.

A source of other interesting results is the following lemma by F. Riesz (1918, pp. 75–76).

2.5-4 F. Riesz's Lemma. *Let Y and Z be subspaces of a normed space X (of any dimension), and suppose that Y is closed and is a proper subset of Z. Then for every real number θ in the interval* $(0, 1)$ *there is a $z \in Z$ such that*

$$\|z\| = 1, \qquad \|z - y\| \geqq \theta \text{ for all } y \in Y.$$

Proof. We consider any $v \in Z - Y$ and denote its distance from Y by a, that is (Fig. 19),

$$a = \inf_{y \in Y} \|v - y\|.$$

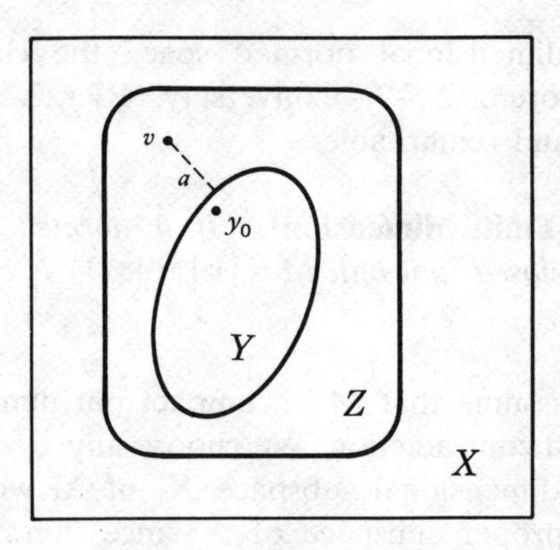

Fig. 19. Notations in the proof of Riesz's lemma

Clearly, $a>0$ since Y is closed. We now take any $\theta \in (0, 1)$. By the definition of an infimum there is a $y_0 \in Y$ such that

$$a \leqq \|v-y_0\| \leqq \frac{a}{\theta} \tag{1}$$

(note that $a/\theta > a$ since $0<\theta<1$). Let

$$z = c(v-y_0) \qquad \text{where} \qquad c = \frac{1}{\|v-y_0\|}.$$

Then $\|z\|=1$, and we show that $\|z-y\| \geqq \theta$ for every $y \in Y$. We have

$$\begin{aligned} \|z-y\| &= \|c(v-y_0)-y\| \\ &= c\,\|v-y_0-c^{-1}y\| \\ &= c\,\|v-y_1\| \end{aligned}$$

where

$$y_1 = y_0 + c^{-1}y.$$

The form of y_1 shows that $y_1 \in Y$. Hence $\|v-y_1\| \geqq a$, by the definition of a. Writing c out and using (1), we obtain

$$\|z-y\| = c\,\|v-y_1\| \geqq ca = \frac{a}{\|v-y_0\|} \geqq \frac{a}{a/\theta} = \theta.$$

Since $y \in Y$ was arbitrary, this completes the proof. ∎

In a finite dimensional normed space the closed unit ball is compact by Theorem 2.5-3. Conversely, Riesz's lemma gives the following useful and remarkable

2.5-5 Theorem (Finite dimension). *If a normed space X has the property that the closed unit ball $M=\{x \mid \|x\| \leqq 1\}$ is compact, then X is finite dimensional.*

Proof. We assume that M is compact but $\dim X=\infty$, and show that this leads to a contradiction. We choose any x_1 of norm 1. This x_1 generates a one dimensional subspace X_1 of X, which is closed (cf. 2.4-3) and is a proper subspace of X since $\dim X=\infty$. By Riesz's lemma there is an $x_2 \in X$ of norm 1 such that

$$\|x_2-x_1\| \geqq \theta=\frac{1}{2}.$$

The elements x_1, x_2 generate a two dimensional proper closed subspace X_2 of X. By Riesz's lemma there is an x_3 of norm 1 such that for all $x \in X_2$ we have

$$\|x_3-x\| \geqq \frac{1}{2}.$$

In particular,

$$\|x_3-x_1\| \geqq \frac{1}{2},$$

$$\|x_3-x_2\| \geqq \frac{1}{2}.$$

Proceeding by induction, we obtain a sequence (x_n) of elements $x_n \in M$ such that

$$\|x_m-x_n\| \geqq \frac{1}{2} \qquad (m \neq n).$$

Obviously, (x_n) cannot have a convergent subsequence. This contradicts the compactness of M. Hence our assumption $\dim X=\infty$ is false, and $\dim X<\infty$. ∎

This theorem has various applications. We shall use it in Chap. 8 as a basic tool in connection with so-called compact operators.

Compact sets are important since they are "well-behaved": they have several basic properties similar to those of finite sets and not shared by noncompact sets. In connection with continuous mappings a fundamental property is that compact sets have compact images, as follows.

2.5-6 Theorem (Continuous mapping). *Let X and Y be metric spaces and $T: X \longrightarrow Y$ a continuous mapping* (cf. 1.3-3). *Then the image of a compact subset M of X under T is compact.*

Proof. By the definition of compactness it suffices to show that every sequence (y_n) in the image $T(M) \subset Y$ contains a subsequence which converges in $T(M)$. Since $y_n \in T(M)$, we have $y_n = Tx_n$ for some $x_n \in M$. Since M is compact, (x_n) contains a subsequence (x_{n_k}) which converges in M. The image of (x_{n_k}) is a subsequence of (y_n) which converges in $T(M)$ by 1.4-8 because T is continuous. Hence $T(M)$ is compact. ∎

From this theorem we conclude that the following property, well-known from calculus for continuous functions, carries over to metric spaces.

2.5-7 Corollary (Maximum and minimum). *A continuous mapping T of a compact subset M of a metric space X into $\mathbf{R}$ assumes a maximum and a minimum at some points of M.*

Proof. $T(M) \subset \mathbf{R}$ is compact by Theorem 2.5-6 and closed and bounded by Lemma 2.5-2 [applied to $T(M)$], so that $\inf T(M) \in T(M)$, $\sup T(M) \in T(M)$, and the inverse images of these two points consist of points of M at which Tx is minimum or maximum, respectively. ∎

Problems

1. Show that $\mathbf{R}^n$ and $\mathbf{C}^n$ are not compact.

2. Show that a discrete metric space X (cf. 1.1-8) consisting of infinitely many points is not compact.

3. Give examples of compact and noncompact curves in the plane $\mathbf{R}^2$.

4. Show that for an infinite subset M in the space s (cf. 2.2-8) to be compact, it is necessary that there are numbers $\gamma_1, \gamma_2, \cdots$ such that for all $x = (\xi_k(x)) \in M$ we have $|\xi_k(x)| \leqq \gamma_k$. (It can be shown that the condition is also sufficient for the compactness of M.)'

5. (Local compactness) A metric space X is said to be *locally compact* if every point of X has a compact neighborhood. Show that **R** and **C** and, more generally, $\mathbf{R}^n$ and $\mathbf{C}^n$ are locally compact.

6. Show that a compact metric space X is locally compact.

7. If $\dim Y < \infty$ in Riesz's lemma 2.5-4, show that one can even choose $\theta = 1$.

8. In Prob. 7, Sec. 2.4, show directly (without using 2.4-5) that there is an $a > 0$ such that $a\|x\|_2 \leqq \|x\|$. (Use 2.5-7.)

9. If X is a compact metric space and $M \subset X$ is closed, show that M is compact.

10. Let X and Y be metric spaces, X compact, and $T: X \longrightarrow Y$ bijective and continuous. Show that T is a homeomorphism (cf. Prob. 5, Sec. 1.6).

2.6 Linear Operators

In calculus we consider the real line **R** and real-valued functions on **R** (or on a subset of **R**). Obviously, any such function is a mapping[5] of its domain into **R**. In functional analysis we consider more general spaces, such as metric spaces and normed spaces, and mappings of these spaces.

In the case of vector spaces and, in particular, normed spaces, a mapping is called an **operator**.

Of special interest are operators which "preserve" the two algebraic operations of vector space, in the sense of the following definition.

2.6-1 Definition (Linear operator). A *linear operator* T is an operator such that

(i) the domain $\mathcal{D}(T)$ of T is a vector space and the range $\mathcal{R}(T)$ lies in a vector space over the same field,

[5] Some familiarity with the concept of a mapping and simple related concepts is assumed, but a review is included in A1.2; cf. Appendix 1.

(*ii*) for all $x, y \in \mathfrak{D}(T)$ and scalars α,

$$T(x+y) = Tx + Ty \tag{1}$$
$$T(\alpha x) = \alpha Tx.$$ ∎

Observe the **notation;** we write Tx instead of $T(x)$; this simplification is standard in functional analysis. Furthermore, ***for the remainder of the book we shall use the following notations.***

$\mathfrak{D}(T)$ denotes the domain of T.

$\mathfrak{R}(T)$ denotes the range of T.

$\mathcal{N}(T)$ denotes the null space of T.

By definition, the **null space** of T is the set of all $x \in \mathfrak{D}(T)$ such that $Tx = 0$. (Another word for null space is "kernel." We shall not adopt this term since we must reserve the word "kernel" for another purpose in the theory of integral equations.)

We should also say something about the use of arrows in connection with operators. Let $\mathfrak{D}(T) \subset X$ and $\mathfrak{R}(T) \subset Y$, where X and Y are vector spaces, both real or both complex. Then T is an operator *from* (or mapping *of*) $\mathfrak{D}(T)$ **onto** $\mathfrak{R}(T)$, written

$$T\colon\ \mathfrak{D}(T) \longrightarrow \mathfrak{R}(T),$$

or from $\mathfrak{D}(T)$ *into* Y, written

$$T\colon\ \mathfrak{D}(T) \longrightarrow Y.$$

If $\mathfrak{D}(T)$ is the whole space X, then—and only then—we write

$$T\colon\ X \longrightarrow Y.$$

Clearly, (1) is equivalent to

$$T(\alpha x + \beta y) = \alpha Tx + \beta Ty. \tag{2}$$

By taking $\alpha = 0$ in (1) we obtain the following formula which we shall need many times in our further work:

$$T0 = 0. \tag{3}$$

Formula (1) expresses the fact that a linear operator T is a **homomorphism** of a vector space (its domain) into another vector space, that is, T preserves the two operations of vector space, in the following sense. In (1) on the left we first apply a vector space operation (addition or multiplication by scalars) and then map the resulting vector into Y, whereas on the right we first map x and y into Y and then perform the vector space operations in Y, the outcome being the same. This property makes linear operators important. In turn, vector spaces are important in functional analysis mainly because of the linear operators defined on them.

We shall now consider some basic examples of linear operators and invite the reader to verify the linearity of the operator in each case.

Examples

2.6-2 Identity operator. The *identity operator* $I_X\colon X \longrightarrow X$ is defined by $I_X x = x$ for all $x \in X$. We also write simply I for I_X; thus, $Ix = x$.

2.6-3 Zero operator. The *zero operator* $0\colon X \longrightarrow Y$ is defined by $0x = 0$ for all $x \in X$.

2.6-4 Differentiation. Let X be the vector space of all polynomials on $[a, b]$. We may define a linear operator T on X by setting

$$Tx(t) = x'(t)$$

for every $x \in X$, where the prime denotes differentiation with respect to t. This operator T maps X onto itself.

2.6-5 Integration. A linear operator T from $C[a, b]$ into itself can be defined by

$$Tx(t) = \int_a^t x(\tau)\, d\tau \qquad t \in [a, b].$$

2.6-6 Multiplication by t. Another linear operator from $C[a, b]$ into itself is defined by

$$Tx(t) = tx(t).$$

T plays a role in physics (quantum theory), as we shall see in Chap. 11.

2.6-7 Elementary vector algebra. The *cross product* with one factor kept fixed defines a linear operator $T_1\colon \mathbf{R}^3 \longrightarrow \mathbf{R}^3$. Similarly, the *dot product* with one fixed factor defines a linear operator $T_2\colon \mathbf{R}^3 \longrightarrow \mathbf{R}$, say,

$$T_2 x = x \cdot a = \xi_1\alpha_1 + \xi_2\alpha_2 + \xi_3\alpha_3$$

where $a = (\alpha_j) \in \mathbf{R}^3$ is fixed.

2.6-8 Matrices. A *real matrix* $A = (\alpha_{jk})$ with r rows and n columns defines an operator $T\colon \mathbf{R}^n \longrightarrow \mathbf{R}^r$ by means of

$$y = Ax$$

where $x = (\xi_j)$ has n components and $y = (\eta_j)$ has r components and both vectors are written as column vectors because of the usual convention of matrix multiplication; writing $y = Ax$ out, we have

$$\begin{bmatrix} \eta_1 \\ \eta_2 \\ \cdot \\ \cdot \\ \cdot \\ \eta_r \end{bmatrix} = \begin{bmatrix} \alpha_{11} & \alpha_{12} & \cdots & \alpha_{1n} \\ \alpha_{21} & \alpha_{22} & \cdots & \alpha_{2n} \\ \cdot & \cdot & \cdots & \cdot \\ \cdot & \cdot & \cdots & \cdot \\ \cdot & \cdot & \cdots & \cdot \\ \alpha_{r1} & \alpha_{r2} & \cdots & \alpha_{rn} \end{bmatrix} \begin{bmatrix} \xi_1 \\ \xi_2 \\ \cdot \\ \cdot \\ \cdot \\ \xi_n \end{bmatrix}$$

T is linear because matrix multiplication is a linear operation. If A were complex, it would define a linear operator from $\mathbf{C}^n$ into $\mathbf{C}^r$. A detailed discussion of the role of matrices in connection with linear operators follows in Sec. 2.9. ∎

In these examples we can easily verify that the ranges and null spaces of the linear operators are vector spaces. This fact is typical. Let us prove it, thereby observing how the linearity is used in simple proofs. The theorem itself will have various applications in our further work.

2.6-9 Theorem (Range and null space). *Let T be a linear operator. Then:*

(a) *The range $\mathscr{R}(T)$ is a vector space.*

(b) *If* $\dim \mathscr{D}(T) = n < \infty$, *then* $\dim \mathscr{R}(T) \leqq n$.

(c) *The null space $\mathscr{N}(T)$ is a vector space.*

Proof. **(a)** We take any $y_1, y_2 \in \mathscr{R}(T)$ and show that $\alpha y_1 + \beta y_2 \in \mathscr{R}(T)$ for any scalars α, β. Since $y_1, y_2 \in \mathscr{R}(T)$, we have $y_1 = Tx_1$, $y_2 = Tx_2$ for some $x_1, x_2 \in \mathscr{D}(T)$, and $\alpha x_1 + \beta x_2 \in \mathscr{D}(T)$ because $\mathscr{D}(T)$ is a vector space. The linearity of T yields

$$T(\alpha x_1 + \beta x_2) = \alpha Tx_1 + \beta Tx_2 = \alpha y_1 + \beta y_2.$$

Hence $\alpha y_1 + \beta y_2 \in \mathscr{R}(T)$. Since $y_1, y_2 \in \mathscr{R}(T)$ were arbitrary and so were the scalars, this proves that $\mathscr{R}(T)$ is a vector space.

(b) We choose $n+1$ elements $y_1, \cdots, y_{n+1}$ of $\mathscr{R}(T)$ in an arbitrary fashion. Then we have $y_1 = Tx_1, \cdots, y_{n+1} = Tx_{n+1}$ for some $x_1, \cdots, x_{n+1}$ in $\mathscr{D}(T)$. Since $\dim \mathscr{D}(T) = n$, this set $\{x_1, \cdots, x_{n+1}\}$ must be linearly dependent. Hence

$$\alpha_1 x_1 + \cdots + \alpha_{n+1} x_{n+1} = 0$$

for some scalars $\alpha_1, \cdots, \alpha_{n+1}$, not all zero. Since T is linear and $T0 = 0$, application of T on both sides gives

$$T(\alpha_1 x_1 + \cdots + \alpha_{n+1} x_{n+1}) = \alpha_1 y_1 + \cdots + \alpha_{n+1} y_{n+1} = 0.$$

This shows that $\{y_1, \cdots, y_{n+1}\}$ is a linearly dependent set because the α_j's are not all zero. Remembering that this subset of $\mathscr{R}(T)$ was chosen in an arbitrary fashion, we conclude that $\mathscr{R}(T)$ has no linearly independent subsets of $n+1$ or more elements. By the definition this means that $\dim \mathscr{R}(T) \leqq n$.

(c) We take any $x_1, x_2 \in \mathscr{N}(T)$. Then $Tx_1 = Tx_2 = 0$. Since T is linear, for any scalars α, β we have

$$T(\alpha x_1 + \beta x_2) = \alpha Tx_1 + \beta Tx_2 = 0.$$

This shows that $\alpha x_1 + \beta x_2 \in \mathscr{N}(T)$. Hence $\mathscr{N}(T)$ is a vector space. ∎

An immediate consequence of part (*b*) of the proof is worth noting:

Linear operators preserve linear dependence.

Let us turn to the inverse of a linear operator. We first remember that a mapping $T: \mathscr{D}(T) \longrightarrow Y$ is said to be **injective** or **one-to-one** if different points in the domain have different images, that is, if for any $x_1, x_2 \in \mathcal{D}(T)$,

(4) $$x_1 \neq x_2 \qquad \Longrightarrow \qquad Tx_1 \neq Tx_2;$$

equivalently,

(4*) $$Tx_1 = Tx_2 \qquad \Longrightarrow \qquad x_1 = x_2.$$

In this case there exists the mapping

(5) $$\begin{aligned} T^{-1}\colon \mathcal{R}(T) &\longrightarrow \mathcal{D}(T) \\ y_0 &\longmapsto x_0 \qquad\qquad (y_0 = Tx_0) \end{aligned}$$

which maps every $y_0 \in \mathcal{R}(T)$ onto that $x_0 \in \mathcal{D}(T)$ for which $Tx_0 = y_0$. See Fig. 20. The mapping T^{-1} is called the **inverse**[6] of T.

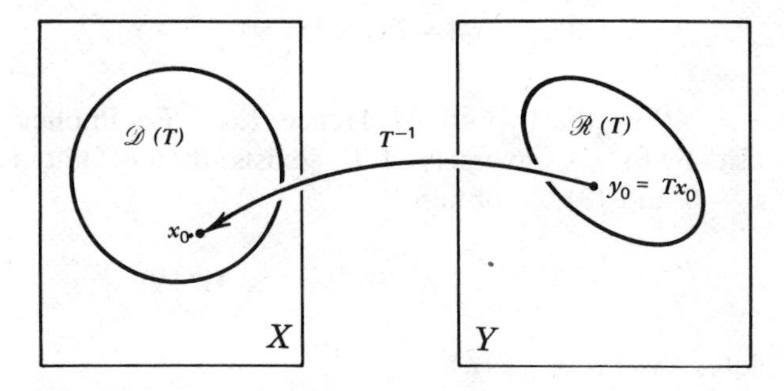

Fig. 20. Notations in connection with the inverse of a mapping; cf. (5)

From (5) we clearly have

$$T^{-1}Tx = x \qquad \text{for all } x \in \mathcal{D}(T)$$

$$TT^{-1}y = y \qquad \text{for all } y \in \mathcal{R}(T).$$

In connection with linear operators on vector spaces the situation is as follows. The inverse of a linear operator exists if and only if the null space of the operator consists of the zero vector only. More

[6] The reader may wish to review the terms "surjective" and "bijective" in A1.2, Appendix 1, which also contains a remark on the use of the term "inverse."

precisely, we have the following useful criterion which we shall apply quite often.

2.6-10 Theorem (Inverse operator). *Let X, Y be vector spaces, both real or both complex. Let $T: \mathcal{D}(T) \longrightarrow Y$ be a linear operator with domain $\mathcal{D}(T) \subset X$ and range $\mathcal{R}(T) \subset Y$. Then:*

(a) *The inverse $T^{-1}: \mathcal{R}(T) \longrightarrow \mathcal{D}(T)$ exists if and only if*

$$Tx = 0 \qquad \Longrightarrow \qquad x = 0.$$

(b) *If T^{-1} exists, it is a linear operator.*

(c) *If* $\dim \mathcal{D}(T) = n < \infty$ *and T^{-1} exists, then* $\dim \mathcal{R}(T) = \dim \mathcal{D}(T)$.

Proof. **(a)** Suppose that $Tx = 0$ implies $x = 0$. Let $Tx_1 = Tx_2$. Since T is linear,

$$T(x_1 - x_2) = Tx_1 - Tx_2 = 0,$$

so that $x_1 - x_2 = 0$ by the hypothesis. Hence $Tx_1 = Tx_2$ implies $x_1 = x_2$, and T^{-1} exists by (4*). Conversely, if T^{-1} exists, then (4*) holds. From (4*) with $x_2 = 0$ and (3) we obtain

$$Tx_1 = T0 = 0 \qquad \Longrightarrow \qquad x_1 = 0.$$

This completes the proof of (a).

(b) We assume that T^{-1} exists and show that T^{-1} is linear. The domain of T^{-1} is $\mathcal{R}(T)$ and is a vector space by Theorem 2.6-9(a). We consider any $x_1, x_2 \in \mathcal{D}(T)$ and their images

$$y_1 = Tx_1 \qquad \text{and} \qquad y_2 = Tx_2.$$

Then

$$x_1 = T^{-1}y_1 \qquad \text{and} \qquad x_2 = T^{-1}y_2.$$

T is linear, so that for any scalars α and β we have

$$\alpha y_1 + \beta y_2 = \alpha Tx_1 + \beta Tx_2 = T(\alpha x_1 + \beta x_2).$$

Since $x_j = T^{-1}y_j$, this implies

$$T^{-1}(\alpha y_1 + \beta y_2) = \alpha x_1 + \beta x_2 = \alpha T^{-1} y_1 + \beta T^{-1} y_2$$

and proves that T^{-1} is linear.

(c) We have $\dim \mathscr{R}(T) \leqq \dim \mathscr{D}(T)$ by Theorem 2.6-9(*b*), and $\dim \mathscr{D}(T) \leqq \dim \mathscr{R}(T)$ by the same theorem applied to T^{-1}. ∎

We finally mention a useful formula for the inverse of the composite of linear operators. (The reader may perhaps know this formula for the case of square matrices.)

2.6-11 Lemma (Inverse of product). *Let $T: X \longrightarrow Y$ and $S: Y \longrightarrow Z$ be bijective linear operators, where X, Y, Z are vector spaces (see* Fig. 21). *Then the inverse $(ST)^{-1}: Z \longrightarrow X$ of the product (the composite) ST exists, and*

$$(ST)^{-1} = T^{-1}S^{-1}. \tag{6}$$

Proof. The operator $ST: X \longrightarrow Z$ is bijective, so that $(ST)^{-1}$ exists. We thus have

$$ST(ST)^{-1} = I_Z$$

where I_Z is the identity operator on Z. Applying S^{-1} and using $S^{-1}S = I_Y$ (the identity operator on Y), we obtain

$$S^{-1}ST(ST)^{-1} = T(ST)^{-1} = S^{-1}I_Z = S^{-1}.$$

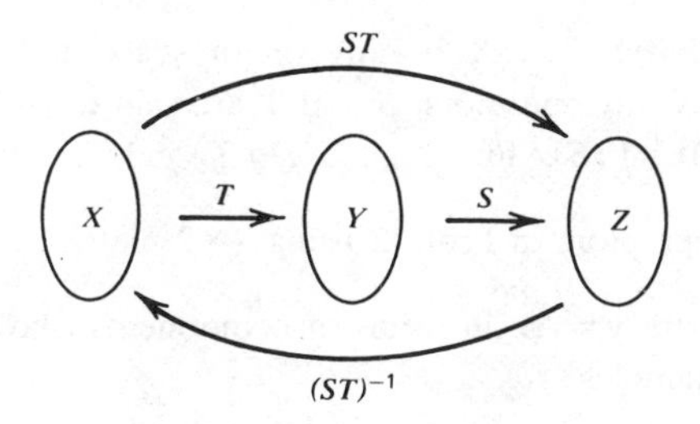

Fig. 21. Notations in Lemma 2.6-11

Applying T^{-1} and using $T^{-1}T = I_X$, we obtain the desired result

$$T^{-1}T(ST)^{-1} = (ST)^{-1} = T^{-1}S^{-1}.$$

This completes the proof. ∎

Problems

1. Show that the operators in 2.6-2, 2.6-3 and 2.6-4 are linear.

2. Show that the operators $T_1, \cdots, T_4$ from $\mathbf{R}^2$ into $\mathbf{R}^2$ defined by

$$(\xi_1, \xi_2) \longmapsto (\xi_1, 0)$$

$$(\xi_1, \xi_2) \longmapsto (0, \xi_2)$$

$$(\xi_1, \xi_2) \longmapsto (\xi_2, \xi_1)$$

$$(\xi_1, \xi_2) \longmapsto (\gamma\xi_1, \gamma\xi_2)$$

respectively, are linear, and interpret these operators geometrically.

3. What are the domain, range and null space of T_1, T_2, T_3 in Prob. 2?

4. What is the null space of T_4 in Prob. 2? Of T_1 and T_2 in 2.6-7? Of T in 2.6-4?

5. Let $T: X \longrightarrow Y$ be a linear operator. Show that the image of a subspace V of X is a vector space, and so is the inverse image of a subspace W of Y.

6. If the product (the composite) of two linear operators exists, show that it is linear.

7. (Commutativity) Let X be any vector space and $S: X \longrightarrow X$ and $T: X \longrightarrow X$ any operators. S and T are said to *commute* if $ST = TS$, that is, $(ST)x = (TS)x$ for all $x \in X$. Do T_1 and T_3 in Prob. 2 commute?

8. Write the operators in Prob. 2 using 2×2 matrices.

9. In 2.6-8, write $y = Ax$ in terms of components, show that T is linear and give examples.

10. Formulate the condition in 2.6-10(a) in terms of the null space of T.

11. Let X be the vector space of all complex 2×2 matrices and define $T: X \longrightarrow X$ by $Tx = bx$, where $b \in X$ is fixed and bx denotes the usual product of matrices. Show that T is linear. Under what condition does T^{-1} exist?

12. Does the inverse of T in 2.6-4 exist?

13. Let $T: \mathcal{D}(T) \longrightarrow Y$ be a linear operator whose inverse exists. If $\{x_1, \cdots, x_n\}$ is a linearly independent set in $\mathcal{D}(T)$, show that the set $\{Tx_1, \cdots, Tx_n\}$ is linearly independent.

14. Let $T: X \longrightarrow Y$ be a linear operator and $\dim X = \dim Y = n < \infty$. Show that $\mathcal{R}(T) = Y$ if and only if T^{-1} exists.

15. Consider the vector space X of all real-valued functions which are defined on $\mathbf{R}$ and have derivatives of all orders everywhere on $\mathbf{R}$. Define $T: X \longrightarrow X$ by $y(t) = Tx(t) = x'(t)$. Show that $\mathcal{R}(T)$ is all of X but T^{-1} does not exist. Compare with Prob. 14 and comment.

2.7 Bounded and Continuous Linear Operators

The reader may have noticed that in the whole last section we did not make any use of norms. We shall now again take norms into account, in the following basic definition.

2.7-1 Definition (Bounded linear operator). Let X and Y be normed spaces and $T: \mathcal{D}(T) \longrightarrow Y$ a linear operator, where $\mathcal{D}(T) \subset X$. The operator T is said to be *bounded* if there is a real number c such that for all $x \in \mathcal{D}(T)$,

$$\|Tx\| \leqq c\|x\|. \tag{1}$$ ∎

In (1) the norm on the left is that on Y, and the norm on the right is that on X. For simplicity we have denoted both norms by the same symbol $\|\cdot\|$, without danger of confusion. Distinction by subscripts ($\|x\|_0$, $\|Tx\|_1$, etc.) seems unnecessary here. Formula (1) shows that a bounded linear operator maps bounded sets in $\mathcal{D}(T)$ onto bounded sets in Y. This motivates the term "bounded operator."

Warning. Note that our present use of the word "bounded" is different from that in calculus, where a bounded function is one whose range is a bounded set. Unfortunately, both terms are standard. But there is little danger of confusion.

What is the smallest possible c such that (1) still holds for all nonzero $x \in \mathscr{D}(T)$? [We can leave out $x=0$ since $Tx=0$ for $x=0$ by (3), Sec. 2.6.] By division,

$$\frac{\|Tx\|}{\|x\|} \leqq c \qquad (x \neq 0)$$

and this shows that c must be at least as big as the supremum of the expression on the left taken over $\mathscr{D}(T)-\{0\}$. Hence the answer to our question is that the smallest possible c in (1) is that supremum. This quantity is denoted by $\|T\|$; thus

$$\|T\| = \sup_{\substack{x \in \mathscr{D}(T) \\ x \neq 0}} \frac{\|Tx\|}{\|x\|}. \tag{2}$$

$\|T\|$ is called the **norm** of the operator T. If $\mathscr{D}(T)=\{0\}$, we define $\|T\|=0$; in this (relatively uninteresting) case, $T=0$ since $T0=0$ by (3), Sec. 2.6.

Note that (1) with $c=\|T\|$ is

$$\|Tx\| \leqq \|T\| \|x\|. \tag{3}$$

This formula will be applied quite frequently.

Of course, we should justify the use of the term "norm" in the present context. This will be done in the following lemma.

2.7-2 Lemma (Norm). *Let T be a bounded linear operator as defined in* 2.7-1. *Then:*

(a) *An alternative formula for the norm of T is*

$$\|T\| = \sup_{\substack{x \in \mathscr{D}(T) \\ \|x\|=1}} \|Tx\|. \tag{4}$$

(b) *The norm defined by* (2) *satisfies* (N1) *to* (N4) *in Sec.* 2.2.

Proof. **(a)** We write $\|x\|=a$ and set $y=(1/a)x$, where $x\neq 0$. Then $\|y\|=\|x\|/a=1$, and since T is linear, (2) gives

$$\|T\|=\sup_{\substack{x\in\mathscr{D}(T)\\ x\neq 0}}\frac{1}{a}\|Tx\|=\sup_{\substack{x\in\mathscr{D}(T)\\ x\neq 0}}\left\|T\left(\frac{1}{a}x\right)\right\|=\sup_{\substack{y\in\mathscr{D}(T)\\ \|y\|=1}}\|Ty\|.$$

Writing x for y on the right, we have (4).

(b) (N1) is obvious, and so is $\|0\|=0$. From $\|T\|=0$ we have $Tx=0$ for all $x\in\mathscr{D}(T)$, so that $T=0$. Hence (N2) holds. Furthermore, (N3) is obtained from

$$\sup_{\|x\|=1}\|\alpha Tx\|=\sup_{\|x\|=1}|\alpha|\,\|Tx\|=|\alpha|\sup_{\|x\|=1}\|Tx\|$$

where $x\in\mathscr{D}(T)$. Finally, (N4) follows from

$$\sup_{\|x\|=1}\|(T_1+T_2)x\|=\sup_{\|x\|=1}\|T_1x+T_2x\|\leqq\sup_{\|x\|=1}\|T_1x\|+\sup_{\|x\|=1}\|T_2x\|;$$

here, $x\in\mathscr{D}(T)$. ∎

Before we consider general properties of bounded linear operators, let us take a look at some typical examples, so that we get a better feeling for the concept of a bounded linear operator.

Examples

2.7-3 Identity operator. The identity operator $I: X\longrightarrow X$ on a normed space $X\neq\{0\}$ is bounded and has norm $\|I\|=1$. Cf. 2.6-2.

2.7-4 Zero operator. The zero operator $0: X\longrightarrow Y$ on a normed space X is bounded and has norm $\|0\|=0$. Cf. 2.6-3.

2.7-5 Differentiation operator. Let X be the normed space of all polynomials on $J=[0,1]$ with norm given $\|x\|=\max|x(t)|$, $t\in J$. A differentiation operator T is defined on X by

$$Tx(t)=x'(t)$$

where the prime denotes differentiation with respect to t. This operator is linear but not bounded. Indeed, let $x_n(t)=t^n$, where $n\in\mathbf{N}$. Then $\|x_n\|=1$ and

$$Tx_n(t)=x_n'(t)=nt^{n-1}$$

so that $\|Tx_n\|=n$ and $\|Tx_n\|/\|x_n\|=n$. Since $n\in\mathbf{N}$ is arbitrary, this shows that there is no fixed number c such that $\|Tx_n\|/\|x_n\|\leqq c$. From this and (1) we conclude that T is not bounded.

Since differentiation is an important operation, our result seems to indicate that unbounded operators are also of practical importance. This is indeed the case, as we shall see in Chaps. 10 and 11, after a detailed study of the theory and application of bounded operators, which are simpler than unbounded ones.

2.7-6 Integral operator. We can define an integral operator $T\colon C[0,1]\longrightarrow C[0,1]$ by

$$y=Tx \qquad \text{where} \qquad y(t)=\int_0^1 k(t,\tau)x(\tau)\,d\tau.$$

Here k is a given function, which is called the *kernel* of T and is assumed to be continuous on the closed square $G=J\times J$ in the $t\tau$-plane, where $J=[0,1]$. This operator is linear.

T is bounded.

To prove this, we first note that the continuity of k on the closed square implies that k is bounded, say, $|k(t,\tau)|\leqq k_0$ for all $(t,\tau)\in G$, where k_0 is a real number. Furthermore,

$$|x(t)|\leqq \max_{t\in J}|x(t)|=\|x\|.$$

Hence

$$\begin{aligned}\|y\|=\|Tx\|&=\max_{t\in J}\left|\int_0^1 k(t,\tau)x(\tau)\,d\tau\right|\\&\leqq \max_{t\in J}\int_0^1 |k(t,\tau)|\,|x(\tau)|\,d\tau\\&\leqq k_0\,\|x\|.\end{aligned}$$

The result is $\|Tx\| \leqq k_0 \|x\|$. This is (1) with $c = k_0$. Hence T is bounded.

2.7-7 Matrix. A real matrix $A = (\alpha_{jk})$ with r rows and n columns defines an operator $T\colon \mathbf{R}^n \longrightarrow \mathbf{R}^r$ by means of

$$y = Ax \tag{5}$$

where $x = (\xi_j)$ and $y = (\eta_j)$ are column vectors with n and r components, respectively, and we used matrix multiplication, as in 2.6-8. In terms of components, (5) becomes

$$\eta_j = \sum_{k=1}^{n} \alpha_{jk}\xi_k \qquad (j = 1, \cdots, r). \tag{5'}$$

T is linear because matrix multiplication is a linear operation.

T is bounded.

To prove this, we first remember from 2.2-2 that the norm on $\mathbf{R}^n$ is given by

$$\|x\| = \left(\sum_{m=1}^{n} \xi_m^{\,2}\right)^{1/2};$$

similarly for $y \in \mathbf{R}^r$. From (5') and the Cauchy-Schwarz inequality (11) in Sec. 1.2 we thus obtain

$$\begin{aligned} \|Tx\|^2 = \sum_{j=1}^{r} \eta_j^{\,2} &= \sum_{j=1}^{r} \left[\sum_{k=1}^{n} \alpha_{jk}\xi_k\right]^2 \\ &\leqq \sum_{j=1}^{r} \left[\left(\sum_{k=1}^{n} \alpha_{jk}^{\,2}\right)^{1/2} \left(\sum_{m=1}^{n} \xi_m^{\,2}\right)^{1/2}\right]^2 \\ &= \|x\|^2 \sum_{j=1}^{r} \sum_{k=1}^{n} \alpha_{jk}^{\,2}. \end{aligned}$$

Noting that the double sum in the last line does not depend on x, we can write our result in the form

$$\|Tx\|^2 \leqq c^2 \|x\|^2 \qquad \text{where} \qquad c^2 = \sum_{j=1}^{r} \sum_{k=1}^{n} \alpha_{jk}^{\,2}.$$

This gives (1) and completes the proof that T is bounded. ∎

The role of matrices in connection with linear operators will be studied in a separate section (Sec. 2.9). Boundedness is typical; it is an essential simplification which we always have in the finite dimensional case, as follows.

2.7-8 Theorem (Finite dimension). *If a normed space X is finite dimensional, then every linear operator on X is bounded.*

Proof. Let $\dim X = n$ and $\{e_1, \cdots, e_n\}$ a basis for X. We take any $x = \sum \xi_j e_j$ and consider any linear operator T on X. Since T is linear,

$$\|Tx\| = \left\| \sum \xi_j Te_j \right\| \leqq \sum |\xi_j| \, \|Te_j\| \leqq \max_k \|Te_k\| \sum |\xi_j|$$

(summations from 1 to n). To the last sum we apply Lemma 2.4-1 with $\alpha_j = \xi_j$ and $x_j = e_j$. Then we obtain

$$\sum |\xi_j| \leqq \frac{1}{c} \left\| \sum \xi_j e_j \right\| = \frac{1}{c} \|x\|.$$

Together,

$$\|Tx\| \leqq \gamma \|x\| \qquad \text{where} \qquad \gamma = \frac{1}{c} \max_k \|Te_k\|.$$

From this and (1) we see that T is bounded. ∎

We shall now consider important general properties of bounded linear operators.

Operators are mappings, so that the definition of continuity (cf. 1.3-3) applies to them. It is a fundamental fact that for a *linear* operator, continuity and boundedness become equivalent concepts. The details are as follows.

Let $T\colon \mathfrak{D}(T) \longrightarrow Y$ be any operator, not necessarily linear, where $\mathfrak{D}(T) \subset X$ and X and Y are normed spaces. By Def. 1.3-3, the operator T is *continuous at an* $x_0 \in \mathfrak{D}(T)$ if for every $\varepsilon > 0$ there is a $\delta > 0$ such that

$$\|Tx - Tx_0\| < \varepsilon \qquad \text{for all } x \in \mathfrak{D}(T) \text{ satisfying} \qquad \|x - x_0\| < \delta.$$

T is *continuous* if T is continuous at every $x \in \mathfrak{D}(T)$.

Now, if T is linear, we have the remarkable

2.7-9 Theorem (Continuity and boundedness). *Let $T: \mathcal{D}(T) \longrightarrow Y$ be a linear*[7] *operator, where $\mathcal{D}(T) \subset X$ and X, Y are normed spaces. Then:*

(a) *T is continuous if and only if T is bounded.*

(b) *If T is continuous at a single point, it is continuous.*

Proof. **(a)** For $T=0$ the statement is trivial. Let $T \neq 0$. Then $\|T\| \neq 0$. We assume T to be bounded and consider any $x_0 \in \mathcal{D}(T)$. Let any $\varepsilon > 0$ be given. Then, since T is linear, for every $x \in \mathcal{D}(T)$ such that

$$\|x - x_0\| < \delta \qquad \text{where} \qquad \delta = \frac{\varepsilon}{\|T\|}$$

we obtain

$$\|Tx - Tx_0\| = \|T(x - x_0)\| \leqq \|T\| \, \|x - x_0\| < \|T\| \delta = \varepsilon.$$

Since $x_0 \in \mathcal{D}(T)$ was arbitrary, this shows that T is continuous.

Conversely, assume that T is continuous at an arbitrary $x_0 \in \mathcal{D}(T)$. Then, given any $\varepsilon > 0$, there is a $\delta > 0$ such that

$$\|Tx - Tx_0\| \leqq \varepsilon \qquad \text{for all } x \in \mathcal{D}(T) \text{ satisfying} \qquad \|x - x_0\| \leqq \delta. \tag{6}$$

We now take any $y \neq 0$ in $\mathcal{D}(T)$ and set

$$x = x_0 + \frac{\delta}{\|y\|} y. \qquad \text{Then} \qquad x - x_0 = \frac{\delta}{\|y\|} y.$$

Hence $\|x - x_0\| = \delta$, so that we may use (6). Since T is linear, we have

$$\|Tx - Tx_0\| = \|T(x - x_0)\| = \left\| T\left(\frac{\delta}{\|y\|} y\right) \right\| = \frac{\delta}{\|y\|} \|Ty\|$$

[7] **Warning.** Unfortunately, continuous linear operators are called "linear operators" by some authors. We shall not adopt this terminology; in fact, there are linear operators of practical importance which are not continuous. A first example is given in 2.7-5 and further operators of that type will be considered in Chaps. 10 and 11.

and (6) implies

$$\frac{\delta}{\|y\|}\|Ty\| \leqq \varepsilon. \qquad \text{Thus} \qquad \|Ty\| \leqq \frac{\varepsilon}{\delta}\|y\|.$$

This can be written $\|Ty\| \leqq c\|y\|$, where $c = \varepsilon/\delta$, and shows that T is bounded.

(b) Continuity of T at a point implies boundedness of T by the second part of the proof of (a), which in turn implies continuity of T by (a). ∎

2.7-10 Corollary (Continuity, null space). *Let T be a bounded linear operator. Then:*

(a) $x_n \longrightarrow x$ *[where* $x_n, x \in \mathcal{D}(T)$*] implies* $Tx_n \longrightarrow Tx$.

(b) *The null space* $\mathcal{N}(T)$ *is closed.*

Proof. **(a)** follows from Theorems 2.7-9(a) and 1.4-8 or directly from (3) because, as $n \longrightarrow \infty$,

$$\|Tx_n - Tx\| = \|T(x_n - x)\| \leqq \|T\|\|x_n - x\| \longrightarrow 0.$$

(b) For every $x \in \overline{\mathcal{N}(T)}$ there is a sequence (x_n) in $\mathcal{N}(T)$ such that $x_n \longrightarrow x$; cf. 1.4-6(a). Hence $Tx_n \longrightarrow Tx$ by part (a) of this Corollary. Also $Tx = 0$ since $Tx_n = 0$, so that $x \in \mathcal{N}(T)$. Since $x \in \overline{\mathcal{N}(T)}$ was arbitrary, $\mathcal{N}(T)$ is closed. ∎

It is worth noting that the range of a bounded linear operator may not be closed. Cf. Prob. 6.

The reader may give the simple proof of another useful formula, namely,

$$\text{(7)} \qquad \|T_1T_2\| \leqq \|T_1\|\|T_2\|, \qquad \|T^n\| \leqq \|T\|^n \qquad (n \in \mathbf{N})$$

valid for bounded linear operators $T_2\colon X \longrightarrow Y$, $T_1\colon Y \longrightarrow Z$ and $T\colon X \longrightarrow X$, where X, Y, Z are normed spaces.

Operators are mappings, and some concepts related to mappings[8] have been discussed, notably the domain, range and null space of an

[8] A review of some of these concepts is given in A1.2; cf. Appendix 1.

operator. Two further concepts (restriction and extension) will now be added. We could have done this earlier, but we prefer to do it here, where we can immediately give an interesting application (Theorem 2.7-11, below). Let us begin by defining equality of operators as follows.

Two operators T_1 and T_2 are defined to be **equal**, written

$$T_1 = T_2,$$

if they have the same domain $\mathfrak{D}(T_1) = \mathfrak{D}(T_2)$ and if $T_1x = T_2x$ for all $x \in \mathfrak{D}(T_1) = \mathfrak{D}(T_2)$.

The **restriction** of an operator $T: \mathfrak{D}(T) \longrightarrow Y$ to a subset $B \subset \mathfrak{D}(T)$ is denoted by

$$T|_B$$

and is the operator defined by

$$T|_B: B \longrightarrow Y, \qquad\qquad T|_B x = Tx \text{ for all } x \in B.$$

An **extension** of T to a set $M \supset \mathfrak{D}(T)$ is an operator

$$\tilde{T}: M \longrightarrow Y \qquad \text{such that} \qquad \tilde{T}|_{\mathfrak{D}(T)} = T,$$

that is, $\tilde{T}x = Tx$ for all $x \in \mathfrak{D}(T)$. [Hence T is the restriction of $\tilde{T}$ to $\mathfrak{D}(T)$.]

If $\mathfrak{D}(T)$ is a proper subset of M, then a given T has many extensions. Of practical interest are usually those extensions which preserve some basic property, for instance linearity (if T happens to be linear) or boundedness (if $\mathfrak{D}(T)$ lies in a normed space and T is bounded). The following important theorem is typical in that respect. It concerns an extension of a bounded linear operator T to the closure $\overline{\mathfrak{D}(T)}$ of the domain such that the extended operator is again bounded and linear, and even has the same norm. This includes the case of an extension from a dense set in a normed space X to all of X. It also includes the case of an extension from a normed space X to its completion (cf. 2.3-2).

2.7-11 Theorem (Bounded linear extension). *Let*

$$T: \mathfrak{D}(T) \longrightarrow Y$$

be a bounded linear operator, where $\mathfrak{D}(T)$ lies in a normed space X and Y is a Banach space. Then T has an extension

$$\tilde{T}: \overline{\mathfrak{D}(T)} \longrightarrow Y$$

where $\tilde{T}$ is a bounded linear operator of norm

$$\|\tilde{T}\| = \|T\|.$$

Proof. We consider any $x \in \overline{\mathfrak{D}(T)}$. By Theorem 1.4-6($a$) there is a sequence (x_n) in $\mathfrak{D}(T)$ such that $x_n \longrightarrow x$. Since T is linear and bounded, we have

$$\|Tx_n - Tx_m\| = \|T(x_n - x_m)\| \leqq \|T\| \|x_n - x_m\|.$$

This shows that (Tx_n) is Cauchy because (x_n) converges. By assumption, Y is complete, so that (Tx_n) converges, say,

$$Tx_n \longrightarrow y \in Y.$$

We define $\tilde{T}$ by

$$\tilde{T}x = y.$$

We show that this definition is independent of the particular choice of a sequence in $\mathfrak{D}(T)$ converging to x. Suppose that $x_n \longrightarrow x$ and $z_n \longrightarrow x$. Then $v_m \longrightarrow x$, where (v_m) is the sequence

$$(x_1, z_1, x_2, z_2, \cdots).$$

Hence (Tv_m) converges by 2.7-10(a), and the two subsequences (Tx_n) and (Tz_n) of (Tv_m) must have the same limit. This proves that $\tilde{T}$ is uniquely defined at every $x \in \overline{\mathfrak{D}(T)}$.

Clearly, $\tilde{T}$ is linear and $\tilde{T}x = Tx$ for every $x \in \mathfrak{D}(T)$, so that $\tilde{T}$ is an extension of T. We now use

$$\|Tx_n\| \leqq \|T\| \|x_n\|$$

and let $n \longrightarrow \infty$. Then $Tx_n \longrightarrow y = \tilde{T}x$. Since $x \longmapsto \|x\|$ defines a continuous mapping (cf. Sec. 2.2), we thus obtain

$$\|\tilde{T}x\| \leqq \|T\| \|x\|.$$

Hence $\tilde{T}$ is bounded and $\|\tilde{T}\| \leqq \|T\|$. Of course, $\|\tilde{T}\| \geqq \|T\|$ because the norm, being defined by a supremum, cannot decrease in an extension. Together we have $\|\tilde{T}\| = \|T\|$. ∎

Problems

1. Prove (7).

2. Let X and Y be normed spaces. Show that a linear operator $T: X \longrightarrow Y$ is bounded if and only if T maps bounded sets in X into bounded sets in Y.

3. If $T \neq 0$ is a bounded linear operator, show that for any $x \in \mathfrak{D}(T)$ such that $\|x\| < 1$ we have the strict inequality $\|Tx\| < \|T\|$.

4. Give a direct proof of 2.7-9(*b*), without using 2.7-9(*a*).

5. Show that the operator $T: l^\infty \longrightarrow l^\infty$ defined by $y = (\eta_j) = Tx$, $\eta_j = \xi_j / j$, $x \doteq (\xi_j)$, is linear and bounded.

6. **(Range)** Show that the range $\mathfrak{R}(T)$ of a bounded linear operator $T: X \longrightarrow Y$ need not be closed in Y. *Hint.* Use T in Prob. 5.

7. **(Inverse operator)** Let T be a bounded linear operator from a normed space X onto a normed space Y. If there is a positive b such that

$$\|Tx\| \geqq b\|x\| \qquad \text{for all } x \in X,$$

show that then $T^{-1}: Y \longrightarrow X$ exists and is bounded.

8. Show that the inverse $T^{-1}: \mathfrak{R}(T) \longrightarrow X$ of a bounded linear operator $T: X \longrightarrow Y$ need not be bounded. *Hint.* Use T in Prob. 5.

9. Let $T: C[0, 1] \longrightarrow C[0, 1]$ be defined by

$$y(t) = \int_0^t x(\tau)\, d\tau.$$

Find $\mathfrak{R}(T)$ and $T^{-1}: \mathfrak{R}(T) \longrightarrow C[0, 1]$. Is T^{-1} linear and bounded?

10. On $C[0, 1]$ define S and T by

$$y(s)=s\int_0^1 x(t)\,dt, \qquad\qquad y(s)=sx(s),$$

respectively. Do S and T commute? Find $\|S\|$, $\|T\|$, $\|ST\|$ and $\|TS\|$.

11. Let X be the normed space of all bounded real-valued functions on $\mathbf{R}$ with norm defined by

$$\|x\|=\sup_{t\in\mathbf{R}} |x(t)|,$$

and let $T\colon X \longrightarrow X$ be defined by

$$y(t)=Tx(t)=x(t-\Delta)$$

where $\Delta>0$ is a constant. (This is a model of a *delay line*, which is an electric device whose output y is a delayed version of the input x, the time delay being Δ; see Fig. 22.) Is T linear? Bounded?

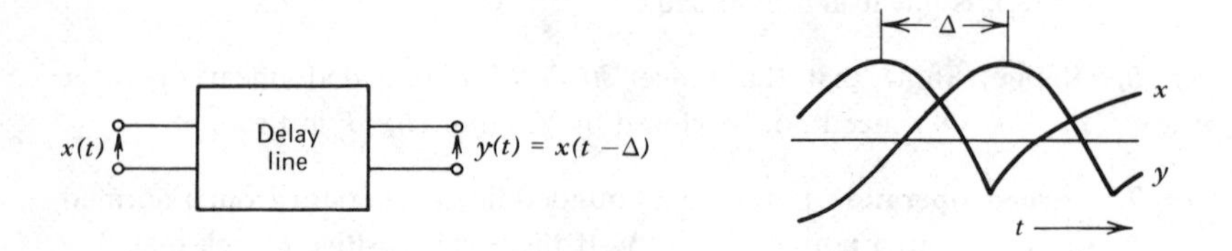

Fig. 22. Electric delay line

12. **(Matrices)** From 2.7-7 we know that an $r\times n$ matrix $A=(\alpha_{jk})$ defines a linear operator from the vector space X of all ordered n-tuples of numbers into the vector space Y of all ordered r-tuples of numbers. Suppose that any norm $\|\cdot\|_1$ is given on X and any norm $\|\cdot\|_2$ is given on Y. Remember from Prob. 10, Sec. 2.4, that there are various norms on the space Z of all those matrices (r and n fixed). A norm $\|\cdot\|$ on Z is said to be *compatible* with $\|\cdot\|_1$ and $\|\cdot\|_2$ if

$$\|Ax\|_2 \leqq \|A\|\,\|x\|_1.$$

Show that the norm defined by

$$\|A\| = \sup_{\substack{x \in X \\ x \neq 0}} \frac{\|Ax\|_2}{\|x\|_1}$$

is compatible with $\|\cdot\|_1$ and $\|\cdot\|_2$. This norm is often called the *natural norm* defined by $\|\cdot\|_1$ and $\|\cdot\|_2$. If we choose $\|x\|_1 = \max_j |\xi_j|$ and $\|y\|_2 = \max_j |\eta_j|$, show that the natural norm is

$$\|A\| = \max_j \sum_{k=1}^{n} |\alpha_{jk}|.$$

13. Show that in 2.7-7 with $r = n$, a compatible norm is defined by

$$\|A\| = \left(\sum_{j=1}^{n} \sum_{k=1}^{n} \alpha_{jk}^2 \right)^{1/2},$$

but for $n > 1$ this is *not* the natural norm defined by the Euclidean norm on $\mathbf{R}^n$.

14. If in Prob. 12 we choose

$$\|x\|_1 = \sum_{k=1}^{n} |\xi_k|, \qquad \|y\|_2 = \sum_{j=1}^{r} |\eta_j|,$$

show that a compatible norm is defined by

$$\|A\| = \max_k \sum_{j=1}^{r} |\alpha_{jk}|.$$

15. Show that for $r = n$, the norm in Prob. 14 is the natural norm corresponding to $\|\cdot\|_1$ and $\|\cdot\|_2$ as defined in that problem.

2.8 Linear Functionals

A **functional** is an operator whose range lies on the real line $\mathbf{R}$ or in the complex plane $\mathbf{C}$. And *functional analysis* was initially the analysis of functionals. The latter appear so frequently that special notations are used. We denote functionals by lowercase letters $f, g, h, \cdots$, the domain of f by $\mathfrak{D}(f)$, the range by $\mathfrak{R}(f)$ and the value of f at an $x \in \mathfrak{D}(f)$ by $f(x)$, with parentheses.

Functionals are operators, so that previous definitions apply. We shall need in particular the following two definitions because most of the functionals to be considered will be linear and bounded.

2.8-1 Definition (Linear functional). A *linear functional* f is a linear operator with domain in a vector space X and range in the scalar field K of X; thus,

$$f\colon \mathfrak{D}(f) \longrightarrow K,$$

where $K = \mathbf{R}$ if X is real and $K = \mathbf{C}$ if X is complex. ∎

2.8-2 Definition (Bounded linear functional). A *bounded linear functional* f is a bounded linear operator (cf. Def. 2.7-1) with range in the scalar field of the normed space X in which the domain $\mathfrak{D}(f)$ lies. Thus there exists a real number c such that for all $x \in \mathfrak{D}(f)$,

$$|f(x)| \leqq c\,\|x\|. \tag{1}$$

Furthermore, the *norm* of f is [cf. (2) in Sec. 2.7]

$$\|f\| = \sup_{\substack{x \in \mathfrak{D}(f) \\ x \neq 0}} \frac{|f(x)|}{\|x\|} \tag{2a}$$

or

$$\|f\| = \sup_{\substack{x \in \mathfrak{D}(f) \\ \|x\| = 1}} |f(x)|. \tag{2b}$$ ∎

Formula (3) in Sec. 2.7 now implies

$$|f(x)| \leqq \|f\|\,\|x\|, \tag{3}$$

and a special case of Theorem 2.7-9 is

2.8-3 Theorem (Continuity and boundedness). *A linear functional f with domain $\mathfrak{D}(f)$ in a normed space is continuous if and only if f is bounded.*

Examples

2.8-4 Norm. The *norm* $\|\cdot\|: X \longrightarrow \mathbf{R}$ on a normed space $(X, \|\cdot\|)$ is a functional on X which is not linear.

2.8-5 Dot product. The familiar *dot product* with one factor kept fixed defines a functional $f: \mathbf{R}^3 \longrightarrow \mathbf{R}$ by means of

$$f(x) = x \cdot a = \xi_1\alpha_1 + \xi_2\alpha_2 + \xi_3\alpha_3,$$

where $a = (\alpha_j) \in \mathbf{R}^3$ is fixed.

f is linear. f is bounded. In fact,

$$|f(x)| = |x \cdot a| \leqq \|x\| \|a\|,$$

so that $\|f\| \leqq \|a\|$ follows from (2b) if we take the supremum over all x of norm one. On the other hand, by taking $x = a$ and using (3) we obtain

$$\|f\| \geqq \frac{|f(a)|}{\|a\|} = \frac{\|a\|^2}{\|a\|} = \|a\|.$$

Hence the norm of f is $\|f\| = \|a\|$.

2.8-6 Definite integral. The *definite integral* is a number if we consider it for a single function, as we do in calculus most of the time. However, the situation changes completely if we consider that integral for all functions in a certain function space. Then the integral becomes a functional on that space, call it f. As a space let us choose $C[a, b]$; cf. 2.2-5. Then f is defined by

$$f(x) = \int_a^b x(t)\, dt \qquad x \in C[a, b].$$

f is linear. We prove that f is bounded and has norm $\|f\| = b - a$.

In fact, writing $J = [a, b]$ and remembering the norm on $C[a, b]$, we obtain

$$|f(x)| = \left| \int_a^b x(t)\, dt \right| \leqq (b-a) \max_{t \in J} |x(t)| = (b-a)\|x\|.$$

Taking the supremum over all x of norm 1, we obtain $\|f\| \leqq b-a$. To get $\|f\| \geqq b-a$, we choose the particular $x = x_0 = 1$, note that $\|x_0\| = 1$ and use (3):

$$\|f\| \geqq \frac{|f(x_0)|}{\|x_0\|} = |f(x_0)| = \int_a^b dt = b-a.$$

2.8-7 Space $C[a, b]$. Another practically important functional on $C[a, b]$ is obtained if we choose a fixed $t_0 \in J = [a, b]$ and set

$$f_1(x) = x(t_0) \qquad\qquad x \in C[a, b].$$

f_1 is linear. f_1 is bounded and has norm $\|f_1\| = 1$. In fact, we have

$$|f_1(x)| = |x(t_0)| \leqq \|x\|,$$

and this implies $\|f_1\| \leqq 1$ by (2). On the other hand, for $x_0 = 1$ we have $\|x_0\| = 1$ and obtain from (3)

$$\|f_1\| \geqq |f_1(x_0)| = 1.$$

2.8-8 Space l^2. We can obtain a linear functional f on the Hilbert space l^2 (cf. 1.2-3) by choosing a fixed $a = (\alpha_j) \in l^2$ and setting

$$f(x) = \sum_{j=1}^{\infty} \xi_j \alpha_j$$

where $x = (\xi_j) \in l^2$. This series converges absolutely and f is bounded, since the Cauchy-Schwarz inequality (11) in Sec. 1.2 gives (summation over j from 1 to ∞)

$$|f(x)| = \left|\sum \xi_j \alpha_j\right| \leqq \sum |\xi_j \alpha_j| \leqq \sqrt{\sum |\xi_j|^2} \sqrt{\sum |\alpha_j|^2} = \|x\| \, \|a\|. \qquad \blacksquare$$

It is of basic importance that the set of all linear functionals defined on a vector space X can itself be made into a vector space. This space is denoted by X^* and is called the **algebraic**[9] **dual space** of X. Its algebraic operations of vector space are defined in a natural way

[9] Note that this definition does not involve a norm. The so-called *dual space* X' consisting of all *bounded* linear functionals on X will be considered in Sec. 2.10.

as follows. The *sum* f_1+f_2 of two functionals f_1 and f_2 is the functional s whose value at every $x \in X$ is

$$s(x)=(f_1+f_2)(x)=f_1(x)+f_2(x);$$

the *product* αf of a scalar α and a functional f is the functional p whose value at $x \in X$ is

$$p(x)=(\alpha f)(x)=\alpha f(x).$$

Note that this agrees with the usual way of adding functions and multiplying them by constants.

We may go a step further and consider the algebraic dual $(X^*)^*$ of X^*, whose elements are the linear functionals defined on X^*. We denote $(X^*)^*$ by X^{**} and call it the **second algebraic dual space** of X.

Why do we consider X^{**}? The point is that we can obtain an interesting and important relation between X and X^{**}, as follows. We choose the notations:

Space	General element	Value at a point
X	x	—
X^*	f	$f(x)$
X^{**}	g	$g(f)$

We can obtain a $g \in X^{**}$, which is a linear functional defined on X^*, by choosing a *fixed* $x \in X$ and setting

$$\textbf{(4)} \qquad g(f)=g_x(f)=f(x) \qquad (x \in X \text{ fixed, } f \in X^* \text{ variable}).$$

The subscript x is a little reminder that we got g by the use of a certain $x \in X$. The reader should observe carefully that here f is the variable whereas x is fixed. Keeping this in mind, he should not have difficulties in understanding our present consideration.

g_x as defined by (4) is linear. This can be seen from

$$g_x(\alpha f_1+\beta f_2)=(\alpha f_1+\beta f_2)(x)=\alpha f_1(x)+\beta f_2(x)=\alpha g_x(f_1)+\beta g_x(f_2).$$

Hence g_x is an element of X^{**}, by the definition of X^{**}.

To each $x \in X$ there corresponds a $g_x \in X^{**}$. This defines a mapping

$$C: X \longrightarrow X^{**}$$

$$x \longmapsto g_x.$$

C is called the **canonical mapping** of X into X^{**}.

C is linear since its domain is a vector space and we have

$$\begin{aligned}
(C(\alpha x+\beta y))(f) &= g_{\alpha x+\beta y}(f) \\
&= f(\alpha x+\beta y) \\
&= \alpha f(x)+\beta f(y) \\
&= \alpha g_x(f)+\beta g_y(f) \\
&= \alpha (Cx)(f)+\beta (Cy)(f).
\end{aligned}$$

C is also called the *canonical embedding* of X into X^{**}. To understand and motivate this term, we first explain the concept of "isomorphism," which is of general interest.

In our work we are concerned with various spaces. Common to all of them is that they consist of a set, call it X, and a "structure" defined on X. For a metric space, this is the metric. For a vector space, the two algebraic operations form the structure. And for a normed space the structure consists of those two algebraic operations and the norm.

Given two spaces X and $\tilde{X}$ of the same kind (for instance, two vector spaces), it is of interest to know whether X and $\tilde{X}$ are "essentially identical," that is, whether they differ at most by the nature of their points. Then we can regard X and $\tilde{X}$ as identical—as two copies of the same "abstract" space—whenever the structure is the primary object of study, whereas the nature of the points does not matter. This situation occurs quite often. It suggests the concept of an **isomorphism.** By definition, this is a bijective mapping of X onto $\tilde{X}$ which preserves the structure.

Accordingly, an *isomorphism* T *of a metric space* $X=(X, d)$ onto a metric space $\tilde{X}=(\tilde{X}, \tilde{d})$ is a bijective mapping which preserves distance, that is, for all $x, y \in X$,

$$\tilde{d}(Tx, Ty)=d(x, y).$$

$\tilde{X}$ is then called *isomorphic* with X. This is nothing new to us but merely another name for a bijective isometry as introduced in Def. 1.6-1. New is the following.

An *isomorphism* T *of a vector space* X onto a vector space $\tilde{X}$ over the same field is a bijective mapping which preserves the two algebraic operations of vector space; thus, for all $x, y \in X$ and scalars α,

$$T(x+y) = Tx + Ty, \qquad T(\alpha x) = \alpha Tx,$$

that is, $T: X \longrightarrow \tilde{X}$ is a bijective linear operator. $\tilde{X}$ is then called *isomorphic* with X, and X and $\tilde{X}$ are called *isomorphic vector spaces.*

Isomorphisms for normed spaces are vector space isomorphisms which also preserve norms. Details follow in Sec. 2.10 where we need such isomorphisms. At present we can apply vector space isomorphisms as follows.

It can be shown that the canonical mapping C is injective. Since C is linear (see before), it is an isomorphism of X onto the range $\mathcal{R}(C) \subset X^{**}$.

If X is isomorphic with a subspace of a vector space Y, we say that X is **embeddable** in Y. Hence X is embeddable in X^{**}, and C is also called the *canonical embedding* of X into X^{**}.

If C is surjective (hence bijective), so that $\mathcal{R}(C) = X^{**}$, then X is said to be **algebraically reflexive.** We shall prove in the next section that if X is finite dimensional, then X is algebraically reflexive.

A similar discussion involving norms and leading to the concept of *reflexivity* of a *normed* space will be presented later (in Sec. 4.6), after the development of suitable tools (in particular, the famous Hahn-Banach theorem).

Problems

1. Show that the functionals in 2.8-7 and 2.8-8 are linear.

2. Show that the functionals defined on $C[a, b]$ by

$$f_1(x) = \int_a^b x(t)y_0(t)\,dt \qquad (y_0 \in C[a, b])$$

$$f_2(x) = \alpha x(a) + \beta x(b) \qquad (\alpha, \beta \text{ fixed})$$

are linear and bounded.

3. Find the norm of the linear functional f defined on $C[-1, 1]$ by

$$f(x)=\int_{-1}^{0} x(t)\, dt-\int_{0}^{1} x(t)\, dt.$$

4. Show that

$$f_1(x)=\max_{t\in J} x(t)$$

$$J=[a, b]$$

$$f_2(x)=\min_{t\in J} x(t)$$

define functionals on $C[a, b]$. Are they linear? Bounded?

5. Show that on any sequence space X we can define a linear functional f by setting $f(x)=\xi_n$ (n fixed), where $x=(\xi_j)$. Is f bounded if $X=l^\infty$?

6. (Space $C'[a, b]$) The space $C^1[a, b]$ or $C'[a, b]$ is the normed space of all continuously differentiable functions on $J=[a, b]$ with norm defined by

$$\|x\|=\max_{t\in J} |x(t)|+\max_{t\in J} |x'(t)|.$$

Show that the axioms of a norm are satisfied. Show that $f(x)=x'(c)$, $c=(a+b)/2$, defines a bounded linear functional on $C'[a, b]$. Show that f is not bounded, considered as a functional on the subspace of $C[a, b]$ which consists of all continuously differentiable functions.

7. If f is a bounded linear functional on a complex normed space, is $\bar{f}$ bounded? Linear? (The bar denotes the complex conjugate.)

8. (Null space) The *null space* $N(M^*)$ of a set $M^*\subset X^*$ is defined to be the set of all $x\in X$ such that $f(x)=0$ for all $f\in M^*$. Show that $N(M^*)$ is a vector space.

9. Let $f\neq 0$ be any linear functional on a vector space X and x_0 any fixed element of $X-\mathcal{N}(f)$, where $\mathcal{N}(f)$ is the null space of f. Show that any $x\in X$ has a unique representation $x=\alpha x_0+y$, where $y\in \mathcal{N}(f)$.

10. Show that in Prob. 9, two elements $x_1, x_2\in X$ belong to the same element of the quotient space $X/\mathcal{N}(f)$ if and only if $f(x_1)=f(x_2)$; show that codim $\mathcal{N}(f)=1$. (Cf. Sec. 2.1, Prob. 14.)

11. Show that two linear functionals $f_1\neq 0$ and $f_2\neq 0$ which are defined on the same vector space and have the same null space are proportional.

12. (Hyperplane) If Y is a subspace of a vector space X and codim $Y=1$ (cf. Sec. 2.1, Prob. 14), then every element of X/Y is called a *hyperplane parallel to* Y. Show that for any linear functional $f \neq 0$ on X, the set $H_1 = \{x \in X \mid f(x) = 1\}$ is a hyperplane parallel to the null space $\mathcal{N}(f)$ of f.

13. If Y is a subspace of a vector space X and f is a linear functional on X such that $f(Y)$ is not the whole scalar field of X, show that $f(y)=0$ for all $y \in Y$.

14. Show that the norm $\|f\|$ of a bounded linear functional $f \neq 0$ on a normed space X can be interpreted geometrically as the reciprocal of the distance $\tilde{d} = \inf\{\|x\| \mid f(x) = 1\}$ of the hyperplane $H_1 = \{x \in X \mid f(x) = 1\}$ from the orgin.

15. (Half space) Let $f \neq 0$ be a bounded linear functional on a real normed space X. Then for any scalar c we have a hyperplane $H_c = \{x \in X \mid f(x) = c\}$, and H_c determines the two *half spaces*

$$X_{c1} = \{x \mid f(x) \leqq c\} \qquad \text{and} \qquad X_{c2} = \{x \mid f(x) \geqq c\}.$$

Show that the closed unit ball lies in X_{c1} where $c = \|f\|$, but for no $\varepsilon > 0$, the half space X_{c1} with $c = \|f\| - \varepsilon$ contains that ball.

2.9 Linear Operators and Functionals on Finite Dimensional Spaces

Finite dimensional vector spaces are simpler than infinite dimensional ones, and it is natural to ask what simplification this entails with respect to linear operators and functionals defined on such a space. This is the question to be considered, and the answer will clarify the role of (finite) matrices in connection with linear operators as well as the structure of the algebraic dual X^* (Sec. 2.8) of a finite dimensional vector space X.

Linear operators on finite dimensional vector spaces can be represented in terms of matrices, as explained below. In this way, matrices become the most important tools for studying linear operators in the finite dimensional case. In this connection we should also remember Theorem 2.7-8 to understand the full significance of our present consideration. The details are as follows.

Let X and Y be finite dimensional vector spaces over the same field and $T: X \longrightarrow Y$ a linear operator. We choose a basis $E = \{e_1, \cdots, e_n\}$ for X and a basis $B = \{b_1, \cdots, b_r\}$ for Y, with the vectors arranged in a definite order which we keep fixed. Then every $x \in X$ has a unique representation

$$x = \xi_1 e_1 + \cdots + \xi_n e_n. \tag{1}$$

Since T is linear, x has the image

$$y = Tx = T\left(\sum_{k=1}^{n} \xi_k e_k\right) = \sum_{k=1}^{n} \xi_k T e_k. \tag{2}$$

Since the representation (1) is unique, we have our first result:

T is uniquely determined if the images $y_k = Te_k$ of the n basis vectors $e_1, \cdots, e_n$ are prescribed.

Since y and $y_k = Te_k$ are in Y, they have unique representations of the form

$$\text{(3)} \quad \begin{aligned} &\text{(a)} & y &= \sum_{j=1}^{r} \eta_j b_j \\ &\text{(b)} & Te_k &= \sum_{j=1}^{r} \tau_{jk} b_j. \end{aligned}$$

Substitution into (2) gives

$$y = \sum_{j=1}^{r} \eta_j b_j = \sum_{k=1}^{n} \xi_k T e_k = \sum_{k=1}^{n} \xi_k \sum_{j=1}^{r} \tau_{jk} b_j = \sum_{j=1}^{r} \left(\sum_{k=1}^{n} \tau_{jk} \xi_k\right) b_j.$$

Since the b_j's form a linearly independent set, the coefficients of each b_j on the left and on the right must be the same, that is,

$$\eta_j = \sum_{k=1}^{n} \tau_{jk} \xi_k \qquad j = 1, \cdots, r. \tag{4}$$

This yields our next result:

The image $y = Tx = \sum \eta_j b_j$ of $x = \sum \xi_k e_k$ can be obtained from (4).

Note the unusual position of the summation index j of τ_{jk} in (3b), which is necessary in order to arrive at the usual position of the summation index in (4).

The coefficients in (4) form a matrix

$$T_{EB} = (\tau_{jk})$$

with r rows and n columns. If a basis E for X and a basis B for Y are given, with the elements of E and B arranged in some definite order (which is arbitrary but fixed), then the matrix T_{EB} is uniquely determined by the linear operator T. We say that the matrix T_{EB} **represents** the operator T with respect to those bases.

By introducing the column vectors $\tilde{x} = (\xi_k)$ and $\tilde{y} = (\eta_j)$ we can write (4) in matrix notation:

(4′) $$\tilde{y} = T_{EB}\tilde{x}.$$

Similarly, (3b) can also be written in matrix notation

(3b′) $$Te = T_{EB}{}^{\mathsf{T}} b$$

where Te is the column vector with components $Te_1, \cdots, Te_n$ (which are themselves vectors) and b is the column vector with components $b_1, \cdots, b_r$, and we have to use the transpose $T_{EB}{}^{\mathsf{T}}$ of T_{EB} because in (3b) we sum over j, which is the first subscript, whereas in (4) we sum over k, which is the second subscript.

Our consideration shows that a linear operator T determines a uniqe matrix representing T with respect to a given basis for X and a given basis for Y, where the vectors of each of the bases are assumed to be arranged in a fixed order. Conversely, any matrix with r rows and n columns determines a linear operator which it represents with respect to given bases for X and Y. (Cf. also 2.6-8 and 2.7-7.)

Let us now turn to **linear functionals** on X, where $\dim X = n$ and $\{e_1, \cdots, e_n\}$ is a basis for X, as before. These functionals constitute the algebraic dual space X^* of X, as we know from the previous section. For every such functional f and every $x = \sum \xi_j e_j \in X$ we have

(5a) $$f(x) = f\left(\sum_{j=1}^{n} \xi_j e_j\right) = \sum_{j=1}^{n} \xi_j f(e_j) = \sum_{j=1}^{n} \xi_j \alpha_j$$

where

(5b) $$\alpha_j = f(e_j) \qquad j = 1, \cdots, n,$$

and f is uniquely determined by its values α_j at the n basis vectors of X.

Conversely, every n-tuple of scalars $\alpha_1, \cdots, \alpha_n$ determines a linear functional on X by (5). In particular, let us take the n-tuples

$$\begin{array}{cccccc} (1, & 0, & 0, & \cdots & 0, & 0) \\ (0, & 1, & 0, & \cdots & 0, & 0) \\ \cdot & \cdot & \cdot & \cdots & \cdot & \cdot \\ (0, & 0, & 0, & \cdots & 0, & 1). \end{array}$$

By (5) this gives n functionals, which we denote by $f_1, \cdots, f_n$, with values

(6) $$f_k(e_j) = \delta_{jk} = \begin{cases} 0 & \text{if } j \neq k, \\ 1 & \text{if } j = k; \end{cases}$$

that is, f_k has the value 1 at the kth basis vector and 0 at the $n-1$ other basis vectors. δ_{jk} is called the *Kronecker delta.* $\{f_1, \cdots, f_n\}$ is called the **dual basis** of the basis $\{e_1, \cdots, e_n\}$ for X. This is justified by the following theorem.

2.9-1 Theorem (Dimension of X^*). *Let X be an n-dimensional vector space and $E = \{e_1, \cdots, e_n\}$ a basis for X. Then $F = \{f_1, \cdots, f_n\}$ given by (6) is a basis for the algebraic dual X^* of X, and* $\dim X^* = \dim X = n$.

Proof. F is a linearly independent set since

(7) $$\sum_{k=1}^{n} \beta_k f_k(x) = 0 \qquad (x \in X)$$

with $x = e_j$ gives

$$\sum_{k=1}^{n} \beta_k f_k(e_j) = \sum_{k=1}^{n} \beta_k \delta_{jk} = \beta_j = 0,$$

so that all the β_k's in (7) are zero. We show that every $f \in X^*$ can be represented as a linear combination of the elements of F in a unique way. We write $f(e_j) = \alpha_j$ as in (5b). By (5a),

$$f(x) = \sum_{j=1}^{n} \xi_j \alpha_j$$

for every $x \in X$. On the other hand, by (6) we obtain

$$f_j(x) = f_j(\xi_1 e_1 + \cdots + \xi_n e_n) = \xi_j.$$

Together,

$$f(x) = \sum_{j=1}^{n} \alpha_j f_j(x).$$

Hence the unique representation of the arbitrary linear functional f on X in terms of the functionals $f_1, \cdots, f_n$ is

$$f = \alpha_1 f_1 + \cdots + \alpha_n f_n. \qquad \blacksquare$$

To prepare for an interesting application of this theorem, we first prove the following lemma. (A similar lemma for arbitrary normed spaces will be given later, in 4.3-4.)

2.9-2 Lemma (Zero vector). *Let X be a finite dimensional vector space. If $x_0 \in X$ has the property that $f(x_0) = 0$ for all $f \in X^*$, then $x_0 = 0$.*

Proof. Let $\{e_1, \cdots, e_n\}$ be a basis for X and $x_0 = \sum \xi_{0j} e_j$. Then (5) becomes

$$f(x_0) = \sum_{j=1}^{n} \xi_{0j} \alpha_j.$$

By assumption this is zero for every $f \in X^*$, that is, for every choice of $\alpha_1, \cdots, \alpha_n$. Hence all ξ_{0j} must be zero. $\blacksquare$

Using this lemma, we can now obtain

2.9-3 Theorem (Algebraic reflexivity). *A finite dimensional vector space is algebraically reflexive.*

Proof. The canonical mapping $C: X \longrightarrow X^{**}$ considered in the previous section is linear. $Cx_0 = 0$ means that for all $f \in X^*$ we have

$$(Cx_0)(f) = g_{x_0}(f) = f(x_0) = 0,$$

by the definition of C. This implies $x_0 = 0$ by Lemma 2.9-2. Hence from Theorem 2.6-10 it follows that the mapping C has an inverse $C^{-1}: \mathcal{R}(C) \longrightarrow X$, where $\mathcal{R}(C)$ is the range of C. We also have $\dim \mathcal{R}(C) = \dim X$ by the same theorem. Now by Theorem 2.9-1,

$$\dim X^{**} = \dim X^* = \dim X.$$

Together, $\dim \mathcal{R}(C) = \dim X^{**}$. Hence $\mathcal{R}(C) = X^{**}$ because $\mathcal{R}(C)$ is a vector space (cf. 2.6-9) and a proper subspace of X^{**} has dimension less than $\dim X^{**}$, by Theorem 2.1-8. By the definition, this proves algebraic reflexivity. ∎

Problems

1. Determine the null space of the operator $T: \mathbf{R}^3 \longrightarrow \mathbf{R}^2$ represented by

$$\begin{bmatrix} 1 & 3 & 2 \\ -2 & 1 & 0 \end{bmatrix}.$$

2. Let $T: \mathbf{R}^3 \longrightarrow \mathbf{R}^3$ be defined by $(\xi_1, \xi_2, \xi_3) \longmapsto (\xi_1, \xi_2, -\xi_1 - \xi_2)$. Find $\mathcal{R}(T)$, $\mathcal{N}(T)$ and a matrix which represents T.

3. Find the dual basis of the basis $\{(1, 0, 0), (0, 1, 0), (0, 0, 1)\}$ for $\mathbf{R}^3$.

4. Let $\{f_1, f_2, f_3\}$ be the dual basis of $\{e_1, e_2, e_3\}$ for $\mathbf{R}^3$, where $e_1 = (1, 1, 1)$, $e_2 = (1, 1, -1)$, $e_3 = (1, -1, -1)$. Find $f_1(x)$, $f_2(x)$, $f_3(x)$, where $x = (1, 0, 0)$.

5. If f is a linear functional on an n-dimensional vector space X, what dimension can the null space $\mathcal{N}(f)$ have?

6. Find a basis for the null space of the functional f defined on $\mathbf{R}^3$ by $f(x) = \xi_1 + \xi_2 - \xi_3$, where $x = (\xi_1, \xi_2, \xi_3)$.

7. Same task as in Prob. 6, if $f(x) = \alpha_1\xi_1 + \alpha_2\xi_2 + \alpha_3\xi_3$, where $\alpha_1 \neq 0$.

8. If Z is an $(n-1)$-dimensional subspace of an n-dimensional vector space X, show that Z is the null space of a suitable linear functional f on X, which is uniquely determined to within a scalar multiple.

9. Let X be the vector space of all real polynomials of a real variable and of degree less than a given n, together with the polynomial $x=0$ (whose degree is left undefined in the usual discussion of degree). Let $f(x)=x^{(k)}(a)$, the value of the kth derivative (k fixed) of $x \in X$ at a fixed $a \in \mathbf{R}$. Show that f is a linear functional on X.

10. Let Z be a proper subspace of an n-dimensional vector space X, and let $x_0 \in X-Z$. Show that there is a linear functional f on X such that $f(x_0)=1$ and $f(x)=0$ for all $x \in Z$.

11. If x and y are different vectors in a finite dimensional vector space X, show that there is a linear functional f on X such that $f(x) \neq f(y)$.

12. If $f_1, \cdots, f_p$ are linear functionals on an n-dimensional vector space X, where $p<n$, show that there is a vector $x \neq 0$ in X such that $f_1(x)=0, \cdots, f_p(x)=0$. What consequences does this result have with respect to linear equations?

13. (Linear extension) Let Z be a proper subspace of an n-dimensional vector space X, and let f be a linear functional on Z. Show that f can be *extended linearly* to X, that is, there is a linear functional $\tilde{f}$ on X such that $\tilde{f}|_Z = f$.

14. Let the functional f on $\mathbf{R}^2$ be defined by $f(x)=4\xi_1-3\xi_2$, where $x=(\xi_1, \xi_2)$. Regard $\mathbf{R}^2$ as the subspace of $\mathbf{R}^3$ given by $\xi_3=0$. Determine all linear extensions $\tilde{f}$ of f from $\mathbf{R}^2$ to $\mathbf{R}^3$.

15. Let $Z \subset \mathbf{R}^3$ be the subspace represented by $\xi_2=0$ and let f on Z be defined by $f(x)=(\xi_1-\xi_3)/2$. Find a linear extension $\tilde{f}$ of f to $\mathbf{R}^3$ such that $\tilde{f}(x_0)=k$ (a given constant), where $x_0=(1, 1, 1)$. Is $\tilde{f}$ unique?

2.10 Normed Spaces of Operators. Dual Space

In Sec. 2.7 we defined the concept of a bounded linear operator and illustrated it by basic examples which gave the reader a first impression of the importance of these operators. In the present section our goal is as follows. We take any two normed spaces X and Y (both real or both complex) and consider the set

$$B(X, Y)$$

consisting of all bounded linear operators from X into Y, that is, each such operator is defined on all of X and its range lies in Y. We want to show that $B(X, Y)$ can itself be made into a normed space.[10]

The whole matter is quite simple. First of all, $B(X, Y)$ becomes a vector space if we define the sum T_1+T_2 of two operators T_1, $T_2 \in B(X, Y)$ in a natural way by

$$(T_1+T_2)x = T_1x + T_2x$$

and the product αT of $T \in B(X, Y)$ and a scalar α by

$$(\alpha T)x = \alpha Tx.$$

Now we remember Lemma 2.7-2(*b*) and have at once the desired result:

2.10-1 Theorem (Space *B*(*X*, *Y*)). *The vector space $B(X, Y)$ of all bounded linear operators from a normed space X into a normed space Y is itself a normed space with norm defined by*

$$\|T\| = \sup_{\substack{x \in X \\ x \neq 0}} \frac{\|Tx\|}{\|x\|} = \sup_{\substack{x \in X \\ \|x\|=1}} \|Tx\|. \tag{1}$$

In what case will $B(X, Y)$ be a Banach space? This is a central question, which is answered in the following theorem. It is remarkable that the condition in the theorem does not involve X; that is, X may or may not be complete:

2.10-2 Theorem (Completeness). *If Y is a Banach space, then $B(X, Y)$ is a Banach space.*

Proof. We consider an arbitrary Cauchy sequence (T_n) in $B(X, Y)$ and show that (T_n) converges to an operator $T \in B(X, Y)$.

[10] B in $B(X, Y)$ suggests "bounded." Another notation for $B(X, Y)$ is $L(X, Y)$, where L suggests "linear." Both notations are common. We use $B(X, Y)$ throughout.

Since (T_n) is Cauchy, for every $\varepsilon > 0$ there is an N such that

$$\|T_n - T_m\| < \varepsilon \qquad (m, n > N).$$

For all $x \in X$ and $m, n > N$ we thus obtain [cf. (3) in Sec. 2.7]

$$\|T_n x - T_m x\| = \|(T_n - T_m)x\| \leqq \|T_n - T_m\| \, \|x\| < \varepsilon \|x\|. \tag{2}$$

Now for any fixed x and given $\tilde{\varepsilon}$ we may choose $\varepsilon = \varepsilon_x$ so that $\varepsilon_x \|x\| < \tilde{\varepsilon}$. Then from (2) we have $\|T_n x - T_m x\| < \tilde{\varepsilon}$ and see that $(T_n x)$ is Cauchy in Y. Since Y is complete, $(T_n x)$ converges, say, $T_n x \longrightarrow y$. Clearly, the limit $y \in Y$ depends on the choice of $x \in X$. This defines an operator $T\colon X \longrightarrow Y$, where $y = Tx$. The operator T is linear since

$$\lim T_n(\alpha x + \beta z) = \lim (\alpha T_n x + \beta T_n z) = \alpha \lim T_n x + \beta \lim T_n z.$$

We prove that T is bounded and $T_n \longrightarrow T$, that is, $\|T_n - T\| \longrightarrow 0$.

Since (2) holds for every $m > N$ and $T_m x \longrightarrow Tx$, we may let $m \longrightarrow \infty$. Using the continuity of the norm, we then obtain from (2) for every $n > N$ and all $x \in X$

$$\|T_n x - Tx\| = \|T_n x - \lim_{m \to \infty} T_m x\| = \lim_{m \to \infty} \|T_n x - T_m x\| \leqq \varepsilon \|x\|. \tag{3}$$

This shows that $(T_n - T)$ with $n > N$ is a bounded linear operator. Since T_n is bounded, $T = T_n - (T_n - T)$ is bounded, that is, $T \in B(X, Y)$. Furthermore, if in (3) we take the supremum over all x of norm 1, we obtain

$$\|T_n - T\| \leqq \varepsilon \qquad (n > N).$$

Hence $\|T_n - T\| \longrightarrow 0$. ∎

This theorem has an important consequence with respect to the dual space X' of X, which is defined as follows.

2.10-3 Definition (Dual space X'). Let X be a normed space. Then the set of all bounded linear functionals on X constitutes a normed space with norm defined by

$$\|f\| = \sup_{\substack{x \in X \\ x \neq 0}} \frac{|f(x)|}{\|x\|} = \sup_{\substack{x \in X \\ \|x\|=1}} |f(x)| \tag{4}$$

[cf. (2) in Sec. 2.8] which is called the *dual space*[11] of X and is denoted by X'. ∎

Since a linear functional on X maps X into $\mathbf{R}$ or $\mathbf{C}$ (the scalar field of X), and since $\mathbf{R}$ or $\mathbf{C}$, taken with the usual metric, is complete, we see that X' is $B(X, Y)$ with the complete space $Y = \mathbf{R}$ or $\mathbf{C}$. Hence Theorem 2.10-2 is applicable and implies the basic

2.10-4 Theorem (Dual space). *The dual space X' of a normed space X is a Banach space (whether or not X is).*

It is a fundamental principle of functional analysis that investigations of spaces are often combined with those of the dual spaces. For this reason it is worthwhile to consider some of the more frequently occurring spaces and find out what their duals look like. In this connection the concept of an isomorphism will be helpful in understanding the present discussion. Remembering our consideration in Sec. 2.8, we give the following definition.

An **isomorphism** of a normed space X onto a normed space $\tilde{X}$ is a bijective linear operator $T: X \longrightarrow \tilde{X}$ which preserves the norm, that is, for all $x \in X$,

$$\|Tx\| = \|x\|.$$

(Hence T is isometric.) X is then called *isomorphic* with $\tilde{X}$, and X and $\tilde{X}$ are called *isomorphic normed spaces.*—From an abstract point of view, X and $\tilde{X}$ are then identical, the isomorphism merely amounting to renaming of the elements (attaching a "tag" T to each point).

Our first example shows that the dual space of $\mathbf{R}^n$ is isomorphic with $\mathbf{R}^n$; we express this more briefly by saying that the dual space of $\mathbf{R}^n$ *is* $\mathbf{R}^n$; similarly for the other examples.

[11] Other terms are *dual, adjoint space* and *conjugate space.* Remember from Sec. 2.8 that the *algebraic* dual space X^* of X is the vector space of *all* linear functionals on X.

Examples

2.10-5 Space $\mathbf{R}^n$. *The dual space of $\mathbf{R}^n$ is $\mathbf{R}^n$.*

Proof. We have $\mathbf{R}^{n\prime}=\mathbf{R}^{n*}$ by Theorem 2.7-8, and every $f \in \mathbf{R}^{n*}$ has a representation (5), Sec. 2.9:

$$f(x)=\sum \xi_k\gamma_k \qquad\qquad \gamma_k=f(e_k)$$

(sum from 1 to n). By the Cauchy-Schwarz inequality (Sec. 1.2),

$$|f(x)| \leqq \sum |\xi_k\gamma_k| \leqq \left(\sum \xi_j^2\right)^{1/2}\left(\sum \gamma_k^2\right)^{1/2} = \|x\| \left(\sum \gamma_k^2\right)^{1/2}.$$

Taking the supremum over all x of norm 1 we obtain

$$\|f\| \leqq \left(\sum \gamma_k^2\right)^{1/2}.$$

However, since for $x=(\gamma_1, \cdots, \gamma_n)$ equality is achieved in the Cauchy-Schwarz inequality, we must in fact have

$$\|f\| = \left(\sum_{k=1}^{n} \gamma_k^2\right)^{1/2}.$$

This proves that the norm of f is the Euclidean norm, and $\|f\|=\|c\|$, where $c=(\gamma_k)\in\mathbf{R}^n$. Hence the mapping of $\mathbf{R}^{n\prime}$ onto $\mathbf{R}^n$ defined by $f \longmapsto c=(\gamma_k)$, $\gamma_k=f(e_k)$, is norm preserving and, since it is linear and bijective, it is an isomorphism. ∎

2.10-6 Space l^1. *The dual space of l^1 is l^∞.*

Proof. A Schauder basis (Sec. 2.3) for l^1 is (e_k), where $e_k=(\delta_{kj})$ has 1 in the kth place and zeros otherwise. Then every $x \in l^1$ has a unique representation

$$x=\sum_{k=1}^{\infty} \xi_k e_k. \tag{5}$$

We consider any $f \in l^{1\prime}$, where $l^{1\prime}$ is the dual space of l^1. Since f is linear and bounded,

$$f(x)=\sum_{k=1}^{\infty} \xi_k\gamma_k \qquad\qquad \gamma_k=f(e_k) \tag{6}$$

where the numbers $\gamma_k = f(e_k)$ are uniquely determined by f. Also $\|e_k\| = 1$ and

$$(7) \qquad |\gamma_k| = |f(e_k)| \leqq \|f\| \, \|e_k\| = \|f\|, \qquad \sup_k |\gamma_k| \leqq \|f\|.$$

Hence $(\gamma_k) \in l^\infty$.

On the other hand, for every $b = (\beta_k) \in l^\infty$ we can obtain a corresponding bounded linear functional g on l^1. In fact, we may define g on l^1 by

$$g(x) = \sum_{k=1}^{\infty} \xi_k \beta_k$$

where $x = (\xi_k) \in l^1$. Then g is linear, and boundedness follows from

$$|g(x)| \leqq \sum |\xi_k \beta_k| \leqq \sup_j |\beta_j| \sum |\xi_k| = \|x\| \sup_j |\beta_j|$$

(sum from 1 to ∞). Hence $g \in l^{1\prime}$.

We finally show that the norm of f is the norm on the space l^∞. From (6) we have

$$|f(x)| = \left| \sum \xi_k \gamma_k \right| \leqq \sup_j |\gamma_j| \sum |\xi_k| = \|x\| \sup_j |\gamma_j|.$$

Taking the supremum over all x of norm 1, we see that

$$\|f\| \leqq \sup_j |\gamma_j|.$$

From this and (7),

$$(8) \qquad \|f\| = \sup_j |\gamma_j|,$$

which is the norm on l^∞. Hence this formula can be written $\|f\| = \|c\|_\infty$, where $c = (\gamma_j) \in l^\infty$. It shows that the bijective linear mapping of $l^{1\prime}$ onto l^∞ defined by $f \longmapsto c = (\gamma_j)$ is an isomorphism. ∎

2.10-7 Space l^p. *The dual space of l^p is l^q; here, $1 < p < +\infty$ and q is the conjugate of p, that is, $1/p + 1/q = 1$.*

Proof. A Schauder basis for l^p is (e_k), where $e_k = (\delta_{kj})$ as in the preceding example. Then every $x \in l^p$ has a unique representation

$$x = \sum_{k=1}^{\infty} \xi_k e_k. \tag{9}$$

We consider any $f \in l^{p\prime}$, where $l^{p\prime}$ is the dual space of l^p. Since f is linear and bounded,

$$f(x) = \sum_{k=1}^{\infty} \xi_k \gamma_k \qquad \gamma_k = f(e_k). \tag{10}$$

Let q be the conjugate of p (cf. 1.2-3) and consider $x_n = (\xi_k^{(n)})$ with

$$\xi_k^{(n)} = \begin{cases} |\gamma_k|^q/\gamma_k & \text{if } k \leqq n \text{ and } \gamma_k \neq 0, \\ 0 & \text{if } k > n \text{ or } \gamma_k = 0. \end{cases} \tag{11}$$

By substituting this into (10) we obtain

$$f(x_n) = \sum_{k=1}^{\infty} \xi_k^{(n)} \gamma_k = \sum_{k=1}^{n} |\gamma_k|^q.$$

We also have, using (11) and $(q-1)p = q$,

$$\begin{aligned} f(x_n) \leqq \|f\| \, \|x_n\| &= \|f\| \Big(\sum |\xi_k^{(n)}|^p \Big)^{1/p} \\ &= \|f\| \Big(\sum |\gamma_k|^{(q-1)p} \Big)^{1/p} \\ &= \|f\| \Big(\sum |\gamma_k|^q \Big)^{1/p} \end{aligned}$$

(sum from 1 to n). Together,

$$f(x_n) = \sum |\gamma_k|^q \leqq \|f\| \Big(\sum |\gamma_k|^q \Big)^{1/p}.$$

Dividing by the last factor and using $1 - 1/p = 1/q$, we get

$$\Big(\sum_{k=1}^{n} |\gamma_k|^q \Big)^{1-1/p} = \Big(\sum_{k=1}^{n} |\gamma_k|^q \Big)^{1/q} \leqq \|f\|.$$

Since n is arbitrary, letting $n \longrightarrow \infty$, we obtain

$$\left(\sum_{k=1}^{\infty} |\gamma_k|^q\right)^{1/q} \leqq \|f\|. \tag{12}$$

Hence $(\gamma_k) \in l^q$.

Conversely, for any $b = (\beta_k) \in l^q$ we can get a corresponding bounded linear functional g on l^p. In fact, we may define g on l^p by setting

$$g(x) = \sum_{k=1}^{\infty} \xi_k \beta_k$$

where $x = (\xi_k) \in l^p$. Then g is linear, and boundedness follows from the Hölder inequality (10), Sec. 1.2. Hence $g \in l^{p\prime}$.

We finally prove that the norm of f is the norm on the space l^q. From (10) and the Hölder inequality we have

$$\begin{aligned} |f(x)| = \left|\sum \xi_k \gamma_k\right| &\leqq \left(\sum |\xi_k|^p\right)^{1/p} \left(\sum |\gamma_k|^q\right)^{1/q} \\ &= \|x\| \left(\sum |\gamma_k|^q\right)^{1/q} \end{aligned}$$

(sum from 1 to ∞); hence by taking the supremum over all x of norm 1 we obtain

$$\|f\| \leqq \left(\sum |\gamma_k|^q\right)^{1/q}.$$

From (12) we see that the equality sign must hold, that is,

$$\|f\| = \left(\sum_{k=1}^{\infty} |\gamma_k|^q\right)^{1/q}. \tag{13}$$

This can be written $\|f\| = \|c\|_q$, where $c = (\gamma_k) \in l^q$ and $\gamma_k = f(e_k)$. The mapping of $l^{p\prime}$ onto l^q defined by $f \longmapsto c$ is linear and bijective, and from (13) we see that it is norm preserving, so that it is an isomorphism. ∎

What is the significance of these and similar examples? In applications it is frequently quite useful to know the general form of bounded linear functionals on spaces of practical importance, and many spaces have been investigated in that respect. Our examples give general representations of bounded linear functionals on $\mathbf{R}^n$, l^1 and l^p with $p>1$. The space $C[a, b]$ will be considered later, in Sec. 4.4, since this will require additional tools (in particular the so-called Hahn-Banach theorem).

Furthermore, remembering the discussion of the second algebraic dual space X^{**} in Sec. 2.8, we may ask whether it is worthwhile to consider $X''=(X')'$, the second dual space of X. The answer is in the affirmative, but we have to postpone this discussion until Sec. 4.6 where we develop suitable tools for obtaining substantial results in that direction. At present let us turn to matters which are somewhat simpler, namely, to inner product and Hilbert spaces. We shall see that these are special normed spaces which are of great importance in applications.

Problems

1. What is the zero element of the vector space $B(X, Y)$? The inverse of a $T \in B(X, Y)$ in the sense of Def. 2.1-1?

2. The operators and functionals considered in the text are defined on the entire space X. Show that without that assumption, in the case of functionals we still have the following theorem. If f and g are bounded linear functionals with domains in a normed space X, then for any nonzero scalars α and β the linear combination $h=\alpha f+\beta g$ is a bounded linear functional with domain $\mathfrak{D}(h)=\mathfrak{D}(f)\cap\mathfrak{D}(g)$.

3. Extend the theorem in Prob. 2 to bounded linear operators T_1 and T_2.

4. Let X and Y be normed spaces and $T_n: X \longrightarrow Y$ $(n=1, 2, \cdots)$ bounded linear operators. Show that convergence $T_n \longrightarrow T$ implies that for every $\varepsilon>0$ there is an N such that for all $n>N$ and all x in any given closed ball we have $\|T_n x - Tx\|<\varepsilon$.

5. Show that 2.8-5 is in agreement with 2.10-5.

6. If X is the space of ordered n-tuples of real numbers and $\|x\|=\max_i |\xi_i|$, where $x=(\xi_1, \cdots, \xi_n)$, what is the corresponding norm on the dual space X'?

7. What conclusion can we draw from 2.10-6 with respect to the space X of all ordered n-tuples of real numbers?

8. Show that the dual space of the space c_0 is l^1. (Cf. Prob. 1 in Sec. 2.3.)

9. Show that a linear functional f on a vector space X is uniquely determined by its values on a Hamel basis for X. (Cf. Sec. 2.1.)

10. Let X and $Y \neq \{0\}$ be normed spaces, where $\dim X = \infty$. Show that there is at least one unbounded linear operator $T: X \longrightarrow Y$. (Use a Hamel basis.)

11. If X is a normed space and $\dim X = \infty$, show that the dual space X' is not identical with the algebraic dual space X^*.

12. (Completeness) The examples in the text can be used to prove completeness of certain spaces. How? For what spaces?

13. (Annihilator) Let $M \neq \varnothing$ be any subset of a normed space X. The *annihilator* M^a of M is defined to be the set of all bounded linear functionals on X which are zero everywhere on M. Thus M^a is a subset of the dual space X' of X. Show that M^a is a vector subspace of X' and is closed. What are X^a and $\{0\}^a$?

14. If M is an m-dimensional subspace of an n-dimensional normed space X, show that M^a is an $(n-m)$-dimensional subspace of X'. Formulate this as a theorem about solutions of a system of linear equations.

15. Let $M = \{(1, 0, -1), (1, -1, 0), (0, 1, -1)\} \subset \mathbf{R}^3$. Find a basis for M^a.

CHAPTER 3

INNER PRODUCT SPACES. HILBERT SPACES

In a normed space we can add vectors and multiply vectors by scalars, just as in elementary vector algebra. Furthermore, the norm on such a space generalizes the elementary concept of the length of a vector. However, what is still missing in a general normed space, and what we would like to have if possible, is an analogue of the familiar dot product

$$a \cdot b = \alpha_1\beta_1 + \alpha_2\beta_2 + \alpha_3\beta_3$$

and resulting formulas, notably

$$|a| = \sqrt{a \cdot a}$$

and the condition for orthogonality (perpendicularity)

$$a \cdot b = 0$$

which are important tools in many applications. Hence the question arises whether the dot product and orthogonality can be generalized to arbitrary vector spaces. In fact, this can be done and leads to *inner product spaces* and complete inner product spaces, called *Hilbert spaces.*

Inner product spaces are special normed spaces, as we shall see. Historically they are older than general normed spaces. Their theory is richer and retains many features of Euclidean space, a central concept being orthogonality. In fact, inner product spaces are probably the most natural generalization of Euclidean space, and the reader should note the great harmony and beauty of the concepts and proofs in this field. The whole theory was initiated by the work of D. Hilbert (1912) on integral equations. The currently used geometrical notation and terminology is analogous to that of Euclidean geometry and was coined by E. Schmidt (1908), who followed a suggestion of G. Kowalewski (as he mentioned on p. 56 of his paper). These spaces have been, up to now, the most useful spaces in practical applications of functional analysis.

Important concepts, brief orientation about main content

An *inner product space* X (Def. 3.1-1) is a vector space with an *inner product* $\langle x, y\rangle$ defined on it. The latter generalizes the dot product of vectors in three dimensional space and is used to define

(I) a *norm* $\|\cdot\|$ by $\|x\| = \langle x, x\rangle^{1/2}$,

(II) *orthogonality* by $\langle x, y\rangle = 0$.

A *Hilbert space* H is a complete inner product space. The theory of inner product and Hilbert spaces is richer than that of general normed and Banach spaces. Distinguishing features are

(i) representations of H as a direct sum of a closed subspace and its *orthogonal complement* (cf. 3.3-4),

(ii) *orthonormal sets and sequences* and corresponding representations of elements of H (cf. Secs. 3.4, 3.5),

(iii) the *Riesz representation* 3.8-1 of bounded linear functionals by inner products,

(iv) the *Hilbert-adjoint operator* T^* of a bounded linear operator T (cf. 3.9-1).

Orthonormal sets and sequences are truly interesting only if they are total (Sec. 3.6). Hilbert-adjoint operators can be used to define classes of operators (*self-adjoint, unitary, normal;* cf. Sec. 3.10) which are of great importance in applications.

3.1 Inner Product Space. Hilbert Space

The spaces to be considered in this chapter are defined as follows.

3.1-1 Definition (Inner product space, Hilbert space). An *inner product space* (or *pre-Hilbert space*) is a vector space X with an inner product defined on X. A *Hilbert space* is a complete inner product space (complete in the metric defined by the inner product; cf. (2), below). Here, an **inner product** *on* X is a mapping of $X \times X$ into the scalar field K of X; that is, with every pair of vectors x and y there is associated a scalar which is written

$$\langle x, y\rangle$$

and is called the *inner product*[1] of x and y, such that for all vectors x, y, z and scalars α we have

(IP1) $$\langle x+y, z\rangle = \langle x, z\rangle + \langle y, z\rangle$$

(IP2) $$\langle \alpha x, y\rangle = \alpha\langle x, y\rangle$$

(IP3) $$\langle x, y\rangle = \overline{\langle y, x\rangle}$$

(IP4) $$\langle x, x\rangle \geqq 0$$ $$\langle x, x\rangle = 0 \quad \Longleftrightarrow \quad x = 0.$$

An inner product on X defines a *norm* on X given by

(1) $$\|x\| = \sqrt{\langle x, x\rangle} \qquad (\geqq 0)$$

and a *metric* on X given by

(2) $$d(x, y) = \|x-y\| = \sqrt{\langle x-y, x-y\rangle}.$$ ∎

Hence *inner product spaces are normed spaces, and Hilbert spaces are Banach spaces.*

In (IP3), the bar denotes complex conjugation. Consequently, if X is a *real* vector space, we simply have

$$\langle x, y\rangle = \langle y, x\rangle \qquad (\textit{Symmetry}).$$

The proof that (1) satisfies the axioms (N1) to (N4) of a norm (cf. Sec. 2.2) will be given at the beginning of the next section.

From (IP1) to (IP3) we obtain the formula

(a) $$\langle \alpha x + \beta y, z\rangle = \alpha\langle x, z\rangle + \beta\langle y, z\rangle$$

(3) (b) $$\langle x, \alpha y\rangle = \bar{\alpha}\langle x, y\rangle$$

(c) $$\langle x, \alpha y + \beta z\rangle = \bar{\alpha}\langle x, y\rangle + \bar{\beta}\langle x, z\rangle$$

[1] Or *scalar product*, but this must not be confused with the product of a vector by a scalar in a vector space.

The notation $\langle\ ,\ \rangle$ for the inner product is quite common. In an elementary text such as the present one it may have the advantage over another popular notation, $(\ ,\ )$, that it excludes confusion with ordered pairs (components of a vector, elements of a product space, arguments of functions depending on two variables, etc.).

which we shall use quite often. (3a) shows that the inner product is linear in the first factor. Since in (3c) we have complex conjugates $\bar{\alpha}$ and $\bar{\beta}$ on the right, we say that the inner product is *conjugate linear* in the second factor. Expressing both properties together, we say that the inner product is *sesquilinear.* This means "$1\frac{1}{2}$ times linear" and is motivated by the fact that "conjugate linear" is also known as "semilinear" (meaning "halflinear"), a less suggestive term which we shall not use.

The reader may show by a simple straightforward calculation that a norm on an inner product space satisfies the important **parallelogram equality**

$$\|x+y\|^2+\|x-y\|^2=2(\|x\|^2+\|y\|^2). \tag{4}$$

This name is suggested by elementary geometry, as we see from Fig. 23 if we remember that the norm generalizes the elementary concept of the length of a vector (cf. Sec. 2.2). It is quite remarkable that such an equation continues to hold in our present much more general setting.

We conclude that if a norm does not satisfy (4), it cannot be obtained from an inner product by the use of (1). Such norms do exist; examples will be given below. Without risking misunderstandings we may thus say:

Not all normed spaces are inner product spaces.

Before we consider examples, let us define the concept of orthogonality, which is basic in the whole theory. We know that if the dot product of two vectors in three dimensional spaces is zero, the vectors are orthogonal, that is, they are perpendicular or at least one of them is the zero vector. This suggests and motivates the following

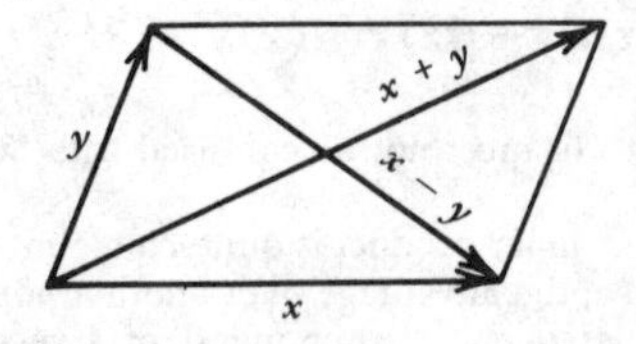

Fig. 23. Parallelogram with sides x and y in the plane

3.1-2 Definition (Orthogonality). An element x of an inner product space X is said to be *orthogonal* to an element $y \in X$ if

$$\langle x, y\rangle = 0.$$

We also say that *x and y are orthogonal*, and we write $x \perp y$. Similarly, for subsets $A, B \subset X$ we write $x \perp A$ if $x \perp a$ for all $a \in A$, and $A \perp B$ if $a \perp b$ for all $a \in A$ and all $b \in B$. ∎

Examples

3.1-3 Euclidean space $\mathbf{R}^n$. The space $\mathbf{R}^n$ is a Hilbert space with inner product defined by

$$\langle x, y\rangle = \xi_1\eta_1 + \cdots + \xi_n\eta_n \tag{5}$$

where $x = (\xi_j) = (\xi_1, \cdots, \xi_n)$ and $y = (\eta_j) = (\eta_1, \cdots, \eta_n)$.

In fact, from (5) we obtain

$$\|x\| = \langle x, x\rangle^{1/2} = (\xi_1{}^2 + \cdots + \xi_n)^{1/2}$$

and from this the Euclidean metric defined by

$$d(x, y) = \|x - y\| = \langle x - y, x - y\rangle^{1/2} = [(\xi_1 - \eta_1)^2 + \cdots + (\xi_n - \eta_n)^2]^{1/2};$$

cf. 2.2-2. Completeness was shown in 1.5-1.

If $n = 3$, formula (5) gives the usual dot product

$$\langle x, y\rangle = x \cdot y = \xi_1\eta_1 + \xi_2\eta_2 + \xi_3\eta_3$$

of $x = (\xi_1, \xi_2, \xi_3)$ and $y = (\eta_1, \eta_2, \eta_3)$, and the orthogonality

$$\langle x, y\rangle = x \cdot y = 0$$

agrees with the elementary concept of perpendicularity.

3.1-4 Unitary space $\mathbf{C}^n$. The space $\mathbf{C}^n$ defined in 2.2-2 is a Hilbert space with inner product given by

$$\langle x, y\rangle = \xi_1\bar{\eta}_1 + \cdots + \xi_n\bar{\eta}_n. \tag{6}$$

In fact, from (6) we obtain the norm defined by

$$\|x\| = (\xi_1\bar{\xi}_1 + \cdots + \xi_n\bar{\xi}_n)^{1/2} = (|\xi_1|^2 + \cdots + |\xi_n|^2)^{1/2}.$$

Here we also see why we have to take complex conjugates $\bar{\eta}_j$ in (6); this entails $\langle y, x\rangle = \overline{\langle x, y\rangle}$, which is (IP3), so that $\langle x, x\rangle$ is real.

3.1-5 Space $L^2[a, b]$. The norm in Example 2.2-7 is defined by

$$\|x\| = \left(\int_a^b x(t)^2 \, dt\right)^{1/2}$$

and can be obtained from the inner product defined by

$$\langle x, y\rangle = \int_a^b x(t)y(t) \, dt. \tag{7}$$

In Example 2.2-7 the functions were assumed to be real-valued, for simplicity. In connection with certain applications it is advantageous to remove that restriction and consider *complex-valued* functions (keeping $t \in [a, b]$ real, as before). These functions form a complex vector space, which becomes an inner product space if we define

$$\langle x, y\rangle = \int_a^b x(t)\overline{y(t)} \, dt. \tag{7*}$$

Here the bar denotes the complex conjugate. It has the effect that (IP3) holds, so that $\langle x, x\rangle$ is still real. This property is again needed in connection with the norm, which is now defined by

$$\|x\| = \left(\int_a^b |x(t)|^2 \, dt\right)^{1/2}$$

because $x(t)\overline{x(t)} = |x(t)|^2$.

The completion of the metric space corresponding to (7) is the real space $L^2[a, b]$; cf. 2.2-7. Similarly, the completion of the metric space corresponding to (7*) is called the *complex space* $L^2[a, b]$. We shall see in the next section that the inner product can be extended from an inner product space to its completion. Together with our present discussion this implies that $L^2[a, b]$ is a Hilbert space.

3.1-6 Hilbert sequence space l^2. The space l^2 (cf. 2.2-3) is a Hilbert space with inner product defined by

$$\langle x, y\rangle = \sum_{j=1}^{\infty} \xi_j \bar{\eta}_j. \tag{8}$$

Convergence of this series follows from the Cauchy-Schwarz inequality (11), Sec. 1.2, and the fact that $x, y \in l^2$, by assumption. We see that (8) generalizes (6). The norm is defined by

$$\|x\| = \langle x, x\rangle^{1/2} = \left(\sum_{j=1}^{\infty} |\xi_j|^2\right)^{1/2}.$$

Completeness was shown in 1.5-4.

l^2 is the prototype of a Hilbert space. It was introduced and investigated by D. Hilbert (1912) in his work on integral equations. An axiomatic definition of Hilbert space was not given until much later, by J. von Neumann (1927), pp. 15–17, in a paper on the mathematical foundation of quantum mechanics. Cf. also J. von Neumann (1929–30), pp. 63–66, and M. H. Stone (1932), pp. 3–4. That definition included separability, a condition which was later dropped from the definition when H. Löwig (1934), F. Rellich (1934) and F. Riesz (1934) showed that for most parts of the theory that condition was an unnecessary restriction. (These papers are listed in Appendix 3.)

3.1-7 Space l^p. *The space l^p with $p \neq 2$ is not an inner product space, hence not a Hilbert space.*

Proof. Our statement means that the norm of l^p with $p \neq 2$ cannot be obtained from an inner product. We prove this by showing that the norm does not satisfy the parallelogram equality (4). In fact, let us take $x = (1, 1, 0, 0, \cdots) \in l^p$ and $y = (1, -1, 0, 0, \cdots) \in l^p$ and calculate

$$\|x\| = \|y\| = 2^{1/p}, \qquad \|x+y\| = \|x-y\| = 2.$$

We now see that (4) is not satisfied if $p \neq 2$.

l^p is complete (cf. 1.5-4). Hence *l^p with $p \neq 2$ is a Banach space which is not a Hilbert space.* The same holds for the space in the next example.

3.1-8 Space $C[a, b]$. *The space $C[a, b]$ is not an inner product space, hence not a Hilbert space.*

Proof. We show that the norm defined by

$$\|x\| = \max_{t \in J} |x(t)| \qquad J = [a, b]$$

cannot be obtained from an inner product since this norm does not satisfy the parallelogram equality (4). Indeed, if we take $x(t) = 1$ and $y(t) = (t-a)/(b-a)$, we have $\|x\| = 1$, $\|y\| = 1$ and

$$x(t) + y(t) = 1 + \frac{t-a}{b-a}$$

$$x(t) - y(t) = 1 - \frac{t-a}{b-a}.$$

Hence $\|x+y\| = 2$, $\|x-y\| = 1$ and

$$\|x+y\|^2 + \|x-y\|^2 = 5 \qquad \text{but} \qquad 2(\|x\|^2 + \|y\|^2) = 4.$$

This completes the proof. ∎

We finally mention the following interesting fact. We know that to an inner product there corresponds a norm which is given by (1). It is remarkable that, conversely, we can "rediscover" the inner product from the corresponding norm. In fact, the reader may verify by straightforward calculation that for a real inner product space we have

$$\langle x, y \rangle = \tfrac{1}{4}(\|x+y\|^2 - \|x-y\|^2) \tag{9}$$

and for a complex inner product space we have

$$\begin{aligned} \operatorname{Re} \langle x, y \rangle &= \tfrac{1}{4}(\|x+y\|^2 - \|x-y\|^2) \\ \operatorname{Im} \langle x, y \rangle &= \tfrac{1}{4}(\|x+iy\|^2 - \|x-iy\|^2). \end{aligned} \tag{10}$$

Formula (10) is sometimes called the **polarization identity.**

Problems

1. Prove (4).

2. **(Pythagorean theorem)** If $x \perp y$ in an inner product space X, show that (Fig. 24)

$$\|x+y\|^2 = \|x\|^2 + \|y\|^2.$$

Extend the formula to m mutually orthogonal vectors.

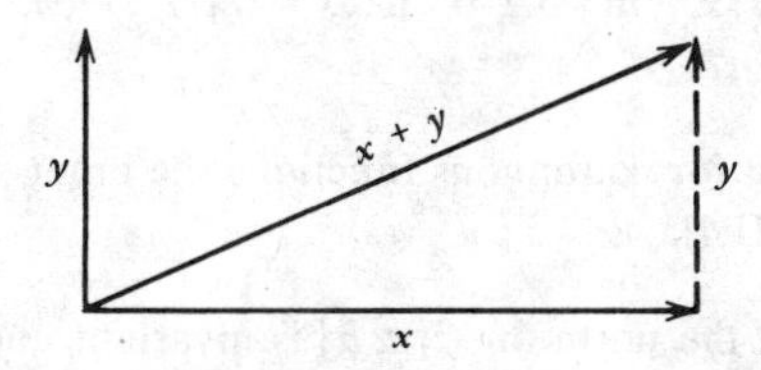

Fig. 24. Illustration of the Pythagorean theorem in the plane

3. If X in Prob. 2 is real, show that, conversely, the given relation implies that $x \perp y$. Show that this may not hold if X is complex. Give examples.

4. If an inner product space X is real, show that the condition $\|x\| = \|y\|$ implies $\langle x+y, x-y\rangle = 0$. What does this mean geometrically if $X = \mathbf{R}^2$? What does the condition imply if X is complex?

5. **(Apollonius' identity)** Verify by direct calculation that for any elements in an inner product space,

$$\|z-x\|^2 + \|z-y\|^2 = \tfrac{1}{2}\|x-y\|^2 + 2\left\|z - \tfrac{1}{2}(x+y)\right\|^2.$$

Show that this identity can also be obtained from the parallelogram equality.

6. Let $x \neq 0$ and $y \neq 0$. (*a*) If $x \perp y$, show that $\{x, y\}$ is a linearly independent set. (*b*) Extend the result to mutually orthogonal nonzero vectors $x_1, \cdots, x_m$.

7. If in an inner product space, $\langle x, u\rangle = \langle x, v\rangle$ for all x, show that $u = v$.

8. Prove (9).

9. Prove (10).

10. Let z_1 and z_2 denote complex numbers. Show that $\langle z_1, z_2\rangle = z_1\bar{z}_2$ defines an inner product, which yields the usual metric on the complex plane. Under what condition do we have orthogonality?

11. Let X be the vector space of all ordered pairs of complex numbers. Can we obtain the norm defined on X by

$$\|x\| = |\xi_1| + |\xi_2| \qquad [x = (\xi_1, \xi_2)]$$

from an inner product?

12. What is $\|x\|$ in 3.1-6 if $x = (\xi_1, \xi_2, \cdots)$, where (*a*) $\xi_n = 2^{-n/2}$, (*b*) $\xi_n = 1/n$?

13. Verify that for continuous functions the inner product in 3.1-5 satisfies (IP1) to (IP4).

14. Show that the norm on $C[a, b]$ is invariant under a linear transformation $t = \alpha\tau + \beta$. Use this to prove the statement in 3.1-8 by mapping $[a, b]$ onto $[0, 1]$ and then considering the functions defined by $\tilde{x}(\tau) = 1$, $\tilde{y}(\tau) = \tau$, where $\tau \in [0, 1]$.

15. If X is a finite dimensional vector space and (e_j) is a basis for X, show that an inner product on X is completely determined by its values $\gamma_{jk} = \langle e_j, e_k\rangle$. Can we choose such scalars γ_{jk} in a completely arbitrary fashion?

3.2 Further Properties of Inner Product Spaces

First of all, we should verify that (1) in the preceding section defines a norm:

(N1) and (N2) in Sec. 2.2 follow from (IP4). Furthermore, (N3) is obtained by the use of (IP2) and (IP3); in fact,

$$\|\alpha x\|^2 = \langle \alpha x, \alpha x\rangle = \alpha\bar{\alpha}\langle x, x\rangle = |\alpha|^2 \|x\|^2.$$

Finally, (N4) is included in

3.2-1 Lemma (Schwarz inequality, triangle inequality). *An inner product and the corresponding norm satisfy the Schwarz inequality and the triangle inequality as follows.*

(a) *We have*

(1) $$|\langle x, y\rangle| \leqq \|x\| \, \|y\|$$ **(Schwarz inequality)**

where the equality sign holds if and only if $\{x, y\}$ *is a linearly dependent set.*

(b) *That norm also satisfies*

(2) $$\|x+y\| \leqq \|x\| + \|y\|$$ **(Triangle inequality)**

where the equality sign holds if and only if[2] $y=0$ *or* $x=cy$ (c *real and* $\geqq 0$).

Proof. **(a)** If $y=0$, then (1) holds since $\langle x, 0\rangle = 0$. Let $y \neq 0$. For every scalar α we have

$$0 \leqq \|x-\alpha y\|^2 = \langle x-\alpha y, x-\alpha y\rangle$$

$$= \langle x, x\rangle - \bar{\alpha}\langle x, y\rangle - \alpha[\langle y, x\rangle - \bar{\alpha}\langle y, y\rangle].$$

We see that the expression in the brackets $[\cdots]$ is zero if we choose $\bar{\alpha} = \langle y, x\rangle / \langle y, y\rangle$. The remaining inequality is

$$0 \leqq \langle x, x\rangle - \frac{\langle y, x\rangle}{\langle y, y\rangle}\langle x, y\rangle = \|x\|^2 - \frac{|\langle x, y\rangle|^2}{\|y\|^2};$$

here we used $\langle y, x\rangle = \overline{\langle x, y\rangle}$. Multiplying by $\|y\|^2$, transferring the last term to the left and taking square roots, we obtain (1).

Equality holds in this derivation if and only if $y=0$ or $0 = \|x-\alpha y\|^2$, hence $x - \alpha y = 0$, so that $x = \alpha y$, which shows linear dependence.

(b) We prove (2). We have

$$\|x+y\|^2 = \langle x+y, x+y\rangle = \|x\|^2 + \langle x, y\rangle + \langle y, x\rangle + \|y\|^2.$$

By the Schwarz inequality,

$$|\langle x, y\rangle| = |\langle y, x\rangle| \leqq \|x\| \, \|y\|.$$

[2] Note that this condition for equality is perfectly "symmetric" in x and y since $x=0$ is included in $x=cy$ (for $c=0$) and so is $y=kx$, $k=1/c$ (for $c>0$).

By the triangle inequality for numbers we thus obtain

$$\begin{aligned}\|x+y\|^2 &\leqq \|x\|^2+2\,|\langle x, y\rangle|+\|y\|^2\\ &\leqq \|x\|^2+2\,\|x\|\,\|y\|+\|y\|^2\\ &=(\|x\|+\|y\|)^2.\end{aligned}$$

Taking square roots on both sides, we have (2).

Equality holds in this derivation if and only if

$$\langle x, y\rangle+\langle y, x\rangle=2\,\|x\|\,\|y\|.$$

The left-hand side is 2 Re $\langle x, y\rangle$, where Re denotes the real part. From this and (1),

$$\text{Re}\,\langle x, y\rangle=\|x\|\,\|y\|\geqq|\langle x, y\rangle|. \tag{3}$$

Since the real part of a complex number cannot exceed the absolute value, we must have equality, which implies linear dependence by part (a), say, $y=0$ or $x=cy$. We show that c is real and $\geqq 0$. From (3) with the equality sign we have Re $\langle x, y\rangle=|\langle x, y\rangle|$. But if the real part of a complex number equals the absolute value, the imaginary part must be zero. Hence $\langle x, y\rangle=\text{Re}\,\langle x, y\rangle\geqq 0$ by (3), and $c\geqq 0$ follows from

$$0\leqq\langle x, y\rangle=\langle cy, y\rangle=c\,\|y\|^2. \qquad \blacksquare$$

The Schwarz inequality (1) is quite important and will be used in proofs over and over again. Another frequently used property is the continuity of the inner product:

3.2-2 Lemma (Continuity of inner product). *If in an inner product space, $x_n \longrightarrow x$ and $y_n \longrightarrow y$, then $\langle x_n, y_n\rangle \longrightarrow \langle x, y\rangle$.*

Proof. Subtracting and adding a term, using the triangle inequality for numbers and, finally, the Schwarz inequality, we obtain

$$\begin{aligned}|\langle x_n, y_n\rangle-\langle x, y\rangle| &= |\langle x_n, y_n\rangle-\langle x_n, y\rangle+\langle x_n, y\rangle-\langle x, y\rangle|\\ &\leqq |\langle x_n, y_n-y\rangle|+|\langle x_n-x, y\rangle|\\ &\leqq \|x_n\|\,\|y_n-y\|+\|x_n-x\|\,\|y\| \quad\longrightarrow\quad 0\end{aligned}$$

since $y_n-y \longrightarrow 0$ and $x_n-x \longrightarrow 0$ as $n \longrightarrow \infty$. $\blacksquare$

As a first application of this lemma, let us prove that every inner product space can be completed. The completion is a Hilbert space and is unique except for isomorphisms. Here the definition of an isomorphism is as follows (as suggested by our discussion in Sec. 2.8).

An **isomorphism** T of an inner product space X onto an inner product space $\tilde{X}$ over the same field is a bijective linear operator $T: X \longrightarrow \tilde{X}$ which preserves the inner product, that is, for all $x, y \in X$,

$$\langle Tx, Ty \rangle = \langle x, y \rangle,$$

where we denoted inner products on X and $\tilde{X}$ by the same symbol, for simplicity. $\tilde{X}$ is then called *isomorphic* with X, and X and $\tilde{X}$ are called *isomorphic inner product spaces.* Note that the bijectivity and linearity guarantees that T is a vector space isomorphism of X onto $\tilde{X}$, so that T preserves the whole structure of inner product space. T is also an isometry of X onto $\tilde{X}$ because distances in X and $\tilde{X}$ are determined by the norms defined by the inner products on X and $\tilde{X}$.

The theorem about the completion of an inner product space can now be stated as follows.

3.2-3 Theorem (Completion). *For any inner product space X there exists a Hilbert space H and an isomorphism A from X onto a dense subspace $W \subset H$. The space H is unique except for isomorphisms.*

Proof. By Theorem 2.3-2 there exists a Banach space H and an isometry A from X onto a subspace W of H which is dense in H. For reasons of continuity, under such an isometry, sums and scalar multiples of elements in X and W correspond to each other, so that A is even an isomorphism of X onto W, both regarded as normed spaces. Lemma 3.2-2 shows that we can define an inner product on H by setting

$$\langle \hat{x}, \hat{y} \rangle = \lim_{n \to \infty} \langle x_n, y_n \rangle,$$

the notations being as in Theorem 2.3-2 (and 1.6-2), that is, (x_n) and (y_n) are representatives of $\hat{x} \in H$ and $\hat{y} \in H$, respectively. Taking (9) and (10), Sec. 3.1, into account, we see that A is an isomorphism of X onto W, both regarded as inner product spaces.

Theorem 2.3-2 also guarantees that H is unique except for isometries, that is, two completions H and $\tilde{H}$ of X are related by an isometry $T: H \longrightarrow \tilde{H}$. Reasoning as in the case of A, we conclude that T must be an isomorphism of the Hilbert space H onto the Hilbert space $\tilde{H}$. ∎

A **subspace** Y of an inner product space X is defined to be a vector subspace of X (cf. Sec. 2.1) taken with the inner product on X restricted to $Y \times Y$.

Similarly, a **subspace** Y of a Hilbert space H is defined to be a subspace of H, regarded as an inner product space. Note that Y need not be a Hilbert space because Y may not be complete. In fact, from Theorems 2.3-1 and 2.4-2 we immediately have the statements (*a*) and (*b*) in the following theorem.

3.2-4 Theorem (Subspace). *Let Y be a subspace of a Hilbert space H. Then:*

(a) *Y is complete if and only if Y is closed in H.*

(b) *If Y is finite dimensional, then Y is complete.*

(c) *If H is separable, so is Y. More generally, every subset of a separable inner product space is separable.*

The simple proof of (*c*) is left to the reader.

Problems

1. What is the Schwarz inequality in $\mathbf{R}^2$ or $\mathbf{R}^3$? Give another proof of it in these cases.

2. Give examples of subspaces of l^2.

3. Let X be the inner product space consisting of the polynomial $x=0$ (cf. the remark in Prob. 9, Sec. 2.9) and all real polynomials in t, of degree not exceeding 2, considered for real $t \in [a, b]$, with inner product defined by (7), Sec. 3.1. Show that X is complete. Let Y consist of all $x \in X$ such that $x(a)=0$. Is Y a subspace of X? Do all $x \in X$ of degree 2 form a subspace of X?

4. Show that $y \perp x_n$ and $x_n \longrightarrow x$ together imply $x \perp y$.

5. Show that for a sequence (x_n) in an inner product space the conditions $\|x_n\| \longrightarrow \|x\|$ and $\langle x_n, x\rangle \longrightarrow \langle x, x\rangle$ imply convergence $x_n \longrightarrow x$.

6. Prove the statement in Prob. 5 for the special case of the complex plane.

7. Show that in an inner product space, $x \perp y$ if and only if we have $\|x+\alpha y\|=\|x-\alpha y\|$ for all scalars α. (See Fig. 25.)

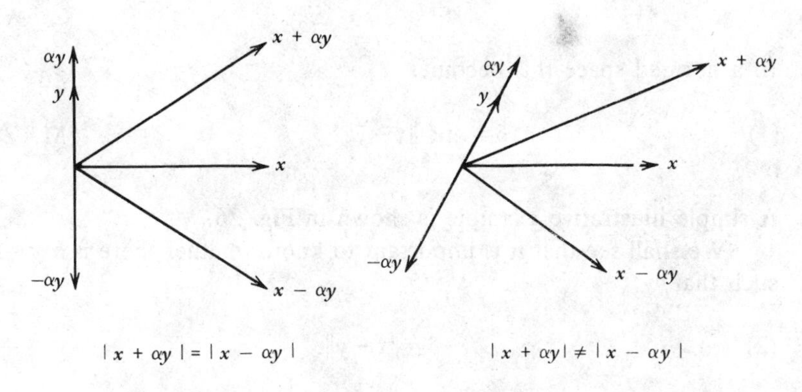

Fig. 25. Illustration of Prob. 7 in the Euclidean plane $\mathbf{R}^2$

8. Show that in an inner product space, $x \perp y$ if and only if $\|x+\alpha y\| \geqq \|x\|$ for all scalars α.

9. Let V be the vector space of all continuous complex-valued functions on $J=[a, b]$. Let $X_1=(V, \|\cdot\|_\infty)$, where $\|x\|_\infty = \max_{t \in J} |x(t)|$; and let $X_2=(V, \|\cdot\|_2)$, where

$$\|x\|_2 = \langle x, x\rangle^{1/2}, \qquad \langle x, y\rangle = \int_a^b x(t)\overline{y(t)}\, dt.$$

Show that the identity mapping $x \longmapsto x$ of X_1 onto X_2 is continuous. (It is not a homeomorphism. X_2 is not complete.)

10. (Zero operator) Let $T\colon X \longrightarrow X$ be a bounded linear operator on a complex inner product space X. *If* $\langle Tx, x\rangle = 0$ for all $x \in X$, show that $T=0$.

Show that this does not hold in the case of a *real* inner product space. *Hint.* Consider a rotation of the Euclidean plane.

3.3 Orthogonal Complements and Direct Sums

In a metric space X, the *distance* δ from an element $x \in X$ to a nonempty subset $M \subset X$ is defined to be

$$\delta = \inf_{\tilde{y} \in M} d(x, \tilde{y}) \qquad (M \neq \emptyset).$$

In a normed space this becomes

$$\delta = \inf_{\tilde{y} \in M} \|x - \tilde{y}\| \qquad (M \neq \emptyset). \tag{1}$$

A simple illustrative example is shown in Fig. 26.

We shall see that it is important to know whether there is a $y \in M$ such that

$$\delta = \|x - y\|, \tag{2}$$

that is, intuitively speaking, a point $y \in M$ which is closest to the given x, and if such an element exists, whether it is unique. This is an *existence and uniqueness problem.* It is of fundamental importance, theoretically as well as in applications, for instance, in connection with approximations of functions.

Figure 27 illustrates that even in a very simple space such as the Euclidean plane $\mathbf{R}^2$, there may be no y satisfying (2), or precisely one such y, or more than one y. And we may expect that other spaces, in particular infinite dimensional ones, will be much more complicated in that respect. For general normed spaces this is the case (as we shall see in Chap. 6), but for Hilbert spaces the situation remains relatively

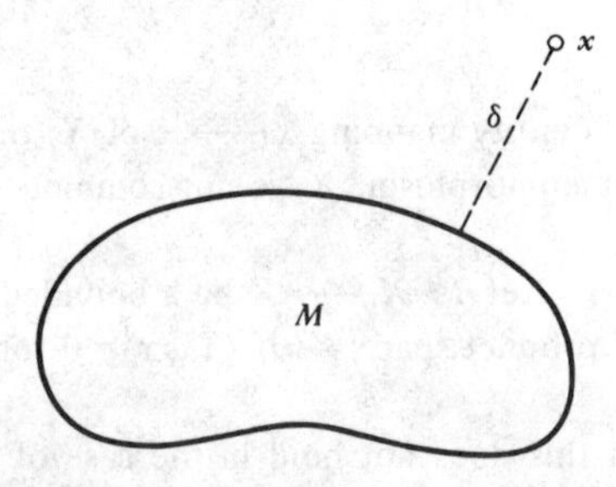

Fig. 26. Illustration of (1) in the case of the plane $\mathbf{R}^2$

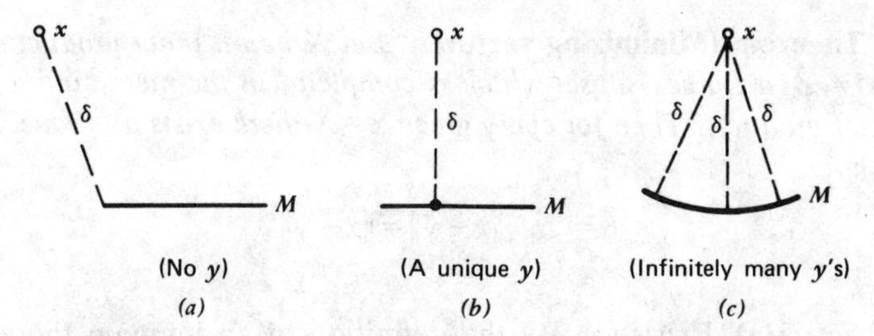

Fig. 27. Existence and uniqueness of points $y \in M$ satisfying (2), where the given $M \subset \mathbf{R}^2$ is an open segment [in (a) and (b)] and a circular arc [in (c)]

simple. This fact is surprising and has various theoretical and practical consequences. It is one of the main reasons why the theory of Hilbert spaces is simpler than that of general Banach spaces.

To consider that existence and uniqueness problem for Hilbert spaces and to formulate the key theorem (3.3-1, below), we need two related concepts, which are of general interest, as follows.

The **segment** joining two given elements x and y of a vector space X is defined to be the set of all $z \in X$ of the form

$$z = \alpha x + (1-\alpha)y \qquad (\alpha \in \mathbf{R}, 0 \leqq \alpha \leqq 1).$$

A subset M of X is said to be **convex** if for every $x, y \in M$ the segment joining x and y is contained in M. Figure 28 shows a simple example.

For instance, every subspace Y of X is convex, and the intersection of convex sets is a convex set.

We can now provide the main tool in this section:

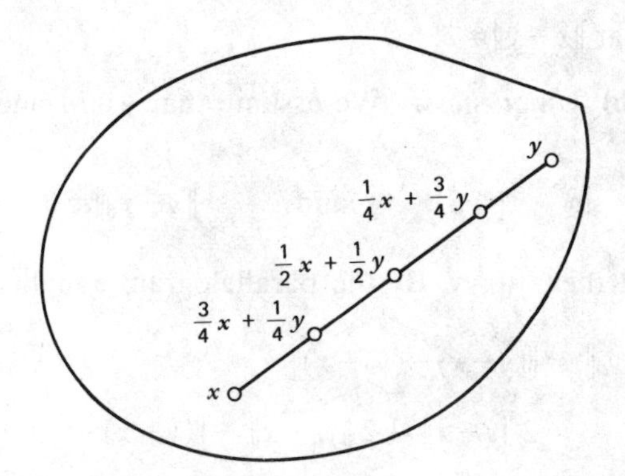

Fig. 28. Illustrative example of a segment in a convex set

3.3-1 Theorem (Minimizing vector). *Let X be an inner product space and $M \neq \varnothing$ a convex subset which is complete (in the metric induced by the inner product). Then for every given $x \in X$ there exists a unique $y \in M$ such that*

$$\delta = \inf_{\tilde{y} \in M} \|x - \tilde{y}\| = \|x - y\|. \tag{3}$$

Proof. **(a)** *Existence.* By the definition of an infimum there is a sequence (y_n) in M such that

$$\delta_n \longrightarrow \delta \qquad \text{where} \qquad \delta_n = \|x - y_n\|. \tag{4}$$

We show that (y_n) is Cauchy. Writing $y_n - x = v_n$, we have $\|v_n\| = \delta_n$ and

$$\|v_n + v_m\| = \|y_n + y_m - 2x\| = 2\,\|\tfrac{1}{2}(y_n + y_m) - x\| \geqq 2\delta$$

because M is convex, so that $\frac{1}{2}(y_n + y_m) \in M$. Furthermore, we have $y_n - y_m = v_n - v_m$. Hence by the parallelogram equality,

$$\begin{aligned}\|y_n - y_m\|^2 = \|v_n - v_m\|^2 &= -\|v_n + v_m\|^2 + 2(\|v_n\|^2 + \|v_m\|^2) \\ &\leqq -(2\delta)^2 + 2(\delta_n{}^2 + \delta_m{}^2),\end{aligned}$$

and (4) implies that (y_n) is Cauchy. Since M is complete, (y_n) converges, say, $y_n \longrightarrow y \in M$. Since $y \in M$, we have $\|x - y\| \geqq \delta$. Also, by (4),

$$\|x - y\| \leqq \|x - y_n\| + \|y_n - y\| = \delta_n + \|y_n - y\| \longrightarrow \delta.$$

This shows that $\|x - y\| = \delta$.

(b) *Uniqueness.* We assume that $y \in M$ and $y_0 \in M$ both satisfy

$$\|x - y\| = \delta \qquad \text{and} \qquad \|x - y_0\| = \delta$$

and show that then $y_0 = y$. By the parallelogram equality,

$$\begin{aligned}\|y - y_0\|^2 &= \|(y - x) - (y_0 - x)\|^2 \\ &= 2\,\|y - x\|^2 + 2\,\|y_0 - x\|^2 - \|(y - x) + (y_0 - x)\|^2 \\ &= 2\delta^2 + 2\delta^2 - 2^2\,\|\tfrac{1}{2}(y + y_0) - x\|^2.\end{aligned}$$

On the right, $\frac{1}{2}(y+y_0)\in M$, so that

$$\|\tfrac{1}{2}(y+y_0)-x\|\geqq \delta.$$

This implies that the right-hand side is less than or equal to $2\delta^2+2\delta^2-4\delta^2=0$. Hence we have the inequality $\|y-y_0\|\leqq 0$. Clearly, $\|y-y_0\|\geqq 0$, so that we must have equality, and $y_0=y$. ∎

Turning from arbitrary convex sets to subspaces, we obtain a lemma which generalizes the familiar idea of elementary geometry that the unique point y in a given subspace Y closest to a given x is found by "dropping a perpendicular from x to Y."

3.3-2 Lemma (Orthogonality). *In Theorem* 3.3-1, *let M be a complete subspace Y and $x\in X$ fixed. Then $z=x-y$ is orthogonal to Y.*

Proof. If $z\perp Y$ were false, there would be a $y_1\in Y$ such that

$$\langle z, y_1\rangle=\beta\neq 0. \tag{5}$$

Clearly, $y_1\neq 0$ since otherwise $\langle z, y_1\rangle=0$. Furthermore, for any scalar α,

$$\begin{aligned}\|z-\alpha y_1\|^2&=\langle z-\alpha y_1, z-\alpha y_1\rangle\\ &=\langle z, z\rangle-\bar\alpha\langle z, y_1\rangle-\alpha[\langle y_1, z\rangle-\bar\alpha\langle y_1, y_1\rangle]\\ &=\langle z, z\rangle-\bar\alpha\beta-\alpha[\bar\beta-\bar\alpha\langle y_1, y_1\rangle].\end{aligned}$$

The expression in the brackets $[\cdots]$ is zero if we choose

$$\bar\alpha=\frac{\bar\beta}{\langle y_1, y_1\rangle}.$$

From (3) we have $\|z\|=\|x-y\|=\delta$, so that our equation now yields

$$\|z-\alpha y_1\|^2=\|z\|^2-\frac{|\beta|^2}{\langle y_1, y_1\rangle}<\delta^2.$$

But this is impossible because we have

$$z-\alpha y_1=x-y_2 \qquad \text{where} \qquad y_2=y+\alpha y_1\in Y,$$

so that $\|z-\alpha y_1\|\geqq\delta$ by the definition of δ. Hence (5) cannot hold, and the lemma is proved. ∎

Our goal is a representation of a Hilbert space as a direct sum which is particularly simple and suitable because it makes use of orthogonality. To understand the situation and the problem, let us first introduce the concept of a direct sum. This concept makes sense for any vector space and is defined as follows.

3.3-3 Definition (Direct sum). A vector space X is said to be the *direct sum* of two subspaces Y and Z of X, written

$$X = Y \oplus Z,$$

if each $x \in X$ has a unique representation

$$x = y + z \qquad\qquad y \in Y,\ z \in Z.$$

Then Z is called an *algebraic complement* of Y in X and vice versa, and Y, Z is called a *complementary pair* of subspaces in X. ∎

For example, $Y = \mathbf{R}$ is a subspace of the Euclidean plane $\mathbf{R}^2$. Clearly, Y has infinitely many algebraic complements in $\mathbf{R}^2$, each of which is a real line. But most convenient is a complement that is perpendicular. We make use of this fact when we choose a Cartesian coordinate system. In $\mathbf{R}^3$ the situation is the same in principle.

Similarly, in the case of a general Hilbert space H, the main interest concerns representations of H as a direct sum of a closed subspace Y and its **orthogonal complement**

$$Y^\perp = \{z \in H \mid z \perp Y\},$$

which is the set of all vectors orthogonal to Y. This gives our main result in this section, which is sometimes called the *projection theorem*, for reasons to be explained after the proof.

3.3-4 Theorem (Direct sum). *Let Y be any closed subspace of a Hilbert space H. Then*

$$H = Y \oplus Z \qquad\qquad Z = Y^\perp. \tag{6}$$

Proof. Since H is complete and Y is closed, Y is complete by Theorem 1.4-7. Since Y is convex, Theorem 3.3-1 and Lemma 3.3-2 imply that for every $x \in H$ there is a $y \in Y$ such that

$$x = y + z \qquad\qquad z \in Z = Y^{\perp}. \tag{7}$$

To prove uniqueness, we assume that

$$x = y + z = y_1 + z_1$$

where $y, y_1 \in Y$ and $z, z_1 \in Z$. Then $y - y_1 = z_1 - z$. Since $y - y_1 \in Y$ whereas $z_1 - z \in Z = Y^{\perp}$, we see that $y - y_1 \in Y \cap Y^{\perp} = \{0\}$. This implies $y = y_1$. Hence also $z = z_1$. ∎

y in (7) is called the **orthogonal projection** of x on Y (or, briefly, the *projection* of x on Y). This term is motivated by elementary geometry. [For instance, we can take $H = \mathbf{R}^2$ and project any point $x = (\xi_1, \xi_2)$ on the ξ_1-axis, which then plays the role of Y; the projection is $y = (\xi_1, 0)$.]

Equation (7) defines a mapping

$$P\colon H \longrightarrow Y$$
$$x \longmapsto y = Px.$$

P is called the (orthogonal) **projection** (or *projection operator*) of H onto Y. See Fig. 29. Obviously, P is a bounded linear operator. P

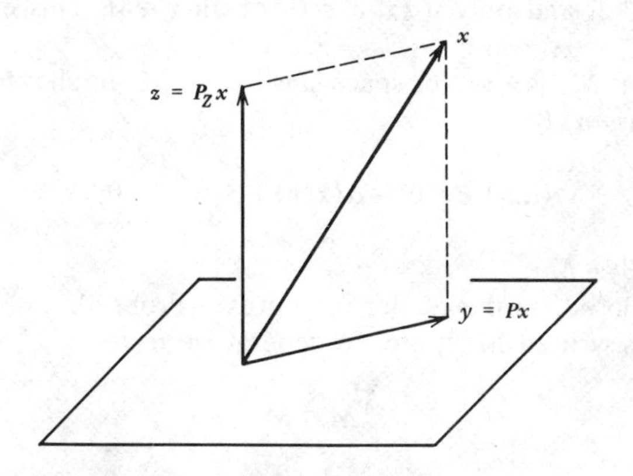

Fig. 29. Notation in connection with Theorem 3.3-4 and formula (9)

maps

$$H \text{ onto } Y,$$
$$Y \text{ onto itself},$$
$$Z = Y^{\perp} \text{ onto } \{0\},$$

and is **idempotent,** that is,

$$P^2 = P;$$

thus, for every $x \in H$,

$$P^2x = P(Px) = Px.$$

Hence $P|_Y$ is the identity operator on Y. And for $Z = Y^{\perp}$ our discussion yields

3.3-5 Lemma (Null space). *The orthogonal complement $Y^{\perp}$ of a closed subspace Y of a Hilbert space H is the null space $\mathcal{N}(P)$ of the orthogonal projection P of H onto Y.*

An orthogonal complement is a special annihilator, where, by definition, the *annihilator* $M^{\perp}$ of a set $M \neq \varnothing$ in an inner product space X is the set[3]

$$M^{\perp} = \{x \in X \mid x \perp M\}.$$

Thus, $x \in M^{\perp}$ if and only if $\langle x, v\rangle = 0$ for all $v \in M$. This explains the name.

Note that $M^{\perp}$ is a vector space since $x, y \in M^{\perp}$ implies for all $v \in M$ and all scalars α, β

$$\langle \alpha x + \beta y, v\rangle = \alpha\langle x, v\rangle + \beta\langle y, v\rangle = 0,$$

hence $\alpha x + \beta y \in M^{\perp}$.

$M^{\perp}$ is closed, as the reader may prove (Prob. 8).

$(M^{\perp})^{\perp}$ is written $M^{\perp\perp}$, etc. In general we have

$$M \subset M^{\perp\perp} \qquad (8^*)$$

[3] This causes no conflict with Prob. 13, Sec. 2.10, as we shall see later (in Sec. 3.8).

because

$$x \in M \quad \Longrightarrow \quad x \perp M^{\perp} \quad \Longrightarrow \quad x \in (M^{\perp})^{\perp}.$$

But for closed subspaces we even have

3.3-6 Lemma (Closed subspace). *If Y is a closed subspace of a Hilbert space H, then*

(8)
$$Y = Y^{\perp\perp}.$$

Proof. $Y \subset Y^{\perp\perp}$ by (8*). We show $Y \supset Y^{\perp\perp}$. Let $x \in Y^{\perp\perp}$. Then $x = y + z$ by 3.3-4, where $y \in Y \subset Y^{\perp\perp}$ by (8*). Since $Y^{\perp\perp}$ is a vector space and $x \in Y^{\perp\perp}$ by assumption, we also have $z = x - y \in Y^{\perp\perp}$, hence $z \perp Y^{\perp}$. But $z \in Y^{\perp}$ by 3.3-4. Together $z \perp z$, hence $z = 0$, so that $x = y$, that is, $x \in Y$. Since $x \in Y^{\perp\perp}$ was arbitrary, this proves $Y \supset Y^{\perp\perp}$. ∎

(8) is the main reason for the use of *closed* subspaces in the present context. Since $Z^{\perp} = Y^{\perp\perp} = Y$, formula (6) can also be written

$$H = Z \oplus Z^{\perp}.$$

It follows that $x \longmapsto z$ defines a projection (Fig. 29)

(9)
$$P_Z \colon H \longrightarrow Z$$

of H onto Z, whose properties are quite similar to those of the projection P considered before.

Theorem 3.3-4 readily implies a characterization of sets in Hilbert spaces whose span is dense, as follows.

3.3-7 Lemma (Dense set). *For any subset $M \neq \varnothing$ of a Hilbert space H, the span of M is dense in H if and only if $M^{\perp} = \{0\}$.*

Proof. (*a*) Let $x \in M^{\perp}$ and assume $V = \text{span } M$ to be dense in H. Then $x \in \bar{V} = H$. By Theorem 1.4-6(*a*) there is a sequence (x_n) in V such that $x_n \longrightarrow x$. Since $x \in M^{\perp}$ and $M^{\perp} \perp V$, we have $\langle x_n, x \rangle = 0$. The continuity of the inner product (cf. Lemma 3.2-2) implies that $\langle x_n, x \rangle \longrightarrow \langle x, x \rangle$. Together, $\langle x, x \rangle = \|x\|^2 = 0$, so that $x = 0$. Since $x \in M^{\perp}$ was arbitrary, this shows that $M^{\perp} = \{0\}$.

(*b*) Conversely, suppose that $M^{\perp}=\{0\}$. If $x \perp V$, then $x \perp M$, so that $x \in M^{\perp}$ and $x=0$. Hence $V^{\perp}=\{0\}$. Noting that V is a subspace of H, we thus obtain $\bar{V}=H$ from 3.3-4 with $Y=\bar{V}$. ∎

Problems

1. Let H be a Hilbert space, $M \subset H$ a convex subset, and (x_n) a sequence in M such that $\|x_n\| \longrightarrow d$, where $d=\inf_{x \in M} \|x\|$. Show that (x_n) converges in H. Give an illustrative example in $\mathbf{R}^2$ or $\mathbf{R}^3$.

2. Show that the subset $M=\{y=(\eta_j) \mid \sum \eta_j=1\}$ of complex space $\mathbf{C}^n$ (cf. 3.1-4) is complete and convex. Find the vector of minimum norm in M.

3. (*a*) Show that the vector space X of all real-valued continuous functions on $[-1, 1]$ is the direct sum of the set of all even continuous functions and the set of all odd continuous functions on $[-1, 1]$. (*b*) Give examples of representations of $\mathbf{R}^3$ as a direct sum (*i*) of a subspace and its orthogonal complement, (*ii*) of any complementary pair of subspaces.

4. (*a*) Show that the conclusion of Theorem 3.3-1 also holds if X is a Hilbert space and $M \subset X$ is a closed subspace. (*b*) How could we use Appolonius' identity (Sec. 3.1, Prob. 5) in the proof of Theorem 3.3-1?

5. Let $X=\mathbf{R}^2$. Find $M^{\perp}$ if M is (*a*) $\{x\}$, where $x=(\xi_1, \xi_2) \neq 0$, (*b*) a linearly independent set $\{x_1, x_2\} \subset X$.

6. Show that $Y=\{x \mid x=(\xi_j) \in l^2,\ \xi_{2n}=0,\ n \in \mathbf{N}\}$ is a closed subspace of l^2 and find $Y^{\perp}$. What is $Y^{\perp}$ if $Y=\operatorname{span}\{e_1, \cdots, e_n\} \subset l^2$, where $e_j=(\delta_{jk})$?

7. Let A and $B \supset A$ be nonempty subsets of an inner product space X. Show that

$$(a)\ A \subset A^{\perp\perp}, \qquad (b)\ B^{\perp} \subset A^{\perp}, \qquad (c)\ A^{\perp\perp\perp}=A^{\perp}.$$

8. Show that the annihilator $M^{\perp}$ of a set $M \neq \varnothing$ in an inner product space X is a closed subspace of X.

9. Show that a subspace Y of a Hilbert space H is closed in H if and only if $Y=Y^{\perp\perp}$.

10. If $M \neq \varnothing$ is any subset of a Hilbert space H, show that $M^{\perp\perp}$ is the smallest closed subspace of H which contains M, that is, $M^{\perp\perp}$ is contained in any closed subspace $Y \subset H$ such that $Y \supset M$.

3.4 Orthonormal Sets and Sequences

Orthogonality of elements as defined in Sec. 3.1 plays a basic role in inner product and Hilbert spaces. A first impression of this fact was given in the preceding section. Of particular interest are sets whose elements are orthogonal in pairs. To understand this, let us remember a familiar situation in Euclidean space $\mathbf{R}^3$. In the space $\mathbf{R}^3$, a set of that kind is the set of the three unit vectors in the positive directions of the axes of a rectangular coordinate system; call these vectors e_1, e_2, e_3. These vectors form a basis for $\mathbf{R}^3$, so that every $x \in \mathbf{R}^3$ has a unique representation (Fig. 30)

$$x = \alpha_1 e_1 + \alpha_2 e_2 + \alpha_3 e_3.$$

Now we see a great advantage of the orthogonality. Given x, we can readily determine the unknown coefficients α_1, α_2, α_3 by taking inner products (dot products). In fact, to obtain α_1, we must multiply that representation of x by e_1, that is,

$$\langle x, e_1 \rangle = \alpha_1 \langle e_1, e_1 \rangle + \alpha_2 \langle e_2, e_1 \rangle + \alpha_3 \langle e_3, e_1 \rangle = \alpha_1,$$

and so on. In more general inner product spaces there are similar and other possibilities for the use of orthogonal and orthonormal sets and

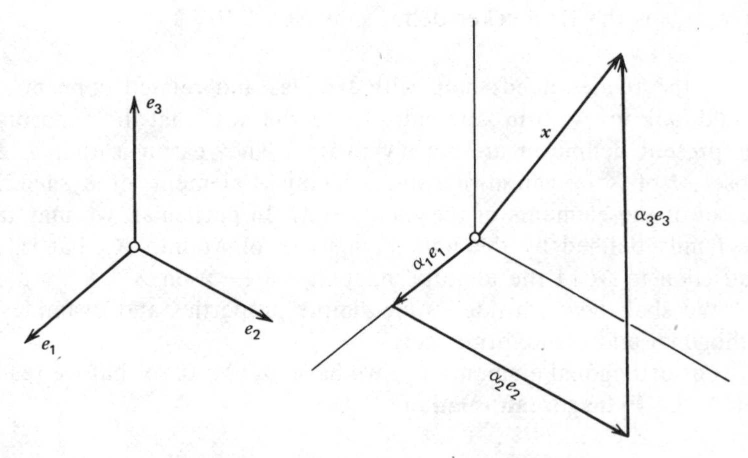

Fig. 30. Orthonormal set $\{e_1, e_2, e_3\}$ in $\mathbf{R}^3$ and representation $x = \alpha_1 e_1 + \alpha_2 e_2 + \alpha_3 e_3$

sequences, as we shall explain. In fact, the application of such sets and sequences makes up quite a substantial part of the whole theory of inner product and Hilbert spaces. Let us begin our study of this situation by introducing the necessary concepts.

3.4-1 Definition (Orthonormal sets and sequences). *An orthogonal set* M in an inner product space X is a subset $M \subset X$ whose elements are pairwise orthogonal. An *orthonormal set* $M \subset X$ is an orthogonal set in X whose elements have norm 1, that is, for all $x, y \in M$,

$$\text{(1)} \qquad \langle x, y\rangle = \begin{cases} 0 & \text{if } x \neq y \\ 1 & \text{if } x = y. \end{cases}$$

If an orthogonal or orthonormal set M is countable, we can arrange it in a sequence (x_n) and call it an *orthogonal* or *orthonormal sequence*, respectively.

More generally, an indexed set, or *family*, (x_α), $\alpha \in I$, is called *orthogonal* if $x_\alpha \perp x_\beta$ for all $\alpha, \beta \in I$, $\alpha \neq \beta$. The family is called *orthonormal* if it is orthogonal and all x_α have norm 1, so that for all $\alpha, \beta \in I$ we have

$$\text{(2)} \qquad \langle x_\alpha, x_\beta\rangle = \delta_{\alpha\beta} = \begin{cases} 0 & \text{if } \alpha \neq \beta, \\ 1 & \text{if } \alpha = \beta. \end{cases}$$

Here, $\delta_{\alpha\beta}$ is the Kronecker delta, as in Sec. 2.9. ∎

If the reader needs help with families and related concepts, he should look up A1.3 in Appendix 1. He will note that the concepts in our present definition are closely related. The reason is that to any subset M of X we can always find a family of elements of X such that the set of the elements of the family is M. In particular, we may take the family defined by the *natural injection* of M into X, that is, the restriction to M of the identity mapping $x \longmapsto x$ on X.

We shall now consider some simple properties and examples of orthogonal and orthonormal sets.

For orthogonal elements x, y we have $\langle x, y\rangle = 0$, so that we readily obtain the **Pythagorean relation**

$$\text{(3)} \qquad \|x+y\|^2 = \|x\|^2 + \|y\|^2.$$

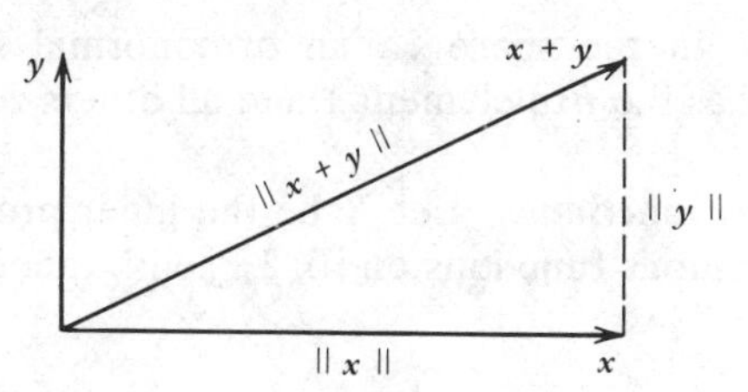

Fig. 31. Pythagorean relation (3) in $\mathbf{R}^2$

Figure 31 shows a familiar example.—More generally, if $\{x_1, \cdots, x_n\}$ is an orthogonal set, then

(4)
$$\|x_1 + \cdots + x_n\|^2 = \|x_1\|^2 + \cdots + \|x_n\|^2.$$

In fact, $\langle x_j, x_k \rangle = 0$ if $j \neq k$; consequently,

$$\left\| \sum_j x_j \right\|^2 = \left\langle \sum_j x_j, \sum_k x_k \right\rangle = \sum_j \sum_k \langle x_j, x_k \rangle = \sum_j \langle x_j, x_j \rangle = \sum_j \|x_j\|^2$$

(summations from 1 to n). We also note

3.4-2 Lemma (Linear independence). *An orthonormal set is linearly independent.*

Proof. Let $\{e_1, \cdots, e_n\}$ be orthonormal and consider the equation

$$\alpha_1 e_1 + \cdots + \alpha_n e_n = 0.$$

Multiplication by a fixed e_j gives

$$\left\langle \sum_k \alpha_k e_k, e_j \right\rangle = \sum_k \alpha_k \langle e_k, e_j \rangle = \alpha_j \langle e_j, e_j \rangle = \alpha_j = 0$$

and proves linear independence for any finite orthonormal set. This also implies linear independence if the given orthonormal set is infinite, by the definition of linear independence in Sec. 2.1. ∎

Examples

3.4-3 Euclidean space $\mathbf{R}^3$. In the space $\mathbf{R}^3$, the three unit vectors $(1, 0, 0)$, $(0, 1, 0)$, $(0, 0, 1)$ in the direction of the three axes of a rectangular coordinate system form an orthonormal set. See Fig. 30.

3.4-4 Space l^2. In the space l^2, an orthonormal sequence is (e_n), where $e_n = (\delta_{nj})$ has the nth element 1 and all others zero. (Cf. 3.1-6.)

3.4-5 Continuous functions. Let X be the inner product space of all real-valued continuous functions on $[0, 2\pi]$ with inner product defined by

$$\langle x, y\rangle = \int_0^{2\pi} x(t)y(t)\,dt$$

(cf. 3.1-5). An orthogonal sequence in X is (u_n), where

$$u_n(t) = \cos nt \qquad\qquad n = 0, 1, \cdots.$$

Another orthogonal sequence in X is (v_n), where

$$v_n(t) = \sin nt \qquad\qquad n = 1, 2, \cdots.$$

In fact, by integration we obtain

$$\text{(5)} \qquad \langle u_m, u_n\rangle = \int_0^{2\pi} \cos mt \cos nt\,dt = \begin{cases} 0 & \text{if } m \neq n \\ \pi & \text{if } m = n = 1, 2, \cdots \\ 2\pi & \text{if } m = n = 0 \end{cases}$$

and similarly for (v_n). Hence an orthonormal sequence is (e_n), where

$$e_0(t) = \frac{1}{\sqrt{2\pi}}, \qquad e_n(t) = \frac{u_n(t)}{\|u_n\|} = \frac{\cos nt}{\sqrt{\pi}} \qquad (n = 1, 2, \cdots).$$

From (v_n) we obtain the orthonormal sequence $(\tilde{e}_n)$, where

$$\tilde{e}_n(t) = \frac{v_n(t)}{\|v_n\|} = \frac{\sin nt}{\sqrt{\pi}} \qquad (n = 1, 2, \cdots).$$

Note that we even have $u_m \perp v_n$ for all m and n. (Proof?) These sequences appear in *Fourier series*, as we shall discuss in the next section. Our examples are sufficient to give us a first impression of what is going on. Further orthonormal sequences of practical importance are included in a later section (Sec. 3.7). ∎

A great advantage of orthonormal sequences over arbitrary linearly independent sequences is the following. If we know that a given x can be represented as a linear combination of some elements of an orthonormal sequence, then the orthonormality makes the actual determination of the coefficients very easy. In fact, if $(e_1, e_2, \cdots)$ is an orthonormal sequence in an inner product space X and we have $x \in \operatorname{span}\{e_1, \cdots, e_n\}$, where n is fixed, then by the definition of the span (Sec. 2.1),

$$x = \sum_{k=1}^{n} \alpha_k e_k, \tag{6}$$

and if we take the inner product by a fixed e_j, we obtain

$$\langle x, e_j \rangle = \left\langle \sum \alpha_k e_k, e_j \right\rangle = \sum \alpha_k \langle e_k, e_j \rangle = \alpha_j.$$

With these coefficients, (6) becomes

$$x = \sum_{k=1}^{n} \langle x, e_k \rangle e_k. \tag{7}$$

This shows that the determination of the unknown coefficients in (6) is simple. Another advantage of orthonormality becomes apparent if in (6) and (7) we want to add another term $\alpha_{n+1} e_{n+1}$, to take care of an

$$\tilde{x} = x + \alpha_{n+1} e_{n+1} \in \operatorname{span}\{e_1, \cdots, e_{n+1}\};$$

then we need to calculate only one more coefficient since the other coefficients remain unchanged.

More generally, if we consider any $x \in X$, not necessarily in $Y_n = \operatorname{span}\{e_1, \cdots, e_n\}$, we can define $y \in Y_n$ by setting

$$y = \sum_{k=1}^{n} \langle x, e_k \rangle e_k, \tag{8a}$$

where n is fixed, as before, and then define z by setting

$$x = y + z, \tag{8b}$$

that is, $z=x-y$. We want to show that $z \perp y$. To really understand what is going on, note the following. *Every* $y \in Y_n$ is a linear combination

$$y=\sum_{k=1}^{n} \alpha_k e_k .$$

Here $\alpha_k=\langle y, e_k\rangle$, as follows from what we discussed right before. Our claim is that for the particular choice $\alpha_k=\langle x, e_k\rangle$, $k=1, \cdots, n$, we shall obtain a y such that $z=x-y \perp y$.

To prove this, we first note that, by the orthonormality,

$$\|y\|^2=\left\langle \sum \langle x, e_k\rangle e_k, \sum \langle x, e_m\rangle e_m \right\rangle=\sum |\langle x, e_k\rangle|^2. \tag{9}$$

Using this, we can now show that $z \perp y$:

$$\begin{aligned}\langle z, y\rangle=\langle x-y, y\rangle&=\langle x, y\rangle-\langle y, y\rangle \\ &=\left\langle x, \sum \langle x, e_k\rangle e_k\right\rangle-\|y\|^2 \\ &=\sum \langle x, e_k\rangle \overline{\langle x, e_k\rangle}-\sum |\langle x, e_k\rangle|^2 \\ &=0.\end{aligned}$$

Hence the Pythagorean relation (3) gives

$$\|x\|^2=\|y\|^2+\|z\|^2. \tag{10}$$

By (9) it follows that

$$\|z\|^2=\|x\|^2-\|y\|^2=\|x\|^2-\sum |\langle x, e_k\rangle|^2. \tag{11}$$

Since $\|z\| \geqq 0$, we have for every $n=1, 2, \cdots$

$$\sum_{k=1}^{n} |\langle x, e_k\rangle|^2 \leqq \|x\|^2. \tag{12*}$$

These sums have nonnegative terms, so that they form a monotone increasing sequence. This sequence converges because it is bounded by $\|x\|^2$. This is the sequence of the partial sums of an infinite series, which thus converges. Hence (12*) implies

3.4-6 Theorem (Bessel inequality). *Let* (e_k) *be an orthonormal sequence in an inner product space* X. *Then for every* $x \in X$

$$\sum_{k=1}^{\infty} |\langle x, e_k \rangle|^2 \leqq \|x\|^2 \qquad \textbf{(Bessel inequality).} \tag{12}$$

The inner products $\langle x, e_k \rangle$ in (12) are called the **Fourier coefficients** of x with respect to the orthonormal sequence (e_k).

Note that if X is finite dimensional, then every orthonormal set in X must be finite because it is linearly independent by 3.4-2. Hence in (12) we then have a finite sum.

We have seen that orthonormal sequences are very convenient to work with. The remaining practical problem is how to obtain an orthonormal sequence if an arbitrary linearly independent sequence is given. This is accomplished by a constructive procedure, the **Gram-Schmidt process** for orthonormalizing a linearly independent sequence (x_j) in an inner product space. The resulting orthonormal sequence (e_j) has the property that for every n,

$$\operatorname{span}\{e_1, \cdots, e_n\} = \operatorname{span}\{x_1, \cdots, x_n\}.$$

The process is as follows.

1st step. The first element of (e_k) is

$$e_1 = \frac{1}{\|x_1\|} x_1.$$

2nd step. x_2 can be written

$$x_2 = \langle x_2, e_1 \rangle e_1 + v_2.$$

Then (Fig. 32)

$$v_2 = x_2 - \langle x_2, e_1 \rangle e_1$$

is not the zero vector since (x_j) is linearly independent; also $v_2 \perp e_1$ since $\langle v_2, e_1 \rangle = 0$, so that we can take

$$e_2 = \frac{1}{\|v_2\|} v_2.$$

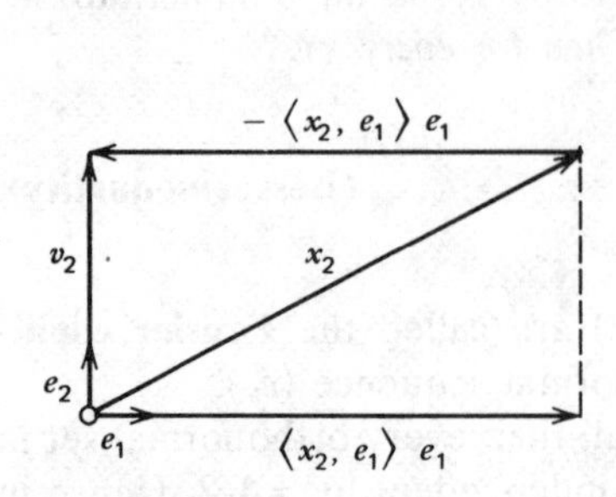

Fig. 32. Gram-Schmidt process, 2nd step

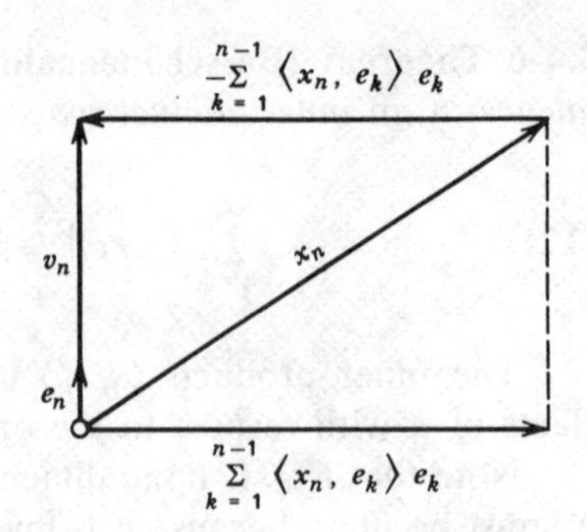

Fig. 33. Gram-Schmidt process, nth step

3rd step. The vector

$$v_3 = x_3 - \langle x_3, e_1\rangle e_1 - \langle x_3, e_2\rangle e_2$$

is not the zero vector, and $v_3 \perp e_1$ as well as $v_3 \perp e_2$. We take

$$e_3 = \frac{1}{\|v_3\|} v_3.$$

***n*th step.** The vector (see Fig. 33)

$$v_n = x_n - \sum_{k=1}^{n-1} \langle x_n, e_k\rangle e_k \tag{13}$$

is not the zero vector and is orthogonal to $e_1, \cdots, e_{n-1}$. From it we obtain

$$e_n = \frac{1}{\|v_n\|} v_n. \tag{14}$$

These are the general formulas for the Gram-Schmidt process, which was designed by E. Schmidt (1907). Cf. also J. P. Gram (1883). Note that the sum which is subtracted on the right-hand side of (13) is the projection of x_n on span $\{e_1, \cdots, e_{n-1}\}$. In other words, in each step we subtract from x_n its "components" in the directions of the previously orthogonalized vectors. This gives v_n, which is then multiplied by $1/\|v_n\|$, so that we get a vector of norm one. v_n cannot be the zero vector for any n. In fact, if n were the smallest subscript for which $v_n = 0$, then (13) shows that x_n would be a linear combination of $e_1, \cdots, e_{n-1}$, hence a linear combination of $x_1, \cdots, x_{n-1}$, contradicting the assumption that $\{x_1, \cdots, x_n\}$ is linearly independent.

Problems

1. Show that an inner product space of finite dimension n has a basis $\{b_1, \cdots, b_n\}$ of orthonormal vectors. (The infinite dimensional case will be considered in Sec. 3.6.)

2. How can we interpret (12*) geometrically in $\mathbf{R}^r$, where $r \geqq n$?

3. Obtain the Schwarz inequality (Sec. 3.2) from (12*).

4. Give an example of an $x \in l^2$ such that we have strict inequality in (12).

5. If (e_k) is an orthonormal sequence in an inner product space X, and $x \in X$, show that $x - y$ with y given by

$$y = \sum_{k=1}^{n} \alpha_k e_k \qquad\qquad \alpha_k = \langle x, e_k \rangle$$

is orthogonal to the subspace $Y_n = \operatorname{span}\{e_1, \cdots e_n\}$.

6. (Minimum property of Fourier coefficients) Let $\{e_1, \cdots, e_n\}$ be an orthonormal set in an inner product space X, where n is fixed. Let $x \in X$ be any fixed element and $y = \beta_1 e_1 + \cdots + \beta_n e_n$. Then $\|x - y\|$ depends on $\beta_1, \cdots, \beta_n$. Show by direct calculation that $\|x - y\|$ is minimum if and only if $\beta_j = \langle x, e_j \rangle$, where $j = 1, \cdots, n$.

7. Let (e_k) be any orthonormal sequence in an inner product space X. Show that for any $x, y \in X$,

$$\sum_{k=1}^{\infty} |\langle x, e_k \rangle \langle y, e_k \rangle| \leqq \|x\| \, \|y\|.$$

8. Show that an element x of an inner product space X cannot have "too many" Fourier coefficients $\langle x, e_k \rangle$ which are "big"; here, (e_k) is a given orthonormal sequence; more precisely, show that the number n_m of $\langle x, e_k \rangle$ such that $|\langle x, e_k \rangle| > 1/m$ must satisfy $n_m < m^2 \|x\|^2$.

9. Orthonormalize the first three terms of the sequence $(x_0, x_1, x_2, \cdots)$, where $x_j(t) = t^j$, on the interval $[-1, 1]$, where

$$\langle x, y \rangle = \int_{-1}^{1} x(t)y(t)\,dt.$$

10. Let $x_1(t) = t^2$, $x_2(t) = t$ and $x_3(t) = 1$. Orthonormalize x_1, x_2, x_3, in this order, on the interval $[-1, 1]$ with respect to the inner product given in Prob. 9. Compare with Prob. 9 and comment.

3.5 Series Related to Orthonormal Sequences and Sets

There are some facts and questions that arise in connection with the Bessel inequality. In this section we first motivate the term "Fourier coefficients," then consider infinite series related to orthonormal sequences, and finally take a first look at orthonormal sets which are uncountable.

3.5-1 Example (Fourier series). A *trigonometric series* is a series of the form

$$a_0 + \sum_{k=1}^{\infty} (a_k \cos kt + b_k \sin kt). \tag{1*}$$

A real-valued function x on $\mathbf{R}$ is said to be *periodic* if there is a positive number p (called a *period* of x) such that $x(t+p) = x(t)$ for all $t \in \mathbf{R}$.

Let x be of period 2π and continuous. By definition, the *Fourier series* of x is the trigonometric series (1*) with coefficients a_k and b_k given by the *Euler formulas*

$$\begin{aligned} a_0 &= \frac{1}{2\pi} \int_0^{2\pi} x(t)\,dt \\ a_k &= \frac{1}{\pi} \int_0^{2\pi} x(t) \cos kt\,dt \qquad k = 1, 2, \cdots, \\ b_k &= \frac{1}{\pi} \int_0^{2\pi} x(t) \sin kt\,dt \qquad k = 1, 2, \cdots. \end{aligned} \tag{2}$$

These coefficients are called the *Fourier coefficients of x*.

If the Fourier series of x converges for each t and has the sum $x(t)$, then we write

$$x(t) = a_0 + \sum_{k=1}^{\infty} (a_k \cos kt + b_k \sin kt). \tag{1}$$

Since x is periodic of period 2π, in (2) we may replace the interval of integration $[0, 2\pi]$ by any other interval of length 2π, for instance $[-\pi, \pi]$.

Fourier series first arose in connection with physical problems considered by D. Bernoulli (vibrating string, 1753) and J. Fourier (heat conduction, 1822). These series help to represent complicated periodic phenomena in terms of simple periodic functions (cosine and sine). They have various physical applications in connection with differential equations (vibrations, heat conduction, potential problems, etc.).

From (2) we see that the determination of Fourier coefficients requires integration. To help those readers who have not seen Fourier series before, we consider as an illustration (see Fig. 34)

$$x(t) = \begin{cases} t & \text{if } -\pi/2 \leqq t < \pi/2 \\ \pi - t & \text{if } \pi/2 \leqq t < 3\pi/2 \end{cases}$$

and $x(t+2\pi) = x(t)$. From (2) we obtain $a_k = 0$ for $k = 0, 1, \cdots$ and, choosing $[-\pi/2, 3\pi/2]$ as a convenient interval of integration and integrating by parts,

$$\begin{aligned} b_k &= \frac{1}{\pi} \int_{-\pi/2}^{\pi/2} t \sin kt \, dt + \frac{1}{\pi} \int_{\pi/2}^{3\pi/2} (\pi - t) \sin kt \, dt \\ &= -\frac{1}{\pi k} [t \cos kt] \Big|_{-\pi/2}^{\pi/2} + \frac{1}{\pi k} \int_{-\pi/2}^{\pi/2} \cos kt \, dt \\ &\quad - \frac{1}{\pi k} [(\pi - t) \cos kt] \Big|_{\pi/2}^{3\pi/2} - \frac{1}{\pi k} \int_{\pi/2}^{3\pi/2} \cos kt \, dt \\ &= \frac{4}{\pi k^2} \sin \frac{k\pi}{2}, \qquad k = 1, 2, \cdots. \end{aligned}$$

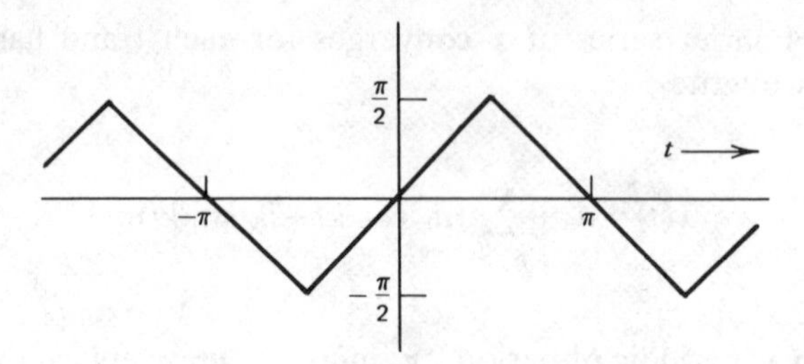

Fig. 34. Graph of the periodic function x, of period 2π, given by $x(t)=t$ if $t\in[-\pi/2, \pi/2)$ and $x(t)=\pi-t$ if $t\in[\pi/2, 3\pi/2)$

Hence (1) takes the form

$$x(t)=\frac{4}{\pi}\left(\sin t-\frac{1}{3^2}\sin 3t+\frac{1}{5^2}\sin 5t-+\cdots\right).$$

The reader may graph the first three partial sums and compare them with the graph of x in Fig. 34.

Returning to general Fourier series, we may ask how these series fit into our terminology and formalism introduced in the preceding section. Obviously, the cosine and sine functions in (1) are those of the sequences (u_k) and (v_k) in 3.4-5, that is

$$u_k(t)=\cos kt, \qquad v_k(t)=\sin kt.$$

Hence we may write (1) in the form

$$x(t)=a_0u_0(t)+\sum_{k=1}^{\infty}[a_ku_k(t)+b_kv_k(t)]. \tag{3}$$

We multiply (3) by a fixed u_j and integrate over t from 0 to 2π. This means that we take the inner product by u_j as defined in 3.4-5. We assume that termwise integration is permissible (uniform convergence would suffice) and use the orthogonality of (u_k) and (v_k) as well as the fact that $u_j \perp v_k$ for all j, k. Then we obtain

$$\begin{aligned}\langle x, u_j\rangle &= a_0\langle u_0, u_j\rangle+\sum[a_k\langle u_k, u_j\rangle+b_k\langle v_k, u_j\rangle]\\ &= a_j\langle u_j, u_j\rangle\\ &= a_j\,\|u_j\|^2=\begin{cases}2\pi a_0 & \text{if } j=0\\ \pi a_j & \text{if } j=1,2,\cdots,\end{cases}\end{aligned}$$

cf. (5), Sec. 3.4. Similarly, if we multiply (3) by v_j and proceed as before, we arrive at

$$\langle x, v_j\rangle = b_j \|v_j\|^2 = \pi b_j$$

where $j = 1, 2, \cdots$. Solving for a_j and b_j and using the orthonormal sequences (e_j) and $(\tilde{e}_j)$, where $e_j = \|u_j\|^{-1} u_j$ and $\tilde{e}_j = \|v_j\|^{-1} v_j$, we obtain

$$\begin{aligned} a_j &= \frac{1}{\|u_j\|^2}\langle x, u_j\rangle = \frac{1}{\|u_j\|}\langle x, e_j\rangle, \\ b_j &= \frac{1}{\|v_j\|^2}\langle x, v_j\rangle = \frac{1}{\|v_j\|}\langle x, \tilde{e}_j\rangle. \end{aligned} \tag{4}$$

This is identical with (2). It shows that in (3),

$$a_k u_k(t) = \frac{1}{\|u_k\|}\langle x, e_k\rangle u_k(t) = \langle x, e_k\rangle e_k(t)$$

and similarly for $b_k v_k(t)$. Hence we may write the Fourier series (1) in the form

$$x = \langle x, e_0\rangle e_0 + \sum_{k=1}^{\infty} [\langle x, e_k\rangle e_k + \langle x, \tilde{e}_k\rangle \tilde{e}_k]. \tag{5}$$

This justifies the term "Fourier coefficients" in the preceding section.

Concluding this example, we mention that the reader can find an introduction to Fourier series in W. Rogosinski (1959); cf. also R. V. Churchill (1963), pp. 77–112 and E. Kreyszig (1972), pp. 377–407. ∎

Our example concerns infinite series and raises the question how we can extend the consideration to other orthonormal sequences and what we can say about the convergence of corresponding series.

Given any orthonormal sequence (e_k) in a Hilbert space H, we may consider series of the form

$$\sum_{k=1}^{\infty} \alpha_k e_k \tag{6}$$

where $\alpha_1, \alpha_2, \cdots$ are any scalars. As defined in Sec. 2.3, such a series *converges* and has the *sum* s if there exists an $s \in H$ such that the sequence (s_n) of the partial sums

$$s_n = \alpha_1 e_1 + \cdots + \alpha_n e_n$$

converges to s, that is, $\|s_n - s\| \longrightarrow 0$ as $n \longrightarrow \infty$.

3.5-2 Theorem (Convergence). *Let (e_k) be an orthonormal sequence in a Hilbert space H. Then:*

(a) *The series (6) converges (in the norm on H) if and only if the following series converges:*

$$\sum_{k=1}^{\infty} |\alpha_k|^2. \tag{7}$$

(b) *If (6) converges, then the coefficients α_k are the Fourier coefficients $\langle x, e_k\rangle$, where x denotes the sum of (6); hence in this case, (6) can be written*

$$x = \sum_{k=1}^{\infty} \langle x, e_k\rangle e_k. \tag{8}$$

(c) *For any $x \in H$, the series (6) with $\alpha_k = \langle x, e_k\rangle$ converges (in the norm of H).*

Proof. **(a)** Let

$$s_n = \alpha_1 e_1 + \cdots + \alpha_n e_n \qquad \text{and} \qquad \sigma_n = |\alpha_1|^2 + \cdots + |\alpha_n|^2.$$

Then, because of the orthonormality, for any m and $n > m$,

$$\|s_n - s_m\|^2 = \|\alpha_{m+1} e_{m+1} + \cdots + \alpha_n e_n\|^2$$

$$= |\alpha_{m+1}|^2 + \cdots + |\alpha_n|^2 = \sigma_n - \sigma_m.$$

Hence (s_n) is Cauchy in H if and only if (σ_n) is Cauchy in $\mathbf{R}$. Since H and $\mathbf{R}$ are complete, the first statement of the theorem follows.

(b) Taking the inner product of s_n and e_j and using the orthonormality, we have

$$\langle s_n, e_j\rangle = \alpha_j \quad \text{for } j = 1, \cdots, k \qquad (k \leqq n \text{ and fixed}).$$

By assumption, $s_n \longrightarrow x$. Since the inner product is continuous (cf. Lemma 3.2-2),

$$\alpha_j = \langle s_n, e_j \rangle \quad \longrightarrow \quad \langle x, e_j \rangle \qquad (j \leqq k).$$

Here we can take k ($\leqq n$) as large as we please because $n \longrightarrow \infty$, so that we have $\alpha_j = \langle x, e_j \rangle$ for every $j = 1, 2, \cdots$.

(c) From the Bessel inequality in Theorem 3.4-6 we see that the series

$$\sum_{k=1}^{\infty} |\langle x, e_k \rangle|^2$$

converges. From this and (a) we conclude that (c) must hold. ∎

If an orthonormal family (e_κ), $\kappa \in I$, in an inner product space X is uncountable (since the index set I is uncountable), we can still form the Fourier coefficients $\langle x, e_\kappa \rangle$ of an $x \in X$, where $\kappa \in I$. Now we use (12*), Sec. 3.4, to conclude that for each fixed $m = 1, 2, \cdots$ the number of Fourier coefficients such that $|\langle x, e_\kappa \rangle| > 1/m$ must be finite. This proves the remarkable

3.5-3 Lemma (Fourier coefficients). *Any x in an inner product space X can have at most countably many nonzero Fourier coefficients $\langle x, e_\kappa \rangle$ with respect to an orthonormal family (e_κ), $\kappa \in I$, in X.*

Hence with any fixed $x \in H$ we can associate a series similar to (8),

$$\sum_{\kappa \in I} \langle x, e_\kappa \rangle e_\kappa \tag{9}$$

and we can arrange the e_κ with $\langle x, e_\kappa \rangle \neq 0$ in a sequence $(e_1, e_2, \cdots)$, so that (9) takes the form (8). Convergence follows from Theorem 3.5-2. We show that the sum does not depend on the order in which those e_κ are arranged in a sequence.

Proof. Let (w_m) be a rearrangement of (e_n). By definition this means that there is a bijective mapping $n \longmapsto m(n)$ of $\mathbf{N}$ onto itself such that corresponding terms of the two sequences are equal, that is,

$w_{m(n)} = e_n$. We set

$$\alpha_n = \langle x, e_n \rangle, \qquad \beta_m = \langle x, w_m \rangle$$

and

$$x_1 = \sum_{n=1}^{\infty} \alpha_n e_n, \qquad x_2 = \sum_{m=1}^{\infty} \beta_m w_m.$$

Then by Theorem 3.5-2(*b*),

$$\alpha_n = \langle x, e_n \rangle = \langle x_1, e_n \rangle, \qquad \beta_m = \langle x, w_m \rangle = \langle x_2, w_m \rangle.$$

Since $e_n = w_{m(n)}$, we thus obtain

$$\begin{aligned} \langle x_1 - x_2, e_n \rangle &= \langle x_1, e_n \rangle - \langle x_2, w_{m(n)} \rangle \\ &= \langle x, e_n \rangle - \langle x, w_{m(n)} \rangle = 0 \end{aligned}$$

and similarly $\langle x_1 - x_2, w_m \rangle = 0$. This implies

$$\begin{aligned} \|x_1 - x_2\|^2 &= \langle x_1 - x_2, \sum \alpha_n e_n - \sum \beta_m w_m \rangle \\ &= \sum \bar{\alpha}_n \langle x_1 - x_2, e_n \rangle - \sum \bar{\beta}_m \langle x_1 - x_2, w_m \rangle = 0. \end{aligned}$$

Consequently, $x_1 - x_2 = 0$ and $x_1 = x_2$. Since the rearrangement (w_m) of (e_n) was arbitrary, this completes the proof. ∎

Problems

1. If (6) converges with sum x, show that (7) has the sum $\|x\|^2$.

2. Derive from (1) and (2) a Fourier series representation of a function $\tilde{x}$ (function of τ) of arbitrary period p.

3. Illustrate with an example that a convergent series $\sum \langle x, e_k \rangle e_k$ need not have the sum x.

4. If (x_j) is a sequence in an inner product space X such that the series $\|x_1\| + \|x_2\| + \cdots$ converges, show that (s_n) is a Cauchy sequence, where $s_n = x_1 + \cdots + x_n$.

5. Show that in a Hilbert space H, convergence of $\sum \|x_j\|$ implies convergence of $\sum x_j$.

6. Let (e_j) be an orthonormal sequence in a Hilbert space H. Show that if

$$x=\sum_{j=1}^{\infty} \alpha_j e_j, \qquad y=\sum_{j=1}^{\infty} \beta_j e_j, \qquad \text{then} \qquad \langle x, y\rangle=\sum_{j=1}^{\infty} \alpha_j \bar{\beta}_j,$$

the series being absolutely convergent.

7. Let (e_k) be an orthonormal sequence in a Hilbert space H. Show that for every $x \in H$, the vector

$$y=\sum_{k=1}^{\infty} \langle x, e_k\rangle e_k$$

exists in H and $x-y$ is orthogonal to every e_k.

8. Let (e_k) be an orthonormal sequence in a Hilbert space H, and let $M=\operatorname{span}(e_k)$. Show that for any $x \in H$ we have $x \in \bar{M}$ if and only if x can be represented by (6) with coefficients $\alpha_k=\langle x, e_k\rangle$.

9. Let (e_n) and $(\tilde{e}_n)$ be orthonormal sequences in a Hilbert space H, and let $M_1=\operatorname{span}(e_n)$ and $M_2=\operatorname{span}(\tilde{e}_n)$. Using Prob. 8, show that $\bar{M}_1=\bar{M}_2$ if and only if

$$\text{(a)} \quad e_n=\sum_{m=1}^{\infty} \alpha_{nm} \tilde{e}_m, \qquad \text{(b)} \quad \tilde{e}_n=\sum_{m=1}^{\infty} \bar{\alpha}_{mn} e_m, \qquad \alpha_{nm}=\langle e_n, \tilde{e}_m\rangle.$$

10. Work out the details of the proof of Lemma 3.5-3.

3.6 Total Orthonormal Sets and Sequences

The truly interesting orthonormal sets in inner product spaces and Hilbert spaces are those which consist of "sufficiently many" elements so that every element in space can be represented or sufficiently accurately approximated by the use of those orthonormal sets. In finite dimensional (n-dimensional) spaces the situation is simple; all we need is an orthonormal set of n elements. The question is what can be done to take care of infinite dimensional spaces, too. Relevant concepts are as follows.

3.6-1 Definition (Total orthonormal set). A *total set* (or *fundamental set*) in a normed space X is a subset $M \subset X$ whose span is dense in X (cf. 1.3-5). Accordingly, an orthonormal set (or sequence or family) in an inner product space X which is total in X is called a *total orthonormal set*[4] (or sequence or family, respectively) in X. ∎

M is total in X if and only if

$$\overline{\operatorname{span} M} = X.$$

This is obvious from the definition.

A total orthonormal family in X is sometimes called an *orthonormal basis* for X. However, it is important to note that this is not a basis, in the sense of algebra, for X as a vector space, unless X is finite dimensional.

In every Hilbert space $H \neq \{0\}$ there exists a total orthonormal set.

For a finite dimensional H this is clear. For an infinite dimensional separable H (cf. 1.3-5) it follows from the Gram-Schmidt process by (ordinary) induction. For a nonseparable H a (nonconstructive) proof results from Zorn's lemma, as we shall see in Sec. 4.1 where we introduce and explain the lemma for another purpose.

All total orthonormal sets in a given Hilbert space $H \neq \{0\}$ have the same cardinality. The latter is called the *Hilbert dimension* or *orthogonal dimension* of H. (If $H = \{0\}$, this dimension is defined to be 0.)

For a finite dimensional H the statement is clear since then the Hilbert dimension is the dimension in the sense of algebra. For an infinite dimensional separable H the statement will readily follow from Theorem 3.6-4 (below) and for a general H the proof would require somewhat more advanced tools from set theory; cf. E. Hewitt and K. Stromberg (1969), p. 246.

[4] Sometimes a *complete* orthonormal set, but we use "complete" only in the sense of Def. 1.4-3; this is preferable since we then avoid the use of the same word in connection with two entirely different concepts. [Moreover, some authors mean by "completeness" of an orthonormal set M the property expressed by (1) in Theorem 3.6-2. We do not adopt this terminology either.]

The following theorem shows that a total orthonormal set cannot be augmented to a more extensive orthonormal set by the adjunction of new elements.

3.6-2 Theorem (Totality). *Let M be a subset of an inner product space X. Then:*

(a) *If M is total in X, then there does not exist a nonzero $x \in X$ which is orthogonal to every element of M; briefly,*

$$x \perp M \quad \Longrightarrow \quad x = 0. \tag{1}$$

(b) *If X is complete, that condition is also sufficient for the totality of M in X.*

Proof. **(a)** Let H be the completion of X; cf. 3.2-3. Then X, regarded as a subspace of H, is dense in H. By assumption, M is total in X, so that span M is dense in X, hence dense in H. Lemma 3.3-7 now implies that the orthogonal complement of M in H is $\{0\}$. A fortiori, if $x \in X$ and $x \perp M$, then $x = 0$.

(b) If X is a Hilbert space and M satisfies that condition, so that $M^{\perp} = \{0\}$, then Lemma 3.3-7 implies that M is total in X. ▮

The completeness of X in (*b*) is essential. If X is not complete, there may not exist an orthonormal set $M \subset X$ such that M is total in X. An example was given by J. Dixmier (1953). Cf. also N. Bourbaki (1955), p. 155.

Another important criterion for totality can be obtained from the Bessel inequality (cf. 3.4-6). For this purpose we consider any given orthonormal set M in a Hilbert space H. From Lemma 3.5-3 we know that each fixed $x \in H$ has at most countably many nonzero Fourier coefficients, so that we can arrange these coefficients in a sequence, say, $\langle x, e_1 \rangle, \langle x, e_2 \rangle, \cdots$. The Bessel inequality is (cf. 3.4-6)

$$\sum_k |\langle x, e_k \rangle|^2 \leqq \|x\|^2 \qquad \textbf{(Bessel inequality)} \tag{2}$$

where the left-hand side is an infinite series or a finite sum. With the equality sign this becomes

$$\sum_k |\langle x, e_k \rangle|^2 = \|x\|^2 \tag{3}$$

(Parseval relation)

and yields another criterion for totality:

3.6-3 Theorem (Totality). *An orthonormal set M in a Hilbert space H is total in H if and only if for all $x \in H$ the Parseval relation* (3) *holds (summation over all nonzero Fourier coefficients of x with respect to M).*

Proof. (*a*) If M is not total, by Theorem 3.6-2 there is a nonzero $x \perp M$ in H. Since $x \perp M$, in (3) we have $\langle x, e_k \rangle = 0$ for all k, so that the left-hand side in (3) is zero, whereas $\|x\|^2 \neq 0$. This shows that (3) does not hold. Hence if (3) holds for all $x \in H$, then M must be total in H.

(*b*) Conversely, assume M to be total in H. Consider any $x \in H$ and its nonzero Fourier coefficients (cf. 3.5-3) arranged in a sequence $\langle x, e_1 \rangle, \langle x, e_2 \rangle, \cdots$, or written in some definite order if there are only finitely many of them. We now define y by

$$y = \sum_k \langle x, e_k \rangle e_k, \tag{4}$$

noting that in the case of an infinite series, convergence follows from Theorem 3.5-2. Let us show that $x - y \perp M$. For every e_j occurring in (4) we have, using the orthonormality,

$$\langle x - y, e_j \rangle = \langle x, e_j \rangle - \sum_k \langle x, e_k \rangle \langle e_k, e_j \rangle = \langle x, e_j \rangle - \langle x, e_j \rangle = 0.$$

And for every $v \in M$ not contained in (4) we have $\langle x, v \rangle = 0$, so that

$$\langle x - y, v \rangle = \langle x, v \rangle - \sum_k \langle x, e_k \rangle \langle e_k, v \rangle = 0 - 0 = 0.$$

Hence $x - y \perp M$, that is, $x - y \in M^\perp$. Since M is total in H, we have $M^\perp = \{0\}$ from 3.3-7. Together, $x - y = 0$, that is, $x = y$. Using (4) and again the orthonormality, we thus obtain (3) from

$$\|x\|^2 = \left\langle \sum_k \langle x, e_k \rangle e_k, \sum_m \langle x, e_m \rangle e_m \right\rangle = \sum_k \langle x, e_k \rangle \overline{\langle x, e_k \rangle}.$$

This completes the proof. ∎

Let us turn to Hilbert spaces which are separable. By Def. 1.3-5 such a space has a countable subset which is dense in the space. Separable Hilbert spaces are simpler than nonseparable ones since they cannot contain uncountable orthonormal sets:

3.6-4 Theorem (Separable Hilbert spaces). *Let H be a Hilbert space. Then:*

(a) *If H is separable, every orthonormal set in H is countable.*

(b) *If H contains an orthonormal sequence which is total in H, then H is separable.*

Proof. **(a)** Let H be separable, B any dense set in H and M any orthonormal set. Then any two distinct elements x and y of M have distance $\sqrt{2}$ since

$$\|x-y\|^2 = \langle x-y, x-y \rangle = \langle x, x \rangle + \langle y, y \rangle = 2.$$

Hence spherical neighborhoods N_x of x and N_y of y of radius $\sqrt{2}/3$ are disjoint. Since B is dense in H, there is a $b \in B$ in N_x and a $\tilde{b} \in B$ in N_y and $b \neq \tilde{b}$ since $N_x \cap N_y = \varnothing$. Hence if M were uncountable, we would have uncountably many such pairwise disjoint spherical neighborhoods (for each $x \in M$ one of them), so that B would be uncountable. Since B was any dense set, this means that H would not contain a dense set which is countable, contradicting separability. From this we conclude that M must be countable.

(b) Let (e_k) be a total orthonormal sequence in H and A the set of all linear combinations

$$\gamma_1^{(n)} e_1 + \cdots + \gamma_n^{(n)} e_n \qquad n = 1, 2, \cdots$$

where $\gamma_k^{(n)} = a_k^{(n)} + i b_k^{(n)}$ and $a_k^{(n)}$ and $b_k^{(n)}$ are rational (and $b_k^{(n)} = 0$ if H is real). Clearly, A is countable. We prove that A is dense in H by showing that for every $x \in H$ and $\varepsilon > 0$ there is a $v \in A$ such that $\|x - v\| < \varepsilon$.

Since the sequence (e_k) is total in H, there is an n such that $Y_n = \operatorname{span}\{e_1, \cdots, e_n\}$ contains a point whose distance from x is less than $\varepsilon/2$. In particular, $\|x - y\| < \varepsilon/2$ for the orthogonal projection y of x on Y_n, which is given by [cf. (8), Sec. 3.4]

$$y = \sum_{k=1}^{n} \langle x, e_k \rangle e_k.$$

Hence we have

$$\left\| x - \sum_{k=1}^{n} \langle x, e_k \rangle e_k \right\| < \frac{\varepsilon}{2}.$$

Since the rationals are dense on $\mathbf{R}$, for each $\langle x, e_k \rangle$ there is a $\gamma_k^{(n)}$ (with rational real and imaginary parts) such that

$$\left\| \sum_{k=1}^{n} [\langle x, e_k \rangle - \gamma_k^{(n)}] e_k \right\| < \frac{\varepsilon}{2}.$$

Hence $v \in A$ defined by

$$v = \sum_{k=1}^{n} \gamma_k^{(n)} e_k$$

satisfies

$$\begin{aligned} \|x - v\| &= \|x - \sum \gamma_k^{(n)} e_k\| \\ &\leqq \|x - \sum \langle x, e_k \rangle e_k\| + \|\sum \langle x, e_k \rangle e_k - \sum \gamma_k^{(n)} e_k\| \\ &< \frac{\varepsilon}{2} + \frac{\varepsilon}{2} = \varepsilon. \end{aligned}$$

This proves that A is dense in H, and since A is countable, H is separable. ∎

For using Hilbert spaces in applications one must know what total orthonormal set or sets to choose in a specific situation and how to investigate properties of the elements of such sets. For certain function spaces this problem will be considered in the next section, which includes special functions of practical interest that arise in this context and have been investigated in very great detail. To conclude this section, let us point out that our present discussion has some further consequences which are of basic importance and can be formulated in terms of isomorphisms of Hilbert spaces. For this purpose we first remember from Sec. 3.2 the following.

An **isomorphism** of a Hilbert space H onto a Hilbert space $\tilde{H}$ over the same field is a bijective linear operator $T\colon H \longrightarrow \tilde{H}$ such that for all $x, y \in H$,

$$\langle Tx, Ty\rangle = \langle x, y\rangle. \tag{5}$$

H and $\tilde{H}$ are then called *isomorphic Hilbert spaces.* Since T is linear, it preserves the vector space structure, and (5) shows that T is isometric. From this and the bijectivity of T it follows that H and $\tilde{H}$ are algebraically as well as metrically indistinguishable; they are essentially the same, except for the nature of their elements, so that we may think of $\tilde{H}$ as being essentially H with a "tag" T attached to each vector x. Or we may regard H and $\tilde{H}$ as two copies (models) of the same abstract space, just as we often do in the case of n-dimensional Euclidean space.

Most exciting in this discussion is the fact that for each Hilbert dimension (cf. at the beginning of this section) there is just one abstract real Hilbert space and just one abstract complex Hilbert space. In other words, two abstract Hilbert spaces over the same field are distinguished only by their Hilbert dimension, a situation which generalizes that in the case of Euclidean spaces. This is the meaning of the following theorem.

3.6-5 Theorem (Isomorphism and Hilbert dimension). *Two Hilbert spaces H and $\tilde{H}$, both real or both complex, are isomorphic if and only if they have the same Hilbert dimension.*

Proof. (*a*) If H is isomorphic with $\tilde{H}$ and $T\colon H \longrightarrow \tilde{H}$ is an isomorphism, then (5) shows that orthonormal elements in H have orthonormal images under T. Since T is bijective, we thus conclude that T maps every total orthonormal set in H onto a total orthonormal set in $\tilde{H}$. Hence H and $\tilde{H}$ have the same Hilbert dimension.

(*b*) Conversely, suppose that H and $\tilde{H}$ have the same Hilbert dimension. The case $H=\{0\}$ and $\tilde{H}=\{0\}$ is trivial. Let $H\neq\{0\}$. Then $\tilde{H}\neq\{0\}$, and any total orthonormal sets M in H and $\tilde{M}$ in $\tilde{H}$ have the same cardinality, so that we can index them by the same index set $\{k\}$ and write $M=(e_k)$ and $\tilde{M}=(\tilde{e}_k)$.

To show that H and $\tilde{H}$ are isomorphic, we construct an isomorphism of H onto $\tilde{H}$. For every $x \in H$ we have

$$x=\sum_k \langle x, e_k\rangle e_k \tag{6}$$

where the right-hand side is a finite sum or an infinite series (cf. 3.5-3), and $\sum_k |\langle x, e_k\rangle|^2<\infty$ by the Bessel inequality. Defining

$$\tilde{x}=Tx=\sum_k \langle x, e_k\rangle \tilde{e}_k \tag{7}$$

we thus have convergence by 3.5-2, so that $\tilde{x} \in \tilde{H}$. The operator T is linear since the inner product is linear with respect to the first factor. T is isometric, because by first using (7) and then (6) we obtain

$$\|\tilde{x}\|^2=\|Tx\|^2=\sum_k |\langle x, e_k\rangle|^2=\|x\|^2.$$

From this and (9), (10) in Sec. 3.1 we see that T preserves the inner product. Furthermore, isometry implies injectivity. In fact, if $Tx=Ty$, then

$$\|x-y\|=\|T(x-y)\|=\|Tx-Ty\|=0,$$

so that $x=y$ and T is injective by 2.6-10.

We finally show that T is surjective. Given any

$$\tilde{x}=\sum_k \alpha_k \tilde{e}_k$$

in $\tilde{H}$, we have $\sum |\alpha_k|^2<\infty$ by the Bessel inequality. Hence

$$\sum_k \alpha_k e_k$$

is a finite sum or a series which converges to an $x \in H$ by 3.5-2, and $\alpha_k=\langle x, e_k\rangle$ by the same theorem. We thus have $\tilde{x}=Tx$ by (7). Since $\tilde{x} \in \tilde{H}$ was arbitrary, this shows that T is surjective. ∎

Problems

1. If F is an orthonormal basis in an inner product space X, can we represent every $x \in X$ as a linear combination of elements of F? (By definition, a linear combination consists of finitely many terms.)

2. Show that if the orthogonal dimension of a Hilbert space H is finite, it equals the dimension of H regarded as a vector space; conversely, if the latter is finite, show that so is the former.

3. From what theorem of elementary geometry does (3) follow in the case of Euclidean n-space?

4. Derive from (3) the following formula (which is often called the *Parseval relation*).
$$\langle x, y\rangle = \sum_{k} \langle x, e_k\rangle \overline{\langle y, e_k\rangle}.$$

5. Show that an orthonormal family (e_κ), $\kappa \in I$, in a Hilbert space H is total if and only if the relation in Prob. 4 holds for every x and y in H.

6. Let H be a separable Hilbert space and M a countable dense subset of H. Show that H contains a total orthonormal sequence which can be obtained from M by the Gram-Schmidt process.

7. Show that if a Hilbert space H is separable, the existence of a total orthonormal set in H can be proved without the use of Zorn's lemma.

8. Show that for any orthonormal sequence F in a separable Hilbert space H there is a total orthonormal sequence $\tilde{F}$ which contains F.

9. Let M be a total set in an inner product space X. If $\langle v, x\rangle = \langle w, x\rangle$ for all $x \in M$, show that $v = w$.

10. Let M be a subset of a Hilbert space H, and let $v, w \in H$. Suppose that $\langle v, x\rangle = \langle w, x\rangle$ for all $x \in M$ implies $v = w$. If this holds for all $v, w \in H$, show that M is total in H.

3.7 Legendre, Hermite and Laguerre Polynomials

The theory of Hilbert spaces has applications to various solid topics in analysis. In the present section we discuss some total orthogonal and orthonormal sequences which are used quite frequently in connection with practical problems (for instance, in quantum mechanics, as we shall see in Chap. 11). Properties of these sequences have been investigated in great detail. A standard reference is A. Erdélyi et al. (1953–55) listed in Appendix 3.

The present section is optional.

3.7-1 Legendre polynomials. The inner product space X of all continuous real-valued functions on $[-1, 1]$ with inner product defined by

$$\langle x, y\rangle = \int_{-1}^{1} x(t)y(t)\, dt$$

can be completed according to Theorem 3.2-3. This gives a Hilbert space which is denoted by $L^2[-1, 1]$; cf. also Example 3.1-5.

We want to obtain a total orthonormal sequence in $L^2[-1, 1]$ which consists of functions that are easy to handle. Polynomials are of this type, and we shall succeed by a very simple idea. We start from the powers $x_0, x_1, x_2, \cdots$ where

$$(1) \qquad x_0(t) = 1, \quad x_1(t) = t, \quad \cdots, \quad x_j(t) = t^j, \cdots \qquad t \in [-1, 1].$$

This sequence is linearly independent. (Proof?) Applying the Gram-Schmidt process (Sec. 3.4), we obtain an orthonormal sequence (e_n). Each e_n is a polynomial since in the process we take linear combinations of the x_j's. The degree of e_n is n, as we shall see.

(e_n) is total in $L^2[-1, 1]$.

Proof. By Theorem 3.2-3 the set $W = A(X)$ is dense in $L^2[-1, 1]$. Hence for any fixed $x \in L^2[-1, 1]$ and given $\varepsilon > 0$ there is a continuous function y defined on $[-1, 1]$ such that

$$\|x - y\| < \frac{\varepsilon}{2}.$$

For this y there is a polynomial z such that for all $t \in [-1, 1]$,

$$|y(t) - z(t)| < \frac{\varepsilon}{2\sqrt{2}}.$$

This follows from the Weierstrass approximation theorem to be proved in Sec. 4.11 and implies

$$\|y-z\|^2=\int_{-1}^{1}|y(t)-z(t)|^2\,dt<2\left(\frac{\varepsilon}{2\sqrt{2}}\right)^2=\frac{\varepsilon^2}{4}.$$

Together, by the triangle inequality,

$$\|x-z\|\leqq\|x-y\|+\|y-z\|<\varepsilon.$$

The definition of the Gram-Schmidt process shows that, by (1), we have $z\in\operatorname{span}\{e_0,\cdots,e_m\}$ for sufficiently large m. Since $x\in L^2[-1,1]$ and $\varepsilon>0$ were arbitrary, this proves totality of (e_n). ∎

For practical purposes one needs explicit formulas. We claim that

(2a) $$e_n(t)=\sqrt{\frac{2n+1}{2}}\,P_n(t)\qquad\qquad n=0,1,\cdots$$

where

(2b) $$P_n(t)=\frac{1}{2^n n!}\frac{d^n}{dt^n}[(t^2-1)^n].$$

P_n is called the *Legendre polynomial of order n.* Formula (2b) is called *Rodrigues' formula.* The square root in (2a) has the effect that $P_n(1)=1$, a property which we shall not prove since we do not need it.

By applying the binomial theorem to $(t^2-1)^n$ and differentiating the result n times term by term we obtain from (2b)

(2c) $$P_n(t)=\sum_{j=0}^{N}(-1)^j\frac{(2n-2j)!}{2^n j!\,(n-j)!\,(n-2j)!}\,t^{n-2j}$$

where $N=n/2$ if n is even and $N=(n-1)/2$ if n is odd. Hence (Fig. 35)

(2*)
$$P_0(t)=1 \qquad\qquad P_1(t)=t$$
$$P_2(t)=\tfrac{1}{2}(3t^2-1) \qquad\qquad P_3(t)=\tfrac{1}{2}(5t^3-3t)$$
$$P_4(t)=\tfrac{1}{8}(35t^4-30t^2+3) \qquad\qquad P_5(t)=\tfrac{1}{8}(63t^5-70t^3+15t)$$

etc.

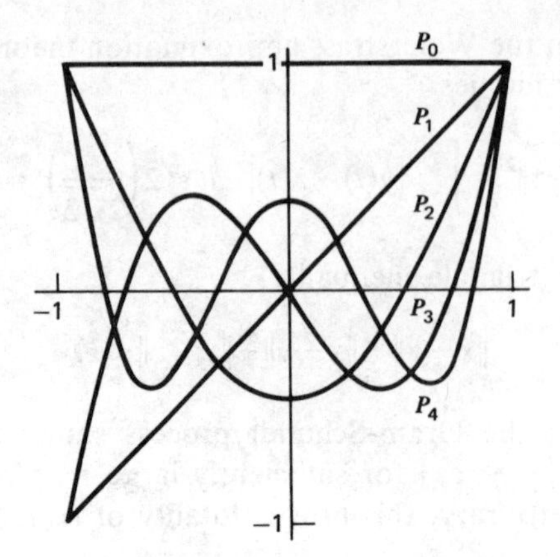

Fig. 35. Legendre polynomials

Proof of (2a) *and* (2b). In part (*a*) we show that (2b) implies

(3)
$$\|P_n\| = \left[\int_{-1}^{1} P_n^{\,2}(t)\,dt\right]^{1/2} = \sqrt{\frac{2}{2n+1}}$$

so that e_n in (2a) comes out with the correct norm, which is 1. In part (*b*) we prove that (P_n) is an orthogonal sequence in the space $L^2[-1, 1]$. This suffices to establish (2a) and (2b) for the following reason. We denote the right-hand side of (2a) at first by $y_n(t)$. Then y_n is a polynomial of degree *n*, and those parts (*a*) and (*b*) imply that (y_n) is an orthonormal sequence in $L^2[-1, 1]$. Let

$$Y_n = \operatorname{span}\{e_0, \cdots, e_n\} = \operatorname{span}\{x_0, \cdots, x_n\} = \operatorname{span}\{y_0, \cdots, y_n\};$$

here the second equality sign follows from the algorithm of the Gram–Schmidt process and the last equality sign from $\dim Y_n = n+1$ together with the linear independence of $\{y_0, \cdots, y_n\}$ stated in 3.4-2. Hence y_n has a representation

(4)
$$y_n = \sum_{j=0}^{n} \alpha_j e_j.$$

Now by the orthogonality,

$$y_n \perp Y_{n-1} = \operatorname{span}\{y_0, \cdots, y_{n-1}\} = \operatorname{span}\{e_0, \cdots, e_{n-1}\}.$$

This implies that for $k=0, \cdots, n-1$ we have

$$0=\langle y_n, e_k\rangle=\sum_{j=0}^{n} \alpha_j\langle e_j, e_k\rangle=\alpha_k.$$

Hence (4) reduces to $y_n=\alpha_n e_n$. Here $|\alpha_n|=1$ since $\|y_n\|=\|e_n\|=1$. Actually, $\alpha_n=+1$ or -1 since both y_n and e_n are real. Now $y_n(t)>0$ for sufficiently large t since the coefficient of t^n in (2c) is positive. Also $e_n(t)>0$ for sufficiently large t, as can be seen from $x_n(t)=t^n$ and (13) and (14) in Sec. 3.4. Hence $\alpha=+1$ and $y_n=e_n$, which establishes (2a) with P_n given by (2b).

This altogether shows that after the presentation of the aforementioned parts (*a*) and (*b*) the proof will be complete.

(a) We derive (3) from (2b). We write $u=t^2-1$. The function u^n and its derivatives $(u^n)', \cdots, (u^n)^{(n-1)}$ are zero at $t=\pm 1$, and $(u^n)^{(2n)}=(2n)!$. Integrating n times by parts, we thus obtain from (2b)

$$\begin{aligned}
(2^n n!)^2\,\|P_n\|^2 &= \int_{-1}^{1} (u^n)^{(n)}(u^n)^{(n)}\,dt \\
&= (u^n)^{(n-1)}(u^n)^{(n)}\Big|_{-1}^{1} - \int_{-1}^{1} (u^n)^{(n-1)}(u^n)^{(n+1)}\,dt \\
&= \cdots \\
&= (-1)^n(2n)!\int_{-1}^{1} u^n\,dt \\
&= 2(2n)!\int_0^1 (1-t^2)^n\,dt \\
&= 2(2n)!\int_0^{\pi/2} \cos^{2n+1}\tau\,d\tau \qquad (t=\sin\tau) \\
&= \frac{2^{2n+1}(n!)^2}{2n+1}.
\end{aligned}$$

Division by $(2^n n!)^2$ yields (3).

(b) We show that $\langle P_m, P_n\rangle=0$ where $0\leqq m<n$. Since P_m is a polynomial, it suffices to prove that $\langle x_m, P_n\rangle=0$ for $m<n$, where

x_m is defined by (1). This result is obtained by m integrations by parts:

$$\begin{aligned} 2^n n!\,\langle x_m, P_n\rangle &= \int_{-1}^{1} t^m (u^n)^{(n)}\,dt \\ &= t^m (u^n)^{(n-1)}\Big|_{-1}^{1} - m\int_{-1}^{1} t^{m-1}(u^n)^{(n-1)}\,dt \\ &= \cdots \\ &= (-1)^m m! \int_{-1}^{1} (u^n)^{(n-m)}\,dt \\ &= (-1)^m m!\,(u^n)^{(n-m-1)}\Big|_{-1}^{1} = 0. \end{aligned}$$

This completes the proof of (2a) and (2b). ∎

The Legendre polynomials are solutions of the important *Legendre differential equation*

$$(1-t^2)P_n'' - 2tP_n' + n(n+1)P_n = 0, \tag{5}$$

and (2c) can also be obtained by applying the power series method to (5).

Furthermore, a total orthonormal sequence in the space $L^2[a, b]$ is (q_n), where

$$q_n = \frac{1}{\|p_n\|}\,p_n, \qquad p_n(t) = P_n(s), \qquad s = 1 + 2\,\frac{t-b}{b-a}. \tag{6}$$

The proof follows if we note that $a \leqq t \leqq b$ corresponds to $-1 \leqq s \leqq 1$ and the orthogonality is preserved under this linear transformation $t \longmapsto s$.

We thus have a total orthonormal sequence in $L^2[a, b]$ for any compact interval $[a, b]$. Theorem 3.6-4 thus implies:

The real space $L^2[a, b]$ is separable.

3.7-2 Hermite polynomials. Further spaces of practical interest are $L^2(-\infty, +\infty)$, $L^2[a, +\infty)$ and $L^2(-\infty, b]$. These are not taken care of by 3.7-1. Since the intervals of integration are infinite, the powers

$x_0, x_1, \cdots$ in 3.7-1 alone would not help. But if we multiply each of them by a simple function which decreases sufficiently rapidly, we can hope to obtain integrals which are finite. Exponential functions with a suitable exponent seem to be a natural choice.

We consider the real space $L^2(-\infty, +\infty)$. The inner product is given by

$$\langle x, y\rangle = \int_{-\infty}^{+\infty} x(t)y(t)\,dt.$$

We apply the Gram-Schmidt process to the sequence of functions defined by

$$w(t) = e^{-t^2/2}, \qquad tw(t), \qquad t^2w(t), \cdots.$$

The factor 1/2 in the exponent is purely conventional and has no deeper meaning. These functions are elements of $L^2(-\infty, +\infty)$. In fact, they are bounded on $\mathbf{R}$, say, $|t^n w(t)| \leqq k_n$ for all t; thus,

$$\left|\int_{-\infty}^{+\infty} t^m e^{-t^2/2}\, t^n e^{-t^2/2}\, dt\right| \leqq k_{m+n} \int_{-\infty}^{+\infty} e^{-t^2/2}\, dt = k_{m+n}\sqrt{2\pi}.$$

The Gram-Schmidt process gives the orthonormal sequence (e_n), where (Fig. 36)

(7a) $$e_n(t) = \frac{1}{(2^n n!\sqrt{\pi})^{1/2}}\, e^{-t^2/2} H_n(t)$$

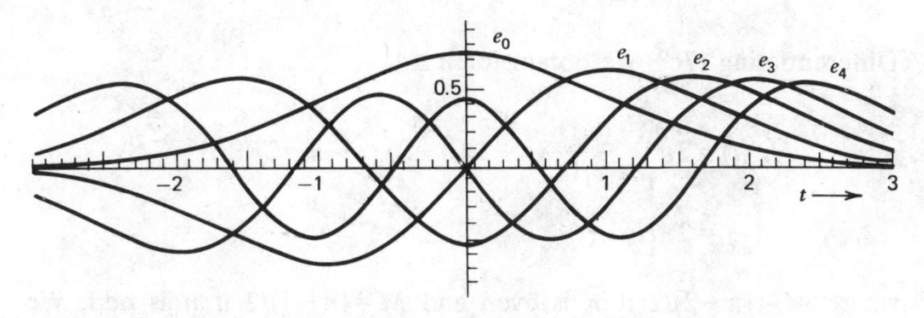

Fig. 36. Functions e_n in (7a) involving Hermite polynomials

and

$$\textbf{(7b)}\qquad H_0(t)=1, \qquad H_n(t)=(-1)^n e^{t^2}\frac{d^n}{dt^n}(e^{-t^2}) \qquad n=1,2,\cdots.$$

H_n is called the *Hermite polynomial of order n.*

Performing the differentiations indicated in (7b), we obtain

$$\textbf{(7c)}\qquad H_n(t)=n!\sum_{j=0}^{N}(-1)^j\frac{2^{n-2j}}{j!\,(n-2j)!}t^{n-2j}$$

where $N=n/2$ if n is even and $N=(n-1)/2$ if n is odd. Note that this can also be written, when $n=2,3,\cdots,$

$$(7c')\qquad H_n(t)=\sum_{j=0}^{N}\frac{(-1)^j}{j!}n(n-1)\cdots(n-2j+1)(2t)^{n-2j}.$$

Explicit expressions for the first few Hermite polynomials are

$$(7^*)\qquad \begin{aligned} &H_0(t)=1 && H_1(t)=2t\\ &H_2(t)=4t^2-2 && H_3(t)=8t^3-12t\\ &H_4(t)=16t^4-48t^2+12 && H_5(t)=32t^5-160t^3+120t. \end{aligned}$$

The sequence (e_n) *defined by* (7a) *and* (7b) *is orthonormal.*

Proof. (7a) and (7b) show that we must prove

$$(8)\qquad \int_{-\infty}^{+\infty} e^{-t^2}H_m(t)H_n(t)\,dt=\begin{cases}0 & \text{if } m\neq n\\ 2^n n!\sqrt{\pi} & \text{if } m=n.\end{cases}$$

Differentiating (7c′), we obtain for $n\geqq 1$

$$\begin{aligned} H_n'(t)&=2n\sum_{j=0}^{M}\frac{(-1)^j}{j!}(n-1)(n-2)\cdots(n-2j)(2t)^{n-1-2j}\\ &=2nH_{n-1}(t)\end{aligned}$$

where $M=(n-2)/2$ if n is even and $M=(n-1)/2$ if n is odd. We apply this formula to H_m, assume $m\leqq n$, denote the exponential function in (8) by v, for simplicity, and integrate m times by parts. Then, by (7b),

$$\begin{aligned}
(-1)^n \int_{-\infty}^{+\infty} e^{-t^2} H_m(t)H_n(t)\,dt &= \int_{-\infty}^{+\infty} H_m(t)v^{(n)}\,dt \\
&= H_m(t)v^{(n-1)}\Big|_{-\infty}^{+\infty} - \int_{-\infty}^{+\infty} 2mH_{m-1}(t)v^{(n-1)}\,dt \\
&= -2m\int_{-\infty}^{+\infty} H_{m-1}(t)v^{(n-1)}\,dt \\
&= \cdots \\
&= (-1)^m 2^m m! \int_{-\infty}^{+\infty} H_0(t)v^{(n-m)}\,dt.
\end{aligned}$$

Here $H_0(t)=1$. If $m<n$, integrating once more, we obtain 0 since v and its derivatives approach zero as $t \longrightarrow +\infty$ or $t \longrightarrow -\infty$. This proves orthogonality of (e_n). We prove (8) for $m=n$, which entails $\|e_n\|=1$ by (7a). If $m=n$, for the last integral, call it J, we obtain

$$J=\int_{-\infty}^{+\infty} e^{-t^2}\,dt=\sqrt{\pi}.$$

This is a familiar result. To verify it, consider J^2, use polar coordinates r, θ and $ds\,dt = r\,dr\,d\theta$, finding

$$\begin{aligned}
J^2 = \int_{-\infty}^{+\infty} e^{-s^2}\,ds \int_{-\infty}^{+\infty} e^{-t^2}\,dt &= \int_{-\infty}^{+\infty}\int_{-\infty}^{+\infty} e^{-(s^2+t^2)}\,ds\,dt \\
&= \int_0^{2\pi}\int_0^{+\infty} e^{-r^2}\,r\,dr\,d\theta \\
&= 2\pi\cdot\tfrac{1}{2}=\pi.
\end{aligned}$$

This proves (8), hence the orthonormality of (e_n). ∎

Classically speaking, one often expresses (8) by saying that the H_n's form an orthogonal sequence with respect to the *weight function* w^2, where w is the function defined at the beginning.

It can be shown that (e_n) *defined by* (7a), (7b) *is total in the real space* $L^2(-\infty, +\infty)$. *Hence this space is separable.* (Cf. 3.6-4.)

We finally mention that the Hermite polynomials H_n satisfy the *Hermite differential equation*

$$H_n'' - 2tH_n' + 2nH_n = 0. \tag{9}$$

Warning. Unfortunately, the terminology in the literature is not unique. In fact, the functions He_n defined by

$$He_0(t) = 1, \qquad He_n(t) = (-1)^n e^{t^2/2} \frac{d^n}{dt^n}(e^{-t^2/2}) \qquad n = 1, 2, \cdots$$

are also called "Hermite polynomials" and, to make things worse, are sometimes denoted by H_n.

An application of Hermite polynomials in quantum mechanics will be considered in Sec. 11.3.

3.7-3 Laguerre polynomials. A total orthonormal sequence in $L^2(-\infty, b]$ or $L^2[a, +\infty)$ can be obtained from such a sequence in $L^2[0, +\infty)$ by the transformations $t = b - s$ and $t = s + a$, respectively.

We consider $L^2[0, +\infty)$. Applying the Gram-Schmidt process to the sequence defined by

$$e^{-t/2}, \qquad te^{-t/2}, \qquad t^2 e^{-t/2}, \qquad \cdots$$

we obtain an orthonormal sequence (e_n). It can be shown that (e_n) is total in $L^2[0, +\infty)$ and is given by (Fig. 37)

$$e_n(t) = e^{-t/2} L_n(t) \qquad n = 0, 1, \cdots \tag{10a}$$

where the *Laguerre polynomial of order n* is defined by

$$L_0(t) = 1, \qquad L_n(t) = \frac{e^t}{n!} \frac{d^n}{dt^n}(t^n e^{-t}) \qquad n = 1, 2, \cdots, \tag{10b}$$

that is,

$$L_n(t) = \sum_{j=0}^{n} \frac{(-1)^j}{j!} \binom{n}{j} t^j. \tag{10c}$$

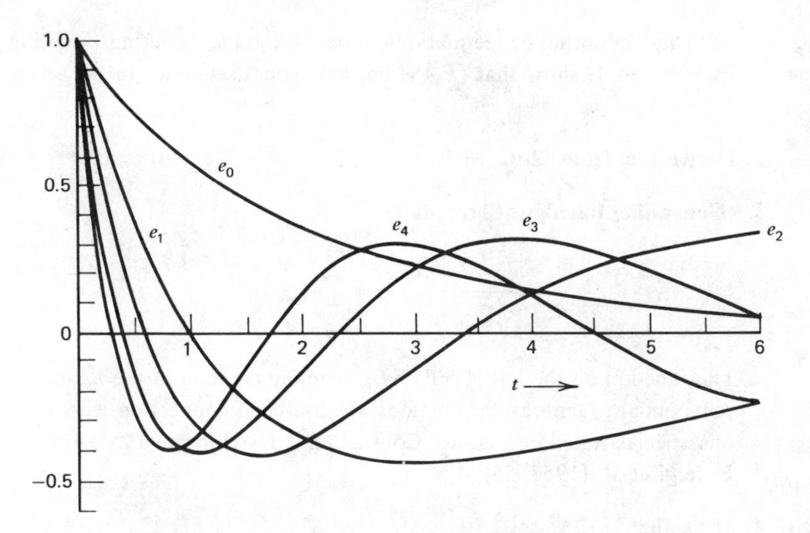

Fig. 37. Functions e_n in (10a) involving Laguerre polynomials

Explicit expressions for the first few Laguerre polynomials are

$$L_0(t) = 1 \qquad L_1(t) = 1 - t$$

$$(10^*) \qquad L_2(t) = 1 - 2t + \tfrac{1}{2}t^2 \qquad L_3(t) = 1 - 3t + \tfrac{3}{2}t^2 - \tfrac{1}{6}t^3$$

$$L_4(t) = 1 - 4t + 3t^2 - \tfrac{2}{3}t^3 + \tfrac{1}{24}t^4.$$

The Laguerre polynomials L_n are solutions of the *Laguerre differential equation*

$$(11) \qquad tL_n'' + (1-t)L_n' + nL_n = 0.$$

For further details, see A. Erdélyi et al. (1953–55); cf. also R. Courant and D. Hilbert (1953–62), vol. I.

Problems

1. Show that the Legendre differential equation can be written

$$[(1-t^2)P_n']' = -n(n+1)P_n.$$

Multiply this by P_m. Multiply the corresponding equation for P_m by

$-P_n$ and add the two equations. Integrating the resulting equation from -1 to 1, show that (P_n) is an orthogonal sequence in the space $L^2[-1, 1]$.

2. Derive (2c) from (2b).

3. (Generating function) Show that

$$\frac{1}{\sqrt{1-2tw+w^2}} = \sum_{n=0}^{\infty} P_n(t)w^n.$$

The function on the left is called a *generating function* of the Legendre polynomials. Generating functions are useful in connection with various special functions; cf. R. Courant and D. Hilbert (1953–62), A. Erdélyi et al. (1953–55).

4. Show that

$$\frac{1}{r} = \frac{1}{\sqrt{r_1^2+r_2^2-2r_1r_2\cos\theta}} = \frac{1}{r_2}\sum_{n=0}^{\infty} P_n(\cos\theta)\left(\frac{r_1}{r_2}\right)^n$$

where r is the distance between given points A_1 and A_2 in $\mathbf{R}^3$, as shown in Fig. 38, and $r_2 > 0$. (This formula is useful in potential theory.)

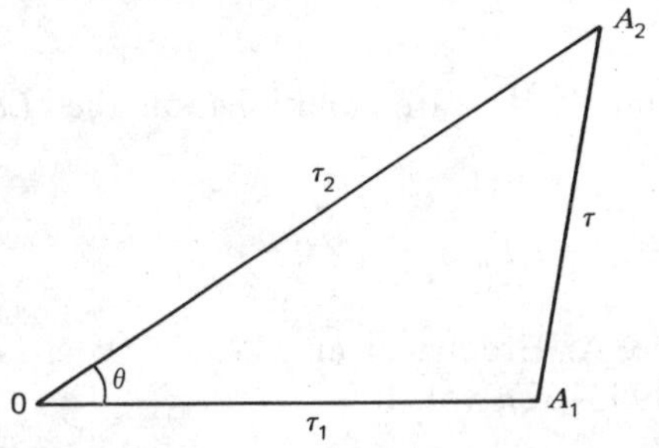

Fig. 38. Problem 4

5. Obtain the Legendre polynomials by the power series method as follows. Substitute $x(t) = c_0 + c_1t + c_2t^2 + \cdots$ into Legendre's equation and show that by determining the coefficients one obtains the solution $x = c_0x_1 + c_1x_2$, where

$$x_1(t) = 1 - \frac{n(n+1)}{2!}t^2 + \frac{(n-2)n(n+1)(n+3)}{4!}t^4 - + \cdots$$

and

$$x_2 = t - \frac{(n-1)(n+2)}{3!}t^3 + \frac{(n-3)(n-1)(n+2)(n+4)}{5!}t^5 - + \cdots .$$

Show that for $n \in \mathbf{N}$, one of these two functions reduces to a polynomial, which agrees with P_n if one chooses $c_n = (2n)!/2^n(n!)^2$ as the coefficient of t^n.

6. **(Generating function)** Show that

$$\exp(2wt - w^2) = \sum_{n=0}^{\infty} \frac{1}{n!} H_n(t) w^n.$$

The function on the left is called a *generating function* of the Hermite polynomials.

7. Using (7b), show that

$$H_{n+1}(t) = 2tH_n(t) - H_n'(t).$$

8. Differentiating the generating function in Prob. 6 with respect to t, show that

$$H_n'(t) = 2nH_{n-1}(t) \qquad (n \geqq 1)$$

and, using Prob. 7, show that H_n satisfies the Hermite differential equation.

9. Solve the differential equation $y'' + (2n+1-t^2)y = 0$ in terms of Hermite polynomials.

10. Using Prob. 8, show that

$$(e^{-t^2} H_n')' = -2ne^{-t^2} H_n.$$

Using this and the method explained in Prob. 1, show that the functions defined by (7a) are orthogonal on $\mathbf{R}$.

11. **(Generating function)** Using (10c), show that

$$\psi(t, w) = \frac{1}{1-w} \exp\left[-\frac{tw}{1-w}\right] = \sum_{n=0}^{\infty} L_n(t) w^n.$$

12. Differentiating ψ in Prob. 11 with respect to w, show that

$$\text{(a)} \qquad (n+1)L_{n+1}(t)-(2n+1-t)L_n(t)+nL_{n-1}(t)=0.$$

Differentiating ψ with respect to t, show that

$$\text{(b)} \qquad L_{n-1}(t)=L'_{n-1}(t)-L'_n(t).$$

13. Using Prob. 12, show that

$$\text{(c)} \qquad tL'_n(t)=nL_n(t)-nL_{n-1}(t).$$

Using this and (b) in Prob. 12, show that L_n satisfies Laguerre's differential equation (11).

14. Show that the functions in (10a) have norm 1.

15. Show that the functions in (10a) constitute an orthogonal sequence in the space $L^2[0, +\infty)$.

3.8 Representation of Functionals on Hilbert Spaces

It is of practical importance to know the general form of bounded linear functionals on various spaces. This was pointed out and explained in Sec. 2.10. For general Banach spaces such formulas and their derivation can sometimes be complicated. However, for a Hilbert space the situation is surprisingly simple:

3.8-1 Riesz's Theorem (Functionals on Hilbert spaces). *Every bounded linear functional f on a Hilbert space H can be represented in terms of the inner product, namely,*

$$f(x)=\langle x, z\rangle \tag{1}$$

where z depends on f, is uniquely determined by f and has norm

$$\|z\|=\|f\|. \tag{2}$$

Proof. We prove that

(a) f has a representation (1),

(b) z in (1) is unique,

(c) formula (2) holds.

The details are as follows.

(a) If $f=0$, then (1) and (2) hold if we take $z=0$. Let $f\neq 0$. To motivate the idea of the proof, let us ask what properties z must have if a representation (1) exists. First of all, $z\neq 0$ since otherwise $f=0$. Second, $\langle x, z\rangle = 0$ for all x for which $f(x)=0$, that is, for all x in the null space $\mathcal{N}(f)$ of f. Hence $z \perp \mathcal{N}(f)$. This suggests that we consider $\mathcal{N}(f)$ and its orthogonal complement $\mathcal{N}(f)^{\perp}$.

$\mathcal{N}(f)$ is a vector space by 2.6-9 and is closed by 2.7-10. Furthermore, $f\neq 0$ implies $\mathcal{N}(f)\neq H$, so that $\mathcal{N}(f)^{\perp}\neq\{0\}$ by the projection theorem 3.3-4. Hence $\mathcal{N}(f)^{\perp}$ contains a $z_0\neq 0$. We set

$$v = f(x)z_0 - f(z_0)x$$

where $x \in H$ is arbitrary. Applying f, we obtain

$$f(v) = f(x)f(z_0) - f(z_0)f(x) = 0.$$

This show that $v \in \mathcal{N}(f)$. Since $z_0 \perp \mathcal{N}(f)$, we have

$$\begin{aligned} 0 = \langle v, z_0\rangle &= \langle f(x)z_0 - f(z_0)x, z_0\rangle \\ &= f(x)\langle z_0, z_0\rangle - f(z_0)\langle x, z_0\rangle. \end{aligned}$$

Noting that $\langle z_0, z_0\rangle = \|z_0\|^2 \neq 0$, we can solve for $f(x)$. The result is

$$f(x) = \frac{f(z_0)}{\langle z_0, z_0\rangle} \langle x, z_0\rangle.$$

This can be written in the form (1), where

$$z = \frac{\overline{f(z_0)}}{\langle z_0, z_0\rangle} z_0.$$

Since $x \in H$ was arbitrary, (1) is proved.

(b) We prove that z in (1) is unique. Suppose that for all $x \in H$,

$$f(x) = \langle x, z_1 \rangle = \langle x, z_2 \rangle.$$

Then $\langle x, z_1 - z_2 \rangle = 0$ for all x. Choosing the particular $x = z_1 - z_2$, we have

$$\langle x, z_1 - z_2 \rangle = \langle z_1 - z_2, z_1 - z_2 \rangle = \|z_1 - z_2\|^2 = 0.$$

Hence $z_1 - z_2 = 0$, so that $z_1 = z_2$, the uniqueness.

(c) We finally prove (2). If $f = 0$, then $z = 0$ and (2) holds. Let $f \neq 0$. Then $z \neq 0$. From (1) with $x = z$ and (3) in Sec. 2.8 we obtain

$$\|z\|^2 = \langle z, z \rangle = f(z) \leqq \|f\| \|z\|.$$

Division by $\|z\| \neq 0$ yields $\|z\| \leqq \|f\|$. It remains to show that $\|f\| \leqq \|z\|$. From (1) and the Schwarz inequality (Sec. 3.2) we see that

$$|f(x)| = |\langle x, z \rangle| \leqq \|x\| \|z\|.$$

This implies

$$\|f\| = \sup_{\|x\|=1} |\langle x, z \rangle| \leqq \|z\|. \qquad \blacksquare$$

The idea of the uniqueness proof in part (*b*) is worth noting for later use:

3.8-2 Lemma (Equality). *If $\langle v_1, w \rangle = \langle v_2, w \rangle$ for all w in an inner product space X, then $v_1 = v_2$. In particular, $\langle v_1, w \rangle = 0$ for all $w \in X$ implies $v_1 = 0$.*

Proof. By assumption, for all w,

$$\langle v_1 - v_2, w \rangle = \langle v_1, w \rangle - \langle v_2, w \rangle = 0.$$

For $w = v_1 - v_2$ this gives $\|v_1 - v_2\|^2 = 0$. Hence $v_1 - v_2 = 0$, so that $v_1 = v_2$. In particular, $\langle v_1, w \rangle = 0$ with $w = v_1$ gives $\|v_1\|^2 = 0$, so that $v_1 = 0$. $\blacksquare$

The practical usefulness of bounded linear functionals on Hilbert spaces results to a large extent from the simplicity of the Riesz representation (1).

Furthermore, (1) is quite important in the theory of operators on Hilbert spaces. In particular, this refers to the Hilbert-adjoint operator T^* of a bounded linear operator T which we shall define in the next section. For this purpose we need a preparation which is of general interest, too. We begin with the following definition.

3.8-3 Definition (Sesquilinear form). Let X and Y be vector spaces over the same field K ($=\mathbf{R}$ or $\mathbf{C}$). Then a *sesquilinear form* (or *sesquilinear functional*) h on $X \times Y$ is a mapping

$$h\colon X \times Y \longrightarrow K$$

such that for all $x, x_1, x_2 \in X$ and $y, y_1, y_2 \in Y$ and all scalars α, β,

$$\begin{aligned} &\text{(a)} && h(x_1+x_2, y) = h(x_1, y) + h(x_2, y) \\ &\text{(b)} && h(x, y_1+y_2) = h(x, y_1) + h(x, y_2) \\ (3)\quad &\text{(c)} && h(\alpha x, y) = \alpha h(x, y) \\ &\text{(d)} && h(x, \beta y) = \bar{\beta} h(x, y). \end{aligned}$$

Hence h is *linear* in the first argument and *conjugate linear* in the second one. If X and Y are real ($K=\mathbf{R}$), then (3d) is simply

$$h(x, \beta y) = \beta h(x, y)$$

and h is called *bilinear* since it is linear in both arguments.

If X and Y are normed spaces and if there is a real number c such that for all x, y

$$(4) \qquad |h(x, y)| \leqq c \, \|x\| \, \|y\|,$$

then h is said to be *bounded*, and the number

$$(5) \qquad \|h\| = \sup_{\substack{x \in X - \{0\} \\ y \in Y - \{0\}}} \frac{|h(x, y)|}{\|x\| \, \|y\|} = \sup_{\substack{\|x\|=1 \\ \|y\|=1}} |h(x, y)|$$

is called the *norm* of h. ∎

For example, the inner product is sesquilinear and bounded. Note that from (4) and (5) we have

$$|h(x, y)| \leqq \|h\| \|x\| \|y\|. \tag{6}$$

The term "sesquilinear" was motivated in Sec. 3.1. In Def. 3.8-3, both words "form" and "functional" are common, the usage of one or the other being largely a matter of individual taste. Perhaps it is slightly preferable to use "form" in this two-variable case and reserve the word "functional" to one-variable cases such as that in Theorem 3.8-1. This is what we shall do.

It is quite interesting that from Theorem 3.8-1 we can obtain a general representation of sesquilinear forms on Hilbert spaces as follows.

3.8-4 Theorem (Riesz representation). *Let H_1, H_2 be Hilbert spaces and*

$$h: H_1 \times H_2 \longrightarrow K$$

a bounded sesquilinear form. Then h has a representation

$$h(x, y) = \langle Sx, y \rangle \tag{7}$$

where $S: H_1 \longrightarrow H_2$ is a bounded linear operator. S is uniquely determined by h and has norm

$$\|S\| = \|h\|. \tag{8}$$

Proof. We consider $\overline{h(x, y)}$. This is linear in y, because of the bar. To make Theorem 3.8-1 applicable, we keep x fixed. Then that theorem yields a representation in which y is variable, say,

$$\overline{h(x, y)} = \langle y, z \rangle.$$

Hence

$$h(x, y) = \langle z, y \rangle. \tag{9}$$

Here $z \in H_2$ is unique but, of course, depends on our fixed $x \in H_1$. It follows that (9) with variable x defines an operator

$$S: H_1 \longrightarrow H_2 \qquad \text{given by} \qquad z = Sx.$$

Substituting $z = Sx$ in (9), we have (7).

S is linear. In fact, its domain is the vector space H_1, and from (7) and the sesquilinearity we obtain

$$\begin{aligned}\langle S(\alpha x_1 + \beta x_2), y\rangle &= h(\alpha x_1 + \beta x_2, y) \\ &= \alpha h(x_1, y) + \beta h(x_2, y) \\ &= \alpha\langle Sx_1, y\rangle + \beta\langle Sx_2, y\rangle \\ &= \langle \alpha Sx_1 + \beta Sx_2, y\rangle\end{aligned}$$

for all y in H_2, so that by Lemma 3.8-2,

$$S(\alpha x_1 + \beta x_2) = \alpha Sx_1 + \beta Sx_2.$$

S is bounded. Indeed, leaving aside the trivial case $S = 0$, we have from (5) and (7)

$$\|h\| = \sup_{\substack{x \neq 0 \\ y \neq 0}} \frac{|\langle Sx, y\rangle|}{\|x\|\,\|y\|} \geqq \sup_{\substack{x \neq 0 \\ Sx \neq 0}} \frac{|\langle Sx, Sx\rangle|}{\|x\|\,\|Sx\|} = \sup_{x \neq 0} \frac{\|Sx\|}{\|x\|} = \|S\|.$$

This proves boundedness. Moreover, $\|h\| \geqq \|S\|$.

We now obtain (8) by noting that $\|h\| \leqq \|S\|$ follows by an application of the Schwarz inequality:

$$\|h\| = \sup_{\substack{x \neq 0 \\ y \neq 0}} \frac{|\langle Sx, y\rangle|}{\|x\|\,\|y\|} \leqq \sup_{x \neq 0} \frac{\|Sx\|\,\|y\|}{\|x\|\,\|y\|} = \|S\|.$$

S is unique. In fact, assuming that there is a linear operator $T: H_1 \longrightarrow H_2$ such that for all $x \in H_1$ and $y \in H_2$ we have

$$h(x, y) = \langle Sx, y\rangle = \langle Tx, y\rangle,$$

we see that $Sx = Tx$ by Lemma 3.8-2 for all $x \in H_1$. Hence $S = T$ by definition. ∎

Problems

1. **(Space $\mathbf{R}^3$)** Show that any linear functional f on $\mathbf{R}^3$ can be represented by a dot product:

$$f(x) = x \cdot z = \xi_1\zeta_1 + \xi_2\zeta_2 + \xi_3\zeta_3.$$

2. **(Space l^2)** Show that every bounded linear functional f on l^2 can be represented in the form

$$f(x) = \sum_{j=1}^{\infty} \xi_j \bar{\zeta}_j \qquad [z = (\zeta_j) \in l^2].$$

3. If z is any fixed element of an inner product space X, show that $f(x) = \langle x, z \rangle$ defines a bounded linear functional f on X, of norm $\|z\|$.

4. Consider Prob. 3. If the mapping $X \longrightarrow X'$ given by $z \longmapsto f$ is surjective, show that X must be a Hilbert space.

5. Show that the dual space of the real space l^2 is l^2. (Use 3.8-1.)

6. Show that Theorem 3.8-1 defines an isometric bijection $T: H \longrightarrow H'$, $z \longmapsto f_z = \langle \cdot, z \rangle$ which is not linear but *conjugate linear,* that is, $\alpha z + \beta v \longmapsto \bar{\alpha} f_z + \bar{\beta} f_v$.

7. Show that the dual space H' of a Hilbert space H is a Hilbert space with inner product $\langle \cdot, \cdot \rangle_1$ defined by

$$\langle f_z, f_v \rangle_1 = \overline{\langle z, v \rangle} = \langle v, z \rangle,$$

where $f_z(x) = \langle x, z \rangle$, etc.

8. Show that any Hilbert space H is isomorphic (cf. Sec. 3.6) with its second dual space $H'' = (H')'$. (This property is called *reflexivity* of H. It will be considered in more detail for normed spaces in Sec. 4.6.)

9. **(Annihilator)** Explain the relation between M^a in Prob. 13, Sec. 2.10, and $M^\perp$ in Sec. 3.3 in the case of a subset $M \neq \varnothing$ of a Hilbert space H.

10. Show that the inner product $\langle \cdot, \cdot \rangle$ on an inner product space X is a bounded sesquilinear form h. What is $\|h\|$ in this case?

11. If X is a vector space and h a sesquilinear form on $X \times X$, show that $f_1(x) = h(x, y_0)$ with fixed y_0 defines a linear functional f_1 on X, and so does $f_2(y) = \overline{h(x_0, y)}$ with fixed x_0.

12. Let X and Y be normed spaces. Show that a bounded sesquilinear form h on $X \times Y$ is jointly continuous in both variables.

13. (Hermitian form) Let X be a vector space over a field K. A *Hermitian sesquilinear form* or, simply, *Hermitian form* h on $X \times X$ is a mapping $h: X \times X \longrightarrow K$ such that for all $x, y, z \in X$ and $\alpha \in K$,

$$h(x+y, z) = h(x, z) + h(y, z)$$

$$h(\alpha x, y) = \alpha h(x, y)$$

$$h(x, y) = \overline{h(y, x)}.$$

What is the last condition if $K = \mathbf{R}$? What condition must be added for h to be an inner product on X?

14. (Schwarz inequality) Let X be a vector space and h a Hermitian form on $X \times X$. This form is said to be *positive semidefinite* if $h(x, x) \geqq 0$ for all $x \in X$. Show that then h satisfies the *Schwarz inequality*

$$|h(x, y)|^2 \leqq h(x, x)h(y, y).$$

15. (Seminorm) If h satisfies the conditions in Prob. 14, show that

$$p(x) = \sqrt{h(x, x)} \qquad (\geqq 0)$$

defines a seminorm on X. (Cf. Prob. 12, Sec. 2.3.)

3.9 Hilbert-Adjoint Operator

The results of the previous section will now enable us to introduce the Hilbert-adjoint operator of a bounded linear operator on a Hilbert space. This operator was suggested by problems in matrices and linear differential and integral equations. We shall see that it also helps to define three important classes of operators (called *self-adjoint*, *unitary* and *normal operators*) which have been studied extensively because they play a key role in various applications.

3.9-1 Definition (Hilbert-adjoint operator T^*). Let $T\colon H_1 \longrightarrow H_2$ be a bounded linear operator, where H_1 and H_2 are Hilbert spaces. Then the *Hilbert-adjoint operator* T^* of T is the operator

$$T^*\colon H_2 \longrightarrow H_1$$

such that[5] for all $x \in H_1$ and $y \in H_2$,

$$\langle Tx, y\rangle = \langle x, T^*y\rangle. \tag{1}$$

∎

Of course, we should first show that this definition makes sense, that is, we should prove that for a given T such a T^* does exist:

3.9-2 Theorem (Existence). *The Hilbert-adjoint operator T^* of T in Def. 3.9-1 exists, is unique and is a bounded linear operator with norm*

$$\|T^*\| = \|T\|. \tag{2}$$

Proof. The formula

$$h(y, x) = \langle y, Tx\rangle \tag{3}$$

defines a sesquilinear form on $H_2 \times H_1$ because the inner product is sesquilinear and T is linear. In fact, conjugate linearity of the form is seen from

$$\begin{aligned} h(y, \alpha x_1 + \beta x_2) &= \langle y, T(\alpha x_1 + \beta x_2)\rangle \\ &= \langle y, \alpha T x_1 + \beta T x_2\rangle \\ &= \bar{\alpha}\langle y, Tx_1\rangle + \bar{\beta}\langle y, Tx_2\rangle \\ &= \bar{\alpha} h(y, x_1) + \bar{\beta} h(y, x_2). \end{aligned}$$

h is bounded. Indeed, by the Schwarz inequality,

$$|h(y, x)| = |\langle y, Tx\rangle| \leqq \|y\| \, \|Tx\| \leqq \|T\| \, \|x\| \, \|y\|.$$

[5] We may denote inner products on H_1 and H_2 by the same symbol since the factors show to which space an inner product refers.

This also implies $\|h\| \leqq \|T\|$. Moreover we have $\|h\| \geqq \|T\|$ from

$$\|h\| = \sup_{\substack{x \neq 0 \\ y \neq 0}} \frac{|\langle y, Tx\rangle|}{\|y\| \|x\|} \geqq \sup_{\substack{x \neq 0 \\ Tx \neq 0}} \frac{|\langle Tx, Tx\rangle|}{\|Tx\| \|x\|} = \|T\|.$$

Together,

$$\|h\| = \|T\|. \tag{4}$$

Theorem 3.8-4 gives a Riesz representation for h; writing T^* for S, we have

$$h(y, x) = \langle T^* y, x\rangle, \tag{5}$$

and we know from that theorem that $T^*: H_2 \longrightarrow H_1$ is a uniquely determined bounded linear operator with norm [cf. (4)]

$$\|T^*\| = \|h\| = \|T\|.$$

This proves (2). Also $\langle y, Tx\rangle = \langle T^* y, x\rangle$ by comparing (3) and (5), so that we have (1) by taking conjugates, and we now see that T^* is in fact the operator we are looking for. ∎

In our study of properties of Hilbert-adjoint operators it will be convenient to make use of the following lemma.

3.9-3 Lemma (Zero operator). *Let X and Y be inner product spaces and $Q: X \longrightarrow Y$ a bounded linear operator. Then:*

(a) *$Q = 0$ if and only if $\langle Qx, y\rangle = 0$ for all $x \in X$ and $y \in Y$.*

(b) *If $Q: X \longrightarrow X$, where X is complex, and $\langle Qx, x\rangle = 0$ for all $x \in X$, then $Q = 0$.*

Proof. **(a)** $Q = 0$ means $Qx = 0$ for all x and implies

$$\langle Qx, y\rangle = \langle 0, y\rangle = 0\langle w, y\rangle = 0.$$

Conversely, $\langle Qx, y\rangle = 0$ for all x and y implies $Qx = 0$ for all x by 3.8-2, so that $Q = 0$ by definition.

(b) By assumption, $\langle Qv, v\rangle = 0$ for every $v = \alpha x + y \in X$, that is,

$$\begin{aligned} 0 &= \langle Q(\alpha x + y), \alpha x + y\rangle \\ &= |\alpha|^2 \langle Qx, x\rangle + \langle Qy, y\rangle + \alpha\langle Qx, y\rangle + \bar{\alpha}\langle Qy, x\rangle. \end{aligned}$$

The first two terms on the right are zero by assumption. $\alpha = 1$ gives

$$\langle Qx, y\rangle + \langle Qy, x\rangle = 0.$$

$\alpha = i$ gives $\bar{\alpha} = -i$ and

$$\langle Qx, y\rangle - \langle Qy, x\rangle = 0.$$

By addition, $\langle Qx, y\rangle = 0$, and $Q = 0$ follows from (*a*). ∎

In part (*b*) of this lemma, it is essential that X be complex. Indeed, the conclusion may not hold if X is real. A counterexample is a rotation Q of the plane $\mathbf{R}^2$ through a right angle. Q is linear, and $Qx \perp x$, hence $\langle Qx, x\rangle = 0$ for all $x \in \mathbf{R}^2$, but $Q \neq 0$. (What about such a rotation in the complex plane?)

We can now list and prove some general properties of Hilbert-adjoint operators which one uses quite frequently in applying these operators.

3.9-4 Theorem (Properties of Hilbert-adjoint operators). *Let H_1, H_2 be Hilbert spaces, $S\colon H_1 \longrightarrow H_2$ and $T\colon H_1 \longrightarrow H_2$ bounded linear operators and α any scalar. Then we have*

(6)

(a) $\langle T^*y, x\rangle = \langle y, Tx\rangle \qquad (x \in H_1, y \in H_2)$

(b) $(S+T)^* = S^* + T^*$

(c) $(\alpha T)^* = \bar{\alpha} T^*$

(d) $(T^*)^* = T$

(e) $\|T^*T\| = \|TT^*\| = \|T\|^2$

(f) $T^*T = 0 \iff T = 0$

(g) $(ST)^* = T^*S^* \qquad$ (assuming $H_2 = H_1$).

Proof. (*a*) From (1) we have (6a):

$$\langle T^*y, x\rangle = \overline{\langle x, T^*y\rangle} = \overline{\langle Tx, y\rangle} = \langle y, Tx\rangle.$$

(*b*) By (1), for all x and y,

$$\begin{aligned}\langle x, (S+T)^*y\rangle &= \langle (S+T)x, y\rangle \\ &= \langle Sx, y\rangle + \langle Tx, y\rangle \\ &= \langle x, S^*y\rangle + \langle x, T^*y\rangle \\ &= \langle x, (S^*+T^*)y\rangle.\end{aligned}$$

Hence $(S+T)^*y = (S^*+T^*)y$ for all y by 3.8-2, which is (6b) by definition.

(*c*) Formula (6c) must not be confused with the formula $T^*(\alpha x) = \alpha T^*x$. It is obtained from the following calculation and subsequent application of Lemma 3.9-3(*a*) to $Q = (\alpha T)^* - \bar{\alpha}T^*$.

$$\begin{aligned}\langle (\alpha T)^*y, x\rangle &= \langle y, (\alpha T)x\rangle \\ &= \langle y, \alpha(Tx)\rangle \\ &= \bar{\alpha}\langle y, Tx\rangle \\ &= \bar{\alpha}\langle T^*y, x\rangle \\ &= \langle \bar{\alpha}T^*y, x\rangle.\end{aligned}$$

(*d*) $(T^*)^*$ is written T^{**} and equals T since for all $x \in H_1$ and $y \in H_2$ we have from (6a) and (1)

$$\langle (T^*)^*x, y\rangle = \langle x, T^*y\rangle = \langle Tx, y\rangle$$

and (6d) follows from Lemma 3.9-3(a) with $Q = (T^*)^* - T$.

(*e*) We see that $T^*T\colon H_1 \longrightarrow H_1$, but $TT^*\colon H_2 \longrightarrow H_2$. By the Schwarz inequality,

$$\|Tx\|^2 = \langle Tx, Tx\rangle = \langle T^*Tx, x\rangle \leqq \|T^*Tx\|\,\|x\| \leqq \|T^*T\|\,\|x\|^2.$$

Taking the supremum over all x of norm 1, we obtain $\|T\|^2 \leqq \|T^*T\|$. Applying (7), Sec. 2.7, and (2), we thus have

$$\|T\|^2 \leqq \|T^*T\| \leqq \|T^*\| \, \|T\| = \|T\|^2.$$

Hence $\|T^*T\| = \|T\|^2$. Replacing T by T^* and using again (2), we also have

$$\|T^{**}T^*\| = \|T^*\|^2 = \|T\|^2.$$

Here $T^{**} = T$ by (6d), so that (6e) is proved.

(*f*) From (6e) we immediately obtain (6f).

(*g*) Repeated application of (1) gives

$$\langle x, (ST)^*y\rangle = \langle (ST)x, y\rangle = \langle Tx, S^*y\rangle = \langle x, T^*S^*y\rangle.$$

Hence $(ST)^*y = T^*S^*y$ by 3.8-2, which is (6g) by definition. ∎

Problems

1. Show that $0^* = 0$, $I^* = I$.

2. Let H be a Hilbert space and $T\colon H \longrightarrow H$ a bijective bounded linear operator whose inverse is bounded. Show that $(T^*)^{-1}$ exists and

$$(T^*)^{-1} = (T^{-1})^*.$$

3. If (T_n) is a sequence of bounded linear operators on a Hilbert space and $T_n \longrightarrow T$, show that $T_n^* \longrightarrow T^*$.

4. Let H_1 and H_2 be Hilbert spaces and $T\colon H_1 \longrightarrow H_2$ a bounded linear operator. If $M_1 \subset H_1$ and $M_2 \subset H_2$ are such that $T(M_1) \subset M_2$, show that $M_1^\perp \supset T^*(M_2^\perp)$.

5. Let M_1 and M_2 in Prob. 4 be closed subspaces. Show that then $T(M_1) \subset M_2$ if and only if $M_1^\perp \supset T^*(M_2^\perp)$.

6. If $M_1 = \mathcal{N}(T) = \{x \mid Tx = 0\}$ in Prob. 4, show that

(*a*) $T^*(H_2) \subset M_1^\perp$, (*b*) $[T(H_1)]^\perp \subset \mathcal{N}(T^*)$, (*c*) $M_1 = [T^*(H_2)]^\perp$.

7. Let T_1 and T_2 be bounded linear operators on a complex Hilbert space H into itself. If $\langle T_1x, x\rangle = \langle T_2x, x\rangle$ for all $x \in H$, show that $T_1 = T_2$.

8. Let $S = I + T^*T\colon H \longrightarrow H$, where T is linear and bounded. Show that $S^{-1}\colon S(H) \longrightarrow H$ exists.

9. Show that a bounded linear operator $T\colon H \longrightarrow H$ on a Hilbert space H has a finite dimensional range if and only if T can be represented in the form

$$Tx = \sum_{j=1}^{n} \langle x, v_j\rangle w_j \qquad [v_j, w_j \in H].$$

10. (Right shift operator) Let (e_n) be a total orthonormal sequence in a separable Hilbert space H and define the *right shift operator* to be the linear operator $T\colon H \longrightarrow H$ such that $Te_n = e_{n+1}$ for $n = 1, 2, \cdots$. Explain the name. Find the range, null space, norm and Hilbert-adjoint operator of T.

3.10 Self-Adjoint, Unitary and Normal Operators

Classes of bounded linear operators of great practical importance can be defined by the use of the Hilbert-adjoint operator as follows.

3.10-1 Definition (Self-adjoint, unitary and normal operators). A bounded linear operator $T\colon H \longrightarrow H$ on a Hilbert space H is said to be

self-adjoint or *Hermitian* if	$T^* = T$,
unitary if T is bijective and	$T^* = T^{-1}$,
normal if	$TT^* = T^*T$. ∎

The Hilbert-adjoint operator T^* of T is defined by (1), Sec. 3.9, that is,

$$\langle Tx, y\rangle = \langle x, T^*y\rangle.$$

If T is self-adjoint, we see that the formula becomes

$$\langle Tx, y\rangle = \langle x, Ty\rangle. \tag{1}$$

If T is self-adjoint or unitary, T is normal.

This can immediately be seen from the definition. Of course, a normal operator need not be self-adjoint or unitary. For example, if $I: H \longrightarrow H$ is the identity operator, then $T = 2iI$ is normal since $T^* = -2iI$ (cf. 3.9-4), so that $TT^* = T^*T = 4I$ but $T^* \neq T$ as well as $T^* \neq T^{-1} = -\frac{1}{2}iI$.

Operators which are not normal will easily result from the next example. Another operator which is not normal is T in Prob. 10, Sec. 3.9, as the reader may prove.

The terms in Def. 3.10-1 are also used in connection with matrices. We want to explain the reason for this and mention some important relations, as follows.

3.10-2 Example (Matrices). We consider $\mathbf{C}^n$ with the inner product defined by (cf. 3.1-4)

$$\langle x, y \rangle = x^{\mathsf{T}} \bar{y}, \tag{2}$$

where x and y are written as column vectors, and T means transposition; thus $x^{\mathsf{T}} = (\xi_1, \cdots, \xi_n)$, and we use the ordinary matrix multiplication.

Let $T: \mathbf{C}^n \longrightarrow \mathbf{C}^n$ be a linear operator (which is bounded by Theorem 2.7-8). A basis for $\mathbf{C}^n$ being given, we can represent T and its Hilbert-adjoint operator T^* by two n-rowed square matrices, say, A and B, respectively.

Using (2) and the familiar rule $(Bx)^{\mathsf{T}} = x^{\mathsf{T}} B^{\mathsf{T}}$ for the transposition of a product, we obtain

$$\langle Tx, y \rangle = (Ax)^{\mathsf{T}} \bar{y} = x^{\mathsf{T}} A^{\mathsf{T}} \bar{y}$$

and

$$\langle x, T^* y \rangle = x^{\mathsf{T}} \bar{B} \bar{y}.$$

By (1), Sec. 3.9, the left-hand sides are equal for all $x, y \in \mathbf{C}^n$. Hence we must have $A^{\mathsf{T}} = \bar{B}$. Consequently,

$$B = \bar{A}^{\mathsf{T}}.$$

Our result is as follows.

If a basis for $\mathbf{C}^n$ is given and a linear operator on $\mathbf{C}^n$ is represented by a certain matrix, then its Hilbert-adjoint operator is represented by the complex conjugate transpose of that matrix.

Consequently, representing matrices are

Hermitian if T is self-adjoint (Hermitian),
unitary if T is unitary,
normal if T is normal.

Similarly, for a linear operator $T: \mathbf{R}^n \longrightarrow \mathbf{R}^n$, representing matrices are:

Real symmetric if T is self-adjoint,
orthogonal if T is unitary.

In this connection, remember the following definitions. A square matrix $A = (\alpha_{jk})$ is said to be:

Hermitian if $\bar{A}^{\mathsf{T}} = A$ (hence $\bar{\alpha}_{kj} = \alpha_{jk}$)
skew-Hermitian if $\bar{A}^{\mathsf{T}} = -A$ (hence $\bar{\alpha}_{kj} = -\alpha_{jk}$)
unitary if $\bar{A}^{\mathsf{T}} = A^{-1}$
normal if $A\bar{A}^{\mathsf{T}} = \bar{A}^{\mathsf{T}}A$.

A *real* square matrix $A = (\alpha_{jk})$ is said to be:

(*Real*) *symmetric* if $A^{\mathsf{T}} = A$ (hence $\alpha_{kj} = \alpha_{jk}$)
(*real*) *skew-symmetric* if $A^{\mathsf{T}} = -A$ (hence $\alpha_{kj} = -\alpha_{jk}$)
orthogonal if $A^{\mathsf{T}} = A^{-1}$.

Hence a real Hermitian matrix is a (real) symmetric matrix. A real skew-Hermitian matrix is a (real) skew-symmetric matrix. A real unitary matrix is an orthogonal matrix. (Hermitian matrices are named after the French mathematician, Charles Hermite, 1822–1901.) ∎

Let us return to linear operators on arbitrary Hilbert spaces and state an important and rather simple criterion for self-adjointness.

3.10-3 Theorem (Self-adjointness). *Let $T: H \longrightarrow H$ be a bounded linear operator on a Hilbert space H. Then:*

(a) *If T is self-adjoint, $\langle Tx, x\rangle$ is real for all $x \in H$.*

(b) *If H is complex and $\langle Tx, x\rangle$ is real for all $x \in H$, the operator T is self-adjoint.*

Proof. **(a)** If T is self-adjoint, then for all x,

$$\overline{\langle Tx, x\rangle} = \langle x, Tx\rangle = \langle Tx, x\rangle.$$

Hence $\langle Tx, x\rangle$ is equal to its complex conjugate, so that it is real.

(b) If $\langle Tx, x\rangle$ is real for all x, then

$$\langle Tx, x\rangle = \overline{\langle Tx, x\rangle} = \overline{\langle x, T^*x\rangle} = \langle T^*x, x\rangle.$$

Hence

$$0 = \langle Tx, x\rangle - \langle T^*x, x\rangle = \langle (T - T^*)x, x\rangle$$

and $T - T^* = 0$ by Lemma 3.9-3(*b*) since H is complex. ∎

In part (*b*) of the theorem it is essential that H be complex. This is clear since for a real H the inner product is real-valued, which makes $\langle Tx, x\rangle$ real without any further assumptions about the linear operator T.

Products (composites[6]) of self-adjoint operators appear quite often in applications, so that the following theorem will be useful.

3.10-4 Theorem (Self-adjointness of product). *The product of two bounded self-adjoint linear operators S and T on a Hilbert space H is self-adjoint if and only if the operators commute,*

$$ST = TS.$$

Proof. By (6g) in the last section and by the assumption,

$$(ST)^* = T^*S^* = TS.$$

Hence

$$ST = (ST)^* \quad \Longleftrightarrow \quad ST = TS.$$

This completes the proof. ∎

Sequences of self-adjoint operators occur in various problems, and for them we have

[6] A review of terms and notations in connection with the composition of mappings is included in A1.2, Appendix 1.

3.10-5 Theorem (Sequences of self-adjoint operators). *Let (T_n) be a sequence of bounded self-adjoint linear operators $T_n: H \longrightarrow H$ on a Hilbert space H. Suppose that (T_n) converges, say,*

$$T_n \longrightarrow T, \qquad \text{that is,} \qquad \|T_n - T\| \longrightarrow 0,$$

where $\|\cdot\|$ is the norm on the space $B(H, H)$; cf. Sec. 2.10. *Then the limit operator T is a bounded self-adjoint linear operator on H.*

Proof. We must show that $T^* = T$. This follows from $\|T - T^*\| = 0$. To prove the latter, we use that, by 3.9-4 and 3.9-2,

$$\|T_n^* - T^*\| = \|(T_n - T)^*\| = \|T_n - T\|$$

and obtain by the triangle inequality in $B(H, H)$

$$\begin{aligned}\|T - T^*\| &\leqq \|T - T_n\| + \|T_n - T_n^*\| + \|T_n^* - T^*\| \\ &= \|T - T_n\| + 0 + \|T_n - T\| \\ &= 2\|T_n - T\| \quad \longrightarrow \quad 0 \qquad (n \longrightarrow \infty).\end{aligned}$$

Hence $\|T - T^*\| = 0$ and $T^* = T$. ∎

These theorems give us some idea about basic properties of self-adjoint linear operators. They will also be helpful in our further work, in particular in the spectral theory of these operators (Chap. 9), where further properties will be discussed.

We now turn to unitary operators and consider some of their basic properties.

3.10-6 Theorem (Unitary operator). *Let the operators $U: H \longrightarrow H$ and $V: H \longrightarrow H$ be unitary; here, H is a Hilbert space. Then:*

(a) *U is isometric* (cf. 1.6-1); *thus $\|Ux\| = \|x\|$ for all $x \in H$;*

(b) *$\|U\| = 1$, provided $H \neq \{0\}$,*

(c) *$U^{-1}(= U^*)$ is unitary,*

(d) *UV is unitary,*

(e) *U is normal.*

Furthermore:

(f) *A bounded linear operator T on a complex Hilbert space H is unitary if and only if T is isometric and surjective.*

Proof. **(a)** can be seen from

$$\|Ux\|^2 = \langle Ux, Ux\rangle = \langle x, U^*Ux\rangle = \langle x, Ix\rangle = \|x\|^2.$$

(b) follows immediately from (*a*).

(c) Since U is bijective, so is U^{-1}, and by 3.9-4,

$$(U^{-1})^* = U^{**} = U = (U^{-1})^{-1}.$$

(d) UV is bijective, and 3.9-4 and 2.6-11 yield

$$(UV)^* = V^*U^* = V^{-1}U^{-1} = (UV)^{-1}.$$

(e) follows from $U^{-1} = U^*$ and $UU^{-1} = U^{-1}U = I$.

(f) Suppose that T is isometric and surjective. Isometry implies injectivity, so that T is bijective. We show that $T^* = T^{-1}$. By the isometry,

$$\langle T^*Tx, x\rangle = \langle Tx, Tx\rangle = \langle x, x\rangle = \langle Ix, x\rangle.$$

Hence

$$\langle (T^*T - I)x, x\rangle = 0$$

and $T^*T - I = 0$ by Lemma 3.9-3(*b*), so that $T^*T = I$. From this,

$$TT^* = TT^*(TT^{-1}) = T(T^*T)T^{-1} = TIT^{-1} = I.$$

Together, $T^*T = TT^* = I$. Hence $T^* = T^{-1}$, so that T is unitary. The converse is clear since T is isometric by (*a*) and surjective by definition. ∎

Note that an isometric operator need not be unitary since it may fail to be surjective. An example is the *right shift operator* $T: l^2 \longrightarrow l^2$ given by

$$(\xi_1, \xi_2, \xi_3, \cdots) \longmapsto (0, \xi_1, \xi_2, \xi_3, \cdots)$$

where $x = (\xi_j) \in l^2$.

Problems

1. If S and T are bounded self-adjoint linear operators on a Hilbert space H and α and β are real, show that $\tilde{T}=\alpha S+\beta T$ is self-adjoint.

2. How could we use Theorem 3.10-3 to prove Theorem 3.10-5 for a complex Hilbert space H?

3. Show that if $T: H \longrightarrow H$ is a bounded self-adjoint linear operator, so is T^n, where n is a positive integer.

4. Show that for any bounded linear operator T on H, the operators

$$T_1=\frac{1}{2}(T+T^*) \qquad \text{and} \qquad T_2=\frac{1}{2i}(T-T^*)$$

are self-adjoint. Show that

$$T=T_1+iT_2, \qquad\qquad T^*=T_1-iT_2.$$

Show uniqueness, that is, $T_1+iT_2=S_1+iS_2$ implies $S_1=T_1$ and $S_2=T_2$; here, S_1 and S_2 are self-adjoint by assumption.

5. On $\mathbf{C}^2$ (cf. 3.1-4) let the operator $T: \mathbf{C}^2 \longrightarrow \mathbf{C}^2$ be defined by $Tx=(\xi_1+i\xi_2, \xi_1-i\xi_2)$, where $x=(\xi_1, \xi_2)$. Find T^*. Show that we have $T^*T=TT^*=2I$. Find T_1 and T_2 as defined in Prob. 4.

6. If $T: H \longrightarrow H$ is a bounded self-adjoint linear operator and $T\neq 0$, then $T^n \neq 0$. Prove this (*a*) for $n=2, 4, 8, 16, \cdots$, (*b*) for every $n\in\mathbf{N}$.

7. Show that the column vectors of a unitary matrix constitute an orthonormal set with respect to the inner product on $\mathbf{C}^n$.

8. Show that an isometric linear operator $T: H \longrightarrow H$ satisfies $T^*T=I$, where I is the identity operator on H.

9. Show that an isometric linear operator $T: H \longrightarrow H$ which is not unitary maps the Hilbert space H onto a proper closed subspace of H.

10. Let X be an inner product space and $T: X \longrightarrow X$ an isometric linear operator. If $\dim X<\infty$, show that T is unitary.

11. **(Unitary equivalence)** Let S and T be linear operators on a Hilbert space H. The operator S is said to be *unitarily equivalent* to T if there

is a unitary operator U on H such that

$$S = UTU^{-1} = UTU^*.$$

If T is self-adjoint, show that S is self-adjoint.

12. Show that T is normal if and only if T_1 and T_2 in Prob. 4 commute. Illustrate part of the situation by two-rowed normal matrices.

13. If $T_n: H \longrightarrow H \ (n = 1, 2, \cdots)$ are normal linear operators and $T_n \longrightarrow T$, show that T is a normal linear operator.

14. If S and T are normal linear operators satisfying $ST^* = T^*S$ and $TS^* = S^*T$, show that their sum $S + T$ and product ST are normal.

15. Show that a bounded linear operator $T: H \longrightarrow H$ on a complex Hilbert space H is normal if and only if $\|T^*x\| = \|Tx\|$ for all $x \in H$. Using this, show that for a normal linear operator,

$$\|T^2\| = \|T\|^2.$$

CHAPTER 4

FUNDAMENTAL THEOREMS FOR NORMED AND BANACH SPACES

This chapter contains, roughly speaking, the basis of the more advanced theory of normed and Banach spaces without which the usefulness of these spaces and their applications would be rather limited. The four important theorems in the chapter are the Hahn-Banach theorem, the uniform boundedness theorem, the open mapping theorem, and the closed graph theorem. These are the cornerstones of the theory of Banach spaces. (The first theorem holds for any normed space.)

Brief orientation about main content

1. *Hahn-Banach theorem* 4.2-1 (variants 4.3-1, 4.3-2). This is an extension theorem for linear functionals on vector spaces. It guarantees that a normed space is richly supplied with linear functionals, so that one obtains an adequate theory of dual spaces as well as a satisfactory theory of adjoint operators (Secs. 4.5, 4.6).

2. *Uniform boundedness theorem* 4.7-3 by Banach and Steinhaus. This theorem gives conditions sufficient for $(\|T_n\|)$ to be bounded, where the T_n's are bounded linear operators from a Banach into a normed space. It has various (simple and deeper) applications in analysis, for instance in connection with Fourier series (cf. 4.7-5), weak convergence (Secs. 4.8, 4.9), summability of sequences (Sec. 4.10), numerical integration (Sec. 4.11), etc.

3. *Open mapping theorem* 4.12-2. This theorem states that a bounded linear operator T from a Banach space onto a Banach space is an open mapping, that is, maps open sets onto open sets. Hence if T is bijective, T^{-1} is continuous ("*bounded inverse theorem*").

4. *Closed graph theorem* 4.13-2. This theorem gives conditions under which a closed linear operator (cf. 4.13-1) is bounded. Closed linear operators are of importance in physical and other applications.

4.1 Zorn's Lemma

We shall need Zorn's lemma in the proof of the fundamental Hahn-Banach theorem, which is an extension theorem for linear functionals and is important for reasons which we shall state when we formulate the theorem. Zorn's lemma has various applications. Two of them will be shown later in this section. The setting for the lemma is a partially ordered set:

4.1-1 Definition (Partially ordered set, chain). A *partially ordered set* is a set M on which there is defined a *partial ordering*, that is, a binary relation which is written $\leqq$ and satisfies the conditions

(PO1) $a \leqq a$ for every $a \in M$. (*Reflexivity*)

(PO2) If $a \leqq b$ and $b \leqq a$, then $a = b$. (*Antisymmetry*)

(PO3) If $a \leqq b$ and $b \leqq c$, then $a \leqq c$. (*Transitivity*)

"Partially"emphasizes that M may contain elements a and b for which neither $a \leqq b$ nor $b \leqq a$ holds. Then a and b are called *incomparable elements.* In contrast, two elements a and b are called *comparable elements* if they satisfy $a \leqq b$ or $b \leqq a$ (or both).

A *totally ordered set* or *chain* is a partially ordered set such that every two elements of the set are comparable. In other words, a chain is a partially ordered set that has no incomparable elements.

An *upper bound* of a subset W of a partially ordered set M is an element $u \in M$ such that

$$x \leqq u \qquad \text{for every } x \in W.$$

(Depending on M and W, such a u may or may not exist.) A *maximal element* of M is an $m \in M$ such that

$$m \leqq x \qquad \text{implies} \qquad m = x.$$

(Again, M may or may not have maximal elements. Note further that a maximal element need not be an upper bound.) ∎

Examples

4.1-2 Real numbers. Let M be the set of all real numbers and let $x \leqq y$ have its usual meaning. M is totally ordered. M has no maximal elements.

4.1-3 Power set. Let $\mathscr{P}(X)$ be the *power set* (set of all subsets) of a given set X and let $A \leqq B$ mean $A \subset B$, that is, A is a subset of B. Then $\mathscr{P}(X)$ is partially ordered. The only maximal element of $\mathscr{P}(X)$ is X.

4.1-4 *n*-tuples of numbers. Let M be the set of all ordered n-tuples $x = (\xi_1, \cdots, \xi_n)$, $y = (\eta_1, \cdots, \eta_n)$, $\cdots$ of real numbers and let $x \leqq y$ mean $\xi_j \leqq \eta_j$ for every $j = 1, \cdots, n$, where $\xi_j \leqq \eta_j$ has its usual meaning. This defines a partial ordering on M.

4.1-5 Positive integers. Let $M = \mathbf{N}$, the set of all positive integers. Let $m \leqq n$ mean that m divides n. This defines a partial ordering on $\mathbf{N}$.

Some further examples are given in the problem set. See also G. Birkhoff (1967).

Using the concepts defined in 4.1-1, we can now formulate Zorn's lemma, which we regard as an axiom.[1]

4.1-6 Zorn's lemma. *Let $M \neq \varnothing$ be a partially ordered set. Suppose that every chain $C \subset M$ has an upper bound. Then M has at least one maximal element.*

Applications

4.1-7 Hamel basis. *Every vector space $X \neq \{0\}$ has a Hamel basis.* (Cf. Sec. 2.1.)

Proof. Let M be the set of all linearly independent subsets of X. Since $X \neq \{0\}$, it has an element $x \neq 0$ and $\{x\} \in M$, so that $M \neq \varnothing$. Set inclusion defines a partial ordering on M; cf. 4.1-3. Every chain $C \subset M$ has an upper bound, namely, the union of all subsets of X which are elements of C. By Zorn's lemma, M has a maximal element B. We show that B is a Hamel basis for X. Let $Y = \operatorname{span} B$. Then Y is a subspace of X, and $Y = X$ since otherwise $B \cup \{z\}$, $z \in X$, $z \notin Y$, would be a linearly independent set containing B as a proper subset, contrary to the maximality of B. ∎

[1] The name "lemma" is for historical reasons. Zorn's lemma can be derived from the **axiom of choice,** which states that for any given set E, there exists a mapping c ("*choice function*") from the power set $\mathscr{P}(E)$ into E such that if $B \subset E$, $B \neq \varnothing$, then $c(B) \in B$. Conversely, this axiom follows from Zorn's lemma, so that Zorn's lemma and the axiom of choice can be regarded as equivalent axioms.

4.1-8 Total orthonormal set. *In every Hilbert space $H \neq \{0\}$ there exists a total orthonormal set.* (Cf. Sec. 3.6.)

Proof. Let M be the set of all orthonormal subsets of H. Since $H \neq \{0\}$, it has an element $x \neq 0$, and an orthonormal subset of H is $\{y\}$, where $y = \|x\|^{-1}x$. Hence $M \neq \varnothing$. Set inclusion defines a partial ordering on M. Every chain $C \subset M$ has an upper bound, namely, the union of all subsets of X which are elements of C. By Zorn's lemma, M has a maximal element F. We prove that F is total in H. Suppose that this is false. Then by Theorem 3.6-2 there exists a nonzero $z \in H$ such that $z \perp F$. Hence $F_1 = F \cup \{e\}$, where $e = \|z\|^{-1}z$, is orthonormal, and F is a proper subset of F_1. This contradicts the maximality of F. ∎

Problems

1. Verify the statements in Example 4.1-3.

2. Let X be the set of all real-valued functions x on the interval $[0, 1]$, and let $x \leqq y$ mean that $x(t) \leqq y(t)$ for all $t \in [0, 1]$. Show that this defines a partial ordering. Is it a total ordering? Does X have maximal elements?

3. Show that the set of all complex numbers $z = x + iy$, $w = u + iv$, $\cdots$ can be partially ordered by defining $z \leqq w$ to mean $x \leqq u$ and $y \leqq v$, where for real numbers, $\leqq$ has its usual meaning.

4. Find all maximal elements of M with respect to the partial ordering in Example 4.1-5, where M is (*a*) $\{2, 3, 4, 8\}$, (*b*) the set of all prime numbers.

5. Prove that a finite partially ordered set A has at least one maximal element.

6. **(Least element, greatest element)** Show that a partially ordered set M can have at most one element a such that $a \leqq x$ for all $x \in M$ and at most one element b such that $x \leqq b$ for all $x \in M$. [If such an a (or b) exists, it is called the *least element* (*greatest element*, respectively) of M.]

7. **(Lower bound)** A *lower bound* of a subset $A \neq \varnothing$ of a partially ordered set M is an $x \in M$ such that $x \leqq y$ for all $y \in A$. Find upper and lower bounds of the subset $A = \{4, 6\}$ in Example 4.1-5.

8. A *greatest lower bound* of a subset $A \neq \emptyset$ of a partially ordered set M is a lower bound x of A such that $l \leqq x$ for any lower bound l of A; we write $x = \text{g.l.b.} A = \inf A$. Similarly, a *least upper bound* y of A, written $y = \text{l.u.b.} A = \sup A$, is an upper bound y of A such that $y \leqq u$ for any upper bound u of A. (a) If A has a g.l.b., show that it is unique. (b) What are g.l.b. $\{A, B\}$ and l.u.b. $\{A, B\}$ in Example 4.1-3?

9. (Lattice) A *lattice* is a partially ordered set M such that any two elements x, y of M have a g.l.b. (written $x \wedge y$) and a l.u.b. (written $x \vee y$). Show that the partially ordered set in Example 4.1-3 is a lattice, where $A \wedge B = A \cap B$ and $A \vee B = A \cup B$.

10. A *minimal element* of a partially ordered set M is an $x \in M$ such that $y \leqq x$ implies $y = x$. Find all minimal elements in Prob. 4(a).

4.2 Hahn-Banach Theorem

The Hahn-Banach theorem is an extension theorem for linear functionals. We shall see in the next section that the theorem guarantees that a normed space is richly supplied with bounded linear functionals and makes possible an adequate theory of dual spaces, which is an essential part of the general theory of normed spaces. In this way the Hahn-Banach theorem becomes one of the most important theorems in connection with bounded linear operators. Furthermore, our discussion will show that the theorem also characterizes the extent to which values of a linear functional can be preassigned. The theorem was discovered by H. Hahn (1927), rediscovered in its present more general form (Theorem 4.2-1) by S. Banach (1929) and generalized to complex vector spaces (Theorem 4.3-1) by H. F. Bohnenblust and A. Sobczyk (1938); cf. the references in Appendix 3.

Generally speaking, in an *extension problem* one considers a mathematical object (for example, a mapping) defined on a subset Z of a given set X and one wants to extend the object from Z to the entire set X in such a way that certain basic properties of the object continue to hold for the extended object.

In the Hahn-Banach theorem, the object to be extended is a linear functional f which is defined on a subspace Z of a vector space X and has a certain boundedness property which will be formulated in terms of a **sublinear functional.** By definition, this is a real-valued functional

p on a vector space X which is **subadditive,** that is,

$$\textbf{(1)} \qquad p(x+y) \leqq p(x)+p(y) \qquad \text{for all } x, y \in X,$$

and **positive-homogeneous,** that is,

$$\textbf{(2)} \qquad p(\alpha x) = \alpha p(x) \qquad \text{for all } \alpha \geqq 0 \text{ in } \mathbf{R} \text{ and } x \in X.$$

(Note that the *norm* on a normed space is such a functional.)

We shall assume that the functional f to be extended is majorized on Z by such a functional p defined on X, and we shall extend f from Z to X without losing the linearity and the majorization, so that the extended functional $\tilde{f}$ on X is still linear and still majorized by p. This is the crux of the theorem. X will be real; a generalization of the theorem that includes complex vector spaces follows in the next section.

4.2-1 Hahn-Banach Theorem (Extension of linear functionals). *Let X be a real vector space and p a sublinear functional on X. Furthermore, let f be a linear functional which is defined on a subspace Z of X and satisfies*

$$(3) \qquad f(x) \leqq p(x) \qquad \text{for all } x \in Z.$$

Then f has a linear extension $\tilde{f}$ from Z to X satisfying

$$(3^*) \qquad \tilde{f}(x) \leqq p(x) \qquad \text{for all } x \in X,$$

that is, $\tilde{f}$ is a linear functional on X, satisfies (3) on X and $\tilde{f}(x) = f(x)$ for every $x \in Z$.*

Proof. Proceeding stepwise, we shall prove:

(*a*) The set E of all linear extensions g of f satisfying $g(x) \leqq p(x)$ on their domain $\mathfrak{D}(g)$ can be partially ordered and Zorn's lemma yields a maximal element $\tilde{f}$ of E.

(*b*) $\tilde{f}$ is defined on the entire space X.

(*c*) An auxiliary relation which was used in (*b*).

We start with part

(a) Let E be the set of all linear extensions g of f which satisfy the condition

$$g(x) \leqq p(x) \qquad \text{for all } x \in \mathfrak{D}(g)$$

Clearly, $E \neq \varnothing$ since $f \in E$. On E we can define a partial ordering by

$$g \leqq h \qquad \text{meaning} \qquad h \text{ is an extension of } g,$$

that is, by definition, $\mathfrak{D}(h) \supset \mathfrak{D}(g)$ and $h(x) = g(x)$ for every $x \in \mathfrak{D}(g)$.

For any chain $C \subset E$ we now define $\hat{g}$ by

$$\hat{g}(x) = g(x) \qquad \text{if } x \in \mathfrak{D}(g) \qquad (g \in C).$$

$\hat{g}$ is a linear functional, the domain being

$$\mathfrak{D}(\hat{g}) = \bigcup_{g \in C} \mathfrak{D}(g),$$

which is a vector space since C is a chain. The definition of $\hat{g}$ is unambiguous. Indeed, for an $x \in \mathfrak{D}(g_1) \cap \mathfrak{D}(g_2)$ with $g_1, g_2 \in C$ we have $g_1(x) = g_2(x)$ since C is a chain, so that $g_1 \leqq g_2$ or $g_2 \leqq g_1$. Clearly, $g \leqq \hat{g}$ for all $g \in C$. Hence $\hat{g}$ is an upper bound of C. Since $C \subset E$ was arbitrary, Zorn's lemma thus implies that E has a maximal element $\tilde{f}$. By the definition of E, this is a linear extension of f which satisfies

$$\tilde{f}(x) \leqq p(x) \qquad x \in \mathfrak{D}(\tilde{f}). \tag{4}$$

(b) We now show that $\mathfrak{D}(\tilde{f})$ is all of X. Suppose that this is false. Then we can choose a $y_1 \in X - \mathfrak{D}(\tilde{f})$ and consider the subspace Y_1 of X spanned by $\mathfrak{D}(\tilde{f})$ and y_1. Note that $y_1 \neq 0$ since $0 \in \mathfrak{D}(\tilde{f})$. Any $x \in Y_1$ can be written

$$x = y + \alpha y_1 \qquad y \in \mathfrak{D}(\tilde{f}).$$

This representation is unique. In fact, $y + \alpha y_1 = \tilde{y} + \beta y_1$ with $\tilde{y} \in \mathfrak{D}(\tilde{f})$ implies $y - \tilde{y} = (\beta - \alpha) y_1$, where $y - \tilde{y} \in \mathfrak{D}(\tilde{f})$ whereas $y_1 \notin \mathfrak{D}(\tilde{f})$, so that the only solution is $y - \tilde{y} = 0$ and $\beta - \alpha = 0$. This means uniqueness.

A functional g_1 on Y_1 is defined by

$$g_1(y + \alpha y_1) = \tilde{f}(y) + \alpha c \tag{5}$$

where c is any real constant. It is not difficult to see that g_1 is linear. Furthermore, for $\alpha = 0$ we have $g_1(y) = \tilde{f}(y)$. Hence g_1 is a *proper extension* of $\tilde{f}$, that is, an extension such that $\mathfrak{D}(\tilde{f})$ is a proper subset of

$\mathfrak{D}(g_1)$. Consequently, if we can prove that $g_1 \in E$ by showing that

$$(6) \qquad g_1(x) \leqq p(x) \qquad \text{for all } x \in \mathfrak{D}(g_1),$$

this will contradict the maximality of $\tilde{f}$, so that $\mathfrak{D}(\tilde{f}) \neq X$ is false and $\mathfrak{D}(\tilde{f}) = X$ is true.

(c) Accordingly, we must finally show that g_1 with a suitable c in (5) satisfies (6).

We consider any y and z in $\mathfrak{D}(\tilde{f})$. From (4) and (1) we obtain

$$\begin{aligned}\tilde{f}(y) - \tilde{f}(z) = \tilde{f}(y-z) &\leqq p(y-z) \\ &= p(y + y_1 - y_1 - z) \\ &\leqq p(y + y_1) + p(-y_1 - z).\end{aligned}$$

Taking the last term to the left and the term $\tilde{f}(y)$ to the right, we have

$$(7) \qquad -p(-y_1 - z) - \tilde{f}(z) \leqq p(y + y_1) - \tilde{f}(y),$$

where y_1 is fixed. Since y does not appear on the left and z not on the right, the inequality continues to hold if we take the supremum over $z \in \mathfrak{D}(\tilde{f})$ on the left (call it m_0) and the infimum over $y \in \mathfrak{D}(\tilde{f})$ on the right, call it m_1. Then $m_0 \leqq m_1$ and for a c with $m_0 \leqq c \leqq m_1$ we have from (7)

$$(8a) \qquad -p(-y_1 - z) - \tilde{f}(z) \leqq c \qquad \text{for all } z \in \mathfrak{D}(\tilde{f})$$

$$(8b) \qquad c \leqq p(y + y_1) - \tilde{f}(y) \qquad \text{for all } y \in \mathfrak{D}(\tilde{f}).$$

We prove (6) first for negative α in (5) and then for positive α. For $\alpha < 0$ we use (8a) with z replaced by $\alpha^{-1} y$, that is,

$$-p\left(-y_1 - \frac{1}{\alpha} y\right) - \tilde{f}\left(\frac{1}{\alpha} y\right) \leqq c.$$

Multiplication by $-\alpha > 0$ gives

$$\alpha p\left(-y_1 - \frac{1}{\alpha} y\right) + \tilde{f}(y) \leqq -\alpha c.$$

From this and (5), using $y+\alpha y_1=x$ (see above), we obtain the desired inequality

$$g_1(x)=\tilde{f}(y)+\alpha c \leqq -\alpha p\left(-y_1-\frac{1}{\alpha}y\right)=p(\alpha y_1+y)=p(x).$$

For $\alpha=0$ we have $x\in\mathcal{D}(\tilde{f})$ and nothing to prove. For $\alpha>0$ we use (8b) with y replaced by $\alpha^{-1}y$ to get

$$c\leqq p\left(\frac{1}{\alpha}y+y_1\right)-\tilde{f}\left(\frac{1}{\alpha}y\right).$$

Multiplication by $\alpha>0$ gives

$$\alpha c\leqq \alpha p\left(\frac{1}{\alpha}y+y_1\right)-\tilde{f}(y)=p(x)-\tilde{f}(y).$$

From this and (5),

$$g_1(x)=\tilde{f}(y)+\alpha c\leqq p(x).$$ ∎

Could we get away without Zorn's lemma? This question is of interest, in particular since the lemma does not give a method of construction. If in (5) we take f instead of $\tilde{f}$, we obtain for each real c a linear extension g_1 of f to the subspace Z_1 spanned by $\mathcal{D}(f)\cup\{y_1\}$, and we can choose c so that $g_1(x)\leqq p(x)$ for all $x\in Z_1$, as may be seen from part (*c*) of the proof with $\tilde{f}$ replaced by f. If $X=Z_1$, we are done. If $X\neq Z_1$, we may take a $y_2\in X-Z_1$ and repeat the process to extend f to Z_2 spanned by Z_1 and y_2, etc. This gives a sequence of subspaces Z_j each containing the preceding, and such that f can be extended linearly from one to the next and the extension g_j satisfies $g_j(x)\leqq p(x)$ for all $x\in Z_j$. If

$$X=\bigcup_{j=1}^{n} Z_j,$$

we are done after n steps, and if

$$X=\bigcup_{j=1}^{\infty} Z_j,$$

we can use ordinary induction. However, if X has no such representation, we do need Zorn's lemma in the proof presented here.

Of course, for special spaces the whole situation may become simpler. Hilbert spaces are of this type, because of the Riesz representation 3.8-1. We shall discuss this fact in the next section.

Problems

1. Show that the absolute value of a linear functional has the properties expressed in (1) and (2).

2. Show that a norm on a vector space X is a sublinear functional on X.

3. Show that $p(x) = \overline{\lim_{n\to\infty}}\, \xi_n$, where $x = (\xi_n) \in l^\infty$, ξ_n real, defines a sublinear functional on l^∞.

4. Show that a sublinear functional p satisfies $p(0) = 0$ and $p(-x) \geqq -p(x)$.

5. **(Convex set)** If p is a sublinear functional on a vector space X, show that $M = \{x \mid p(x) \leqq \gamma,\ \gamma > 0 \text{ fixed}\}$, is a convex set. (Cf. Sec. 3.3.)

6. If a subadditive functional p on a normed space X is continuous at 0 and $p(0) = 0$, show that p is continuous for all $x \in X$.

7. If p_1 and p_2 are sublinear functionals on a vector space X and c_1 and c_2 are positive constants, show that $p = c_1 p_1 + c_2 p_2$ is sublinear on X.

8. If a subadditive functional defined on a normed space X is nonnegative outside a sphere $\{x \mid \|x\| = r\}$, show that it is nonnegative for all $x \in X$.

9. Let p be a sublinear functional on a real vector space X. Let f be defined on $Z = \{x \in X \mid x = \alpha x_0,\ \alpha \in \mathbf{R}\}$ by $f(x) = \alpha p(x_0)$ with fixed $x_0 \in X$. Show that f is a linear functional on Z satisfying $f(x) \leqq p(x)$.

10. If p is a sublinear functional on a real vector space X, show that there exists a linear functional $\tilde{f}$ on X such that $-p(-x) \leqq \tilde{f}(x) \leqq p(x)$.

4.3 Hahn-Banach Theorem for Complex Vector Spaces and Normed Spaces

The Hahn-Banach theorem 4.2-1 concerns *real* vector spaces. A generalization that includes complex vector spaces was obtained by H. F. Bohnenblust and A. Sobczyk (1938):

4.3-1 Hahn-Banach Theorem (Generalized). *Let X be a real or complex vector space and p a real-valued functional on X which is subadditive, that is, for all $x, y \in X$,*

$$p(x+y) \leqq p(x)+p(y) \tag{1}$$

(as in Theorem 4.2-1), and for every scalar α satisfies

$$p(\alpha x) = |\alpha|\, p(x). \tag{2}$$

Furthermore, let f be a linear functional which is defined on a subspace Z of X and satisfies

$$|f(x)| \leqq p(x) \qquad \text{for all } x \in Z. \tag{3}$$

Then f has a linear extension $\tilde{f}$ from Z to X satisfying

$$|\tilde{f}(x)| \leqq p(x) \qquad \text{for all } x \in X. \tag{3*}$$

Proof. **(a)** *Real vector space.* If X is real, the situation is simple. Then (3) implies $f(x) \leqq p(x)$ for all $x \in Z$. Hence by the Hahn-Banach theorem 4.2-1 there is a linear extension $\tilde{f}$ from Z to X such that

$$\tilde{f}(x) \leqq p(x) \qquad \text{for all } x \in X. \tag{4}$$

From this and (2) we obtain

$$-\tilde{f}(x) = \tilde{f}(-x) \leqq p(-x) = |-1|\, p(x) = p(x),$$

that is, $\tilde{f}(x) \geqq -p(x)$. Together with (4) this proves (3*).

(b) *Complex vector space.* Let X be complex. Then Z is a complex vector space, too. Hence f is complex-valued, and we can write

$$f(x) = f_1(x) + if_2(x) \qquad x \in Z$$

where f_1 and f_2 are real-valued. For a moment we regard X and Z as real vector spaces and denote them by X_r and Z_r, respectively; this simply means that we restrict multiplication by scalars to real numbers (instead of complex numbers). Since f is linear on Z and f_1 and f_2 are real-valued, f_1 and f_2 are linear functionals on Z_r. Also $f_1(x) \leqq |f(x)|$ because the real part of a complex number cannot exceed the absolute value. Hence by (3),

$$f_1(x) \leqq p(x) \qquad \text{for all } x \in Z_r.$$

By the Hahn-Banach theorem 4.2-1 there is a linear extension $\tilde{f}_1$ of f_1 from Z_r to X_r such that

$$\tilde{f}_1(x) \leqq p(x) \qquad \text{for all } x \in X_r. \tag{5}$$

This takes care of f_1 and we now turn to f_2. Returning to Z and using $f = f_1 + if_2$, we have for every $x \in Z$

$$i[f_1(x) + if_2(x)] = if(x) = f(ix) = f_1(ix) + if_2(ix).$$

The real parts on both sides must be equal:

$$f_2(x) = -f_1(ix) \qquad x \in Z. \tag{6}$$

Hence if for all $x \in X$ we set

$$\tilde{f}(x) = \tilde{f}_1(x) - i\tilde{f}_1(ix) \qquad x \in X, \tag{7}$$

we see from (6) that $\tilde{f}(x) = f(x)$ on Z. This shows that $\tilde{f}$ is an extension of f from Z to X. Our remaining task is to prove that

(i) $\tilde{f}$ is a linear functional on the *complex* vector space X,

(ii) $\tilde{f}$ satisfies (3*) on X.

That (i) holds can be seen from the following calculation which uses (7) and the linearity of $\tilde{f}_1$ on the real vector space X_r; here $a + ib$ with real a and b is any complex scalar:

$$\begin{aligned}
\tilde{f}((a+ib)x) &= \tilde{f}_1(ax+ibx) - i\tilde{f}_1(iax - bx) \\
&= a\tilde{f}_1(x) + b\tilde{f}_1(ix) - i[a\tilde{f}_1(ix) - b\tilde{f}_1(x)] \\
&= (a+ib)[\tilde{f}_1(x) - i\tilde{f}_1(ix)] \\
&= (a+ib)\tilde{f}(x).
\end{aligned}$$

We prove (ii). For any x such that $\tilde{f}(x) = 0$ this holds since $p(x) \geqq 0$ by (1) and (2); cf. also Prob. 1. Let x be such that $\tilde{f}(x) \neq 0$.

Then we can write, using the polar form of complex quantities,

$$\tilde{f}(x) = |\tilde{f}(x)| e^{i\theta}, \qquad \text{thus} \qquad |\tilde{f}(x)| = \tilde{f}(x) e^{-i\theta} = \tilde{f}(e^{-i\theta} x).$$

Since $|\tilde{f}(x)|$ is real, the last expression is real and thus equal to its real part. Hence by (2),

$$|\tilde{f}(x)| = \tilde{f}(e^{-i\theta} x) = \tilde{f}_1(e^{-i\theta} x) \leqq p(e^{-i\theta} x) = |e^{-i\theta}| p(x) = p(x).$$

This completes the proof. ∎

Although the Hahn-Banach theorem says nothing directly about continuity, a principal application of the theorem deals with bounded linear functionals. This brings us back to normed spaces, which is our main concern. In fact, Theorem 4.3-1 implies the basic

4.3-2 Hahn-Banach Theorem (Normed spaces). *Let f be a bounded linear functional on a subspace Z of a normed space X. Then there exists a bounded linear functional $\tilde{f}$ on X which is an extension of f to X and has the same norm,*

$$\|\tilde{f}\|_X = \|f\|_Z \tag{8}$$

where

$$\|\tilde{f}\|_X = \sup_{\substack{x \in X \\ \|x\|=1}} |\tilde{f}(x)|, \qquad \|f\|_Z = \sup_{\substack{x \in Z \\ \|x\|=1}} |f(x)|$$

(and $\|f\|_Z = 0$ in the trivial case $Z = \{0\}$).

Proof. If $Z = \{0\}$, then $f = 0$, and the extension is $\tilde{f} = 0$. Let $Z \neq \{0\}$. We want to use Theorem 4.3-1. Hence we must first discover a suitable p. For all $x \in Z$ we have

$$|f(x)| \leqq \|f\|_Z \|x\|.$$

This is of the form (3), where

$$p(x) = \|f\|_Z \|x\|. \tag{9}$$

We see that p is defined on all of X. Furthermore, p satisfies (1) on X since by the triangle inequality,

$$p(x+y)=\|f\|_Z\,\|x+y\|\leqq\|f\|_Z\,(\|x\|+\|y\|)=p(x)+p(y).$$

p also satisfies (2) on X because

$$p(\alpha x)=\|f\|_Z\,\|\alpha x\|=|\alpha|\,\|f\|_Z\,\|x\|=|\alpha|\,p(x).$$

Hence we can now apply Theorem 4.3-1 and conclude that there exists a linear functional $\tilde{f}$ on X which is an extension of f and satisfies

$$|\tilde{f}(x)|\leqq p(x)=\|f\|_Z\,\|x\| \qquad\qquad x\in X.$$

Taking the supremum over all $x\in X$ of norm 1, we obtain the inequality

$$\|\tilde{f}\|_X=\sup_{\substack{x\in X\\ \|x\|=1}}|\tilde{f}(x)|\leqq\|f\|_Z.$$

Since under an extension the norm cannot decrease, we also have $\|\tilde{f}\|_X\geqq\|f\|_Z$. Together we obtain (8) and the theorem is proved. ∎

In special cases the situation may become very simple. Hilbert spaces are of this type. Indeed, if Z is a closed subspace of a Hilbert space $X=H$, then f has a Riesz representation 3.8-1, say,

$$f(x)=\langle x, z\rangle \qquad\qquad z\in Z$$

where $\|z\|=\|f\|$. Of course, since the inner product is defined on all of H, this gives at once a linear extension $\tilde{f}$ of f from Z to H, and $\tilde{f}$ has the same norm as f because $\|\tilde{f}\|=\|z\|=\|f\|$ by Theorem 3.8-1. Hence in this case the extension is immediate.

From Theorem 4.3-2 we shall now derive another useful result which, roughly speaking, shows that the dual space X' of a normed space X consists of sufficiently many bounded linear functionals to distinguish between the points of X. This will become essential in connection with adjoint operators (Sec. 4.5) and so-called weak convergence (Sec. 4.8).

4.3-3 Theorem (Bounded linear functionals). *Let X be a normed space and let $x_0 \neq 0$ be any element of X. Then there exists a bounded linear functional $\tilde{f}$ on X such that*

$$\|\tilde{f}\| = 1, \qquad \tilde{f}(x_0) = \|x_0\|.$$

Proof. We consider the subspace Z of X consisting of all elements $x = \alpha x_0$ where α is a scalar. On Z we define a linear functional f by

$$f(x) = f(\alpha x_0) = \alpha \|x_0\|. \tag{10}$$

f is bounded and has norm $\|f\| = 1$ because

$$|f(x)| = |f(\alpha x_0)| = |\alpha| \, \|x_0\| = \|\alpha x_0\| = \|x\|.$$

Theorem 4.3-2 implies that f has a linear extension $\tilde{f}$ from Z to X, of norm $\|\tilde{f}\| = \|f\| = 1$. From (10) we see that $\tilde{f}(x_0) = f(x_0) = \|x_0\|$. ∎

4.3-4 Corollary (Norm, zero vector). *For every x in a normed space X we have*

$$\|x\| = \sup_{\substack{f \in X' \\ f \neq 0}} \frac{|f(x)|}{\|f\|}. \tag{11}$$

Hence if x_0 is such that $f(x_0) = 0$ for all $f \in X'$, then $x_0 = 0$.

Proof. From Theorem 4.3-3 we have, writing x for x_0,

$$\sup_{\substack{f \in X' \\ f \neq 0}} \frac{|f(x)|}{\|f\|} \geqq \frac{|\tilde{f}(x)|}{\|\tilde{f}\|} = \frac{\|x\|}{1} = \|x\|,$$

and from $|f(x)| \leqq \|f\| \, \|x\|$ we obtain

$$\sup_{\substack{f \in X' \\ f \neq 0}} \frac{|f(x)|}{\|f\|} \leqq \|x\|. \qquad ∎$$

Problems

1. **(Seminorm)** Show that (1) and (2) imply $p(0)=0$ and $p(x) \geqq 0$, so that p is a seminorm (cf. Prob. 12, Sec. 2.3).

2. Show that (1) and (2) imply $|p(x)-p(y)| \leqq p(x-y)$.

3. It was shown that $\tilde{f}$ defined by (7) is a linear functional on the *complex* vector space X. Show that for this purpose it suffices to prove that $\tilde{f}(ix)=i\tilde{f}(x)$.

4. Let p be defined on a vector space X and satisfy (1) and (2). Show that for any given $x_0 \in X$ there is a linear functional $\tilde{f}$ on X such that $\tilde{f}(x_0)=p(x_0)$ and $|\tilde{f}(x)| \leqq p(x)$ for all $x \in X$.

5. If X in Theorem 4.3-1 is a normed space and $p(x) \leqq k \|x\|$ for some $k>0$, show that $\|\tilde{f}\| \leqq k$.

6. To illustrate Theorem 4.3-2, consider a functional f on the Euclidean plane $\mathbf{R}^2$ defined by $f(x)=\alpha_1\xi_1+\alpha_2\xi_2$, $x=(\xi_1, \xi_2)$, its linear extensions $\tilde{f}$ to $\mathbf{R}^3$ and the corresponding norms.

7. Give another proof of Theorem 4.3-3 in the case of a Hilbert space.

8. Let X be a normed space and X' its dual space. If $X \neq \{0\}$, show that X' cannot be $\{0\}$.

9. Show that for a separable normed space X, Theorem 4.3-2 can be proved directly, without the use of Zorn's lemma (which was used indirectly, namely, in the proof of Theorem 4.2-1).

10. Obtain the second statement in 4.3-4 directly from 4.3-3.

11. If $f(x)=f(y)$ for every bounded linear functional f on a normed space X, show that $x=y$.

12. To illustrate Theorem 4.3-3, let X be the Euclidean plane $\mathbf{R}^2$ and find the functional $\tilde{f}$.

13. Show that under the assumptions of Theorem 4.3-3 there is a bounded linear functional $\hat{f}$ on X such that $\|\hat{f}\|=\|x_0\|^{-1}$ and $\hat{f}(x_0)=1$.

14. **(Hyperplane)** Show that for any sphere $S(0; r)$ in a normed space X and any point $x_0 \in S(0; r)$ there is a hyperplane $H_0 \ni x_0$ such that the ball $\tilde{B}(0; r)$ lies entirely in one of the two half spaces determined by

H_0. (Cf. Probs. 12, 15, Sec. 2.8.) A simple illustration is shown in Fig. 39.

15. If x_0 in a normed space X is such that $|f(x_0)| \leqq c$ for all $f \in X'$ of norm 1, show that $\|x_0\| \leqq c$.

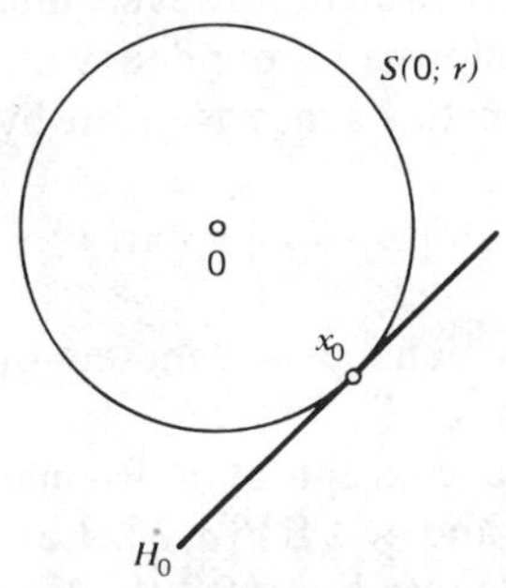

Fig. 39. Illustration of Prob. 14 in the case of the Euclidean plane $\mathbf{R}^2$

4.4 Application to Bounded Linear Functionals on $C[a, b]$

The Hahn-Banach theorem 4.3-2 has many important applications. One of them was considered in the preceding section. Another one will be presented in this section.[2] In fact, we shall use Theorem 4.3-2 for obtaining a general representation formula for bounded linear functionals on $C[a, b]$, where $[a, b]$ is a fixed compact interval. The significance of such general representations of functionals on special spaces was explained at the end of Sec. 2.10. In the present case the representation will be in terms of a Riemann-Stieltjes integral. So let us recall the definition and a few properties of this integral, which is a generalization of the familiar Riemann integral. We begin with the following concept.

A function w defined on $[a, b]$ is said to be of **bounded variation** on $[a, b]$ if its *total variation* $\operatorname{Var}(w)$ on $[a, b]$ is finite, where

$$\operatorname{Var}(w) = \sup \sum_{j=1}^{n} |w(t_j) - w(t_{j-1})|, \tag{1}$$

[2] This section is optional. It will be needed only once (namely, in Sec. 9.9).

the supremum being taken over all *partitions*

$$a = t_0 < t_1 < \cdots < t_n = b \tag{2}$$

of the interval $[a, b]$; here, $n \in \mathbf{N}$ is arbitrary and so is the choice of values $t_1, \ldots, t_{n-1}$ in $[a, b]$ which, however, must satisfy (2).

Obviously, all functions of bounded variation on $[a, b]$ form a vector space. A norm on this space is given by

$$\|w\| = |w(a)| + \mathrm{Var}(w). \tag{3}$$

The normed space thus defined is denoted by $BV[a, b]$, where BV suggests "bounded variation."

We now obtain the concept of a Riemann-Stieltjes integral as follows. Let $x \in C[a, b]$ and $w \in BV[a, b]$. Let P_n be any partition of $[a, b]$ given by (2) and denote by $\eta(P_n)$ the length of a largest interval $[t_{j-1}, t_j]$, that is,

$$\eta(P_n) = \max(t_1 - t_0, \cdots, t_n - t_{n-1}).$$

For every partition P_n of $[a, b]$ we consider the sum

$$s(P_n) = \sum_{j=1}^{n} x(t_j)[w(t_j) - w(t_{j-1})]. \tag{4}$$

There exists a number $\mathcal{J}$ with the property that for every $\varepsilon > 0$ there is a $\delta > 0$ such that

$$\eta(P_n) < \delta \tag{5}$$

implies

$$|\mathcal{J} - s(P_n)| < \varepsilon. \tag{6}$$

$\mathcal{J}$ is called the **Riemann-Stieltjes integral** of x over $[a, b]$ with respect to w and is denoted by

$$\int_a^b x(t)\, dw(t). \tag{7}$$

Hence we can obtain (7) as the limit of the sums (4) for a sequence (P_n) of partitions of $[a, b]$ satisfying $\eta(P_n) \longrightarrow 0$ as $n \longrightarrow \infty$; cf. (5).

Note that for $w(t)=t$, the integral (7) is the familiar Riemann integral of x over $[a, b]$.

Also, if x is continuous on $[a, b]$ and w has a derivative which is integrable on $[a, b]$, then

$$\int_a^b x(t)\,dw(t) = \int_a^b x(t)w'(t)\,dt \tag{8}$$

where the prime denotes differentiation with respect to t.

The integral (7) depends linearly on $x \in C[a, b]$, that is, for all $x_1, x_2 \in C[a, b]$ and scalars α and β we have

$$\int_a^b [\alpha x_1(t) + \beta x_2(t)]\,dw(t) = \alpha \int_a^b x_1(t)\,dw(t) + \beta \int_a^b x_2(t)\,dw(t).$$

The integral also depends linearly on $w \in BV[a, b]$; that is, for all $w_1, w_2 \in BV[a, b]$ and scalars γ and δ we have

$$\int_a^b x(t)\,d(\gamma w_1 + \delta w_2)(t) = \gamma \int_a^b x(t)\,dw_1(t) + \delta \int_a^b x(t)\,dw_2(t).$$

We shall also need the inequality

$$\left| \int_a^b x(t)\,dw(t) \right| \leqq \max_{t \in J} |x(t)| \operatorname{Var}(w), \tag{9}$$

where $J=[a, b]$. We note that this generalizes a familiar formula from calculus. In fact, if $w(t)=t$, then $\operatorname{Var}(w)=b-a$ and (9) takes the form

$$\left| \int_a^b x(t)\,dt \right| \leqq \max_{t \in J} |x(t)|\,(b-a).$$

The representation theorem for bounded linear functionals on $C[a, b]$ by F. Riesz (1909) can now be stated as follows.

4.4-1 Riesz's Theorem (Functionals on $C[a, b]$). *Every bounded linear functional f on $C[a, b]$ can be represented by a Riemann-Stieltjes integral*

$$f(x) = \int_a^b x(t)\,dw(t) \tag{10}$$

where w is of bounded variation on $[a, b]$ *and has the total variation*

$$\text{(11)} \qquad \operatorname{Var}(w) = \|f\|.$$

Proof. From the Hahn-Banach theorem 4.3-2 for normed spaces we see that f has an extension $\tilde{f}$ from $C[a, b]$ to the normed space $B[a, b]$ consisting of all bounded functions on $[a, b]$ with norm defined by

$$\|x\| = \sup_{t \in J} |x(t)| \qquad J = [a, b].$$

Furthermore, by that theorem, the linear functional $\tilde{f}$ is bounded and has the same norm as f, that is,

$$\|\tilde{f}\| = \|f\|.$$

We define the function w needed in (10). For this purpose we consider the function x_t shown in Fig. 40. This function is defined on $[a, b]$ and, by definition, is 1 on $[a, t]$ and 0 otherwise. Clearly, $x_t \in B[a, b]$. We mention that x_t is called the *characteristic function* of the interval $[a, t]$. Using x_t and the functional $\tilde{f}$, we define w on $[a, b]$ by

$$w(a) = 0 \qquad w(t) = \tilde{f}(x_t), \qquad t \in (a, b].$$

We show that this function w is of bounded variation and $\operatorname{Var}(w) \leq \|f\|$.

For a complex quantity we can use the polar form. In fact, setting $\theta = \arg \zeta$, we may write

$$\zeta = |\zeta|\, e(\zeta) \qquad \text{where} \qquad e(\zeta) = \begin{cases} 1 & \text{if } \zeta = 0 \\ e^{i\theta} & \text{if } \zeta \neq 0. \end{cases}$$

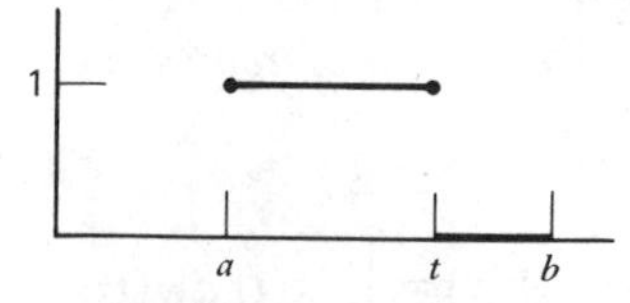

Fig. 40. The function x_t

We see that if $\zeta \neq 0$, then $|\zeta| = \zeta/e^{i\theta} = \zeta e^{-i\theta}$. Hence for any ζ, zero or not, we have

$$|\zeta| = \zeta\, \overline{e(\zeta)}, \tag{12}$$

where the bar indicates complex conjugation, as usual. For simplifying our subsequent formulas we also write

$$\varepsilon_j = \overline{e(w(t_j) - w(t_{j-1}))}$$

and $x_{t_j} = x_j$. In this way we avoid subscripts of subscripts. Then, by (12), for any partition (2) we obtain

$$\begin{aligned}
\sum_{j=1}^{n} |w(t_j) - w(t_{j-1})| &= |\tilde{f}(x_1)| + \sum_{j=2}^{n} |\tilde{f}(x_j) - \tilde{f}(x_{j-1})| \\
&= \varepsilon_1 \tilde{f}(x_1) + \sum_{j=2}^{n} \varepsilon_j [\tilde{f}(x_j) - \tilde{f}(x_{j-1})] \\
&= \tilde{f}\Big(\varepsilon_1 x_1 + \sum_{j=2}^{n} \varepsilon_j [x_j - x_{j-1}]\Big) \\
&\leqq \|\tilde{f}\| \, \Big\| \varepsilon_1 x_1 + \sum_{j=2}^{n} \varepsilon_j [x_j - x_{j-1}] \Big\|.
\end{aligned}$$

On the right, $\|\tilde{f}\| = \|f\|$ (see before) and the other factor $\|\cdots\|$ equals 1 because $|\varepsilon_j| = 1$ and from the definition of the x_j's we see that for each $t \in [a, b]$ only one of the terms $x_1, x_2 - x_1, \cdots$ is not zero (and its norm is 1). On the left we can now take the supremum over all partitions of $[a, b]$. Then we have

$$\operatorname{Var}(w) \leqq \|f\|. \tag{13}$$

Hence w is of bounded variation on $[a, b]$.

We prove (10), where $x \in C[a, b]$. For every partition P_n of the form (2) we define a function, which we denote simply by z_n [instead of $z(P_n)$ or z_{P_n}, say], keeping in mind that z_n depends on P_n, not merely on n. The defining formula is

$$z_n = x(t_0) x_1 + \sum_{j=2}^{n} x(t_{j-1})[x_j - x_{j-1}]. \tag{14}$$

Then $z_n \in B[a, b]$. By the definition of w,

$$\tilde{f}(z_n) = x(t_0)\tilde{f}(x_1) + \sum_{j=2}^{n} x(t_{j-1})[\tilde{f}(x_j) - \tilde{f}(x_{j-1})]$$

$$(15) \qquad = x(t_0)w(t_1) + \sum_{j=2}^{n} x(t_{j-1})[w(t_j) - w(t_{j-1})]$$

$$= \sum_{j=1}^{n} x(t_{j-1})[w(t_j) - w(t_{j-1})],$$

where the last equality follows from $w(t_0) = w(a) = 0$. We now choose any sequence (P_n) of partitions of $[a, b]$ such that $\eta(P_n) \longrightarrow 0$; cf. (5). (Note that the t_j in (15) depend on P_n, a fact which we keep in mind without expressing it by a bulkier notation such as $t_{j,n}$.) As $n \longrightarrow \infty$, the sum on the right-hand side of (15) approaches the integral in (10), and (10) follows, provided $\tilde{f}(z_n) \longrightarrow \tilde{f}(x)$, which equals $f(x)$ since $x \in C[a, b]$.

We prove that $\tilde{f}(z_n) \longrightarrow \tilde{f}(x)$. Remembering the definition of x_t (see Fig. 40), we see that (14) yields $z_n(a) = x(a) \cdot 1$ since the sum in (14) is zero at $t = a$. Hence $z_n(a) - x(a) = 0$. Furthermore, by (14), if $t_{j-1} < t \leqq t_j$, then we obtain $z_n(t) = x(t_{j-1}) \cdot 1$; see Fig. 40. It follows that for those t,

$$|z_n(t) - x(t)| = |x(t_{j-1}) - x(t)|.$$

Consequently, if $\eta(P_n) \longrightarrow 0$, then $\|z_n - x\| \longrightarrow 0$ because x is continuous on $[a, b]$, hence uniformly continuous on $[a, b]$, since $[a, b]$ is compact. The continuity of $\tilde{f}$ now implies that $\tilde{f}(z_n) \longrightarrow \tilde{f}(x)$, and $\tilde{f}(x) = f(x)$, so that (10) is established.

We finally prove (11). From (10) and (9) we have

$$|f(x)| \leqq \max_{t \in J} |x(t)| \operatorname{Var}(w) = \|x\| \operatorname{Var}(w).$$

Taking the supremum over all $x \in C[a, b]$ of norm one, we obtain $\|f\| \leqq \operatorname{Var}(w)$. Together with (13) this yields (11). ∎

We note that w in the theorem is not unique, but can be made unique by imposing the normalizing conditions that w be zero at a and continuous from the right:

$$w(a)=0, \qquad w(t+0)=w(t) \qquad (a<t<b).$$

For details, see A. E. Taylor (1958), pp. 197–200. Cf. also F. Riesz and B. Sz.-Nagy (1955), p. 111.

It is interesting that Riesz's theorem also served later as a starting point of the modern theory of integration. For further historical remarks, see N. Bourbaki (1955), p. 169.

4.5 Adjoint Operator

With a bounded linear operator $T: X \longrightarrow Y$ on a normed space X we can associate the so-called adjoint operator $T^\times$ of T. A motivation for $T^\times$ comes from its usefulness in the solution of equations involving operators, as we shall see in Sec. 8.5; such equations arise, for instance, in physics and other applications. In the present section we define the adjoint operator $T^\times$ and consider some of its properties, including its relation to the Hilbert-adjoint[3] operator T^* defined in Sec. 3.9. It is important to note that our present discussion depends on the Hahn-Banach theorem (via Theorem 4.3-3), and we would not get very far without it.

We consider a bounded linear operator $T: X \longrightarrow Y$, where X and Y are normed spaces, and want to define the adjoint operator $T^\times$ of T. For this purpose we start from any bounded linear functional g on Y. Clearly, g is defined for all $y \in Y$. Setting $y = Tx$, we obtain a functional on X, call it f:

$$\textbf{(1)} \qquad f(x)=g(Tx) \qquad x \in X.$$

f is linear since g and T are linear. f is bounded because

$$|f(x)|=|g(Tx)| \leqq \|g\| \, \|Tx\| \leqq \|g\| \, \|T\| \, \|x\|.$$

[3] In the case of Hilbert spaces the adjoint operator $T^\times$ is *not* identical with the Hilbert-adjoint operator T^* of T (although $T^\times$ and T^* are then related as explained later in this section). The asterisk for the Hilbert-adjoint operator is almost standard. Hence one should *not* denote the adjoint operator by T^*, because it is troublesome to have a notation mean one thing in a Hilbert space and another thing in the theory of general normed spaces. We use $T^\times$ for the adjoint operator. We prefer this over the less perspicuous T' which is also used in the literature.

Taking the supremum over all $x \in X$ of norm one, we obtain the inequality

$$\|f\| \leqq \|g\| \, \|T\|. \tag{2}$$

This shows that $f \in X'$, where X' is the dual space of X defined in 2.10-3. By assumption, $g \in Y'$. Consequently, for *variable* $g \in Y'$, formula (1) defines an operator from Y' into X', which is called the *adjoint operator* of T and is denoted by $T^\times$. Thus we have

$$\begin{array}{c} X \xrightarrow{\ T\ } Y \\ \\ X' \xleftarrow{\ T^\times\ } Y' \end{array} \tag{3}$$

Note carefully that $T^\times$ is an operator defined on Y' whereas the given operator T is defined on X. We summarize:

4.5-1 Definition (Adjoint operator $T^\times$). Let $T: X \longrightarrow Y$ be a bounded linear operator, where X and Y are normed spaces. Then the *adjoint operator* $T^\times: Y' \longrightarrow X'$ of T is defined by

$$f(x) = (T^\times g)(x) = g(Tx) \qquad (g \in Y') \tag{4}$$

where X' and Y' are the dual spaces of X and Y, respectively.

Our first goal is to prove that the adjoint operator has the same norm as the operator itself. This property is basic, as we shall see later. In the proof we shall need Theorem 4.3-3, which resulted from the Hahn-Banach theorem. In this way the Hahn-Banach theorem is vital for establishing a satisfactory theory of adjoint operators, which in turn is an essential part of the general theory of linear operators.

4.5-2 Theorem (Norm of the adjoint operator). *The adjoint operator $T^\times$ in Def.* 4.5-1 *is linear and bounded, and*

$$\|T^\times\| = \|T\|. \tag{5}$$

Proof. The operator $T^\times$ is linear since its domain Y' is a vector space and we readily obtain

$$\begin{aligned}(T^\times(\alpha g_1+\beta g_2))(x)&=(\alpha g_1+\beta g_2)(Tx)\\&=\alpha g_1(Tx)+\beta g_2(Tx)\\&=\alpha(T^\times g_1)(x)+\beta(T^\times g_2)(x).\end{aligned}$$

We prove (5). From (4) we have $f=T^\times g$, and by (2) it follows that

$$\|T^\times g\|=\|f\|\leqq\|g\|\,\|T\|.$$

Taking the supremum over all $g\in Y'$ of norm one, we obtain the inequality

$$\|T^\times\|\leqq\|T\|. \tag{6}$$

Hence to get (5), we must now prove $\|T^\times\|\geqq\|T\|$. Theorem 4.3-3 implies that for every nonzero $x_0\in X$ there is a $g_0\in Y'$ such that

$$\|g_0\|=1 \qquad\text{and}\qquad g_0(Tx_0)=\|Tx_0\|.$$

Here, $g_0(Tx_0)=(T^\times g_0)(x_0)$ by the definition of the adjoint operator $T^\times$. Writing $f_0=T^\times g_0$, we thus obtain

$$\begin{aligned}\|Tx_0\|=g_0(Tx_0)&=f_0(x_0)\\&\leqq\|f_0\|\,\|x_0\|\\&=\|T^\times g_0\|\,\|x_0\|\\&\leqq\|T^\times\|\,\|g_0\|\,\|x_0\|.\end{aligned}$$

Since $\|g_0\|=1$, we thus have for every $x_0\in X$

$$\|Tx_0\|\leqq\|T^\times\|\,\|x_0\|.$$

(This includes $x_0=0$ since $T0=0$.) But always

$$\|Tx_0\|\leqq\|T\|\,\|x_0\|,$$

and here $c = \|T\|$ is the *smallest* constant c such that $\|Tx_0\| \leqq c\|x_0\|$ holds for all $x_0 \in X$. Hence $\|T^\times\|$ cannot be smaller than $\|T\|$, that is, we must have $\|T^\times\| \geqq \|T\|$. This and (6) imply (5). ∎

Let us illustrate the present discussion by matrices representing operators. This will also help the reader in setting up examples of his own.

4.5-3 Example (Matrix). In n-dimensional Euclidean space $\mathbf{R}^n$ a linear operator $T: \mathbf{R}^n \longrightarrow \mathbf{R}^n$ can be represented by matrices, (cf. Sec. 2.9) where such a matrix $T_E = (\tau_{jk})$ depends on the choice of a basis $E = \{e_1, \cdots, e_n\}$ for $\mathbf{R}^n$, whose elements are arranged in some order which is kept fixed. We choose a basis E, regard $x = (\xi_1, \cdots, \xi_n)$, $y = (\eta_1, \cdots, \eta_n)$ as column vectors and employ the usual notation for matrix multiplication. Then

$$(7) \qquad y = T_E x, \qquad \text{in components} \qquad \eta_j = \sum_{k=1}^{n} \tau_{jk}\xi_k,$$

where $j = 1, \cdots, n$. Let $F = \{f_1, \cdots, f_n\}$ be the dual basis of E (cf. Sec. 2.9). This is a basis for $\mathbf{R}^{n\prime}$ (which is also Euclidean n-space, by 2.10-5). Then every $g \in \mathbf{R}^{n\prime}$ has a representation

$$g = \alpha_1 f_1 + \cdots + \alpha_n f_n.$$

Now by the definition of the dual basis we have $f_j(y) = f_j(\sum \eta_k e_k) = \eta_j$. Hence by (7) we obtain

$$g(y) = g(T_E x) = \sum_{j=1}^{n} \alpha_j \eta_j = \sum_{j=1}^{n} \sum_{k=1}^{n} \alpha_j \tau_{jk} \xi_k.$$

Interchanging the order of summation, we can write this in the form

$$(8) \qquad g(T_E x) = \sum_{k=1}^{n} \beta_k \xi_k \qquad \text{where} \qquad \beta_k = \sum_{j=1}^{n} \tau_{jk} \alpha_j.$$

We may regard this as the definition of a functional f on X in terms of g, that is,

$$f(x) = g(T_E x) = \sum_{k=1}^{n} \beta_k \xi_k.$$

Remembering the definition of the adjoint operator, we can write this

$$f = T_E^{\times} g, \qquad \text{in components,} \qquad \beta_k = \sum_{j=1}^{n} \tau_{jk}\alpha_j.$$

Noting that in β_k we sum with respect to the first subscript (so that we sum over all elements of a *column* of T_E), we have the following result.

If T is represented by a matrix T_E, then the adjoint operator $T^{\times}$ is represented by the transpose of T_E.

We mention that this also holds if T is a linear operator from $\mathbf{C}^n$ into $\mathbf{C}^n$. ▮

In working with the adjoint operator, the subsequent formulas (9) to (12) are helpful; the corresponding proofs are left to the reader. Let $S, T \in B(X, Y)$; cf. Sec. 2.10. Then

(9) $$(S+T)^{\times} = S^{\times} + T^{\times}$$

(10) $$(\alpha T)^{\times} = \alpha T^{\times}.$$

Let X, Y, Z be normed spaces and $T \in B(X, Y)$ and $S \in B(Y, Z)$. Then for the adjoint operator of the product ST we have (see Fig. 41)

(11) $$(ST)^{\times} = T^{\times} S^{\times}.$$

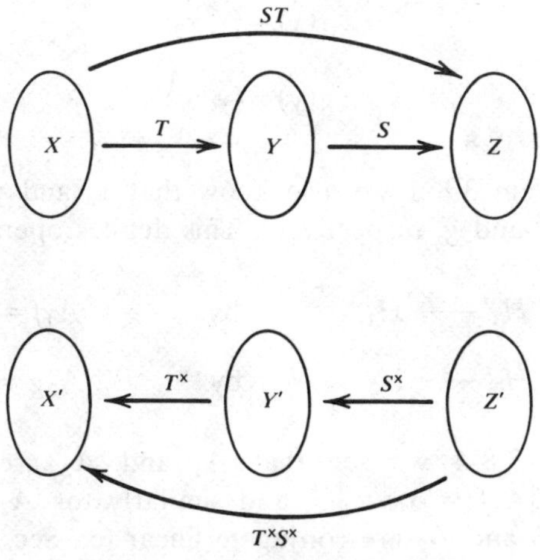

Fig. 41. Illustration of formula (11)

If $T \in B(X, Y)$ and T^{-1} exists and $T^{-1} \in B(Y, X)$, then $(T^{\times})^{-1}$ also exists, $(T^{\times})^{-1} \in B(X', Y')$ and

$$(T^{\times})^{-1} = (T^{-1})^{\times}. \tag{12}$$

Relation between the adjoint operator $T^{\times}$ and the Hilbert-adjoint operator T^*. (Cf. Sec. 3.9.) We show that such a relation exists in the case of a bounded linear operator $T: X \longrightarrow Y$ if X and Y are Hilbert spaces, say $X = H_1$ and $Y = H_2$. In this case we first have (Fig. 42)

$$\begin{array}{c} H_1 \xrightarrow{\quad T \quad} H_2 \\ \\ H_1' \xleftarrow{\quad T^{\times} \quad} H_2' \end{array} \tag{13}$$

where, as before, the adjoint operator $T^{\times}$ of the given operator T is defined by

$$\begin{aligned} &\text{(a)} \qquad T^{\times} g = f \\ &\text{(b)} \qquad g(Tx) = f(x) \end{aligned} \qquad (f \in H_1', \ g \in H_2'). \tag{14}$$

The new feature is that since f and g are functionals on Hilbert spaces, they have Riesz representations (cf. 3.8-1), say,

$$\begin{aligned} &\text{(a)} \qquad f(x) = \langle x, x_0 \rangle \qquad (x_0 \in H_1) \\ &\text{(b)} \qquad g(y) = \langle y, y_0 \rangle \qquad (y_0 \in H_2), \end{aligned} \tag{15}$$

and from Theorem 3.8-1 we also know that x_0 and y_0 are uniquely determined by f and g, respectively. This defines operators

$$A_1: H_1' \longrightarrow H_1 \qquad \text{by} \qquad A_1 f = x_0,$$

$$A_2: H_2' \longrightarrow H_2 \qquad \text{by} \qquad A_2 g = y_0.$$

From Theorem 3.8-1 we see that A_1 and A_2 are bijective and isometric since $\|A_1 f\| = \|x_0\| = \|f\|$, and similarly for A_2. Furthermore, the operators A_1 and A_2 are conjugate linear (cf. Sec. 3.1). In fact, if

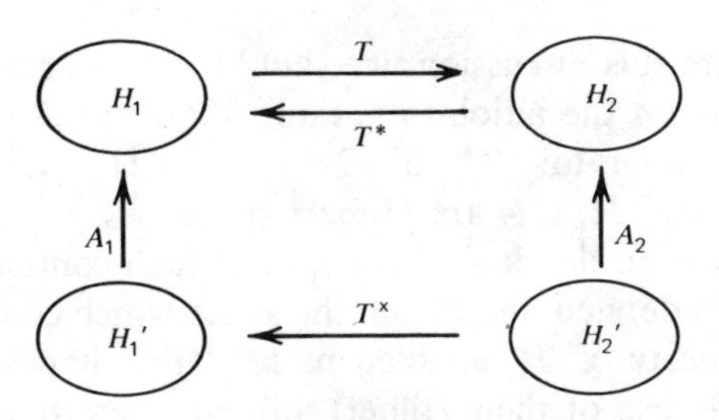

Fig. 42. Operators in formulas (13) and (17)

we write $f_1(x)=\langle x, x_1\rangle$ and $f_2(x)=\langle x, x_2\rangle$, we have for all x and scalars α, β

$$
\begin{aligned}
(\alpha f_1+\beta f_2)(x) &= \alpha f_1(x)+\beta f_2(x) \\
&= \alpha\langle x, x_1\rangle+\beta\langle x, x_2\rangle \qquad (16) \\
&= \langle x, \bar{\alpha}x_1+\bar{\beta}x_2\rangle.
\end{aligned}
$$

By the definition of A_1 this shows conjugate linearity

$$A_1(\alpha f_1+\beta f_2)=\bar{\alpha}A_1f_1+\bar{\beta}A_1f_2.$$

For A_2 the proof is similar.

Composition gives the operator (see Fig. 42)

$$(17)\quad T^*=A_1T^\times A_2^{-1}\colon\ H_2\longrightarrow H_1 \qquad \text{defined by} \qquad T^*y_0=x_0.$$

T^* is linear since it involves *two* conjugate linear mappings, in addition to the linear operator $T^\times$. We prove that T^* is indeed the Hilbert-adjoint operator of T. This is simple since from (14) to (16) we immediately have

$$\langle Tx, y_0\rangle = g(Tx)=f(x)=\langle x, x_0\rangle=\langle x, T^*y_0\rangle,$$

which is (1) in Sec. 3.9, except for the notation. Our result is:

Formula (17) *represents the Hilbert-adjoint operator T^* of a linear operator T on a Hilbert space in terms of the adjoint operator $T^\times$ of T.*

Note further that $\|T^*\|=\|T\|$ (Theorem 3.9-2) now follows immediately from (5) and the isometry of A_1 and A_2. ∎

To complete this discussion, we should also list some of the main differences between the adjoint operator $T^\times$ of $T\colon X \longrightarrow Y$ and the Hilbert-adjoint operator T^* of $T\colon H_1 \longrightarrow H_2$, where X, Y are normed spaces and H_1, H_2 are Hilbert spaces.

$T^\times$ is defined on the dual of the space which contains the range of T, whereas T^* is defined directly on the space which contains the range of T. This property of T^* enabled us to define important classes of operators by the use of their Hilbert-adjoint operators (cf. 3.10-1).

For $T^\times$ we have by (10)

$$(\alpha T)^\times = \alpha T^\times$$

but for T^* we have by 3.9-4

$$(\alpha T)^* = \bar{\alpha} T^*.$$

In the finite dimensional case, $T^\times$ is represented by the transpose of the matrix representing T, whereas T^* is represented by the complex conjugate transpose of that matrix (for details, see 4.5-3 and 3.10-2).

Problems

1. Show that the functional defined by (1) is linear.

2. What are the adjoints of a zero operator 0 and an identity operator I?

3. Prove (9).

4. Prove (10).

5. Prove (11).

6. Show that $(T^n)^\times = (T^\times)^n$.

7. What formula for matrices do we obtain by combining (11) and Example 4.5-3?

8. Prove (12).

9. (Annihilator) Let X and Y be normed spaces, $T\colon X \longrightarrow Y$ a bounded linear operator and $M = \overline{\mathcal{R}(T)}$, the closure of the range of T. Show that (cf. Prob. 13, Sec. 2.10)

$$M^a = \mathcal{N}(T^\times).$$

10. (Annihilator) Let B be a subset of the dual space X' of a normed space X. The *annihilator* ${}^aB$ of B is defined to be

$$ {}^aB = \{x \in X \mid f(x) = 0 \text{ for all } f \in B\}. $$

Show that in Prob. 9,

$$ \mathscr{R}(T) \subset {}^a\mathscr{N}(T^\times). $$

What does this mean with respect to the task of solving an equation $Tx = y$?

4.6 Reflexive Spaces

Algebraic reflexivity of *vector* spaces was discussed in Sec. 2.8. *Reflexivity* of *normed* spaces will be the topic of the present section. But let us first recall what we did in Sec. 2.8. We remember that a vector space X is said to be *algebraically reflexive* if the canonical mapping $C\colon X \longrightarrow X^{**}$ is surjective. Here $X^{**} = (X^*)^*$ is the second algebraic dual space of X and the mapping C is defined by $x \longmapsto g_x$ where

$$ (1) \qquad g_x(f) = f(x) \qquad (f \in X^* \text{ variable}); $$

that is, for any $x \in X$ the image is the linear functional g_x defined by (1). If X is finite dimensional, then X is algebraically reflexive. This was shown in Theorem 2.9-3.

Let us now turn to our actual task. We consider a normed space X, its dual space X' as defined in 2.10-3 and, moreover, the dual space $(X')'$ of X'. This space is denoted by X'' and is called the **second dual space** of X (or *bidual space* of X).

We define a functional g_x on X' by choosing a fixed $x \in X$ and setting

$$ (2) \qquad g_x(f) = f(x) \qquad (f \in X' \text{ variable}). $$

This looks like (1), but note that now f is bounded. And g_x turns out to be bounded, too, since we have the basic

4.6-1 Lemma (Norm of g_x). *For every fixed x in a normed space X, the functional g_x defined by* (2) *is a bounded linear functional on X', so that $g_x \in X''$, and has the norm*

$$\|g_x\| = \|x\|. \tag{3}$$

Proof. Linearity of g_x is known from Sec. 2.8, and (3) follows from (2) and Corollary 4.3-4:

$$\|g_x\| = \sup_{\substack{f \in X' \\ f \neq 0}} \frac{|g_x(f)|}{\|f\|} = \sup_{\substack{f \in X' \\ f \neq 0}} \frac{|f(x)|}{\|f\|} = \|x\|. \tag{4}$$ ∎

To every $x \in X$ there corresponds a unique bounded linear functional $g_x \in X''$ given by (2). This defines a mapping

$$\begin{aligned} C\colon X &\longrightarrow X'' \\ x &\longmapsto g_x. \end{aligned} \tag{5}$$

C is called the **canonical mapping** of X into X''. We show that C is linear and injective and preserves the norm. This can be expressed in terms of an isomorphism of normed spaces as defined in Sec. 2.10:

4.6-2 Lemma (Canonical mapping). *The canonical mapping C given by* (5) *is an isomorphism of the normed space X onto the normed space $\mathcal{R}(C)$, the range of C.*

Proof. Linearity of C is seen as in Sec. 2.8 because

$$g_{\alpha x + \beta y}(f) = f(\alpha x + \beta y) = \alpha f(x) + \beta f(y) = \alpha g_x(f) + \beta g_y(f).$$

In particular, $g_x - g_y = g_{x-y}$. Hence by (3) we obtain

$$\|g_x - g_y\| = \|g_{x-y}\| = \|x - y\|.$$

This shows that C is isometric; it preserves the norm. Isometry implies injectivity. We can also see this directly from our formula. Indeed, if $x \neq y$, then $g_x \neq g_y$ by axiom (N2) in Sec. 2.2. Hence C is bijective, regarded as a mapping onto its range. ∎

X is said to be **embeddable** in a normed space Z if X is isomorphic with a subspace of Z. This is similar to Sec. 2.8, but note that here we are dealing with isomorphisms of normed spaces, that is, vector space isomorphisms which preserve norm (cf. Sec. 2.10). Lemma 4.6-2 shows that X is embeddable in X'', and C is also called the *canonical embedding* of X into X''.

In general, C will not be surjective, so that the range $\mathcal{R}(C)$ will be a *proper* subspace of X''. The surjective case when $\mathcal{R}(C)$ is all of X'' is important enough to give it a name:

4.6-3 Definition (Reflexivity). A normed space X is said to be *reflexive* if

$$\mathcal{R}(C) = X''$$

where $C\colon X \longrightarrow X''$ is the canonical mapping given by (5) and (2). ∎

This concept was introduced by H. Hahn (1927) and called "reflexivity" by E. R. Lorch (1939). Hahn recognized the importance of reflexivity in his study of linear equations in normed spaces which was motivated by integral equations and also contains the Hahn-Banach theorem as well as the earliest investigation of dual spaces.

If X is reflexive, it is isomorphic (hence isometric) with X'', by Lemma 4.6-2. It is interesting that the converse does not generally hold, as R. C. James (1950, 1951) has shown.

Furthermore, completeness does not imply reflexivity, but conversely we have

4.6-4 Theorem (Completeness). *If a normed space X is reflexive, it is complete (hence a Banach space).*

Proof. Since X'' is the dual space of X', it is complete by Theorem 2.10-4. Reflexivity of X means that $\mathcal{R}(C) = X''$. Completeness of X now follows from that of X'' by Lemma 4.6-2. ∎

$\mathbf{R}^n$ is reflexive. This follows directly from 2.10-5. It is typical of any finite dimensional normed space X. Indeed, if $\dim X < \infty$, then every linear functional on X is bounded (cf. 2.7-8), so that $X' = X^*$ and algebraic reflexivity of X (cf. 2.9-3) thus implies

4.6-5 Theorem (Finite dimension). *Every finite dimensional normed space is reflexive.*

l^p with $1<p<+\infty$ is reflexive. This follows from 2.10-7. Similarly, $L^p[a, b]$ with $1<p<+\infty$ is reflexive, as can be shown. It can also be proved that nonreflexive spaces are $C[a, b]$ (cf. 2.2-5), l^1 (proof below), $L^1[a, b]$, l^∞ (cf. 2.2-4) and the subspaces c and c_0 of l^∞, where c is the space of all convergent sequences of scalars and c_0 is the space of all sequences of scalars converging to zero.

4.6-6 Theorem (Hilbert space). *Every Hilbert space H is reflexive.*

Proof. We shall prove surjectivity of the canonical mapping $C\colon H \longrightarrow H''$ by showing that for every $g \in H''$ there is an $x \in H$ such that $g = Cx$. As a preparation we define $A\colon H' \longrightarrow H$ by $Af = z$, where z is given by the Riesz representation $f(x) = \langle x, z\rangle$ in 3.8-1. From 3.8-1 we know that A is bijective and isometric. A is conjugate linear, as we see from (16), Sec. 4.5. Now H' is complete by 2.10-4 and a Hilbert space with inner product defined by

$$\langle f_1, f_2\rangle_1 = \langle Af_2, Af_1\rangle.$$

Note the order of f_1, f_2 on both sides. (IP1) to (IP4) in Sec. 3.1 is readily verified. In particular, (IP2) follows from the conjugate linearity of A:

$$\langle \alpha f_1, f_2\rangle_1 = \langle Af_2, A(\alpha f_1)\rangle = \langle Af_2, \bar{\alpha} Af_1\rangle = \alpha\langle f_1, f_2\rangle_1.$$

Let $g \in H''$ be arbitrary. Let its Riesz representation be

$$g(f) = \langle f, f_0\rangle_1 = \langle Af_0, Af\rangle.$$

We now remember that $f(x) = \langle x, z\rangle$ where $z = Af$. Writing $Af_0 = x$, we thus have

$$\langle Af_0, Af\rangle = \langle x, z\rangle = f(x).$$

Together, $g(f) = f(x)$, that is, $g = Cx$ by the definition of C. Since $g \in H''$ was arbitrary, C is surjective, so that H is reflexive. ∎

Sometimes separability and nonseparability (cf. 1.3-5) can play a role in proofs that certain spaces are not reflexive. This connection between reflexivity and separability is interesting and quite simple. The key is Theorem 4.6-8 (below), which states that separability of X'

implies separability of X (the converse not being generally true). Hence if a normed space X is reflexive, X'' is isomorphic with X by 4.6-2, so that in this case, separability of X implies separability of X'' and, by 4.6-8, the space X' is also separable. From this we have the following result.

A separable normed space X with a nonseparable dual space X' cannot be reflexive.

Example. l^1 is not reflexive.

Proof. l^1 is separable by 1.3-10, but $l^{1\prime} = l^\infty$ is not; cf. 2.10-6 and 1.3-9.

The desired Theorem 4.6-8 will be obtained from the following lemma. A simple illustration of the lemma is shown in Fig. 43.

4.6-7 Lemma (Existence of a functional). *Let Y be a proper closed subspace of a normed space X. Let $x_0 \in X - Y$ be arbitrary and*

$$\delta = \inf_{\tilde{y} \in Y} \|\tilde{y} - x_0\| \tag{6}$$

the distance from x_0 to Y. Then there exists an $\tilde{f} \in X'$ such that

$$\|\tilde{f}\| = 1, \qquad \tilde{f}(y) = 0 \text{ for all } y \in Y, \qquad \tilde{f}(x_0) = \delta. \tag{7}$$

Proof. The idea of the proof is simple. We consider the subspace $Z \subset X$ spanned by Y and x_0, define on Z a bounded linear functional f by

$$f(z) = f(y + \alpha x_0) = \alpha\delta \qquad y \in Y, \tag{8}$$

show that f satisfies (7) and extend f to X by 4.3-2. The details are as follows.

Every $z \in Z = \operatorname{span}(Y \cup \{x_0\})$ has a unique representation

$$z = y + \alpha x_0 \qquad y \in Y.$$

This is used in (8). Linearity of f is readily seen. Also, since Y is closed, $\delta > 0$, so that $f \neq 0$. Now $\alpha = 0$ gives $f(y) = 0$ for all $y \in Y$. For

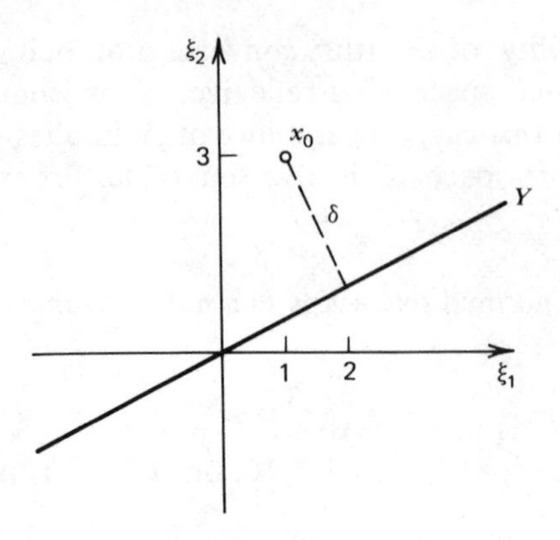

Fig. 43. Illustration of Lemma 4.6-7 for the Euclidean space $X=\mathbf{R}^3$, where Y is represented by $\xi_2=\xi_1/2$, $\xi_3=0$ and $x_0=(1, 3, 0)$, so that $\delta=\sqrt{5}$, $Z=\operatorname{span}(Y\cup\{x_0\})$ is the $\xi_1\xi_2$-plane and $f(z)=(-\xi_1+2\xi_2)/\sqrt{5}$.

$\alpha=1$ and $y=0$ we have $f(x_0)=\delta$.

We show that f is bounded. $\alpha=0$ gives $f(z)=0$. Let $\alpha\neq 0$. Using (6) and noting that $-(1/\alpha)y\in Y$, we obtain

$$\begin{aligned}|f(z)|=|\alpha|\,\delta&=|\alpha|\inf_{\tilde{y}\in Y}\|\tilde{y}-x_0\|\\&\leqq|\alpha|\left\|-\frac{1}{\alpha}y-x_0\right\|\\&=\|y+\alpha x_0\|,\end{aligned}$$

that is, $|f(z)|\leqq\|z\|$. Hence f is bounded and $\|f\|\leqq 1$.

We show that $\|f\|\geqq 1$. By the definition of an infimum, Y contains a sequence (y_n) such that $\|y_n-x_0\|\longrightarrow\delta$. Let $z_n=y_n-x_0$. Then we have $f(z_n)=-\delta$ by (8) with $\alpha=-1$. Also

$$\|f\|=\sup_{\substack{z\in Z\\ z\neq 0}}\frac{|f(z)|}{\|z\|}\geqq\frac{|f(z_n)|}{\|z_n\|}=\frac{\delta}{\|z_n\|}\longrightarrow\frac{\delta}{\delta}=1$$

as $n\longrightarrow\infty$. Hence $\|f\|\geqq 1$, so that $\|f\|=1$. By the Hahn-Banach theorem 4.3-2 for normed spaces we can extend f to X without increasing the norm. ∎

Using this lemma, we shall now obtain the desired

4.6-8 Theorem (Separability). *If the dual space X' of a normed space X is separable, then X itself is separable.*

Proof. We assume that X' is separable. Then the unit sphere $U' = \{f \mid \|f\| = 1\} \subset X'$ also contains a countable dense subset, say, (f_n). Since $f_n \in U'$, we have

$$\|f_n\| = \sup_{\|x\|=1} |f_n(x)| = 1.$$

By the definition of a supremum we can find points $x_n \in X$ of norm 1 such that

$$|f_n(x_n)| \geqq \frac{1}{2}.$$

Let Y be the closure of span (x_n). Then Y is separable because Y has a countable dense subset, namely, the set of all linear combinations of the x_n's with coefficients whose real and imaginary parts are rational.

We show that $Y = X$. Suppose $Y \neq X$. Then, since Y is closed, by Lemma 4.6-7 there exists an $\tilde{f} \in X'$ with $\|\tilde{f}\| = 1$ and $\tilde{f}(y) = 0$ for all $y \in Y$. Since $x_n \in Y$, we have $\tilde{f}(x_n) = 0$ and for all n,

$$\begin{aligned} \frac{1}{2} \leqq |f_n(x_n)| &= |f_n(x_n) - \tilde{f}(x_n)| \\ &= |(f_n - \tilde{f})(x_n)| \\ &\leqq \|f_n - \tilde{f}\| \, \|x_n\|, \end{aligned}$$

where $\|x_n\| = 1$. Hence $\|f_n - \tilde{f}\| \geqq \frac{1}{2}$, but this contradicts the assumption that (f_n) is dense in U' because $\tilde{f}$ is itself in U'; in fact, $\|\tilde{f}\| = 1$. ∎

Problems

1. What are the functionals f and g_x in (2) if $X = \mathbf{R}^n$?

2. Give a simpler proof of Lemma 4.6-7 for the case that X is a Hilbert space.

3. If a normed space X is reflexive, show that X' is reflexive.

4. Show that a Banach space X is reflexive if and only if its dual space X' is reflexive. (*Hint.* It can be shown that a closed subspace of a reflexive Banach space is reflexive. Use this fact, without proving it.)

5. Show that under the assumptions of Lemma 4.6-7 there exists a bounded linear functional h on X such that

$$\|h\| = 1/\delta, \qquad h(y) = 0 \text{ for all } y \in Y, \qquad h(x_0) = 1.$$

6. Show that different closed subspaces Y_1 and Y_2 of a normed space X have different annihilators. (Cf. Sec. 2.10, Prob. 13.)

7. Let Y be a closed subspace of a normed space X such that every $f \in X'$ which is zero everywhere on Y is zero everywhere on the whole space X. Show that then $Y = X$.

8. Let M be any subset of a normed space X. Show that an $x_0 \in X$ is an element of $A = \overline{\text{span } M}$ if and only if $f(x_0) = 0$ for every $f \in X'$ such that $f|_M = 0$.

9. **(Total set)** Show that a subset M of a normed space X is total in X if and only if every $f \in X'$ which is zero everywhere on M is zero everywhere on X.

10. Show that if a normed space X has a linearly independent subset of n elements, so does the dual space X'.

4.7 Category Theorem. Uniform Boundedness Theorem

The uniform boundedness theorem (or *uniform boundedness principle*) by S. Banach and H. Steinhaus (1927) is of great importance. In fact, throughout analysis there are many instances of results related to this theorem, the earliest being an investigation by H. Lebesgue (1909). The uniform boundedness theorem is often regarded as one of the corner stones of functional analysis in normed spaces, the others being the Hahn–Banach theorem (Secs. 4.2, 4.3), the open mapping theorem (Sec. 4.12) and the closed graph theorem (Sec. 4.13). Unlike the Hahn–Banach theorem, the other three of these four theorems require completeness. Indeed, they characterize some of the most important properties of Banach spaces which normed spaces in general may not have.

It is quite interesting to note that we shall obtain all three theorems from a common source. More precisely, we shall prove the so-called Baire's category theorem and derive from it the uniform boundedness theorem (in this section) as well as the open mapping theorem (in Sec. 4.12). The latter will then readily entail the closed graph theorem (in Sec. 4.13).

Baire's category theorem has various other applications in functional analysis and is the main reason why category enters into numerous proofs; cf., for instance, the more advanced books by R. E. Edwards (1965) and J. L. Kelley and I. Namioka (1963).

In Def. 4.7-1 we state the concepts needed for Baire's theorem 4.7-2. Each concept has two names, a new name and an old one given in parentheses. The latter is on the way out because "category" is now being used for an entirely different mathematical purpose (which will not occur in this book).

4.7-1 Definition (Category). A subset M of a metric space X is said to be

(a) *rare* (or *nowhere dense*) *in* X if its closure $\bar{M}$ has no interior points (cf. Sec. 1.3),

(b) *meager* (or *of the first category*) *in* X if M is the union of countably many sets each of which is rare in X,

(c) *nonmeager* (or *of the second category*) *in* X if M is not meager in X. ∎

4.7-2 Baire's Category Theorem (Complete metric spaces). *If a metric space $X \neq \varnothing$ is complete, it is nonmeager in itself.*

Hence if $X \neq \varnothing$ is complete and

$$X = \bigcup_{k=1}^{\infty} A_k \qquad (A_k \text{ closed}) \tag{1}$$

then at least one A_k contains a nonempty open subset.

Proof. The idea of the proof is simple. Suppose the complete metric space $X \neq \varnothing$ were meager in itself. Then

$$X = \bigcup_{k=1}^{\infty} M_k \tag{1*}$$

with each M_k rare in X. We shall construct a Cauchy sequence (p_k) whose limit p (which exists by completeness) is in no M_k, thereby contradicting the representation (1*).

By assumption, M_1 is rare in X, so that, by definition, $\bar{M}_1$ does not contain a nonempty open set. But X does (for instance, X itself). This implies $\bar{M}_1 \neq X$. Hence the complement $\bar{M}_1^{\mathrm{C}} = X - \bar{M}_1$ of $\bar{M}_1$ is not empty and open. We may thus choose a point p_1 in $\bar{M}_1^{\mathrm{C}}$ and an open ball about it, say,

$$B_1 = B(p_1; \varepsilon_1) \subset \bar{M}_1^{\mathrm{C}} \qquad \varepsilon_1 < \tfrac{1}{2}.$$

By assumption, M_2 is rare in X, so that $\bar{M}_2$ does not contain a nonempty open set. Hence it does not contain the open ball $B(p_1; \frac{1}{2}\varepsilon_1)$. This implies that $\bar{M}_2^{\mathrm{C}} \cap B(p_1; \frac{1}{2}\varepsilon_1)$ is not empty and open, so that we may choose an open ball in this set, say,

$$B_2 = B(p_2; \varepsilon_2) \subset \bar{M}_2^{\mathrm{C}} \cap B(p_1; \tfrac{1}{2}\varepsilon_1) \qquad \varepsilon_2 < \tfrac{1}{2}\varepsilon_1.$$

By induction we thus obtain a sequence of balls

$$B_k = B(p_k; \varepsilon_k) \qquad \varepsilon_k < 2^{-k}$$

such that $B_k \cap M_k = \varnothing$ and

$$B_{k+1} \subset B(p_k; \tfrac{1}{2}\varepsilon_k) \subset B_k \qquad k = 1, 2, \cdots.$$

Since $\varepsilon_k < 2^{-k}$, the sequence (p_k) of the centers is Cauchy and converges, say, $p_k \longrightarrow p \in X$ because X is complete by assumption. Also, for every m and $n > m$ we have $B_n \subset B(p_m; \frac{1}{2}\varepsilon_m)$, so that

$$\begin{aligned} d(p_m, p) &\leqq d(p_m, p_n) + d(p_n, p) \\ &< \tfrac{1}{2}\varepsilon_m + d(p_n, p) \quad \longrightarrow \quad \tfrac{1}{2}\varepsilon_m \end{aligned}$$

as $n \longrightarrow \infty$. Hence $p \in B_m$ for every m. Since $B_m \subset \bar{M}_m^{\mathrm{C}}$, we now see that $p \notin M_m$ for every m, so that $p \notin \bigcup M_m = X$. This contradicts $p \in X$. Baire's theorem is proved. ∎

We note that the converse of Baire's theorem is not generally true. An example of an incomplete normed space which is nonmeager in itself is given in N. Bourbaki (1955), Ex. 6, pp. 3–4.

From Baire's theorem we shall now readily obtain the desired uniform boundedness theorem. This theorem states that if X is a Banach space and a sequence of operators $T_n \in B(X, Y)$ is bounded at every point $x \in X$, then the sequence is uniformly bounded. In other words, pointwise boundedness implies boundedness in some stronger sense, namely, uniform boundedness. (The real number c_x in (2), below, will vary in general with x, a fact which we indicate by the subscript x; the essential point is that c_x does not depend on n.)

4.7-3 Uniform Boundedness Theorem. *Let (T_n) be a sequence of bounded linear operators $T_n\colon X \longrightarrow Y$ from a Banach space X into a normed space Y such that $(\|T_n x\|)$ is bounded for every $x \in X$, say,*

$$\|T_n x\| \leqq c_x \qquad n = 1, 2, \cdots, \tag{2}$$

where c_x is a real number. Then the sequence of the norms $\|T_n\|$ is bounded, that is, there is a c such that

$$\|T_n\| \leqq c \qquad n = 1, 2, \cdots. \tag{3}$$

Proof. For every $k \in \mathbf{N}$, let $A_k \subset X$ be the set of all x such that

$$\|T_n x\| \leqq k \qquad \text{for all } n.$$

A_k is closed. Indeed, for any $x \in \bar{A}_k$ there is a sequence (x_j) in A_k converging to x. This means that for every fixed n we have $\|T_n x_j\| \leqq k$ and obtain $\|T_n x\| \leqq k$ because T_n is continuous and so is the norm (cf. Sec. 2.2). Hence $x \in A_k$, and A_k is closed.

By (2), each $x \in X$ belongs to some A_k. Hence

$$X = \bigcup_{k=1}^{\infty} A_k.$$

Since X is complete, Baire's theorem implies that some A_k contains an open ball, say,

$$B_0 = B(x_0; r) \subset A_{k_0}. \tag{4}$$

Let $x \in X$ be arbitrary, not zero. We set

$$z = x_0 + \gamma x \qquad \gamma = \frac{r}{2\,\|x\|}. \tag{5}$$

Then $\|z-x_0\|<r$, so that $z \in B_0$. By (4) and from the definition of A_{k_0} we thus have $\|T_n z\| \leqq k_0$ for all n. Also $\|T_n x_0\| \leqq k_0$ since $x_0 \in B_0$. From (5) we obtain

$$x = \frac{1}{\gamma}(z - x_0).$$

This yields for all n

$$\|T_n x\| = \frac{1}{\gamma}\|T_n(z-x_0)\| \leqq \frac{1}{\gamma}(\|T_n z\| + \|T_n x_0\|) \leqq \frac{4}{r}\|x\|\, k_0.$$

Hence for all n,

$$\|T_n\| = \sup_{\|x\|=1} \|T_n x\| \leqq \frac{4}{r} k_0,$$

which is of the form (3) with $c = 4k_0/r$. ∎

Applications

4.7-4 Space of polynomials. *The normed space X of all polynomials with norm defined by*

$$\text{(6)} \qquad \|x\| = \max_j |\alpha_j| \qquad (\alpha_0, \alpha_1, \cdots \text{ the coefficients of } x)$$

is not complete.

Proof. We construct a sequence of bounded linear operators on X which satisfies (2) but not (3), so that X cannot be complete.

We may write a polynomial $x \neq 0$ of degree N_x in the form

$$x(t) = \sum_{j=0}^{\infty} \alpha_j t^j \qquad (\alpha_j = 0 \text{ for } j > N_x).$$

(For $x = 0$ the degree is not defined in the usual discussion of degree, but this does not matter here.) As a sequence of operators on X we take the sequence of functionals $T_n = f_n$ defined by

$$\text{(7)} \qquad T_n 0 = f_n(0) = 0, \qquad T_n x = f_n(x) = \alpha_0 + \alpha_1 + \cdots + \alpha_{n-1}.$$

f_n is linear. f_n is bounded since $|\alpha_j| \leqq \|x\|$ by (6), so that $|f_n(x)| \leqq n \|x\|$. Furthermore, for each fixed $x \in X$ the sequence $(|f_n(x)|)$ satisfies (2) because a polynomial x of degree N_x has $N_x + 1$ coefficients, so that by (7) we have

$$|f_n(x)| \leqq (N_x + 1) \max_j |\alpha_j| = c_x$$

which is of the form (2).

We now show that (f_n) does not satisfy (3), that is, there is no c such that $\|T_n\| = \|f_n\| \leqq c$ for all n. This we do by choosing particularly disadvantageous polynomials. For f_n we choose x defined by

$$x(t) = 1 + t + \cdots + t^n.$$

Then $\|x\| = 1$ by (6) and

$$f_n(x) = 1 + 1 + \cdots + 1 = n = n \|x\|.$$

Hence $\|f_n\| \geqq |f_n(x)|/\|x\| = n$, so that $(\|f_n\|)$ is unbounded. ∎

4.7-5 Fourier series. From 3.5-1 we remember that the *Fourier series* of a given periodic function x of period 2π is of the form

$$\text{(8)} \qquad \tfrac{1}{2}a_0 + \sum_{m=1}^{\infty} (a_m \cos mt + b_m \sin mt)$$

with the Fourier coefficients of x given by the Euler formulas

$$\text{(9)} \qquad a_m = \frac{1}{\pi} \int_0^{2\pi} x(t) \cos mt \, dt, \qquad b_m = \frac{1}{\pi} \int_0^{2\pi} x(t) \sin mt \, dt.$$

[We wrote $a_0/2$ in (8) to have only two formulas in (9), whereas in 3.5-1 we wrote a_0 and needed three Euler formulas.]

It is well-known that the series (8) may converge even at points where x is discontinuous. (Problem 15 gives a simple example.) This shows that continuity is not necessary for convergence. Surprising enough, continuity is not sufficient either.[4] Indeed, using the uniform boundedness theorem, we can show the following.

[4] Continuity and the existence of the right-hand and left-hand derivatives at a point t_0 is sufficient for convergence at t_0. Cf. W. Rogosinski (1959), p. 70.

There exist real-valued continuous functions whose Fourier series diverge at a given point t_0.

Proof. Let X be the normed space of all real-valued continuous functions of period 2π with norm defined by

$$\|x\| = \max |x(t)|. \tag{10}$$

X is a Banach space, as follows from 1.5-5 with $a=0$ and $b=2\pi$. We may take $t_0=0$, without restricting generality. To prove our statement, we shall apply the uniform boundedness theorem 4.7-3 to $T_n = f_n$ where $f_n(x)$ is the value at $t=0$ of the nth partial sum of the Fourier series of x. Since for $t=0$ the sine terms are zero and the cosine is one, we see from (8) and (9) that

$$\begin{aligned} f_n(x) &= \tfrac{1}{2}a_0 + \sum_{m=1}^{n} a_m \\ &= \frac{1}{\pi}\int_0^{2\pi} x(t)\left[\frac{1}{2} + \sum_{m=1}^{n} \cos mt\right] dt. \end{aligned}$$

We want to determine the function represented by the sum under the integral sign. For this purpose we calculate

$$\begin{aligned} 2\sin\tfrac{1}{2}t \sum_{m=1}^{n} \cos mt &= \sum_{m=1}^{n} 2\sin\tfrac{1}{2}t\cos mt \\ &= \sum_{m=1}^{n} [-\sin(m-\tfrac{1}{2})t + \sin(m+\tfrac{1}{2})t] \\ &= -\sin\tfrac{1}{2}t + \sin(n+\tfrac{1}{2})t, \end{aligned}$$

where the last expression follows by noting that most of the terms drop out in pairs. Dividing this by $\sin\frac{1}{2}t$ and adding 1 on both sides, we have

$$1 + 2\sum_{m=1}^{n} \cos mt = \frac{\sin(n+\frac{1}{2})t}{\sin\frac{1}{2}t}.$$

Consequently, the formula for $f_n(x)$ can be written in the simple form

$$f_n(x) = \frac{1}{2\pi}\int_0^{2\pi} x(t)q_n(t)\,dt, \qquad q_n(t) = \frac{\sin(n+\frac{1}{2})t}{\sin\frac{1}{2}t}. \tag{11}$$

Using this, we can show that the linear functional f_n is bounded. In fact, by (10) and (11),

$$|f_n(x)| \leqq \frac{1}{2\pi} \max |x(t)| \int_0^{2\pi} |q_n(t)|\, dt = \frac{\|x\|}{2\pi} \int_0^{2\pi} |q_n(t)|\, dt.$$

From this we see that f_n is bounded. Furthermore, by taking the supremum over all x of norm one we obtain

$$\|f_n\| \leqq \frac{1}{2\pi} \int_0^{2\pi} |q_n(t)|\, dt.$$

Actually, the equality sign holds, as we shall now prove. For this purpose let us first write

$$|q_n(t)| = y(t) q_n(t)$$

where $y(t) = +1$ at every t at which $q_n(t) \geqq 0$ and $y(t) = -1$ elsewhere. y is not continuous, but for any given $\varepsilon > 0$ it may be modified to a continuous x of norm 1 such that for this x we have

$$\frac{1}{2\pi} \left| \int_0^{2\pi} [x(t) - y(t)] q_n(t)\, dt \right| < \varepsilon.$$

Writing this as two integrals and using (11), we obtain

$$\frac{1}{2\pi} \left| \int_0^{2\pi} x(t) q_n(t)\, dt - \int_0^{2\pi} y(t) q_n(t)\, dt \right| = \left| f_n(x) - \frac{1}{2\pi} \int_0^{2\pi} |q_n(t)|\, dt \right| < \varepsilon.$$

Since $\varepsilon > 0$ was arbitrary and $\|x\| = 1$, this proves the desired formula

$$(12) \qquad \|f_n\| = \frac{1}{2\pi} \int_0^{2\pi} |q_n(t)|\, dt.$$

We finally show that the sequence $(\|f_n\|)$ is unbounded. Substituting into (12) the expression for q_n from (11), using the fact that

$|\sin \frac{1}{2}t| < \frac{1}{2}t$ for $t \in (0, 2\pi]$ and setting $(n+\frac{1}{2})t = v$, we obtain

$$\begin{aligned}
\|f_n\| &= \frac{1}{2\pi} \int_0^{2\pi} \left| \frac{\sin (n+\frac{1}{2})t}{\sin \frac{1}{2}t} \right| dt \\
&> \frac{1}{\pi} \int_0^{2\pi} \frac{|\sin (n+\frac{1}{2})t|}{t} dt \\
&= \frac{1}{\pi} \int_0^{(2n+1)\pi} \frac{|\sin v|}{v} dv \\
&= \frac{1}{\pi} \sum_{k=0}^{2n} \int_{k\pi}^{(k+1)\pi} \frac{|\sin v|}{v} dv \\
&\geqq \frac{1}{\pi} \sum_{k=0}^{2n} \frac{1}{(k+1)\pi} \int_{k\pi}^{(k+1)\pi} |\sin v| \, dv \\
&= \frac{2}{\pi^2} \sum_{k=0}^{2n} \frac{1}{k+1} \longrightarrow \infty \qquad \text{as } n \longrightarrow \infty
\end{aligned}$$

since the harmonic series diverges. Hence $(\|f_n\|)$ is unbounded, so that (3) (with $T_n = f_n$) does not hold. Since X is complete, this implies that (2) cannot hold for all x. Hence there must be an $x \in X$ such that $(|f_n(x)|)$ is unbounded. But by the definition of the f_n's this means that the Fourier series of that x diverges at $t = 0$. ∎

Note that our existence proof does not tell us how to find such a continuous function x whose Fourier series diverges at a t_0. Examples of such functions were given by L. Fejér (1910); one is reproduced in W. Rogosinski (1959), pp. 76–77.

Problems

1. Of what category is the set of all rational numbers (*a*) in **R**, (*b*) in itself (taken with the usual metric)?

2. Of what category is the set of all integers (*a*) in **R**, (*b*) in itself (taken with the metric induced from **R**)?

3. Find all rare sets in a discrete metric space X. (Cf. 1.1-8.)

4. Find a meager dense subset in $\mathbf{R}^2$.

5. Show that a subset M of a metric space X is rare in X if and only if $(\bar{M})^C$ is dense in X.

6. Show that the complement M^C of a meager subset M of a complete metric space X is nonmeager.

7. **(Resonance)** Let X be a Banach space, Y a normed space and $T_n \in B(X, Y)$, $n = 1, 2, \cdots$, such that $\sup_n \|T_n\| = +\infty$. Show that there is an $x_0 \in X$ such that $\sup_n \|T_n x_0\| = +\infty$. [The point x_0 is often called a *point of resonance*, and our problem motivates the term *resonance theorem* for the uniform boundedness theorem.]

8. Show that completeness of X is essential in Theorem 4.7-3 and cannot be omitted. [Consider the subspace $X \subset l^\infty$ consisting of all $x = (\xi_j)$ such that $\xi_j = 0$ for $j \geqq J \in \mathbf{N}$, where J depends on x, and let T_n be defined by $T_n x = f_n(x) = n\xi_n$.]

9. Let $T_n = S^n$, where the operator $S: l^2 \longrightarrow l^2$ is defined by $(\xi_1, \xi_2, \xi_3, \cdots) \longmapsto (\xi_3, \xi_4, \xi_5, \cdots)$. Find a bound for $\|T_n x\|$; find $\lim_{n\to\infty} \|T_n x\|$, $\|T_n\|$ and $\lim_{n\to\infty} \|T_n\|$.

10. **(Space c_0)** Let $y = (\eta_j)$, $\eta_j \in \mathbf{C}$, be such that $\sum \xi_j \eta_j$ converges for every $x = (\xi_j) \in c_0$, where $c_0 \subset l^\infty$ is the subspace of all complex sequences converging to zero. Show that $\sum |\eta_j| < \infty$. (Use 4.7-3.)

11. Let X be a Banach space, Y a normed space and $T_n \in B(X, Y)$ such that $(T_n x)$ is Cauchy in Y for every $x \in X$. Show that $(\|T_n\|)$ is bounded.

12. If, in addition, Y in Prob. 11 is complete, show that $T_n x \longrightarrow Tx$, where $T \in B(X, Y)$.

13. If (x_n) in a Banach space X is such that $(f(x_n))$ is bounded for all $f \in X'$, show that $(\|x_n\|)$ is bounded.

14. If X and Y are Banach spaces and $T_n \in B(X, Y)$, $n = 1, 2, \cdots$, show that equivalent statements are:
(*a*) $(\|T_n\|)$ is bounded,
(*b*) $(\|T_n x\|)$ is bounded for all $x \in X$,
(*c*) $(|g(T_n x)|)$ is bounded for all $x \in X$ and all $g \in Y'$.

15. To illustrate that a Fourier series of a function x may converge even at a point where x is discontinuous, find the Fourier series of

$$x(t) = \begin{cases} 0 & \text{if } -\pi \leqq t < 0 \\ 1 & \text{if } \quad 0 \leqq t < \pi \end{cases} \qquad \text{and} \qquad x(t + 2\pi) = x(t).$$

Graph x and the partial sums s_0, s_1, s_2, s_3, and compare with Fig. 44. Show that at $t = \pm n\pi$ the series has the value 1/2, the arithmetic mean of the right and left limits of x; this behavior is typical of Fourier series.

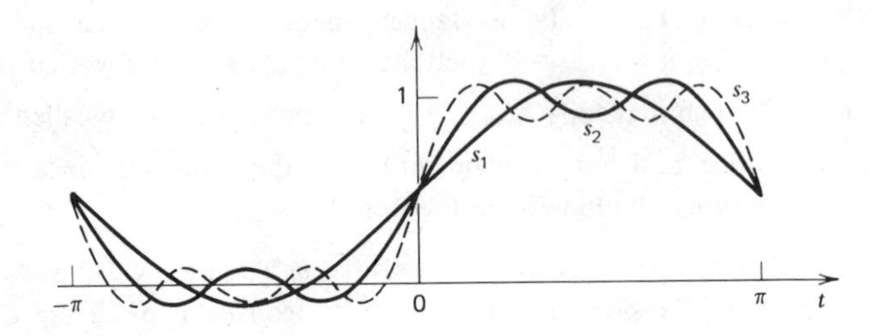

Fig. 44. First three partial sums s_1, s_2, s_3 in Prob. 15

4.8 Strong and Weak Convergence

We know that in calculus one defines different types of convergence (ordinary, conditional, absolute and uniform convergence). This yields greater flexibility in the theory and application of sequences and series. In functional analysis the situation is similar, and one has an even greater variety of possibilities that turn out to be of practical interest. In the present section we are primarily concerned with "weak convergence". This is a basic concept. We present it now since its theory makes essential use of the uniform boundedness theorem discussed in the previous section. In fact, this is one of the major applications of that theorem.

Convergence of sequences of elements in a normed space was defined in Sec. 2.3 and, from now on, will be called *strong convergence*, to distinguish it from "weak convergence" to be introduced shortly. Hence we first state

4.8-1 Definition (Strong convergence). A sequence (x_n) in a normed space X is said to be *strongly convergent* (or *convergent in the norm*) if there is an $x \in X$ such that

$$\lim_{n\to\infty} \|x_n - x\| = 0.$$

This is written

$$\lim_{n\to\infty} x_n = x$$

or simply

$$x_n \longrightarrow x.$$

x is called the *strong limit* of (x_n), and we say that (x_n) *converges strongly to x.* ∎

Weak convergence is defined in terms of bounded linear functionals on X as follows.

4.8-2 Definition (Weak convergence). A sequence (x_n) in a normed space X is said to be *weakly convergent* if there is an $x \in X$ such that for every $f \in X'$,

$$\lim_{n\to\infty} f(x_n) = f(x).$$

This is written

$$x_n \xrightarrow{\ w\ } x$$

or $x_n \rightharpoonup x$. The element x is called the *weak limit* of (x_n), and we say that (x_n) *converges weakly to x.* ∎

Note that weak convergence means convergence of the sequence of numbers $a_n = f(x_n)$ for every $f \in X'$.

Weak convergence has various applications throughout analysis (for instance, in the calculus of variations and the general theory of differential equations). The concept illustrates a basic principle of functional analysis, namely, the fact that the investigation of spaces is often related to that of their dual spaces.

For applying weak convergence one needs to know certain basic properties, which we state in the following lemma. The reader will note that in the proof we use the Hahn–Banach theorem (via 4.3-4 as well as 4.6-1) and the uniform boundedness theorem. This demonstrates the importance of these theorems in connection with weak convergence.

4.8-3 Lemma (Weak convergence). *Let (x_n) be a weakly convergent sequence in a normed space X, say, $x_n \xrightarrow{w} x$. Then:*

(a) *The weak limit x of (x_n) is unique.*

(b) *Every subsequence of (x_n) converges weakly to x.*

(c) *The sequence $(\|x_n\|)$ is bounded.*

Proof. **(a)** Suppose that $x_n \xrightarrow{w} x$ as well as $x_n \xrightarrow{w} y$. Then $f(x_n) \longrightarrow f(x)$ as well as $f(x_n) \longrightarrow f(y)$. Since $(f(x_n))$ is a sequence of numbers, its limit is unique. Hence $f(x) = f(y)$, that is, for every $f \in X'$ we have

$$f(x) - f(y) = f(x - y) = 0.$$

This implies $x - y = 0$ by Corollary 4.3-4 and shows that the weak limit is unique.

(b) follows from the fact that $(f(x_n))$ is a convergent sequence of numbers, so that every subsequence of $(f(x_n))$ converges and has the same limit as the sequence.

(c) Since $(f(x_n))$ is a convergent sequence of numbers, it is bounded, say, $|f(x_n)| \leqq c_f$ for all n, where c_f is a constant depending on f but not on n. Using the canonical mapping $C: X \longrightarrow X''$ (Sec. 4.6), we can define $g_n \in X''$ by

$$g_n(f) = f(x_n) \qquad\qquad f \in X'.$$

(We write g_n instead of g_{x_n}, to avoid subscripts of subscripts.) Then for all n,

$$|g_n(f)| = |f(x_n)| \leqq c_f,$$

that is, the sequence $(|g_n(f)|)$ is bounded for every $f \in X'$. Since X' is complete by 2.10-4, the uniform boundedness theorem 4.7-3 is applicable and implies that $(\|g_n\|)$ is bounded. Now $\|g_n\| = \|x_n\|$ by 4.6-1, so that (c) is proved. ∎

The reader may perhaps wonder why weak convergence does not play a role in calculus. The simple reason is that in finite dimensional normed spaces the distinction between strong and weak convergence disappears completely. Let us prove this fact and also justify the terms "strong" and "weak."

4.8-4 Theorem (Strong and weak convergence). *Let (x_n) be a sequence in a normed space X. Then:*

(a) *Strong convergence implies weak convergence with the same limit.*

(b) *The converse of (a) is not generally true.*

(c) *If* $\dim X < \infty$, *then weak convergence implies strong convergence.*

Proof. **(a)** By definition, $x_n \longrightarrow x$ means $\|x_n - x\| \longrightarrow 0$ and implies that for every $f \in X'$,

$$|f(x_n) - f(x)| = |f(x_n - x)| \leqq \|f\| \, \|x_n - x\| \longrightarrow 0.$$

This shows that $x_n \xrightarrow{w} x$.

(b) can be seen from an orthonormal sequence (e_n) in a Hilbert space H. In fact, every $f \in H'$ has a Riesz representation $f(x) = \langle x, z \rangle$. Hence $f(e_n) = \langle e_n, z \rangle$. Now the Bessel inequality is (cf. 3.4-6)

$$\sum_{n=1}^{\infty} |\langle e_n, z \rangle|^2 \leqq \|z\|^2.$$

Hence the series on the left converges, so that its terms must approach zero as $n \longrightarrow \infty$. This implies

$$f(e_n) = \langle e_n, z \rangle \longrightarrow 0.$$

Since $f \in H'$ was arbitrary, we see that $e_n \xrightarrow{w} 0$. However, (e_n) does not converge strongly because

$$\|e_m - e_n\|^2 = \langle e_m - e_n, e_m - e_n \rangle = 2 \qquad (m \neq n).$$

(c) Suppose that $x_n \xrightarrow{w} x$ and $\dim X = k$. Let $\{e_1, \cdots, e_k\}$ be any basis for X and, say,

$$x_n = \alpha_1^{(n)} e_1 + \cdots + \alpha_k^{(n)} e_k$$

and

$$x = \alpha_1 e_1 + \cdots + \alpha_k e_k.$$

By assumption, $f(x_n) \longrightarrow f(x)$ for every $f \in X'$. We take in particular $f_1, \cdots, f_k$ defined by

$$f_j(e_j) = 1, \qquad f_j(e_m) = 0 \qquad (m \neq j).$$

(We mention that this is the dual basis of $\{e_1, \cdots, e_k\}$; cf. Sec. 2.9.) Then

$$f_j(x_n) = \alpha_j^{(n)}, \qquad f_j(x) = \alpha_j.$$

Hence $f_j(x_n) \longrightarrow f_j(x)$ implies $\alpha_j^{(n)} \longrightarrow \alpha_j$. From this we readily obtain

$$\begin{aligned} \|x_n - x\| &= \left\| \sum_{j=1}^{k} (\alpha_j^{(n)} - \alpha_j) e_j \right\| \\ &\leqq \sum_{j=1}^{k} |\alpha_j^{(n)} - \alpha_j| \, \|e_j\| \longrightarrow 0 \end{aligned}$$

as $n \longrightarrow \infty$. This shows that (x_n) converges strongly to x. ∎

It is interesting to note that there also exist infinite dimensional spaces such that strong and weak convergence are equivalent concepts. An example is l^1, as was shown by I. Schur (1921).

In conclusion let us take a look at weak convergence in two particularly important types of spaces.

Examples

4.8-5 Hilbert space. *In a Hilbert space, $x_n \xrightarrow{w} x$ if and only if $\langle x_n, z\rangle \longrightarrow \langle x, z\rangle$ for all z in the space.*

Proof. Clear by 3.8-1.

4.8-6 Space l^p. *In the space l^p, where $1<p<+\infty$, we have $x_n \xrightarrow{w} x$ if and only if:*

(A) *The sequence $(\|x_n\|)$ is bounded.*

(B) *For every fixed j we have $\xi_j^{(n)} \longrightarrow \xi_j$ as $n \longrightarrow \infty$; here, $x_n = (\xi_j^{(n)})$ and $x = (\xi_j)$.*

Proof. The dual space of l^p is l^q; cf. 2.10-7. A Schauder basis of l^q is (e_n), where $e_n = (\delta_{nj})$ has 1 in the nth place and zeros elsewhere. Span (e_n) is dense in l^q, so that the conclusion results from the following lemma.

4.8-7 Lemma (Weak convergence). *In a normed space X we have $x_n \xrightarrow{w} x$ if and only if:*

(A) *The sequence $(\|x_n\|)$ is bounded.*

(B) *For every element f of a total subset $M \subset X'$ we have $f(x_n) \longrightarrow f(x)$.*

Proof. In the case of weak convergence, (A) follows from Lemma 4.8-3 and (B) is trivial.

Conversely, suppose that (A) and (B) hold. Let us consider any $f \in X'$ and show that $f(x_n) \longrightarrow f(x)$, which means weak convergence, by the definition.

By (A) we have $\|x_n\| \leqq c$ for all n and $\|x\| \leqq c$, where c is sufficiently large. Since M is total in X', for every $f \in X'$ there is a sequence (f_j) in span M such that $f_j \longrightarrow f$. Hence for any given $\varepsilon > 0$ we can find a j such that

$$\|f_j - f\| < \frac{\varepsilon}{3c}.$$

Moreover, since $f_j \in \text{span } M$, by assumption (B) there is an N such that for all $n > N$,

$$|f_j(x_n) - f_j(x)| < \frac{\varepsilon}{3}.$$

Using these two inequalities and applying the triangle inequality, we obtain for $n > N$

$$\begin{aligned} |f(x_n)-f(x)| &\leqq |f(x_n)-f_j(x_n)|+|f_j(x_n)-f_j(x)|+|f_j(x)-f(x)| \\ &< \|f-f_j\| \, \|x_n\|+\frac{\varepsilon}{3}+\|f_j-f\| \, \|x\| \\ &< \frac{\varepsilon}{3c}\, c+\frac{\varepsilon}{3}+\frac{\varepsilon}{3c}\, c=\varepsilon. \end{aligned}$$

Since $f \in X'$ was arbitrary, this shows that the sequence (x_n) converges weakly to x. ∎

Problems

1. **(Pointwise convergence)** If $x_n \in C[a, b]$ and $x_n \xrightarrow{w} x \in C[a, b]$, show that (x_n) is *pointwise convergent* on $[a, b]$, that is, $(x_n(t))$ converges for every $t \in [a, b]$.

2. Let X and Y be normed spaces, $T \in B(X, Y)$ and (x_n) a sequence in X. If $x_n \xrightarrow{w} x_0$, show that $Tx_n \xrightarrow{w} Tx_0$.

3. If (x_n) and (y_n) are sequences in the same normed space X, show that $x_n \xrightarrow{w} x$ and $y_n \xrightarrow{w} y$ implies $x_n + y_n \xrightarrow{w} x + y$ as well as $\alpha x_n \xrightarrow{w} \alpha x$, where α is any scalar.

4. Show that $x_n \xrightarrow{w} x_0$ implies $\varliminf_{n\to\infty} \|x_n\| \geqq \|x_0\|$. (Use Theorem 4.3-3.)

5. If $x_n \xrightarrow{w} x_0$ in a normed space X, show that $x_0 \in \bar{Y}$, where $Y = \operatorname{span}(x_n)$. (Use Lemma 4.6-7.)

6. If (x_n) is a weakly convergent sequence in a normed space X, say, $x_n \xrightarrow{w} x_0$, show that there is a sequence (y_m) of linear combinations of elements of (x_n) which converges strongly to x_0.

7. Show that any closed subspace Y of a normed space X contains the limits of all weakly convergent sequences of its elements.

8. **(Weak Cauchy sequence)** A *weak Cauchy sequence* in a real or complex normed space X is a sequence (x_n) in X such that for every $f \in X'$ the sequence $(f(x_n))$ is Cauchy in **R** or **C**, respectively. [Note that then $\lim_{n\to\infty} f(x_n)$ exists.] Show that a weak Cauchy sequence is bounded.

9. Let A be a set in a normed space X such that every nonempty subset of A contains a weak Cauchy sequence. Show that A is bounded.

10. (Weak completeness) A normed space X is said to be *weakly complete* if each weak Cauchy sequence in X converges weakly in X. If X is reflexive, show that X is weakly complete.

4.9 Convergence of Sequences of Operators and Functionals

Sequences of bounded linear operators and functionals arise frequently in the abstract formulation of concrete situations, for instance in connection with convergence problems of Fourier series or sequences of interpolation polynomials or methods of numerical integration, to name just a few. In such cases one is usually concerned with the convergence of those sequences of operators or functionals, with boundedness of corresponding sequences of norms or with similar properties.

Experience shows that for sequences of *elements* in a normed space, strong and weak convergence as defined in the previous section are useful concepts. For sequences of *operators* $T_n \in B(X, Y)$ three types of convergence turn out to be of theoretical as well as practical value. These are

(1) Convergence in the norm on $B(X, Y)$,

(2) Strong convergence of $(T_n x)$ in Y,

(3) Weak convergence of $(T_n x)$ in Y.

The definitions and terminology are as follows; they were introduced by J. von Neumann (1929–30b).

4.9-1 Definition (Convergence of sequences of operators). Let X and Y be normed spaces. A sequence (T_n) of operators $T_n \in B(X, Y)$ is said to be:

(1) **uniformly operator convergent**[5] if (T_n) converges in the norm on $B(X, Y)$

[5] "Operator" is often omitted from each of the three terms. We retain it for clarity.

(2) **strongly operator convergent** if $(T_n x)$ converges strongly in Y for every $x \in X$,

(3) **weakly operator convergent** if $(T_n x)$ converges weakly in Y for every $x \in X$.

In formulas this means that there is an operator $T: X \longrightarrow Y$ such that

$$(1) \qquad \|T_n - T\| \longrightarrow 0$$

$$(2) \qquad \|T_n x - Tx\| \longrightarrow 0 \qquad \text{for all } x \in X$$

$$(3) \qquad |f(T_n x) - f(Tx)| \longrightarrow 0 \qquad \text{for all } x \in X \text{ and all } f \in Y'$$

respectively. T is called the *uniform, strong* and *weak operator limit* of (T_n), respectively. ∎

We pointed out in the previous section that even in calculus, in a much simpler situation, the use of several concepts of convergence gives greater flexibility. Nevertheless the reader may still be bewildered by the many concepts of convergence we have just introduced. He may ask why it is necessary to have three kinds of convergence for sequences of operators. The answer is that many of the operators that appear in practical problems are given as some sort of limit of simpler operators. And it is important to know what is meant by "some sort" and to know what properties of the limiting operator are implied by the properties of the sequence. Also, at the beginning of an investigation, one does not always know in what sense limits will exist; hence it is useful to have a variety of possibilities. Perhaps in a specific problem one is at first able to establish convergence only in a very "mild" sense, so that one has at least something to start from, and then later develop tools for proving convergence in a stronger sense, which guarantees "better" properties of the limit operator. This is a typical situation, for example in partial differential equations.

It is not difficult to show that

$$(1) \quad \Longrightarrow \quad (2) \quad \Longrightarrow \quad (3)$$

(the limit being the same), but the converse is not generally true, as can be seen from the following examples.

Examples

4.9-2 (Space l^2). In the space l^2 we consider a sequence (T_n), where $T_n\colon l^2 \longrightarrow l^2$ is defined by

$$T_n x = (\underbrace{0, 0, \cdots, 0}_{(n \text{ zeros})}, \xi_{n+1}, \xi_{n+2}, \xi_{n+3}, \cdots);$$

here, $x = (\xi_1, \xi_2, \cdots) \in l^2$. This operator T_n is linear and bounded. Clearly, (T_n) is strongly operator convergent to 0 since $T_n x \longrightarrow 0 = 0x$. However, (T_n) is not uniformly operator convergent since we have $\|T_n - 0\| = \|T_n\| = 1$.

4.9-3 (Space l^2). Another sequence (T_n) of operators $T_n\colon l^2 \longrightarrow l^2$ is defined by

$$T_n x = (\underbrace{0, 0, \cdots, 0}_{(n \text{ zeros})}, \xi_1, \xi_2, \xi_3, \cdots)$$

where $x = (\xi_1, \xi_2, \cdots) \in l^2$. This operator T_n is linear and bounded. We show that (T_n) is weakly operator convergent to 0 but not strongly.

Every bounded linear functional f on l^2 has a Riesz representation 3.8-1, that is, by 3.1-6,

$$f(x) = \langle x, z \rangle = \sum_{j=1}^{\infty} \xi_j \bar{\zeta}_j$$

where $z = (\zeta_j) \in l^2$. Hence, setting $j = n + k$ and using the definition of T_n, we have

$$f(T_n x) = \langle T_n x, z \rangle = \sum_{j=n+1}^{\infty} \xi_{j-n} \bar{\zeta}_j = \sum_{k=1}^{\infty} \xi_k \bar{\zeta}_{n+k}.$$

By the Cauchy–Schwarz inequality in 1.2-3,

$$|f(T_n x)|^2 = |\langle T_n x, z \rangle|^2 \leqq \sum_{k=1}^{\infty} |\xi_k|^2 \sum_{m=n+1}^{\infty} |\zeta_m|^2.$$

The last series is the remainder of a convergent series. Hence the right-hand side approaches 0 as $n \longrightarrow \infty$; thus $f(T_n x) \longrightarrow 0 = f(0x)$. Consequently, (T_n) is weakly operator convergent to 0.

However, (T_n) is not strongly operator convergent because for $x = (1, 0, 0, \cdots)$ we have

$$\|T_m x - T_n x\| = \sqrt{1^2 + 1^2} = \sqrt{2} \qquad (m \neq n).$$ ∎

Linear functionals are linear operators (with range in the scalar field **R** or **C**), so that (1), (2) and (3) apply immediately. However, (2) and (3) now become equivalent, for the following reason. We had $T_n x \in Y$, but we now have $f_n(x) \in \mathbf{R}$ (or **C**). Hence convergence in (2) and (3) now takes place in the finite dimensional (one-dimensional) space **R** (or **C**) and equivalence of (2) and (3) follows from Theorem 4.8-4(*c*). The two remaining concepts are called *strong* and *weak** *convergence* (read "weak star convergence"):

4.9-4 Definition (Strong and weak* convergence of a sequence of functionals). Let (f_n) be a sequence of bounded linear functionals on a normed space X. Then:

(a) *Strong convergence* of (f_n) means that there is an $f \in X'$ such that $\|f_n - f\| \longrightarrow 0$. This is written

$$f_n \longrightarrow f.$$

(b) *Weak* convergence* of (f_n) means that there is an $f \in X'$ such that $f_n(x) \longrightarrow f(x)$ for all $x \in X$. This is written[6]

$$f_n \xrightarrow{w^*} f.$$

f in (*a*) and (*b*) is called the *strong limit* and *weak* limit* of (f_n), respectively. ∎

Returning to operators $T_n \in B(X, Y)$, we ask what can be said about the limit operator $T\colon X \longrightarrow Y$ in (1), (2) and (3).

If the convergence is uniform, $T \in B(X, Y)$; otherwise $\|T_n - T\|$ would not make sense. If the convergence is strong or weak, T is still linear but may be unbounded if X is not complete.

[6] This concept is somewhat more important than *weak convergence* of (f_n), which, by 4.8-2, means that $g(f_n) \longrightarrow g(f)$ for all $g \in X''$. Weak convergence implies weak* convergence, as can be seen by the use of the canonical mapping defined in Sec. 4.6. (Cf. Prob. 4.)

Example. The space X of all sequences $x=(\xi_j)$ in l^2 with only finitely many nonzero terms, taken with the metric on l^2, is not complete. A sequence of bounded linear operators T_n on X is defined by

$$T_nx=(\xi_1, 2\xi_2, 3\xi_3, \cdots, n\xi_n, \xi_{n+1}, \xi_{n+2}, \cdots),$$

so that T_nx has terms $j\xi_j$ if $j \leqq n$ and ξ_j if $j>n$. This sequence (T_n) converges strongly to the unbounded linear operator T defined by $Tx=(\eta_j)$, where $\eta_j=j\xi_j$.

However, if X is complete, the situation illustrated by this example cannot occur since then we have the basic

4.9-5 Lemma (Strong operator convergence). *Let $T_n \in B(X, Y)$, where X is a Banach space and Y a normed space. If (T_n) is strongly operator convergent with limit T, then $T \in B(X, Y)$.*

Proof. Linearity of T follows readily from that of T_n. Since $T_nx \longrightarrow Tx$ for every $x \in X$, the sequence (T_nx) is bounded for every x; cf. 1.4-2. Since X is complete, $(\|T_n\|)$ is bounded by the uniform boundedness theorem, say, $\|T_n\| \leqq c$ for all n. From this, it follows that $\|T_nx\| \leqq \|T_n\| \|x\| \leqq c\|x\|$. This implies $\|Tx\| \leqq c\|x\|$. ∎

A useful criterion for strong operator convergence is

4.9-6 Theorem (Strong operator convergence). *A sequence (T_n) of operators $T_n \in B(X, Y)$, where X and Y are Banach spaces, is strongly operator convergent if and only if:*

(A) The sequence $(\|T_n\|)$ is bounded.
(B) The sequence (T_nx) is Cauchy in Y for every x in a total subset M of X.

Proof. If $T_nx \longrightarrow Tx$ for every $x \in X$, then (A) follows from the uniform boundedness theorem (since X is complete) and (B) is trivial.

Conversely, suppose that (A) and (B) hold, so that, say, $\|T_n\| \leqq c$ for all n. We consider any $x \in X$ and show that (T_nx) converges strongly in Y. Let $\varepsilon>0$ be given. Since span M is dense in X, there is a $y \in \operatorname{span} M$ such that

$$\|x-y\|<\frac{\varepsilon}{3c}.$$

Since $y \in \operatorname{span} M$, the sequence $(T_n y)$ is Cauchy by (B). Hence there is an N such that

$$\|T_n y - T_m y\| < \frac{\varepsilon}{3} \qquad (m, n > N).$$

Using these two inequalities and applying the triangle inequality, we readily see that $(T_n x)$ is Cauchy in Y because for $m, n > N$ we obtain

$$\begin{aligned}\|T_n x - T_m x\| &\leqq \|T_n x - T_n y\| + \|T_n y - T_m y\| + \|T_m y - T_m x\| \\ &< \|T_n\| \|x - y\| + \frac{\varepsilon}{3} + \|T_m\| \|x - y\| \\ &< c\frac{\varepsilon}{3c} + \frac{\varepsilon}{3} + c\frac{\varepsilon}{3c} = \varepsilon.\end{aligned}$$

Since Y is complete, $(T_n x)$ converges in Y. Since $x \in X$ was arbitrary, this proves strong operator convergence of (T_n). ∎

4.9-7 Corollary (Functionals). *A sequence (f_n) of bounded linear functionals on a Banach space X is weak* convergent, the limit being a bounded linear functional on X, if and only if:*

(A) The sequence $(\|f_n\|)$ is bounded.
(B) The sequence $(f_n(x))$ is Cauchy for every x in a total subset M of X.

This has interesting applications. Two of them will be discussed in the next sections.

Problems

1. Show that uniform operator convergence $T_n \longrightarrow T$, $T_n \in B(X, Y)$, implies strong operator convergence with the same limit T.

2. If $S_n, T_n \in B(X, Y)$, and (S_n) and (T_n) are strongly operator convergent with limits S and T, show that $(S_n + T_n)$ is strongly operator convergent with the limit $S + T$.

3. Show that strong operator convergence in $B(X, Y)$ implies weak operator convergence with the same limit.

4. Show that weak convergence in footnote 6 implies weak* convergence. Show that the converse holds if X is reflexive.

5. Strong operator convergence does not imply uniform operator convergence. Illustrate this by considering $T_n = f_n \colon l^1 \longrightarrow \mathbf{R}$, where $f_n(x) = \xi_n$ and $x = (\xi_n)$.

6. Let $T_n \in B(X, Y)$, where $n = 1, 2, \cdots$. To motivate the term "uniform" in Def. 4.9-1, show that $T_n \longrightarrow T$ if and only if for every $\varepsilon > 0$ there is an N, *depending only on* ε, such that for all $n > N$ and all $x \in X$ of norm 1 we have

$$\|T_n x - Tx\| < \varepsilon.$$

7. Let $T_n \in B(X, Y)$, where X is a Banach space. If (T_n) is strongly operator convergent, show that $(\|T_n\|)$ is bounded.

8. Let $T_n \longrightarrow T$, where $T_n \in B(X, Y)$. Show that for every $\varepsilon > 0$ and every closed ball $K \subset X$ there is an N such that $\|T_n x - Tx\| < \varepsilon$ for all $n > N$ and all $x \in K$.

9. Show that $\|T\| \leqq \underline{\lim}_{n \to \infty} \|T_n\|$ in Lemma 4.9-5.

10. Let X be a separable Banach space and $M \subset X'$ a bounded set. Show that every sequence of elements of M contains a subsequence which is weak* convergent to an element of X'.

4.10 Application to Summability of Sequences

Weak* convergence has important applications in the theory of divergent sequences (and series). A divergent sequence has no limit in the usual sense. In that theory, one aims at associating with certain divergent sequences a "limit" in a generalized sense. A procedure for that purpose is called a *summability method*.

For instance, a divergent sequence $x = (\xi_k)$ being given, we may calculate the sequence $y = (\eta_n)$ of the arithmetic means

$$\eta_1 = \xi_1, \qquad \eta_2 = \frac{1}{2}(\xi_1 + \xi_2), \quad \cdots, \quad \eta_n = \frac{1}{n}(\xi_1 + \cdots + \xi_n), \quad \cdots.$$

This is an example of a summability method. If y converges with limit η (in the usual sense), we say that x is *summable* by the present method and has the *generalized limit* η. For instance, if

$$x=(0,1,0,1,0,\cdots) \qquad \text{then} \qquad y=(0,\tfrac{1}{2},\tfrac{1}{3},\tfrac{1}{2},\tfrac{2}{5},\cdots)$$

and x has the generalized limit $\frac{1}{2}$.

A summability method is called a *matrix method* if it can be represented in the form

$$y=Ax$$

where $x=(\xi_k)$ and $y=(\eta_n)$ are written as infinite column vectors and $A=(\alpha_{nk})$ is an infinite matrix; here, n, $k=1, 2, \cdots$. In the formula $y=Ax$ we used matrix multiplication, that is, y has the terms

$$\eta_n=\sum_{k=1}^{\infty}\alpha_{nk}\xi_k \qquad n=1,2,\cdots. \tag{1}$$

The above example illustrates a matrix method. (What is the matrix?)

Relevant terms are as follows. The method given by (1) is briefly called an A-*method* because the corresponding matrix is denoted by A. If the series in (1) all converge and $y=(\eta_n)$ converges in the usual sense, its limit is called the A-*limit* of x, and x is said to be A-*summable.* The set of all A-summable sequences is called the *range* of the A-method.

An A-method is said to be **regular** (or *permanent*) if its range includes all convergent sequences and if for every such sequence the A-limit equals the usual limit, that is, if

$$\xi_k \longrightarrow \xi \qquad \text{implies} \qquad \eta_n \longrightarrow \xi.$$

Obviously, regularity is a rather natural requirement. In fact, a method which is not applicable to certain convergent sequences or alters their limit would be of no practical use. A basic criterion for regularity is as follows.

4.10-1 Toeplitz Limit Theorem (Regular summability methods). *An A-summability method with matrix $A=(\alpha_{nk})$ is regular if and* *only if*

$$\text{(2)} \qquad \lim_{n\to\infty} \alpha_{nk} = 0 \qquad \text{for } k = 1, 2, \cdots$$

$$\text{(3)} \qquad \lim_{n\to\infty} \sum_{k=1}^{\infty} \alpha_{nk} = 1$$

$$\text{(4)} \qquad \sum_{k=1}^{\infty} |\alpha_{nk}| \leqq \gamma \qquad \text{for } n = 1, 2, \cdots$$

where γ is a constant which does not depend on n.

Proof. We show that
(a) (2) to (4) are necessary for regularity,
(b) (2) to (4) are sufficient for regularity.
The details are as follows.

(a) Suppose that the A-method is regular. Let x_k have 1 as the kth term and all other terms zero. For x_k we have $\eta_n = \alpha_{nk}$ in (1). Since x_k is convergent and has the limit 0, this shows that (2) must hold.

Furthermore, $x = (1, 1, 1, \cdots)$ has the limit 1. And from (1) we see that η_n now equals the series in (3). Consequently, (3) must hold.

We prove that (4) is necessary for regularity. Let c be the Banach space of all convergent sequences with norm defined by

$$\|x\| = \sup_j |\xi_j|,$$

cf. 1.5-3. Linear functionals f_{nm} on c are defined by

$$\text{(5)} \qquad f_{nm}(x) = \sum_{k=1}^{m} \alpha_{nk}\xi_k \qquad m, n = 1, 2, \cdots.$$

Each f_{nm} is bounded since

$$|f_{nm}(x)| \leqq \sup_j |\xi_j| \sum_{k=1}^{m} |\alpha_{nk}| = \left(\sum_{k=1}^{m} |\alpha_{nk}| \right) \|x\|.$$

Regularity implies the convergence of the series in (1) for all $x \in c$.

Hence (1) defines linear functionals $f_1, f_2, \cdots$ on c given by

$$\eta_n = f_n(x) = \sum_{k=1}^{\infty} \alpha_{nk}\xi_k \qquad n = 1, 2, \cdots. \tag{6}$$

From (5) we see that $f_{nm}(x) \longrightarrow f_n(x)$ as $m \longrightarrow \infty$ for all $x \in c$. This is weak* convergence, and f_n is bounded by Lemma 4.9-5 (with $T = f_n$). Also, $(f_n(x))$ converges for all $x \in c$, and $(\|f_n\|)$ is bounded by Corollary 4.9-7, say,

$$\|f_n\| \leqq \gamma \qquad \text{for all } n. \tag{7}$$

For an arbitrary fixed $m \in \mathbf{N}$ define

$$\xi_k^{(n,m)} = \begin{cases} |\alpha_{nk}|/\alpha_{nk} & \text{if } k \leqq m \text{ and } \alpha_{nk} \neq 0 \\ 0 & \text{if } k > m \text{ or } \alpha_{nk} = 0. \end{cases}$$

Then we have $x_{nm} = (\xi_k^{(n,m)}) \in c$. Also $\|x_{nm}\| = 1$ if $x_{nm} \neq 0$ and $\|x_{nm}\| = 0$ if $x_{nm} = 0$. Furthermore,

$$f_{nm}(x_{nm}) = \sum_{k=1}^{m} \alpha_{nk}\xi_k^{(n,m)} = \sum_{k=1}^{m} |\alpha_{nk}|$$

for all m. Hence

$$\text{(8)} \quad \begin{aligned} &\text{(a)} \qquad \sum_{k=1}^{m} |\alpha_{nk}| = f_{nm}(x_{nm}) \leqq \|f_{nm}\| \\ &\text{(b)} \qquad \sum_{k=1}^{\infty} |\alpha_{nk}| \leqq \|f_n\|. \end{aligned}$$

This shows that the series in (4) converges, and (4) follows from (7).

(b) We prove that (2) to (4) is sufficient for regularity. We define a linear functional f on c by

$$f(x) = \xi = \lim_{k \to \infty} \xi_k$$

where $x=(\xi_k)\in c$. Boundedness of f can be seen from

$$|f(x)|=|\xi|\leqq \sup_j |\xi_j|=\|x\|.$$

Let $M\subset c$ be the set of all sequences whose terms are equal from some term on, say, $x=(\xi_k)$ where

$$\xi_j=\xi_{j+1}=\xi_{j+2}=\cdots=\xi,$$

and j depends on x. Then $f(x)=\xi$, as above, and in (1) and (6) we obtain

$$\begin{aligned}\eta_n=f_n(x)&=\sum_{k=1}^{j-1}\alpha_{nk}\xi_k+\xi\sum_{k=j}^{\infty}\alpha_{nk}\\&=\sum_{k=1}^{j-1}\alpha_{nk}(\xi_k-\xi)+\xi\sum_{k=1}^{\infty}\alpha_{nk}.\end{aligned}$$

Hence by (2) and (3),

$$\eta_n=f_n(x)\longrightarrow 0+\xi\cdot 1=\xi=f(x) \tag{9}$$

for every $x\in M$.

We want to use Corollary 4.9-7 again. Hence we show next that the set M on which we have the convergence expressed in (9) is dense in c. Let $x=(\xi_k)\in c$ with $\xi_k\longrightarrow\xi$. Then for every $\varepsilon>0$ there is an N such that

$$|\xi_k-\xi|<\varepsilon \qquad\qquad \text{for } k\geqq N.$$

Clearly,

$$\tilde{x}=(\xi_1,\cdots,\xi_{N-1},\xi,\xi,\xi,\cdots)\in M$$

and

$$x-\tilde{x}=(0,\cdots,0,\xi_N-\xi,\xi_{N+1}-\xi,\cdots).$$

It follows that $\|x-\tilde{x}\|\leqq\varepsilon$. Since $x\in c$ was arbitrary, this shows that M is dense in c.

Finally, by (4),

$$|f_n(x)| \leqq \|x\| \sum_{k=1}^{\infty} |\alpha_{nk}| \leqq \gamma \|x\|$$

for every $x \in c$ and all n. Hence $\|f_n\| \leqq \gamma$, that is, $(\|f_n\|)$ is bounded. Furthermore, (9) means convergence $f_n(x) \longrightarrow f(x)$ for all x in the dense set M. By Corollary 4.9-7 this implies weak* convergence $f_n \xrightarrow{w^*} f$. Thus we have shown that if $\xi = \lim \xi_k$ exists, it follows that $\eta_n \longrightarrow \xi$. By definition, this means regularity and the theorem is proved. ▮

Problems

1. **Cesàro's summability method C_1** is defined by

$$\eta_n = \frac{1}{n}(\xi_1 + \cdots + \xi_n) \qquad n = 1, 2, \cdots,$$

that is, one takes arithmetic means. Find the corresponding matrix A.

2. Apply the method C_1 in Prob. 1 to the sequences

$$(1, 0, 1, 0, 1, 0, \cdots) \qquad \text{and} \qquad \left(1, 0, -\frac{1}{4}, -\frac{2}{8}, -\frac{3}{16}, -\frac{4}{32}, \cdots\right).$$

3. In Prob. 1, express (ξ_n) in terms of (η_n). Find (ξ_n) such that $(\eta_n) = (1/n)$.

4. Use the formula in Prob. 3 for obtaining a sequence which is not C_1-summable.

5. **Hölder's summability methods H_p** are defined as follows. H_1 is identical with C_1 in Prob. 1. The method H_2 consists of two successive applications of H_1, that is, we first take the arithmetic means and then again the arithmetic means of those means. H_3 consists of three successive applications of H_1, etc. Apply H_1 and H_2 to the sequence $(1, -3, 5, -7, 9, -11, \cdots)$. Comment.

6. **(Series)** An infinite series is said to be A-*summable* if the sequence of its partial sums is A-summable, and the A-limit of that sequence is called the A-*sum* of the series. Show that $1 + z + z^2 + \cdots$ is C_1-summable for $|z| = 1$, $z \neq 1$, and the C_1-sum is $1/(1-z)$.

7. (Cesàro's C_k-method) Given (ξ_n), let $\sigma_n^{(0)} = \xi_n$ and

$$\sigma_n^{(k)} = \sigma_0^{(k-1)} + \sigma_1^{(k-1)} + \cdots + \sigma_n^{(k-1)} \qquad (k \geqq 1, n = 0, 1, 2, \cdots).$$

If for a fixed $k \in \mathbf{N}$ we have $\eta_n^{(k)} = \sigma_n^{(k)} / \binom{n+k}{k} \longrightarrow \eta$, we say that (ξ_n) is C_k*-summable* and has the C_k*-limit* η. Show that the method has the advantage that $\sigma_n^{(k)}$ can be represented in terms of the ξ_j's in a very simple fashion, namely

$$\sigma_n^{(k)} = \sum_{\nu=0}^{n} \binom{n+k-1-\nu}{k-1} \xi_\nu.$$

8. Euler's method for series associates with a given series

$$\sum_{j=0}^{\infty} (-1)^j a_j \qquad \text{the transformed series} \qquad \sum_{n=0}^{\infty} \frac{\Delta^n a_0}{2^{n+1}}$$

where

$$\Delta^0 a_j = a_j, \qquad \Delta^n a_j = \Delta^{n-1} a_j - \Delta^{n-1} a_{j+1}, \qquad j = 1, 2, \cdots,$$

and $(-1)^j$ is written for convenience (hence the a_j need not be positive). It can be shown that the method is regular, so that the convergence of the given series implies that of the transformed series, the sum being the same. Show that the method gives

$$\ln 2 = 1 - \frac{1}{2} + \frac{1}{3} - \frac{1}{4} + - \cdots = \frac{1}{1 \cdot 2^1} + \frac{1}{2 \cdot 2^2} + \frac{1}{3 \cdot 2^3} + \frac{1}{4 \cdot 2^4} + \cdots.$$

9. Show that Euler's method in Prob. 8 yields

$$\frac{\pi}{4} = \text{arc tan } 1 = 1 - \frac{1}{3} + \frac{1}{5} - \frac{1}{7} + - \cdots = \frac{1}{2}\left(1 + \frac{1}{3} + \frac{1 \cdot 2}{3 \cdot 5} + \frac{1 \cdot 2 \cdot 3}{3 \cdot 5 \cdot 7} + \cdots\right).$$

10. Show that Euler's method yields the following result. Comment.

$$\sum_{n=0}^{\infty} \frac{(-1)^n}{4^n} = \frac{1}{2} \sum_{n=0}^{\infty} \left(\frac{3}{8}\right)^n.$$

4.11 Numerical Integration and Weak* Convergence

Weak* convergence has useful applications to numerical integration, differentiation and interpolation. In this section we consider numerical integration, that is, the problem of obtaining approximate values for a given integral

$$\int_a^b x(t)\,dt.$$

Since the problem is important in applications, various methods have been developed for that purpose, for example the trapezoidal rule, Simpson's rule and more complicated formulas by Newton-Cotes and Gauss. (For a review of some elementary facts, see the problem set at the end of the section.)

The common feature of those and other methods is that we first choose points in $[a, b]$, called *nodes*, and then approximate the unknown value of the integral by a linear combination of the values of x at the nodes. The nodes and the coefficients of that linear combination depend on the method but not on the integrand x. Of course, the usefulness of a method is largely determined by its accuracy, and one may want the accuracy to increase as the number of nodes gets larger.

In this section we shall see that functional analysis can offer help in that respect. In fact, we shall describe a general setting for those methods and consider the problem of convergence as the number of nodes increases.

We shall be concerned with continuous functions. This suggests introducing the Banach space $X = C[a, b]$ of all continuous real-valued functions on $J = [a, b]$, with norm defined by

$$\|x\| = \max_{t \in J} |x(t)|.$$

Then the above definite integral defines a linear functional f on X by means of

$$f(x) = \int_a^b x(t)\,dt. \tag{1}$$

To obtain a formula for numerical integration, we may proceed as in those aforementioned methods. Thus, for each positive integer n we choose $n+1$ real numbers

$$t_0^{(n)}, \cdots, t_n^{(n)} \qquad \text{(called **nodes**)}$$

such that

$$a \leqq t_0^{(n)} < \cdots < t_n^{(n)} \leqq b. \tag{2}$$

Then we choose $n+1$ real numbers

$$\alpha_0^{(n)}, \cdots, \alpha_n^{(n)} \qquad \text{(called **coefficients**)}$$

and define linear functionals f_n on X by setting

$$f_n(x) = \sum_{k=0}^{n} \alpha_k^{(n)} x(t_k^{(n)}) \qquad n = 1, 2, \cdots. \tag{3}$$

This defines a numerical process of integration, the value $f_n(x)$ being an approximation to $f(x)$, where x is given. To find out about the accuracy of the process, we consider the f_n's as follows.

Each f_n is bounded since $|x(t_k^{(n)})| \leqq \|x\|$ by the definition of the norm. Consequently,

$$|f_n(x)| \leqq \sum_{k=0}^{n} |\alpha_k^{(n)}| |x(t_k^{(n)})| \leqq \left(\sum_{k=0}^{n} |\alpha_k^{(n)}| \right) \|x\|. \tag{4}$$

For later use we show that f_n has the norm

$$\|f_n\| = \sum_{k=0}^{n} |\alpha_k^{(n)}|. \tag{5}$$

Indeed, (4) shows that $\|f_n\|$ cannot exceed the right-hand side of (5), and equality follows if we take an $x_0 \in X$ such that $|x_0(t)| \leqq 1$ on J and

$$x_0(t_k^{(n)}) = \operatorname{sgn} \alpha_k^{(n)} = \begin{cases} 1 & \text{if } \alpha_k^{(n)} \geqq 0 \\ -1 & \text{if } \alpha_k^{(n)} < 0 \end{cases}$$

since then $\|x_0\| = 1$ and

$$f_n(x_0) = \sum_{k=0}^{n} \alpha_k^{(n)} \operatorname{sgn} \alpha_k^{(n)} = \sum_{k=0}^{n} |\alpha_k^{(n)}|.$$

For a given $x \in X$, formula (3) yields an approximate value $f_n(x)$ of $f(x)$ in (1). Of course, we are interested in the accuracy, as was mentioned before, and we want it to increase with increasing n. This suggests the following concept.

4.11-1 Definition (Convergence). The numerical process of integration defined by (3) is said to be *convergent* for an $x \in X$ if for that x,

$$f_n(x) \longrightarrow f(x) \qquad (n \longrightarrow \infty), \tag{6}$$

where f is defined by (1). ∎

Furthermore, since exact integration of polynomials is easy, it is natural to make the following

4.11-2 Requirement. *For every n, if x is a polynomial of degree not exceeding n, then*

$$f_n(x) = f(x). \tag{7}$$

∎

Since the f_n's are linear, it suffices to require (7) for the $n+1$ powers defined by

$$x_0(t) = 1, \qquad x_1(t) = t, \qquad \cdots, \qquad x_n(t) = t^n.$$

In fact, then for a polynomial of degree n given by $x(t) = \sum \beta_j t^j$ we obtain

$$f_n(x) = \sum_{j=0}^{n} \beta_j f_n(x_j) = \sum_{j=0}^{n} \beta_j f(x_j) = f(x).$$

We see that we thus have the $n+1$ conditions

$$f_n(x_j) = f(x_j) \qquad j = 0, \cdots, n. \tag{8}$$

We show that these conditions can be fulfilled. $2n+2$ parameters are available, namely, $n+1$ nodes and $n+1$ coefficients. Hence we can choose some of them in an arbitrary fashion. Let us choose the nodes $t_k^{(n)}$, and let us prove that we can then determine those coefficients uniquely.

In (8) we now have $x_j(t_k^{(n)}) = (t_k^{(n)})^j$ so that (8) takes the form

$$\sum_{k=0}^{n} \alpha_k^{(n)}(t_k^{(n)})^j = \int_a^b t^j\,dt = \frac{1}{j+1}(b^{j+1} - a^{j+1}) \tag{9}$$

where $j = 0, \cdots, n$. For each fixed n this is a nonhomogeneous system of $n+1$ linear equations in the $n+1$ unknowns $\alpha_0^{(n)}, \cdots, \alpha_n^{(n)}$. A unique solution exists if the corresponding homogeneous system

$$\sum_{k=0}^{n} (t_k^{(n)})^j \gamma_k = 0 \qquad (j = 0, \cdots, n)$$

has only the trivial solution $\gamma_0 = 0, \cdots, \gamma_n = 0$ or, equally well, if the same holds for the system

$$\sum_{j=0}^{n} (t_k^{(n)})^j \gamma_j = 0 \qquad (k = 0, \cdots, n) \tag{10}$$

whose coefficient matrix is the transpose of the coefficient matrix of the previous system. This holds, since (10) means that the polynomial

$$\sum_{j=0}^{n} \gamma_j t^j$$

which is of degree n, is zero at the $n+1$ nodes, hence it must be identically zero, that is, all its coefficients γ_j are zero. ∎

Our result is that for every choice of nodes satisfying (2) there are uniquely determined coefficients such that 4.11-2 holds; hence the corresponding process is convergent for all polynomials. And we ask what additional conditions we should impose in order that the process is convergent for all real-valued continuous functions on $[a, b]$. A corresponding criterion was given by G. Pólya (1933):

4.11-3 Pólya Convergence Theorem (Numerical integration). *A process of numerical integration* (3) *which satisfies* 4.11-2 *converges for all real-valued continuous functions on* $[a, b]$ *if and only if there is a number* c *such that*

$$\sum_{k=0}^{n} |\alpha_k^{(n)}| \leqq c \qquad \text{for all } n. \tag{11}$$

Proof. The set W of all polynomials with real coefficients is dense in the real space $X = C[a, b]$, by the Weierstrass approximation theorem (proof below), and for every $x \in W$ we have convergence by 4.11-2. From (5) we see that $(\|f_n\|)$ is bounded if and only if (11) holds for some real number c. The theorem now follows from Corollary 4.9-7, since convergence $f_n(x) \longrightarrow f(x)$ for all $x \in X$ is weak* convergence $f_n \xrightarrow{w^*} f$. ∎

It is trivial that in this theorem we may replace the polynomials by any other set which is dense in the real space $C[a, b]$.

Furthermore, in most integration methods the coefficients are all nonnegative. Taking $x = 1$, we then have by 4.11-2

$$f_n(1) = \sum_{k=0}^{n} \alpha_k^{(n)} = \sum_{k=0}^{n} |\alpha_k^{(n)}| = f(1) = \int_a^b dt = b - a,$$

so that (11) holds. This proves

4.11-4 Steklov's Theorem (Numerical integration). *A process of numerical integration* (3) *which satisfies* 4.11-2 *and has nonnegative coefficients* $\alpha_k^{(n)}$ *converges for every continuous function.*

In the proof of 4.11-3 we used the

4.11-5 Weierstrass Approximation Theorem (Polynomials). *The set W of all polynomials with real coefficients is dense in the real space $C[a, b]$.*

Hence for every $x \in C[a, b]$ and given $\varepsilon > 0$ there exists a polynomial p such that $|x(t) - p(t)| < \varepsilon$ for all $t \in [a, b]$.

Proof. Every $x \in C[a, b]$ is uniformly continuous on $J = [a, b]$ since J is compact. Hence for any $\varepsilon > 0$ there is a y whose graph is an arc of a polygon such that

$$\max_{t \in J} |x(t) - y(t)| < \frac{\varepsilon}{3}. \tag{12}$$

We first assume that $x(a) = x(b)$ and $y(a) = y(b)$. Since y is piecewise linear and continuous, its Fourier coefficients have bounds of the form

$|a_0| < k$, $|a_m| < k/m^2$, $|b_m| < k/m^2$. This can be seen by applying integration by parts to the formulas for a_m and b_m (cf. 3.5-1 where we have $[a, b] = [0, 2\pi]$). (Cf. also Prob. 10 at the end of this section.) Hence for the Fourier series of y (representing the periodic extension of y, of period $b - a$), we have, writing $\kappa = 2\pi/(b-a)$ for simplicity,

$$\left| a_0 + \sum_{m=1}^{\infty} (a_m \cos \kappa mt + b_m \sin \kappa mt) \right| \tag{13}$$

$$\leqq 2k\left(1 + \sum_{m=1}^{\infty} \frac{1}{m^2}\right) = 2k\left(1 + \frac{1}{6}\pi^2\right).$$

This shows that the series converges uniformly on J. Together with the continuity of y it implies that that series has the sum y [see, for instance, W. Rogosinski (1959, p. 15)]. Consequently, for the nth partial sum s_n with sufficiently large n,

$$\max_{t \in J} |y(t) - s_n(t)| < \frac{\varepsilon}{3}. \tag{14}$$

The Taylor series of the cosine and sine functions in s_n also converge uniformly on J, so that there is a polynomial p (obtained, for instance, from suitable partial sums of those series) such that

$$\max_{t \in J} |s_n(t) - p(t)| < \frac{\varepsilon}{3}.$$

From this, (12), (14) and

$$|x(t) - p(t)| \leqq |x(t) - y(t)| + |y(t) - s_n(t)| + |s_n(t) - p(t)|$$

we have

$$\max_{t \in J} |x(t) - p(t)| < \varepsilon. \tag{15}$$

This takes care of every $x \in C[a, b]$ such that $x(a) = x(b)$. If $x(a) \neq x(b)$, take $u(t) = x(t) - \gamma(t - a)$ with γ such that $u(a) = u(b)$. Then for u there is a polynomial q satisfying $|u(t) - q(t)| < \varepsilon$ on J. Hence $p(t) = q(t) + \gamma(t - a)$ satisfies (15) because $x - p = u - q$. Since $\varepsilon > 0$ was arbitrary, we have shown that W is dense in $C[a, b]$. ∎

The first proof of the theorem was given by K. Weierstrass (1885), and there are many other proofs, for instance, one by S. N. Bernstein (1912), which yields a uniformly convergent sequence of polynomials ("Bernstein polynomials") explicitly in terms of x. Bernstein's proof can be found in K. Yosida (1971), pp. 8–9.

Problems

1. The **rectangular rule** is (Fig. 45)

$$\int_a^b x(t)\,dt \approx h[x(t_1^*)+\cdots+x(t_n^*)], \qquad h=\frac{b-a}{n}$$

where $t_k^* = a+(k-\frac{1}{2})h$. How is this formula obtained? What are the nodes and the coefficients? How can we obtain error bounds for the approximate value given by the formula?

2. The **trapezoidal rule** is (Fig. 46)

$$\int_{t_0}^{t_1} x(t)\,dt \approx \frac{h}{2}(x_0+x_1), \qquad h=\frac{b-a}{n}$$

or

$$\int_a^b x(t)\,dt \approx h(\tfrac{1}{2}x_0+x_1+\cdots+x_{n-1}+\tfrac{1}{2}x_n)$$

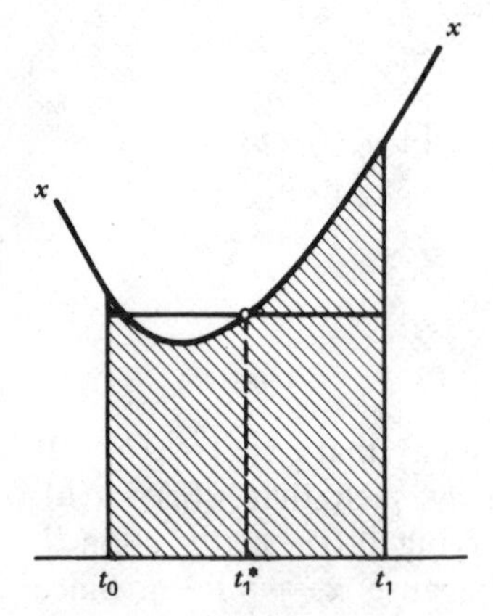

Fig. 45. Rectangular rule

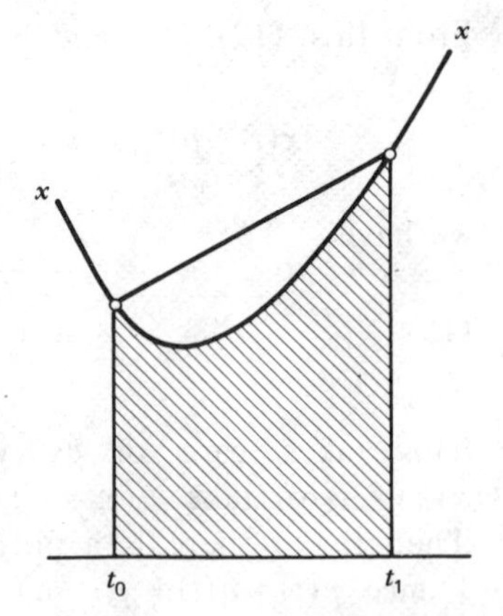

Fig. 46. Trapezoidal rule

where $x_k = x(t_k)$ and $t_k = a + kh$. Explain how the formulas are obtained if we approximate x by a piecewise linear function.

3. Simpson's rule is (Fig. 47)

$$\int_{t_0}^{t_2} x(t)\,dt \approx \frac{h}{3}(x_0 + 4x_1 + x_2) \qquad h = \frac{b-a}{n}$$

or

$$\int_a^b x(t)\,dt \approx \frac{h}{3}(x_0 + 4x_1 + 2x_2 + \cdots + 4x_{n-1} + x_n)$$

where n is even, $x_k = x(t_k)$ and $t_k = a + kh$. Show that these formulas are obtained if we approximate x on $[t_0, t_2]$ by a quadratic polynomial with values at t_0, t_1, t_2 equal to those of x; similarly on $[t_2, t_4]$, etc.

4. Let $f(x) = f_n(x) - \varepsilon_n(x)$ where f_n is the approximation obtained by the trapezoidal rule. Show that for any twice continuously differentiable function x we have the error bounds

$$k_n m_2^* \leqq \varepsilon_n(x) \leqq k_n m_2 \qquad \text{where} \qquad k_n = \frac{(b-a)^3}{12n^2}$$

and m_2 and m_2^* are the maximum and minimum of x'' on $[a, b]$.

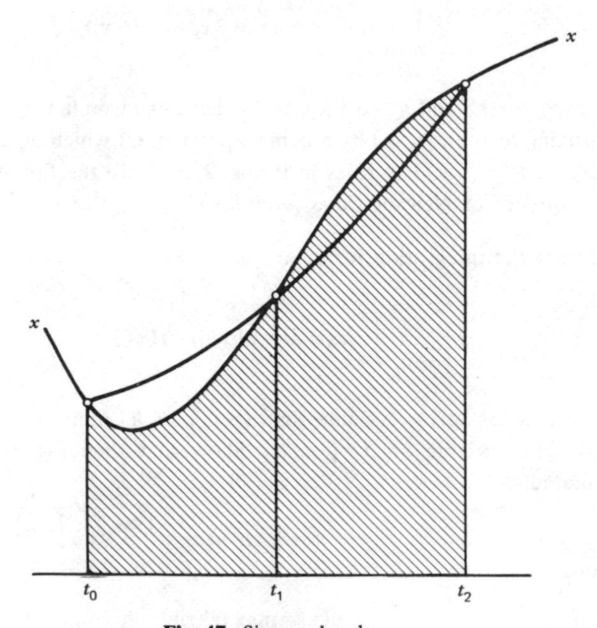

Fig. 47. Simpson's rule

5. Simpson's rule is widely used in practice. To get a feeling for the increase in accuracy, apply both the trapezoidal rule and Simpson's rule with $n = 10$ to the integral

$$I = \int_0^1 e^{-t^2}\, dt$$

and compare the values

$$0.746211 \qquad \text{and} \qquad 0.746825$$

with the actual value 0.746824 (exact to 6D).

t	e^{-t^2}
0	1.000 000
0.1	0.990 050
0.2	0.960 789
0.3	0.913 931
0.4	0.852 144
0.5	0.778 801
0.6	0.697 676
0.7	0.612 626
0.8	0.527 292
0.9	0.444 858
1.0	0.367 879

6. Using Prob. 4, show that bounds for the error of 0.746 211 in Prob. 5 are $-0.001\,667$ and $0.000\,614$, so that

$$0.745\,597 \leqq I \leqq 0.747\,878.$$

7. The **three-eights rule** is

$$\int_{t_0}^{t_3} x(t)\, dt \approx \frac{3h}{8}(x_0 + 3x_1 + 3x_2 + x_3)$$

where $x_k = x(t_k)$ and $t_k = a + kh$. Show that this formula is obtained if we approximate x on $[t_0, t_3]$ by a cubical polynomial which equals x at the nodes t_0, t_1, t_2, t_3. (The rules in Probs. 2, 3, 7 are the first members of the sequence of *Newton-Cotes formulas.*)

8. Consider the integration formula

$$\int_{-h}^{h} x(t)\, dt = 2hx(0) + r(x)$$

where r is the error. Assume that $x \in C^1[-h, h]$, that is, x is continuously differentiable on $J = [-h, h]$. Show that then the error can be estimated

$$|r(x)| \leqq h^2 p(x)$$

where

$$p(x) = \max_{t \in J} |x'(t)|.$$

Show that p is a seminorm on the vector space of those functions. (Cf. Prob. 12, Sec. 2.3.)

9. If x is real analytic, show that

$$\int_{-h}^{h} x(t)\,dt = 2h\left(x(0) + x''(0)\frac{h^2}{3!} + x^{\mathrm{IV}}(0)\frac{h^4}{5!} + \cdots\right). \tag{16}$$

Assume for the integral an approximate expression of the form $2h(\alpha_{-1}x(-h) + \alpha_0 x(0) + \alpha_1 x(h))$ and determine α_{-1}, α_0, α_1 so that as many powers $h, h^2, \cdots$ as possible agree with (16). Show that this gives Simpson's rule

$$\int_{-h}^{h} x(t)\,dt \approx \frac{h}{3}(x(-h) + 4x(0) + x(h)).$$

Why does this derivation show that the rule is exact for cubical polynomials?

10. In the proof of the Weierstrass approximation theorem we used bounds for the Fourier coefficients of a continuous and piecewise linear function. How can those bounds be obtained?

4.12 Open Mapping Theorem

We have discussed the Hahn-Banach theorem and the uniform boundedness theorem and shall now approach the third "big" theorem in this chapter, the open mapping theorem. It will be concerned with open mappings. These are mappings such that the image of every open set is an open set (definition below). Remembering our discussion of the importance of open sets (cf. Sec. 1.3), we understand that open mappings are of general interest. More specifically, the open mapping theorem states conditions under which a bounded linear operator is an open mapping. As in the uniform boundedness theorem we again need completeness, and the present theorem exhibits another reason why Banach spaces are more satisfactory than incomplete normed spaces. The theorem also gives conditions under which the inverse of a bounded linear operator is bounded. The proof of the open mapping theorem will be based on Baire's category theorem stated and explained in Sec. 4.7.

Let us begin by introducing the concept of an open mapping.

4.12-1 Definition (Open mapping). Let X and Y be metric spaces. Then $T\colon \mathscr{D}(T) \longrightarrow Y$ with domain $\mathscr{D}(T) \subset X$ is called an *open mapping* if for every open set in $\mathscr{D}(T)$ the image is an open set in Y. ∎

Note that if a mapping is not surjective, one must take care to distinguish between the assertions that the mapping is open as a mapping from its domain

(*a*) into Y,

(*b*) onto its range.

(*b*) is weaker than (*a*). For instance, if $X \subset Y$, the mapping $x \longmapsto x$ of X into Y is open if and only if X is an open subset of Y, whereas the mapping $x \longmapsto x$ of X onto its range (which is X) is open in any case.

Furthermore, to avoid confusion, we should remember that, by Theorem 1.3-4, a continuous mapping $T\colon X \longrightarrow Y$ has the property that for every open set in Y the inverse image is an open set in X. This does *not* imply that T maps open sets in X onto open sets in Y. For example, the mapping $\mathbf{R} \longrightarrow \mathbf{R}$ given by $t \longmapsto \sin t$ is continuous but maps $(0, 2\pi)$ onto $[-1, 1]$.

4.12-2 Open Mapping Theorem, Bounded Inverse Theorem. *A bounded linear operator T from a Banach space X onto a Banach space Y is an open mapping. Hence if T is bijective, T^{-1} is continuous and thus bounded.*

The proof will readily follow from

4.12-3 Lemma (Open unit ball). *A bounded linear operator T from a Banach space X onto a Banach space Y has the property that the image $T(B_0)$ of the open unit ball $B_0 = B(0; 1) \subset X$ contains an open ball about $0 \in Y$.*

Proof. Proceeding stepwise, we prove:

(*a*) The closure of the image of the open ball $B_1 = B(0; \frac{1}{2})$ contains an open ball B^*.

(*b*) $\overline{T(B_n)}$ contains an open ball V_n about $0 \in Y$, where $B_n = B(0; 2^{-n}) \subset X$.

(*c*) $T(B_0)$ contains an open ball about $0 \in Y$.

The details are as follows.

(a) In connection with subsets $A \subset X$ we shall write αA (α a scalar) and $A + w$ ($w \in X$) to mean

$$\alpha A = \{x \in X \mid x = \alpha a, \ a \in A\} \qquad \text{(1)} \qquad \text{(Fig. 48)}$$

$$A + w = \{x \in X \mid x = a + w, \ a \in A\} \qquad \text{(2)} \qquad \text{(Fig. 49)}$$

and similarly for subsets of Y.

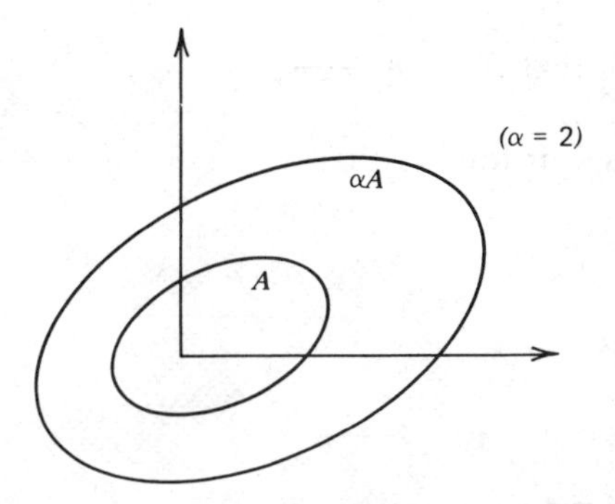

Fig. 48. Illustration of formula (1)

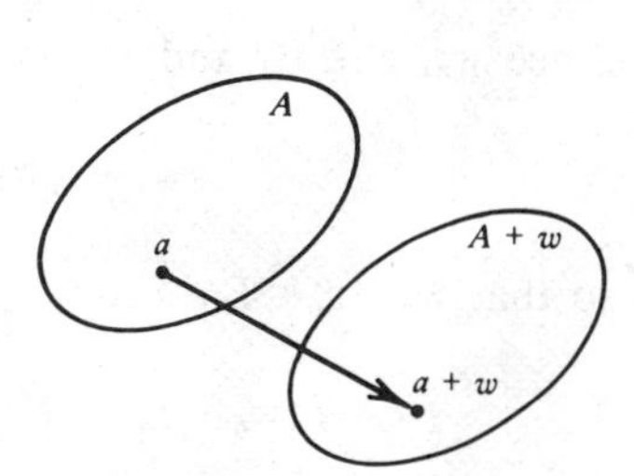

Fig. 49. Illustration of formula (2)

We consider the open ball $B_1 = B(0; \frac{1}{2}) \subset X$. Any fixed $x \in X$ is in kB_1 with real k sufficiently large ($k > 2 \|x\|$). Hence

$$X = \bigcup_{k=1}^{\infty} kB_1.$$

Since T is surjective and linear,

$$Y = T(X) = T\left(\bigcup_{k=1}^{\infty} kB_1\right) = \bigcup_{k=1}^{\infty} kT(B_1) = \bigcup_{k=1}^{\infty} \overline{kT(B_1)}. \qquad \text{(3)}$$

Note that by taking closures we did not add further points to the union since that union was already the whole space Y. Since Y is complete, it is nonmeager in itself, by Baire's category theorem 4.7-2. Hence, noting that (3) is similar to (1) in 4.7-2, we conclude that a $\overline{kT(B_1)}$ must contain some open ball. This implies that $\overline{T(B_1)}$ also contains an open ball, say, $B^* = B(y_0; \varepsilon) \subset \overline{T(B_1)}$. It follows that

$$B^* - y_0 = B(0; \varepsilon) \subset \overline{T(B_1)} - y_0. \qquad \text{(4)}$$

(b) We prove that $B^* - y_0 \subset \overline{T(B_0)}$, where B_0 is given in the theorem. This we do by showing that [cf. (4)]

$$\overline{T(B_1)} - y_0 \subset \overline{T(B_0)}. \tag{5}$$

Let $y \in \overline{T(B_1)} - y_0$. Then $y + y_0 \in \overline{T(B_1)}$, and we remember that $y_0 \in \overline{T(B_1)}$, too. By 1.4-6(*a*) there are

$$u_n = Tw_n \in T(B_1) \qquad \text{such that} \qquad u_n \longrightarrow y + y_0,$$

$$v_n = Tz_n \in T(B_1) \qquad \text{such that} \qquad v_n \longrightarrow y_0.$$

Since w_n, $z_n \in B_1$ and B_1 has radius $\frac{1}{2}$, it follows that

$$\|w_n - z_n\| \leqq \|w_n\| + \|z_n\| < \tfrac{1}{2} + \tfrac{1}{2} = 1,$$

so that $w_n - z_n \in B_0$. From

$$T(w_n - z_n) = Tw_n - Tz_n = u_n - v_n \quad \longrightarrow \quad y$$

we see that $y \in \overline{T(B_0)}$. Since $y \in \overline{T(B_1)} - y_0$ was arbitrary, this proves (5). From (4) we thus have

$$B^* - y_0 = B(0; \varepsilon) \subset \overline{T(B_0)}. \tag{6}$$

Let $B_n = B(0; 2^{-n}) \subset X$. Since T is linear, $\overline{T(B_n)} = 2^{-n}\overline{T(B_0)}$. From (6) we thus obtain

$$V_n = B(0; \varepsilon/2^n) \subset \overline{T(B_n)}. \tag{7}$$

(c) We finally prove that

$$V_1 = B(0; \tfrac{1}{2}\varepsilon) \subset T(B_0)$$

by showing that every $y \in V_1$ is in $T(B_0)$. So let $y \in V_1$. From (7) with $n = 1$ we have $V_1 \subset \overline{T(B_1)}$. Hence $y \in \overline{T(B_1)}$. By 1.4-6(*a*) there must be a $v \in T(B_1)$ close to y, say, $\|y - v\| < \varepsilon/4$. Now $v \in T(B_1)$ implies $v = Tx_1$ for some $x_1 \in B_1$. Hence

$$\|y - Tx_1\| < \frac{\varepsilon}{4}.$$

From this and (7) with $n=2$ we see that $y-Tx_1 \in V_2 \subset \overline{T(B_2)}$. As before we conclude that there is an $x_2 \in B_2$ such that

$$\|(y-Tx_1)-Tx_2\| < \frac{\varepsilon}{8}.$$

Hence $y-Tx_1-Tx_2 \in V_3 \subset \overline{T(B_3)}$, and so on. In the nth step we can choose an $x_n \in B_n$ such that

$$\left\| y - \sum_{k=1}^{n} Tx_k \right\| < \frac{\varepsilon}{2^{n+1}} \qquad (n=1,2,\cdots). \tag{8}$$

Let $z_n = x_1 + \cdots + x_n$. Since $x_k \in B_k$, we have $\|x_k\| < 1/2^k$. This yields for $n > m$

$$\|z_n - z_m\| \leqq \sum_{k=m+1}^{n} \|x_k\| < \sum_{k=m+1}^{\infty} \frac{1}{2^k} \longrightarrow 0$$

as $m \longrightarrow \infty$. Hence (z_n) is Cauchy. (z_n) converges, say, $z_n \longrightarrow x$ because X is complete. Also $x \in B_0$ since B_0 has radius 1 and

$$\sum_{k=1}^{\infty} \|x_k\| \leqslant \sum_{k=2}^{\infty} \frac{1}{2^k} < \frac{1}{2} + \frac{1}{2} = 1 \tag{9}$$

Since T is continuous, $Tz_n \longrightarrow Tx$, and (8) shows that $Tx = y$. Hence $y \in T(B_0)$. ∎

***Proof of Theorem* 4.12-2.** We prove that for every open set $A \subset X$ the image $T(A)$ is open in Y. This we do by showing that for every $y = Tx \in T(A)$ the set $T(A)$ contains an open ball about $y = Tx$.

Let $y = Tx \in T(A)$. Since A is open, it contains an open ball with center x. Hence $A - x$ contains an open ball with center 0; let the radius of the ball be r and set $k = 1/r$, so that $r = 1/k$. Then $k(A-x)$ contains the open unit ball $B(0; 1)$. Lemma 4.12-3 now implies that $T(k(A-x)) = k[T(A) - Tx]$ contains an open ball about 0, and so does $T(A) - Tx$. Hence $T(A)$ contains an open ball about $Tx = y$. Since $y \in T(A)$ was arbitrary, $T(A)$ is open.

Finally, if $T^{-1}: Y \longrightarrow X$ exists, it is continuous by Theorem 1.3-4 because T is open. Since T^{-1} is linear by Theorem 2.6-10, it is bounded by Theorem 2.7-9. ∎

Problems

1. Show that $T: \mathbf{R}^2 \longrightarrow \mathbf{R}$ defined by $(\xi_1, \xi_2) \longmapsto (\xi_1)$ is open. Is the mapping $\mathbf{R}^2 \longmapsto \mathbf{R}^2$ given by $(\xi_1, \xi_2) \longmapsto (\xi_1, 0)$ an open mapping?

2. Show that an open mapping need not map closed sets onto closed sets.

3. Extending (1) and (2), we can define

$$A + B = \{x \in X \mid x = a + b, a \in A, b \in B\},$$

where $A, B \subset X$. To become familiar with this notation find αA, $A + w$, $A + A$, where $A = \{1, 2, 3, 4\}$. Explain Fig. 50.

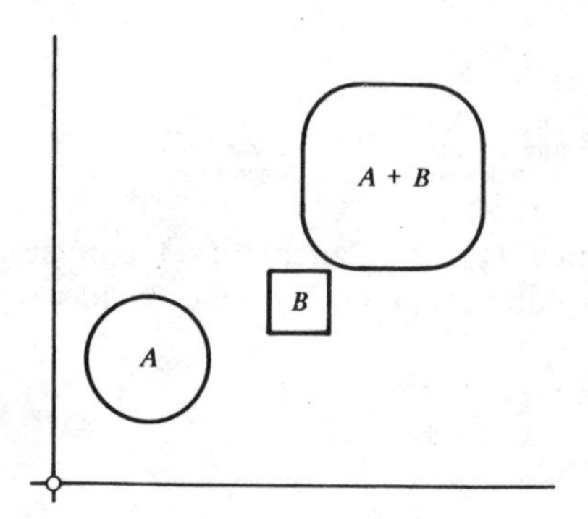

Fig. 50. Sets A, B and $A + B$ in the plane

4. Show that in (9) the inequality is strict.

5. Let X be the normed space whose points are sequences of complex numbers $x = (\xi_j)$ with only finitely many nonzero terms and norm defined by $\|x\| = \sup_j |\xi_j|$. Let $T: X \longrightarrow X$ be defined by

$$y = Tx = \left(\xi_1, \frac{1}{2}\xi_2, \frac{1}{3}\xi_3, \cdots\right).$$

Show that T is linear and bounded but T^{-1} is unbounded. Does this contradict 4.12-2?

6. Let X and Y be Banach spaces and $T: X \longrightarrow Y$ an injective bounded linear operator. Show that $T^{-1}: \mathcal{R}(T) \longrightarrow X$ is bounded if and only if $\mathcal{R}(T)$ is closed in Y.

7. Let $T: X \longrightarrow Y$ be a bounded linear operator, where X and Y are Banach spaces. If T is bijective, show that there are positive real numbers a and b such that $a\|x\| \leqq \|Tx\| \leqq b\|x\|$ for all $x \in X$.

8. (Equivalent norms) Let $\|\cdot\|_1$ and $\|\cdot\|_2$ be norms on a vector space X such that $X_1 = (X, \|\cdot\|_1)$ and $X_2 = (X, \|\cdot\|_2)$ are complete. If $\|x_n\|_1 \longrightarrow 0$ always implies $\|x_n\|_2 \longrightarrow 0$, show that convergence in X_1 implies convergence in X_2 and conversely, and there are positive numbers a and b such that for all $x \in X$,

$$a\|x\|_1 \leqq \|x\|_2 \leqq b\|x\|_1.$$

(Note that then these norms are equivalent; cf. Def. 2.4-4.)

9. Let $X_1 = (X, \|\cdot\|_1)$ and $X_2 = (X, \|\cdot\|_2)$ be Banach spaces. If there is a constant c such that $\|x\|_1 \leqq c\|x\|_2$ for all $x \in X$, show that there is a constant k such that $\|x\|_2 \leqq k\|x\|_1$ for all $x \in X$ (so that the two norms are equivalent; cf. Def. 2.4-4).

10. From Sec. 1.3 we know that the set $\mathcal{T}$ of all open subsets of a metric space X is called a *topology* for X. Consequently, each norm on a vector space X defines a topology for X. If we have two norms on X such that $X_1 = (X, \|\cdot\|_1)$ and $X_2 = (X, \|\cdot\|_2)$ are Banach spaces and the topologies $\mathcal{T}_1$ and $\mathcal{T}_2$ defined by $\|\cdot\|_1$ and $\|\cdot\|_2$ satisfy $\mathcal{T}_1 \supset \mathcal{T}_2$, show that $\mathcal{T}_1 = \mathcal{T}_2$.

4.13 Closed Linear Operators. Closed Graph Theorem

Not all linear operators of practical importance are bounded. For instance, the differential operator in 2.7-5 is unbounded, and in quantum mechanics and other applications one needs unbounded operators quite frequently. However, practically all of the linear operators which the analyst is likely to use are so-called closed linear operators. This makes it worthwhile to give an introduction to these operators. In this section we define closed linear operators on normed spaces and consider some of their properties, in particular in connection with the important closed graph theorem which states sufficient conditions under which a closed linear operator on a Banach space is bounded.

A more detailed study of closed and other unbounded operators in Hilbert spaces will be presented in Chap. 10 and applications to quantum mechanics in Chap. 11.

Let us begin with the definition.

4.13-1 Definition (Closed linear operator). Let X and Y be normed spaces and $T: \mathfrak{D}(T) \longrightarrow Y$ a linear operator with domain $\mathfrak{D}(T) \subset X$. Then T is called a *closed linear operator* if its *graph*

$$\mathscr{G}(T) = \{(x, y) \mid x \in \mathfrak{D}(T), \ y = Tx\}$$

is closed in the normed space $X \times Y$, where the two algebraic operations of a vector space in $X \times Y$ are defined as usual, that is

$$(x_1, y_1) + (x_2, y_2) = (x_1 + x_2, y_1 + y_2)$$

$$\alpha(x, y) = (\alpha x, \alpha y)$$

(α a scalar) and the norm on $X \times Y$ is defined by[7]

$$\|(x, y)\| = \|x\| + \|y\|. \tag{1}$$

∎

Under what conditions will a closed linear operator be bounded? An answer is given by the important

4.13-2 Closed Graph Theorem. *Let X and Y be Banach spaces and $T: \mathfrak{D}(T) \longrightarrow Y$ a closed linear operator, where $\mathfrak{D}(T) \subset X$. Then if $\mathfrak{D}(T)$ is closed in X, the operator T is bounded.*

Proof. We first show that $X \times Y$ with norm defined by (1) is complete. Let (z_n) be Cauchy in $X \times Y$, where $z_n = (x_n, y_n)$. Then for every $\varepsilon > 0$ there is an N such that

$$\|z_n - z_m\| = \|x_n - x_m\| + \|y_n - y_m\| < \varepsilon \qquad (m, n > N). \tag{2}$$

Hence (x_n) and (y_n) are Cauchy in X and Y, respectively, and converge, say, $x_n \longrightarrow x$ and $y_n \longrightarrow y$, because X and Y are complete. This implies that $z_n \longrightarrow z = (x, y)$ since from (2) with $m \longrightarrow \infty$ we have $\|z_n - z\| \leqq \varepsilon$ for $n > N$. Since the Cauchy sequence (z_n) was arbitrary, $X \times Y$ is complete.

[7] For other norms, see Prob. 2.

By assumption, $\mathscr{G}(T)$ is closed in $X \times Y$ and $\mathscr{D}(T)$ is closed in X. Hence $\mathscr{G}(T)$ and $\mathscr{D}(T)$ are complete by 1.4-7. We now consider the mapping

$$P: \mathscr{G}(T) \longrightarrow \mathscr{D}(T)$$
$$(x, Tx) \longmapsto x.$$

P is linear. P is bounded because

$$\|P(x, Tx)\| = \|x\| \leqq \|x\| + \|Tx\| = \|(x, Tx)\|.$$

P is bijective; in fact the inverse mapping is

$$P^{-1}: \mathscr{D}(T) \longrightarrow \mathscr{G}(T)$$
$$x \longmapsto (x, Tx).$$

Since $\mathscr{G}(T)$ and $\mathscr{D}(T)$ are complete, we can apply the bounded inverse theorem 4.12-2 and see that P^{-1} is bounded, say, $\|(x, Tx)\| \leqq b\|x\|$ for some b and all $x \in \mathscr{D}(T)$. Hence T is bounded because

$$\|Tx\| \leqq \|Tx\| + \|x\| = \|(x, Tx)\| \leqq b\,\|x\|$$

for all $x \in \mathscr{D}(T)$. ∎

By definition, $\mathscr{G}(T)$ is closed if and only if $z = (x, y) \in \overline{\mathscr{G}(T)}$ implies $z \in \mathscr{G}(T)$. From Theorem 1.4-6(*a*) we see that $z \in \overline{\mathscr{G}(T)}$ if and only if there are $z_n = (x_n, Tx_n) \in \mathscr{G}(T)$ such that $z_n \longrightarrow z$, hence

$$x_n \longrightarrow x, \qquad Tx_n \longrightarrow y; \tag{3}$$

and $z = (x, y) \in \mathscr{G}(T)$ if and only if $x \in \mathscr{D}(T)$ and $y = Tx$. This proves the following useful criterion which expresses a property that is often taken as a definition of closedness of a linear operator.

4.13-3 Theorem (Closed linear operator). *Let $T: \mathscr{D}(T) \longrightarrow Y$ be a linear operator, where $\mathscr{D}(T) \subset X$ and X and Y are normed spaces. Then T is closed if and only if it has the following property. If $x_n \longrightarrow x$, where $x_n \in \mathscr{D}(T)$, and $Tx_n \longrightarrow y$, then $x \in \mathscr{D}(T)$ and $Tx = y$.*

Note well that this property is different from the following property of a bounded linear operator. If a linear operator T is bounded and thus continuous, and if (x_n) is a sequence in $\mathscr{D}(T)$ which converges in $\mathscr{D}(T)$, then (Tx_n) also converges; cf. 1.4-8. This need not hold for a closed linear operator. However, if T is closed and two sequences (x_n) and $(\tilde{x}_n)$ in the domain of T converge with the same limit and if the corresponding sequences (Tx_n) and $(T\tilde{x}_n)$ both converge, then the latter have the same limit (cf. Prob. 6).

4.13-4 Example (Differential operator). Let $X = C[0, 1]$ and

$$T: \mathscr{D}(T) \longrightarrow X$$

$$x \longmapsto x'$$

where the prime denotes differentiation and $\mathscr{D}(T)$ is the subspace of functions $x \in X$ which have a continuous derivative. Then T is not bounded, but is closed.

Proof. We see from 2.7-5 that T is not bounded. We prove that T is closed by applying Theorem 4.13-3. Let (x_n) in $\mathscr{D}(T)$ be such that both (x_n) and (Tx_n) converge, say,

$$x_n \longrightarrow x \qquad \text{and} \qquad Tx_n = x_n' \longrightarrow y.$$

Since convergence in the norm of $C[0, 1]$ is uniform convergence on $[0, 1]$, from $x_n' \longrightarrow y$ we have

$$\int_0^t y(\tau)\, d\tau = \int_0^t \lim_{n\to\infty} x_n'(\tau)\, d\tau = \lim_{n\to\infty} \int_0^t x_n'(\tau)\, d\tau = x(t) - x(0),$$

that is,

$$x(t) = x(0) + \int_0^t y(\tau)\, d\tau.$$

This shows that $x \in \mathscr{D}(T)$ and $x' = y$. Theorem 4.13-3 now implies that T is closed. ∎

It is worth noting that in this example, $\mathscr{D}(T)$ is not closed in X since T would then be bounded by the closed graph theorem.

Closedness does not imply boundedness of a linear operator. Conversely, boundedness does not imply closedness.

Proof. The first statement is illustrated by 4.13-4 and the second one by the following example. Let $T: \mathcal{D}(T) \longrightarrow \mathcal{D}(T) \subset X$ be the identity operator on $\mathcal{D}(T)$, where $\mathcal{D}(T)$ is a proper dense subspace of a normed space X. Then it is trivial that T is linear and bounded. However, T is not closed. This follows immediately from Theorem 4.13-3 if we take an $x \in X - \mathcal{D}(T)$ and a sequence (x_n) in $\mathcal{D}(T)$ which converges to x. ∎

Our present discussion seems to indicate that in connection with unbounded operators the determination of domains and extension problems may play a basic role. This is in fact so, as we shall see in more detail in Chap. 10. The statement which we have just proved is rather negative in spirit. On the positive side we have

4.13-5 Lemma (Closed operator). *Let $T: \mathcal{D}(T) \longrightarrow Y$ be a bounded linear operator with domain $\mathcal{D}(T) \subset X$, where X and Y are normed spaces. Then:*

(a) *If $\mathcal{D}(T)$ is a closed subset of X, then T is closed.*

(b) *If T is closed and Y is complete, then $\mathcal{D}(T)$ is a closed subset of X.*

Proof. **(a)** If (x_n) is in $\mathcal{D}(T)$ and converges, say, $x_n \longrightarrow x$, and is such that (Tx_n) also converges, then $x \in \overline{\mathcal{D}(T)} = \mathcal{D}(T)$ since $\mathcal{D}(T)$ is closed, and $Tx_n \longrightarrow Tx$ since T is continuous. Hence T is closed by Theorem 4.13-3.

(b) For $x \in \overline{\mathcal{D}(T)}$ there is a sequence (x_n) in $\mathcal{D}(T)$ such that $x_n \longrightarrow x$; cf. 1.4-6. Since T is bounded,

$$\|Tx_n - Tx_m\| = \|T(x_n - x_m)\| \leqq \|T\| \, \|x_n - x_m\|.$$

This shows that (Tx_n) is Cauchy. (Tx_n) converges, say, $Tx_n \longrightarrow y \in Y$ because Y is complete. Since T is closed, $x \in \mathcal{D}(T)$ by 4.13-3 (and $Tx = y$). Hence $\mathcal{D}(T)$ is closed because $x \in \overline{\mathcal{D}(T)}$ was arbitrary. ∎

Problems

1. Prove that (1) defines a norm on $X \times Y$.

2. Other frequently used norms on the product $X \times Y$ of normed spaces

X and Y are defined by

$$\|(x, y)\| = \max\{\|x\|, \|y\|\}$$

and

$$\|(x, y)\|_0 = (\|x\|^2 + \|y\|^2)^{1/2}.$$

Verify that these are norms.

3. Show that the graph $\mathscr{G}(T)$ of a linear operator $T: X \longrightarrow Y$ is a vector subspace of $X \times Y$.

4. If X and Y in Def. 4.13-1 are Banach spaces, show that $V = X \times Y$ with norm defined by (1) is a Banach space.

5. (Inverse) If the inverse T^{-1} of a closed linear operator exists, show that T^{-1} is a closed linear operator.

6. Let T be a closed linear operator. If two sequences (x_n) and $(\tilde{x}_n)$ in $\mathscr{D}(T)$ converge with the same limit x and if (Tx_n) and $(T\tilde{x}_n)$ both converge, show that (Tx_n) and $(T\tilde{x}_n)$ have the same limit.

7. Obtain the second statement in Theorem 4.12-2 from the closed graph theorem.

8. Let X and Y be normed spaces and let $T: X \longrightarrow Y$ be a closed linear operator. (*a*) Show that the image A of a compact subset $C \subset X$ is closed in Y. (*b*) Show that the inverse image B of a compact subset $K \subset Y$ is closed in X. (Cf. Def. 2.5-1.)

9. If $T: X \longrightarrow Y$ is a closed linear operator, where X and Y are normed spaces and Y is compact, show that T is bounded.

10. Let X and Y be normed spaces and X compact. If $T: X \longrightarrow Y$ is a bijective closed linear operator, show that T^{-1} is bounded.

11. (Null space) Show that the null space $\mathscr{N}(T)$ of a closed linear operator $T: X \longrightarrow Y$ is a closed subspace of X.

12. Let X and Y be normed spaces. If $T_1: X \longrightarrow Y$ is a closed linear operator and $T_2 \in B(X, Y)$, show that $T_1 + T_2$ is a closed linear operator.

13. Let T be a closed linear operator with domain $\mathscr{D}(T)$ in a Banach space X and range $\mathscr{R}(T)$ in a normed space Y. If T^{-1} exists and is bounded, show that $\mathscr{R}(T)$ is closed.

14. Assume that the terms of the series $u_1+u_2+\cdots$ are continuously differentiable functions on the interval $J=[0, 1]$ and that the series is uniformly convergent on J and has the sum x. Furthermore, suppose that $u_1'+u_2'+\cdots$ also converges uniformly on J. Show that then x is continuously differentiable on $(0, 1)$ and $x'=u_1'+u_2'+\cdots$.

15. **(Closed extension)** Let $T\colon \mathfrak{D}(T) \longrightarrow Y$ be a linear operator with graph $\mathscr{G}(T)$, where $\mathfrak{D}(T) \subset X$ and X and Y are Banach spaces. Show that T has an extension $\tilde{T}$ which is a closed linear operator with graph $\overline{\mathscr{G}(T)}$ if and only if $\overline{\mathscr{G}(T)}$ does not contain an element of the form $(0, y)$, where $y \neq 0$.

CHAPTER 5

FURTHER APPLICATIONS: BANACH FIXED POINT THEOREM

This chapter is optional. Its material will not be used in the remaining chapters.

Prerequisite is Chap. 1 (but not Chaps. 2 to 4), so that the present chapter can also be studied immediately after Chap. 1 if so desired.

The Banach fixed point theorem is important as a source of existence and uniqueness theorems in different branches of analysis. In this way the theorem provides an impressive illustration of the unifying power of functional analytic methods and of the usefulness of fixed point theorems in analysis.

Brief orientation about main content

The *Banach fixed point theorem* or *contraction theorem* 5.1-2 concerns certain mappings (*contractions*, cf. 5.1-1) of a complete metric space into itself. It states conditions sufficient for the existence and uniqueness of a *fixed point* (point that is mapped onto itself). The theorem also gives an iterative process by which we can obtain approximations to the fixed point and error bounds (cf. 5.1-3). We consider three important fields of application of the theorem, namely,

linear algebraic equations (Sec. 5.2),
ordinary differential equations (Sec. 5.3),
integral equations (Sec. 5.4).

There are other applications (for instance, partial differential equations) whose discussion would require more prerequisites.

5.1 Banach Fixed Point Theorem

A **fixed point** of a mapping $T: X \longrightarrow X$ of a set X into itself is an $x \in X$ which is mapped onto itself (is "kept fixed" by T), that is,

$$Tx = x,$$

the image Tx coincides with x.

For example, a translation has no fixed points, a rotation of the plane has a single fixed point (the center of rotation), the mapping $x \longmapsto x^2$ of $\mathbf{R}$ into itself has two fixed points (0 and 1) and the projection $(\xi_1, \xi_2) \longmapsto \xi_1$ of $\mathbf{R}^2$ onto the ξ_1-axis has infinitely many fixed points (all points of the ξ_1-axis).

The Banach fixed point theorem to be stated below is an existence and uniqueness theorem for fixed points of certain mappings, and it also gives a constructive procedure for obtaining better and better approximations to the fixed point (the solution of the practical problem). This procedure is called an **iteration**. By definition, this is a method such that we choose an arbitrary x_0 in a given set and calculate recursively a sequence $x_0, x_1, x_2, \cdots$ from a relation of the form

$$x_{n+1} = Tx_n \qquad n = 0, 1, 2, \cdots;$$

that is, we choose an arbitrary x_0 and determine successively $x_1 = Tx_0$, $x_2 = Tx_1, \cdots$.

Iteration procedures are used in nearly every branch of applied mathematics, and convergence proofs and error estimates are very often obtained by an application of Banach's fixed point theorem (or more difficult fixed point theorems). Banach's theorem gives sufficient conditions for the existence (and uniqueness) of a fixed point for a class of mappings, called contractions. The definition is as follows.

5.1-1 Definition (Contraction). Let $X = (X, d)$ be a metric space. A mapping $T: X \longrightarrow X$ is called a *contraction on* X if there is a positive real number $\alpha < 1$ such that for all $x, y \in X$

$$\text{(1)} \qquad d(Tx, Ty) \leqq \alpha d(x, y) \qquad (\alpha < 1).$$

Geometrically this means that any points x and y have images that are closer together than those points x and y; more precisely, the ratio $d(Tx, Ty)/d(x, y)$ does not exceed a constant α which is strictly less than 1.

5.1-2 Banach Fixed Point Theorem (Contraction Theorem). *Consider a metric space $X = (X, d)$, where $X \neq \varnothing$. Suppose that X is complete and let $T: X \longrightarrow X$ be a contraction on X. Then T has precisely one fixed point.*

Proof. We construct a sequence (x_n) and show that it is Cauchy, so that it converges in the complete space X, and then we prove that its limit x is a fixed point of T and T has no further fixed points. This is the idea of the proof.

We choose any $x_0 \in X$ and define the "iterative sequence" (x_n) by

$$x_0, \qquad x_1 = Tx_0, \qquad x_2 = Tx_1 = T^2 x_0, \quad \cdots, \quad x_n = T^n x_0, \quad \cdots. \tag{2}$$

Clearly, this is the sequence of the images of x_0 under repeated application of T. We show that (x_n) is Cauchy. By (1) and (2),

$$\begin{aligned} d(x_{m+1}, x_m) &= d(Tx_m, Tx_{m-1}) \\ &\leqq \alpha d(x_m, x_{m-1}) \\ &= \alpha d(Tx_{m-1}, Tx_{m-2}) \\ &\leqq \alpha^2 d(x_{m-1}, x_{m-2}) \\ &\cdots \leqq \alpha^m d(x_1, x_0). \end{aligned} \tag{3}$$

Hence by the triangle inequality and the formula for the sum of a geometric progression we obtain for $n > m$

$$\begin{aligned} d(x_m, x_n) &\leqq d(x_m, x_{m+1}) + d(x_{m+1}, x_{m+2}) + \cdots + d(x_{n-1}, x_n) \\ &\leqq (\alpha^m + \alpha^{m+1} + \cdots + \alpha^{n-1}) d(x_0, x_1) \\ &= \alpha^m \frac{1 - \alpha^{n-m}}{1 - \alpha} d(x_0, x_1). \end{aligned}$$

Since $0 < \alpha < 1$, in the numerator we have $1 - \alpha^{n-m} < 1$. Consequently,

$$d(x_m, x_n) \leqq \frac{\alpha^m}{1 - \alpha} d(x_0, x_1) \qquad (n > m). \tag{4}$$

On the right, $0 < \alpha < 1$ and $d(x_0, x_1)$ is fixed, so that we can make the right-hand side as small as we please by taking m sufficiently large (and $n > m$). This proves that (x_m) is Cauchy. Since X is complete, (x_m) converges, say, $x_m \longrightarrow x$. We show that this limit x is a fixed point of the mapping T.

From the triangle inequality and (1) we have

$$\begin{aligned} d(x, Tx) &\leqq d(x, x_m) + d(x_m, Tx) \\ &\leqq d(x, x_m) + \alpha d(x_{m-1}, x) \end{aligned}$$

and can make the sum in the second line smaller than any preassigned $\varepsilon > 0$ because $x_m \longrightarrow x$. We conclude that $d(x, Tx) = 0$, so that $x = Tx$ by (M2), Sec. 1.1. This shows that x is a fixed point of T.

x is the only fixed point of T because from $Tx = x$ and $T\tilde{x} = \tilde{x}$ we obtain by (1)

$$d(x, \tilde{x}) = d(Tx, T\tilde{x}) \leqq \alpha d(x, \tilde{x})$$

which implies $d(x, \tilde{x}) = 0$ since $\alpha < 1$. Hence $x = \tilde{x}$ by (M2) and the theorem is proved. ∎

5.1-3 Corollary (Iteration, error bounds). *Under the conditions of Theorem* 5.1-2 *the iterative sequence* (2) *with arbitrary $x_0 \in X$ converges to the unique fixed point x of T. Error estimates are the* **prior estimate**

$$d(x_m, x) \leqq \frac{\alpha^m}{1-\alpha} d(x_0, x_1) \tag{5}$$

and the **posterior estimate**

$$d(x_m, x) \leqq \frac{\alpha}{1-\alpha} d(x_{m-1}, x_m). \tag{6}$$

Proof. The first statement is obvious from the previous proof. Inequality (5) follows from (4) by letting $n \longrightarrow \infty$. We derive (6). Taking $m = 1$ and writing y_0 for x_0 and y_1 for x_1, we have from (5)

$$d(y_1, x) \leqq \frac{\alpha}{1-\alpha} d(y_0, y_1).$$

Setting $y_0 = x_{m-1}$, we have $y_1 = Ty_0 = x_m$ and obtain (6). ∎

The prior error bound (5) can be used at the beginning of a calculation for estimating the number of steps necessary to obtain a given accuracy. (6) can be used at intermediate stages or at the end of a calculation. It is at least as accurate as (5) and may be better; cf. Prob. 8.

From the viewpoint of applied mathematics the situation is not yet completely satisfactory because it frequently happens that a mapping T is a contraction not on the entire space X but merely on a subset Y of X. However, if Y is closed, it is complete by Theorem 1.4-7, so that T

has a fixed point x in Y, and $x_m \longrightarrow x$ as before, provided we impose a suitable restriction on the choice of x_0, so that the x_m's remain in Y. A typical and practically useful result of this kind is as follows.

5.1-4 Theorem (Contraction on a ball). *Let T be a mapping of a complete metric space $X=(X, d)$ into itself. Suppose T is a contraction on a closed ball $Y=\{x \mid d(x, x_0) \leqq r\}$, that is, T satisfies (1) for all $x, y \in Y$. Moreover, assume that*

$$d(x_0, Tx_0) < (1-\alpha)r. \tag{7}$$

Then the iterative sequence (2) converges to an $x \in Y$. This x is a fixed point of T and is the only fixed point of T in Y.

Proof. We merely have to show that all x_m's as well as x lie in Y. We put $m=0$ in (4), change n to m and use (7) to get

$$d(x_0, x_m) \leqq \frac{1}{1-\alpha} d(x_0, x_1) < r.$$

Hence all x_m's are in Y. Also $x \in Y$ since $x_m \longrightarrow x$ and Y is closed. The assertion of the theorem now follows from the proof of Banach's theorem 5.1-2. ∎

For later use the reader may give the simple proof of

5.1-5 Lemma (Continuity). *A contraction T on a metric space X is a continuous mapping.*

Problems

1. Give further examples of mappings in elementary geometry which have (*a*) a single fixed point, (*b*) infinitely many fixed points.

2. Let $X=\{x \in \mathbf{R} \mid x \geqq 1\} \subset \mathbf{R}$ and let the mapping $T: X \longrightarrow X$ be defined by $Tx = x/2 + x^{-1}$. Show that T is a contraction and find the smallest α.

3. Illustrate with an example that in Theorem 5.1-2, completeness is essential and cannot be omitted.

4. It is important that in Banach's theorem 5.1-2 the condition (1) cannot be replaced by $d(Tx, Ty) < d(x, y)$ when $x \neq y$. To see this, consider

$X = \{x \mid 1 \leqq x < +\infty\}$, taken with the usual metric of the real line, and $T: X \longrightarrow X$ defined by $x \longmapsto x + x^{-1}$. Show that $|Tx - Ty| < |x - y|$ when $x \neq y$, but the mapping has no fixed points.

5. If $T: X \longrightarrow X$ satisfies $d(Tx, Ty) < d(x, y)$ when $x \neq y$ and T has a fixed point, show that the fixed point is unique; here, (X, d) is a metric space.

6. If T is a contraction, show that T^n $(n \in \mathbf{N})$ is a contraction. If T^n is a contraction for an $n > 1$, show that T need not be a contraction.

7. Prove Lemma 5.1-5.

8. Show that the error bounds given by (5) form a proper monotone decreasing sequence. Show that (6) is at least as good as (5).

9. Show that in the case of Theorem 5.1-4 we have the prior error estimate $d(x_m, x) < \alpha^m r$ and the posterior estimate (6).

10. In analysis, a usual sufficient condition for the convergence of an iteration $x_n = g(x_{n-1})$ is that g be continuously differentiable and

$$|g'(x)| \leqq \alpha < 1.$$

Verify this by the use of Banach's fixed point theorem.

11. To find approximate numerical solutions of a given equation $f(x) = 0$, we may convert the equation to the form $x = g(x)$, choose an initial value x_0 and compute

$$x_n = g(x_{n-1}) \qquad n = 1, 2, \cdots.$$

Suppose that g is continuously differentiable on some interval $J = [x_0 - r, x_0 + r]$ and satisfies $|g'(x)| \leqq \alpha < 1$ on J as well as

$$|g(x_0) - x_0| < (1 - \alpha)r.$$

Show that then $x = g(x)$ has a unique solution x on J, the iterative sequence (x_m) converges to that solution, and one has the error estimates

$$|x - x_m| < \alpha^m r, \qquad |x - x_m| \leqq \frac{\alpha}{1 - \alpha} |x_m - x_{m-1}|.$$

12. Using Banach's theorem 5.1-2, set up an iteration process for solving $f(x)=0$ if f is continuously differentiable on an interval $J=[a, b]$, $f(a)<0, f(b)>0$ and $0<k_1 \leqq f'(x) \leqq k_2$ $(x \in J)$; use $g(x)=x-\lambda f(x)$ with a suitable λ.

13. Consider an iteration process for solving $f(x)=x^3+x-1=0$; proceed as follows. (*a*) Show that one possibility is

$$x_n = g(x_{n-1}) = (1+x_{n-1}^2)^{-1}.$$

Choose $x_0=1$ and perform three steps. Is $|g'(x)|<1$? (Cf. Prob. 10.) Show that the iteration can be illustrated by Fig. 51. (*b*) Estimate the errors by (5). (*c*) We can write $f(x)=0$ in the form $x=1-x^3$. Is this form suitable for iteration? Try $x_0=1$, $x_0=0.5$, $x_0=2$ and see what happens.

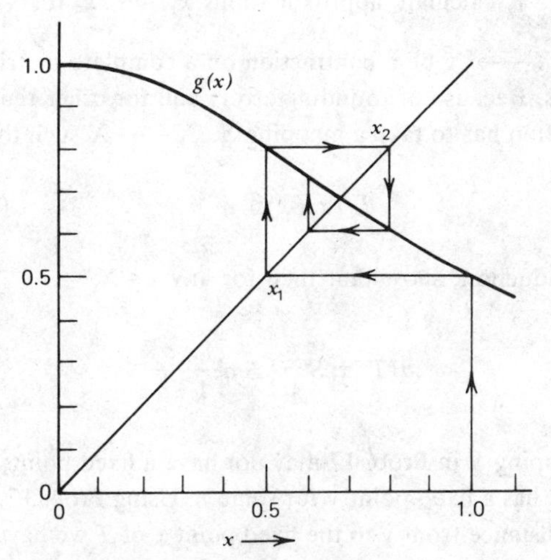

Fig. 51. Iteration in Prob. 13(*a*)

14. Show that another iteration process for the equation in Prob. 13 is

$$x_n = x_{n-1}^{1/2}(1+x_{n-1}^2)^{-1/2}.$$

Choose $x_0=1$. Determine x_1, x_2, x_3. What is the reason for the rapid convergence? (The real root is 0.682 328, 6D.)

15. (Newton's method) Let f be real-valued and twice continuously differentiable on an interval $[a, b]$, and let $\hat{x}$ be a simple zero of f in

(a, b). Show that *Newton's method* defined by

$$x_{n+1} = g(x_n), \qquad g(x_n) = x_n - \frac{f(x_n)}{f'(x_n)}$$

is a contraction in some neighborhood of $\hat{x}$ (so that the iterative sequence converges to $\hat{x}$ for any x_0 sufficiently close to $\hat{x}$).

16. (Square root) Show that an iteration for calculating the square root of a given positive number c is

$$x_{n+1} = g(x_n) = \frac{1}{2}\left(x_n + \frac{c}{x_n}\right)$$

where $n = 0, 1, \cdots$. What condition do we get from Prob. 10? Starting from $x_0 = 1$, calculate approximations $x_1, \cdots, x_4$ for $\sqrt{2}$.

17. Let $T: X \longrightarrow X$ be a contraction on a complete metric space, so that (1) holds. Because of rounding errors and for other reasons, instead of T one often has to take a mapping $S: X \longrightarrow X$ such that for all $x \in X$,

$$d(Tx, Sx) \leqq \eta \qquad (\eta > 0, \text{suitable}).$$

Using induction, show that then for any $x \in X$,

$$d(T^m x, S^m x) \leqq \eta \frac{1-\alpha^m}{1-\alpha} \qquad (m = 1, 2, \cdots).$$

18. The mapping S in Prob. 17 may not have a fixed point; but in practice, S^n often has a fixed point y for some n. Using Prob. 17, show that then for the distance from y to the fixed point x of T we have

$$d(x, y) \leqq \frac{\eta}{1-\alpha}.$$

19. In Prob. 17, let $x = Tx$ and $y_m = S^m y_0$. Using (5) and Prob. 17, show that

$$d(x, y_m) \leqq \frac{1}{1-\alpha} [\eta + \alpha^m d(y_0, Sy_0)].$$

What is the significance of this formula in applications?

20. (Lipschitz condition) A mapping T: $[a, b] \longrightarrow [a, b]$ is said to satisfy a *Lipschitz condition* with a *Lipschitz constant* k on $[a, b]$ if there is a constant k such that for all $x, y \in [a, b]$,

$$|Tx - Ty| \leqq k\,|x - y|.$$

(*a*) Is T a contraction? (*b*) If T is continuously differentiable, show that T satisfies a Lipschitz condition. (*c*) Does the converse of (*b*) hold?

5.2 Application of Banach's Theorem to Linear Equations

Banach's fixed point theorem has important applications to iteration methods for solving systems of linear algebraic equations and yields sufficient conditions for convergence and error bounds.

To understand the situation, we first remember that for solving such a system there are various *direct methods* (methods that would yield the exact solution after finitely many arithmetical operations if the precision—the word length of our computer—were unlimited); a familiar example is Gauss' elimination method (roughly, a systematic version of the elimination taught in school). However, an iteration, or *indirect method*, may be more efficient if the system is special, for instance, if it is *sparse*, that is, if it consists of many equations but has only a small number of nonzero coefficients. (Vibrational problems, networks and difference approximations of partial differential equations often lead to sparse systems.) Moreover, the usual direct methods require about $n^3/3$ arithmetical operations (n = number of equations = number of unknowns), and for large n, rounding errors may become quite large, whereas in an iteration, errors due to roundoff (or even blunders) may be damped out eventually. In fact, iteration methods are frequently used to improve "solutions" obtained by direct methods.

To apply Banach's theorem, we need a complete metric space and a contraction mapping on it. We take the set X of all ordered n-tuples of real numbers, written

$$x = (\xi_1, \cdots, \xi_n), \qquad y = (\eta_1, \cdots, \eta_n), \qquad z = (\zeta_1, \cdots, \zeta_n),$$

etc. On X we define a metric d by

$$(1) \qquad d(x, z) = \max_j |\xi_j - \zeta_j|.$$

$X = (X, d)$ is complete; the simple proof is similar to that in Example 1.5-1.

On X we define $T\colon X \longrightarrow X$ by

$$(2) \qquad y = Tx = Cx + b$$

where $C = (c_{jk})$ is a fixed real $n \times n$ matrix and $b \in X$ a fixed vector. Here and later in this section, all vectors are *column vectors*, because of the usual conventions of matrix multiplication.

Under what condition will T be a contraction? Writing (2) in components, we have

$$\eta_j = \sum_{k=1}^{n} c_{jk}\xi_k + \beta_j \qquad j = 1, \cdots, n,$$

where $b = (\beta_j)$. Setting $w = (\omega_j) = Tz$, we thus obtain from (1) and (2)

$$\begin{aligned} d(y, w) = d(Tx, Tz) &= \max_j |\eta_j - \omega_j| \\ &= \max_j \left| \sum_{k=1}^{n} c_{jk}(\xi_k - \zeta_k) \right| \\ &\leqq \max_i |\xi_i - \zeta_i| \max_j \sum_{k=1}^{n} |c_{jk}| \\ &= d(x, z) \max_j \sum_{k=1}^{n} |c_{jk}|. \end{aligned}$$

We see that this can be written $d(y, w) \leqq \alpha d(x, z)$, where

$$(3) \qquad \alpha = \max_j \sum_{k=1}^{n} |c_{jk}|.$$

Banach's theorem 5.1-2 thus yields

5.2-1 Theorem (Linear equations). *If a system*

$$x = Cx + b \qquad (C = (c_{jk}),\ b \text{ given}) \tag{4}$$

of n linear equations in n unknowns $\xi_1, \cdots, \xi_n$ *(the components of x) satisfies*

$$\sum_{k=1}^{n} |c_{jk}| < 1 \qquad (j = 1, \cdots, n), \tag{5}$$

it has precisely one solution x. This solution can be obtained as the limit of the iterative sequence $(x^{(0)}, x^{(1)}, x^{(2)}, \cdots)$, *where* $x^{(0)}$ *is arbitrary and*

$$x^{(m+1)} = Cx^{(m)} + b \qquad m = 0, 1, \cdots. \tag{6}$$

Error bounds are [cf. (3)]

$$d(x^{(m)}, x) \leqq \frac{\alpha}{1-\alpha}\, d(x^{(m-1)}, x^{(m)}) \leqq \frac{\alpha^m}{1-\alpha}\, d(x^{(0)}, x^{(1)}). \tag{7}$$

Condition (5) is sufficient for convergence. It is a **row sum criterion** because it involves *row sums* obtained by summing the absolute values of the elements in a row of C. If we replaced (1) by other metrics, we would obtain other conditions. Two cases of practical importance are included in Probs. 7 and 8.

How is Theorem 5.2-1 related to methods used in practice? A system of n linear equations in n unknowns is usually written

$$Ax = c, \tag{8}$$

where A is an n-rowed square matrix. Many iterative methods for (8) with $\det A \neq 0$ are such that one writes $A = B - G$ with a suitable nonsingular matrix B. Then (8) becomes

$$Bx = Gx + c$$

or

$$x = B^{-1}(Gx + c).$$

This suggests the iteration (6) where

$$C = B^{-1}G, \qquad b = B^{-1}c. \tag{9}$$

Let us illustrate this by two standard methods, the Jacobi iteration, which is largely of theoretical interest, and the Gauss-Seidel iteration, which is widely used in applied mathematics.

5.2-2 Jacobi iteration. This iteration method is defined by

$$\xi_j^{(m+1)} = \frac{1}{a_{jj}}\left(\gamma_j - \sum_{\substack{k=1 \\ k\neq j}}^{n} a_{jk}\xi_k^{(m)}\right) \qquad j = 1, \cdots, n, \tag{10}$$

where $c = (\gamma_j)$ in (8) and we assume $a_{jj} \neq 0$ for $j = 1, \cdots, n$. This iteration is suggested by solving the jth equation in (8) for ξ_j. It is not difficult to verify that (10) can be written in the form (6) with

$$C = -D^{-1}(A - D), \qquad b = D^{-1}c \tag{11}$$

where $D = \operatorname{diag}(a_{jj})$ is the diagonal matrix whose nonzero elements are those of the principal diagonal of A.

Condition (5) applied to C in (11) is sufficient for the convergence of the Jacobi iteration. Since C in (11) is relatively simple, we can express (5) directly in terms of the elements of A. The result is the row sum criterion for the Jacobi iteration

$$\sum_{\substack{k=1 \\ k\neq j}}^{n} \left|\frac{a_{jk}}{a_{jj}}\right| < 1 \qquad j = 1, \cdots, n, \tag{12}$$

or

$$\sum_{\substack{k=1 \\ k\neq j}}^{n} |a_{jk}| < |a_{jj}| \qquad j = 1, \cdots, n. \tag{12*}$$

This shows that, roughly speaking, convergence is guaranteed if the elements in the principal diagonal of A are sufficiently large.

Note that in the Jacobi iteration some components of $x^{(m+1)}$ may already be available at a certain instant but are not used while the computation of the remaining components is still in progress, that is, all the components of a new approximation are introduced simultaneously at the end of an iterative cycle. We express this fact by saying that the Jacobi iteration is a method of *simultaneous corrections.*

5.2-3 Gauss-Seidel iteration. This is a method of *successive corrections*, in which at every instant all of the latest known components are used. The method is defined by

$$\xi_j^{(m+1)} = \frac{1}{a_{jj}}\left(\gamma_j - \sum_{k=1}^{j-1} a_{jk}\xi_k^{(m+1)} - \sum_{k=j+1}^{n} a_{jk}\xi_k^{(m)}\right), \tag{13}$$

where $j = 1, \cdots, n$ and we again assume $a_{jj} \neq 0$ for all j.

We obtain a matrix form of (13) by writing (Fig. 52)

$$A = -L + D - U$$

where D is as in the Jacobi iteration and L and U are lower and upper triangular, respectively, with principal diagonal elements all zero, the minus signs being a matter of convention and convenience. We now imagine that each equation in (13) is multiplied by a_{jj}. Then we can write the resulting system in the form

$$Dx^{(m+1)} = c + Lx^{(m+1)} + Ux^{(m)}$$

or

$$(D-L)x^{(m+1)} = c + Ux^{(m)}.$$

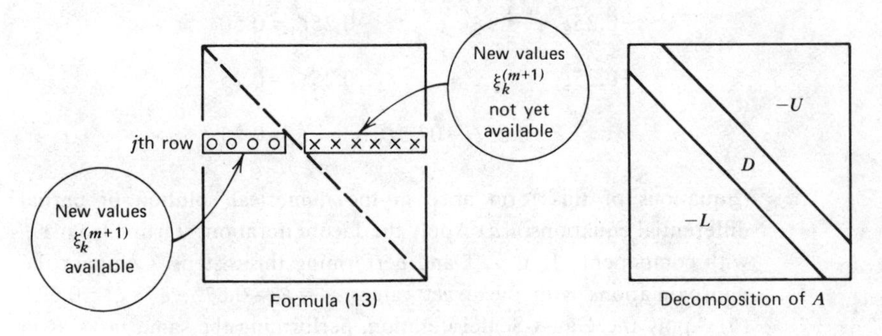

Fig. 52. Explanation of the Gauss-Seidel formulas (13) and (14)

Multiplication by $(D-L)^{-1}$ gives (6) with

$$C=(D-L)^{-1}U, \qquad b=(D-L)^{-1}c. \tag{14}$$

Condition (5) applied to C in (14) is sufficient for the convergence of the Gauss-Seidel iteration. Since C is complicated, the remaining practical problem is to get simpler conditions sufficient for the validity of (5). We mention without proof that (12) is sufficient, but there are better conditions, which the interested reader can find in J. Todd (1962), pp. 494, 495, 500.

Problems

1. Verify (11) and (14).

2. Consider the system

$$\begin{aligned} 5\xi_1 - \xi_2 &= 7 \\ -3\xi_1 + 10\xi_2 &= 24. \end{aligned}$$

(*a*) Determine the exact solution. (*b*) Apply the Jacobi iteration. Does C satisfy (5)? Starting from $x^{(0)}$ with components 1, 1, calculate $x^{(1)}$, $x^{(2)}$ and the error bounds (7) for $x^{(2)}$. Compare these bounds with the actual error of $x^{(2)}$. (*c*) Apply the Gauss-Seidel iteration, performing the same tasks as in (*b*).

3. Consider the system

$$\begin{aligned} \xi_1 - 0.25\xi_2 - 0.25\xi_3 \qquad\qquad &= 0.50 \\ -0.25\xi_1 + \xi_2 \qquad\qquad - 0.25\xi_4 &= 0.50 \\ -0.25\xi_1 \qquad\qquad + \xi_3 - 0.25\xi_4 &= 0.25 \\ -0.25\xi_2 - 0.25\xi_3 + \xi_4 &= 0.25. \end{aligned}$$

(Equations of this form arise in the numerical solution of partial differential equations.) (*a*) Apply the Jacobi iteration, starting from $x^{(0)}$ with components 1, 1, 1, 1 and performing three steps. Compare the approximations with the exact values $\xi_1 = \xi_2 = 0.875$, $\xi_3 = \xi_4 = 0.625$. (*b*) Apply the Gauss-Seidel iteration, performing the same tasks as in (*a*).

4. Gershgorin's theorem states that if λ is an eigenvalue of a square matrix $C=(c_{jk})$, then for some j, where $1 \leqq j \leqq n$,

$$|c_{jj}-\lambda| \leqq \sum_{\substack{k=1 \\ k \neq j}}^{n} |c_{jk}|.$$

(An *eigenvalue* of C is a number λ such that $Cx=\lambda x$ for some $x \neq 0$.) (*a*) Show that (4) can be written $Kx=b$, where $K=I-C$, and Gershgorin's theorem and (5) together imply that K cannot have an eigenvalue 0 (so that K is nonsingular, that is, $\det K \neq 0$, and $Kx=b$ has a unique solution). (*b*) Show that (5) and Gershgorin's theorem imply that C in (6) has spectral radius less than 1. (It can be shown that this is necessary and sufficient for the convergence of the iteration. The *spectral radius* of C is $\max_j |\lambda_j|$, where $\lambda_1, \cdots, \lambda_n$ are the eigenvalues of C.)

5. An example of a system for which the Jacobi iteration diverges whereas the Gauss-Seidel iteration converges is

$$\begin{aligned} 2\xi_1 + \xi_2 + \xi_3 &= 4 \\ \xi_1 + 2\xi_2 + \xi_3 &= 4 \\ \xi_1 + \xi_2 + 2\xi_3 &= 4. \end{aligned}$$

Starting from $x^{(0)}=0$, verify divergence of the Jacobi iteration and perform the first few steps of the Gauss-Seidel iteration to obtain the impression that the iteration seems to converge to the exact solution $\xi_1=\xi_2=\xi_3=1$.

6. It is plausible to think that the Gauss-Seidel iteration is better than the Jacobi iteration in all cases. Actually, the two methods are not comparable. This is surprising. For example, in the case of the system

$$\begin{aligned} \xi_1 \qquad + \xi_3 &= 2 \\ -\xi_1 + \xi_2 \qquad &= 0 \\ \xi_1 + 2\xi_2 - 3\xi_3 &= 0 \end{aligned}$$

the Jacobi iteration converges whereas the Gauss-Seidel iteration diverges. Derive these two facts from the necessary and sufficient conditions stated in Prob. 4(*b*).

7. (Column sum criterion) To the metric in (1) there corresponds the condition (5). If we use on X the metric d_1 defined by

$$d_1(x, z) = \sum_{j=1}^{n} |\xi_j - \zeta_j|,$$

show that instead of (5) we obtain the condition

$$\sum_{j=1}^{n} |c_{jk}| < 1 \qquad (k = 1, \cdots, n). \tag{15}$$

8. (Square sum criterion) To the metric in (1) there corresponds the condition (5). If we use on X the Euclidean metric d_2 defined by

$$d_2(x, z) = \left[\sum_{j=1}^{n} (\xi_j - \zeta_j)^2\right]^{1/2},$$

show that instead of (5) we obtain the condition

$$\sum_{j=1}^{n} \sum_{k=1}^{n} c_{jk}{}^2 < 1. \tag{16}$$

9. (Jacobi iteration) Show that for the Jacobi iteration the sufficient convergence conditions (5), (15) and (16) take the form

$$\sum_{\substack{k=1 \\ k \neq j}}^{n} \frac{|a_{jk}|}{|a_{jj}|} < 1, \qquad \sum_{\substack{j=1 \\ j \neq k}}^{n} \frac{|a_{jk}|}{|a_{jj}|} < 1, \qquad \sum_{\substack{j=1 \\ j \neq k}}^{n} \sum_{k=1}^{n} \frac{a_{jk}{}^2}{a_{jj}{}^2} < 1.$$

10. Find a matrix C which satisfies (5) but neither (15) nor (16).

5.3 Application of Banach's Theorem to Differential Equations

The most interesting applications of Banach's fixed point theorem arise in connection with function spaces. The theorem then yields existence and uniqueness theorems for differential and integral equations, as we shall see.

In fact, in this section let us consider an *explicit ordinary differential equation of the first order*

(1a) $$x' = f(t, x) \qquad (' = d/dt).$$

An **initial value problem** for such an equation consists of the equation and an *initial condition*

(1b) $$x(t_0) = x_0$$

where t_0 and x_0 are given real numbers.

We shall use Banach's theorem to prove the famous Picard's theorem which, while not the strongest of its type that is known, plays a vital role in the theory of ordinary differential equations. The idea of approach is quite simple: (1) will be converted to an integral equation, which defines a mapping T, and the conditions of the theorem will imply that T is a contraction such that its fixed point becomes the solution of our problem.

5.3-1 Picard's Existence and Uniqueness Theorem (Ordinary differential equations). *Let f be continuous on a rectangle* (Fig. 53)

$$R = \{(t, x) \mid |t - t_0| \leqq a, |x - x_0| \leqq b\}$$

and thus bounded on R, say (see Fig. 54)

(2) $$|f(t, x)| \leqq c \qquad \text{for all } (t, x) \in R$$

Suppose that f satisfies a **Lipschitz condition** *on R with respect to its second argument, that is, there is a constant k* (Lipschitz constant) *such*

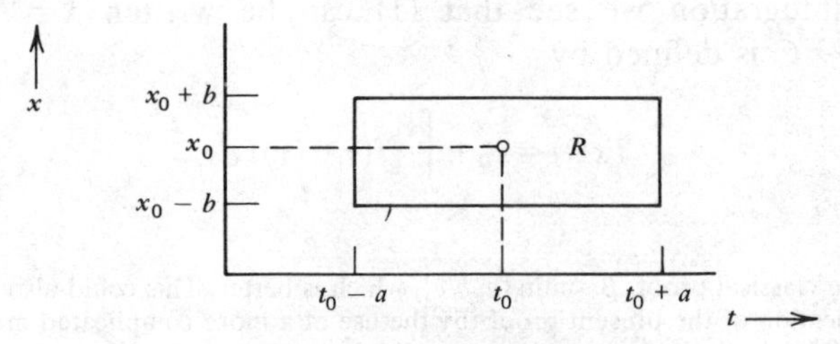

Fig. 53. The rectangle R

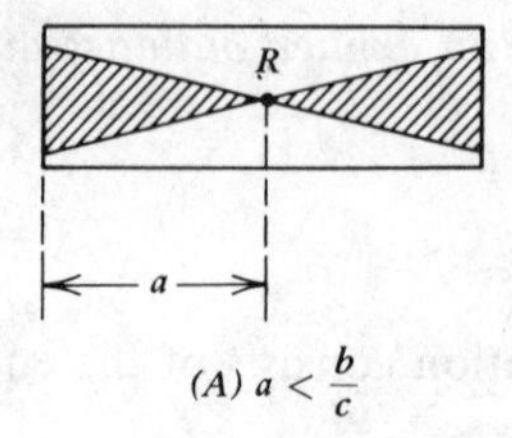

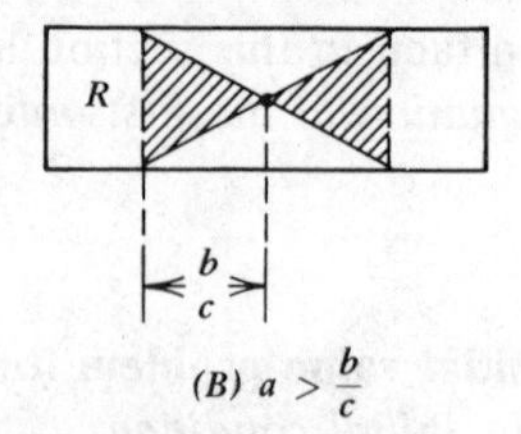

Fig. 54. Geometric illustration of inequality (2) for (*A*) relatively small *c*, (*B*) relatively large *c*. The solution curve must remain in the shaded region bounded by straight lines with slopes $\pm c$.

that for (t, x), $(t, v) \in R$

$$|f(t, x) - f(t, v)| \leqq k\,|x - v|. \tag{3}$$

Then the initial value problem (1) *has a unique solution. This solution exists on an interval* $[t_0 - \beta, t_0 + \beta]$, *where*[1]

$$\beta < \min\left\{a, \frac{b}{c}, \frac{1}{k}\right\}. \tag{4}$$

Proof. Let $C(J)$ be the metric space of all real-valued continuous functions on the interval $J = [t_0 - \beta, t_0 + \beta]$ with metric d defined by

$$d(x, y) = \max_{t \in J} |x(t) - y(t)|.$$

$C(J)$ is complete, as we know from 1.5-5. Let $\tilde{C}$ be the subspace of $C(J)$ consisting of all those functions $x \in C(J)$ that satisfy

$$|x(t) - x_0| \leqq c\beta. \tag{5}$$

It is not difficult to see that $\tilde{C}$ is closed in $C(J)$ (cf. Prob. 6), so that $\tilde{C}$ is complete by 1.4-7.

By integration we see that (1) can be written $x = Tx$, where $T\colon \tilde{C} \longrightarrow \tilde{C}$ is defined by

$$Tx(t) = x_0 + \int_{t_0}^{t} f(\tau, x(\tau))\, d\tau. \tag{6}$$

[1] In the classical proof, $\beta < \min\{a, b/c\}$, which is better. This could also be obtained by a modification of the present proof (by the use of a more complicated metric); cf. A. Bielecki (1956) in the references in Appendix 3.

Indeed, T is defined for all $x \in \tilde{C}$, because $c\beta < b$ by (4), so that if $x \in \tilde{C}$, then $\tau \in J$ and $(\tau, x(\tau)) \in R$, and the integral in (6) exists since f is continuous on R. To see that T maps $\tilde{C}$ into itself, we can use (6) and (2), obtaining

$$|Tx(t) - x_0| = \left| \int_{t_0}^{t} f(\tau, x(\tau))\, d\tau \right| \leqq c\,|t - t_0| \leqq c\beta.$$

We show that T is a contraction on $\tilde{C}$. By the Lipschitz condition (3),

$$\begin{aligned} |Tx(t) - Tv(t)| &= \left| \int_{t_0}^{t} [f(\tau, x(\tau)) - f(\tau, v(\tau))]\, d\tau \right| \\ &\leqq |t - t_0| \max_{\tau \in J} k\, |x(\tau) - v(\tau)| \\ &\leqq k\beta\, d(x, v). \end{aligned}$$

Since the last expression does not depend on t, we can take the maximum on the left and have

$$d(Tx, Tv) \leqq \alpha d(x, v) \qquad \text{where} \qquad \alpha = k\beta.$$

From (4) we see that $\alpha = k\beta < 1$, so that T is indeed a contraction on $\tilde{C}$. Theorem 5.1-2 thus implies that T has a unique fixed point $x \in \tilde{C}$, that is, a continuous function x on J satisfying $x = Tx$. Writing $x = Tx$ out, we have by (6)

$$x(t) = x_0 + \int_{t_0}^{t} f(\tau, x(\tau))\, d\tau. \tag{7}$$

Since $(\tau, x(\tau)) \in R$ where f is continuous, (7) may be differentiated. Hence x is even differentiable and satisfies (1). Conversely, every solution of (1) must satisfy (7). This completes the proof. ∎

Banach's theorem also implies that the solution x of (1) is the limit of the sequence $(x_0, x_1, \cdots)$ obtained by the *Picard iteration*

$$x_{n+1}(t) = x_0 + \int_{t_0}^{t} f(\tau, x_n(\tau))\, d\tau \tag{8}$$

where $n=0, 1, \cdots$. However, the practical usefulness of this way of obtaining approximations to the solution of (1) and corresponding error bounds is rather limited because of the integrations involved.

We finally mention the following. It can be shown that continuity of f is sufficient (but not necessary) for the existence of a solution of the problem (1), but is not sufficient for uniqueness. A Lipschitz condition is sufficient (as Picard's theorem shows), but not necessary. For details, see the book by E. L. Ince (1956), which also contains historical remarks about Picard's theorem (on p. 63) as well as a classical proof, so that the reader may compare our present approach with the classical one.

Problems

1. If the partial derivative $\partial f/\partial x$ of f exists and is continuous on the rectangle R (cf. Picard's theorem), show that f satisfies a Lipschitz condition on R with respect to its second argument.

2. Show that f defined by $f(t, x)=|\sin x|+t$ satisfies a Lipschitz condition on the whole tx-plane with respect to its second argument, but $\partial f/\partial x$ does not exist when $x=0$. What fact does this illustrate?

3. Does f defined by $f(t, x)=|x|^{1/2}$ satisfy a Lipschitz condition?

4. Find all initial conditions such that the initial value problem $tx'=2x$, $x(t_0)=x_0$, has (*a*) no solutions, (*b*) more than one solution, (*c*) precisely one solution.

5. Explain the reasons for the restrictions $\beta < b/c$ and $\beta < 1/k$ in (4).

6. Show that in the proof of Picard's theorem, $\tilde{C}$ is closed in $C(J)$.

7. Show that in Picard's theorem, instead of the constant x_0 we can take any other function $y_0 \in \tilde{C}$, $y_0(t_0)=x_0$, as the initial function of the iteration.

8. Apply the Picard iteration (8) to $x'=1+x^2$, $x(0)=0$. Verify that for x_3, the terms involving $t, t^2, \cdots, t^5$ are the same as those of the exact solution.

9. Show that $x'=3x^{2/3}$, $x(0)=0$, has infinitely many solutions x, given by

$$x(t)=0 \quad \text{if } t<c, \qquad x(t)=(t-c)^3 \quad \text{if } t \geqq c,$$

where $c>0$ is any constant. Does $3x^{2/3}$ on the right satisfy a Lipschitz condition?

10. Show that solutions of the initial value problem

$$x' = |x|^{1/2}, \qquad x(0)=0$$

are $x_1=0$ and x_2, where $x_2(t)=t\,|t|/4$. Does this contradict Picard's theorem? Find further solutions.

5.4 Application of Banach's Theorem to Integral Equations

We finally consider the Banach fixed point theorem as a source of existence and uniqueness theorems for integral equations. An integral equation of the form

$$\textbf{(1)} \qquad x(t) - \mu \int_a^b k(t, \tau) x(\tau)\, d\tau = v(t)$$

is called a **Fredholm equation** *of the second kind.*[2] Here, $[a, b]$ is a given interval. x is a function on $[a, b]$ which is unknown. μ is a parameter. The **kernel** k of the equation is a given function on the square $G=[a, b]\times[a, b]$ shown in Fig. 55, and v is a given function on $[a, b]$.

Integral equations can be considered on various function spaces. In this section we consider (1) on $C[a, b]$, the space of all continuous functions defined on the interval $J=[a, b]$ with metric d given by

$$(2) \qquad d(x, y) = \max_{t\in J} |x(t) - y(t)|;$$

[2] The presence of the term $x(t)$ enables us to apply iteration, as Theorem 5.4-1 shows. An equation without that term is of the form

$$\int_a^b k(t, \tau) x(\tau)\, d\tau = v(t)$$

and is said to be of the *first kind.*

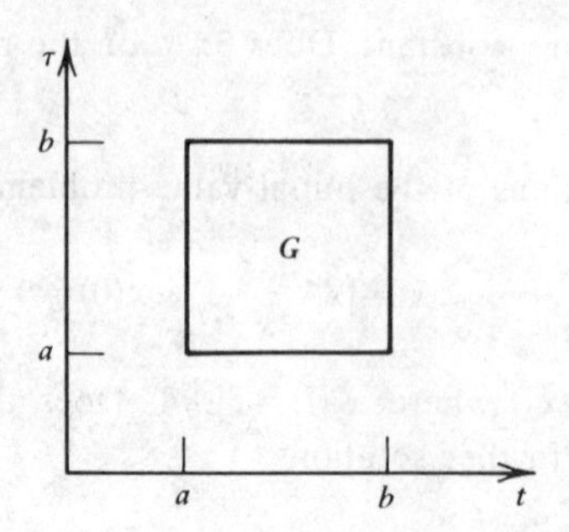

Fig. 55. Domain of definition G of the kernel k in the integral equation (1) in the case of positive a and b

cf. 1.5-5. For the proposed application of Banach's theorem it is important to note that $C[a, b]$ is complete. We assume that $v \in C[a, b]$ and k is continuous on G. Then k is a bounded function on G, say,

$$|k(t, \tau)| \leqq c \qquad \text{for all } (t, \tau) \in G. \tag{3}$$

Obviously, (1) can be written $x = Tx$ where

$$Tx(t) = v(t) + \mu \int_a^b k(t, \tau) x(\tau) \, d\tau. \tag{4}$$

Since v and k are continuous, formula (4) defines an operator $T\colon C[a, b] \longrightarrow C[a, b]$. We now impose a restriction on μ such that T becomes a contraction. From (2) to (4) we have

$$\begin{aligned}
d(Tx, Ty) &= \max_{t \in J} |Tx(t) - Ty(t)| \\
&= |\mu| \max_{t \in J} \left| \int_a^b k(t, \tau)[x(\tau) - y(\tau)] \, d\tau \right| \\
&\leqq |\mu| \max_{t \in J} \int_a^b |k(t, \tau)| \, |x(\tau) - y(\tau)| \, d\tau \\
&\leqq |\mu| \, c \max_{\sigma \in J} |x(\sigma) - y(\sigma)| \int_a^b d\tau \\
&= |\mu| \, c \, d(x, y)(b - a).
\end{aligned}$$

This can be written $d(Tx, Ty) \leqq \alpha\, d(x, y)$, where

$$\alpha = |\mu|\, c(b-a).$$

We see that T becomes a contraction ($\alpha < 1$) if

$$|\mu| < \frac{1}{c(b-a)}. \tag{5}$$

Banach's fixed point theorem 5.1-2 now gives

5.4-1 Theorem (Fredholm integral equation). *Suppose k and v in* (1) *to be continuous on $J \times J$ and $J = [a, b]$, respectively, and assume that μ satisfies* (5) *with c defined in* (3). *Then* (1) *has a unique solution x on J. This function x is the limit of the iterative sequence $(x_0, x_1, \cdots)$, where x_0 is any continuous function on J and for $n = 0, 1, \cdots$,*

$$x_{n+1}(t) = v(t) + \mu \int_a^b k(t, \tau) x_n(\tau)\, d\tau. \tag{6}$$

Fredholm's famous theory of integral equations will be discussed in Chap. 8.

We now consider the **Volterra integral equation**

$$x(t) - \mu \int_a^t k(t, \tau) x(\tau)\, d\tau = v(t). \tag{7}$$

The difference between (1) and (7) is that in (1) the upper limit of integration b is constant, whereas here in (7) it is variable. This is essential. In fact, without any restriction on μ we now get the following existence and uniqueness theorem.

5.4-2 Theorem (Volterra integral equation). *Suppose that v in* (7) *is continuous on $[a, b]$ and the kernel k is continuous on the triangular region R in the $t\tau$-plane given by $a \leqq \tau \leqq t$, $a \leqq t \leqq b$; see Fig. 56. Then* (7) *has a unique solution x on $[a, b]$ for every μ.*

Proof. We see that equation (7) can be written $x = Tx$ with $T\colon C[a, b] \longrightarrow C[a, b]$ defined by

$$Tx(t) = v(t) + \mu \int_a^t k(t, \tau) x(\tau)\, d\tau. \tag{8}$$

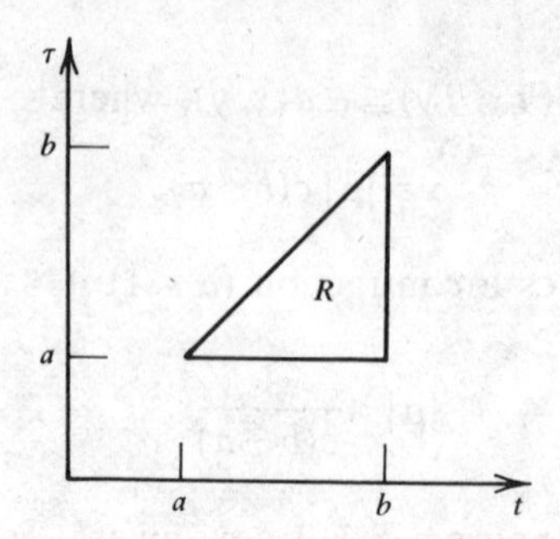

Fig. 56. Triangular region R in Theorem 5.4-2 in the case of positive a and b

Since k is continuous on R and R is closed and bounded, k is a bounded function on R, say,

$$|k(t, \tau)| \leqq c \qquad \text{for all } (t, \tau) \in R.$$

Using (2), we thus obtain for all $x, y \in C[a, b]$

$$\begin{aligned} |Tx(t) - Ty(t)| &= |\mu| \left| \int_a^t k(t, \tau)[x(\tau) - y(\tau)] \, d\tau \right| \\ &\leqq |\mu| \, c \, d(x, y) \int_a^t d\tau \\ &= |\mu| \, c(t-a) \, d(x, y). \end{aligned} \tag{9}$$

We show by induction that

$$|T^m x(t) - T^m y(t)| \leqq |\mu|^m c^m \frac{(t-a)^m}{m!} d(x, y). \tag{10}$$

For $m = 1$ this is (9). Assuming that (10) holds for any m, we obtain from (8)

$$\begin{aligned} |T^{m+1}x(t) - T^{m+1}y(t)| &= |\mu| \left| \int_a^t k(t, \tau)[T^m x(\tau) - T^m y(\tau)] \, d\tau \right| \\ &\leqq |\mu| \, c \int_a^t |\mu|^m \, c^m \frac{(\tau - a)^m}{m!} \, d\tau \, d(x, y) \\ &= |\mu|^{m+1} \, c^{m+1} \frac{(t-a)^{m+1}}{(m+1)!} \, d(x, y), \end{aligned}$$

which completes the inductive proof of (10).

Using $t-a \leqq b-a$ on the right-hand side of (10) and then taking the maximum over $t \in J$ on the left, we obtain from (10)

$$d(T^m x, T^m y) \leqq \alpha_m d(x, y)$$

where

$$\alpha_m = |\mu|^m c^m \frac{(b-a)^m}{m!}.$$

For any fixed μ and sufficiently large m we have $\alpha_m < 1$. Hence the corresponding T^m is a contraction on $C[a, b]$. The assertion of our theorem now follows from

5.4-3 Lemma (Fixed point). *Let $T: X \longrightarrow X$ be a mapping on a complete metric space $X = (X, d)$, and suppose that T^m is a contraction on X for some positive integer m. Then T has a unique fixed point.*

Proof. By assumption, $B = T^m$ is a contraction on X. By Banach's fixed point theorem 5.1-2, this mapping B has a unique fixed point $\hat{x}$, that is, $B\hat{x} = \hat{x}$. Hence $B^n\hat{x} = \hat{x}$. Banach's theorem also implies that for every $x \in X$,

$$B^n x \longrightarrow \hat{x} \qquad \text{as } n \longrightarrow \infty.$$

For the particular $x = T\hat{x}$, since $B^n = T^{nm}$, we thus obtain

$$\begin{aligned} \hat{x} = \lim_{n \to \infty} B^n T\hat{x} &= \lim_{n \to \infty} TB^n\hat{x} \\ &= \lim_{n \to \infty} T\hat{x} \\ &= T\hat{x}. \end{aligned}$$

This shows that $\hat{x}$ is a fixed point of T. Since every fixed point of T is also a fixed point of B, we see that T cannot have more than one fixed point. This completes the proof. ∎

We finally note that a Volterra equation can be regarded as a special Fredholm equation whose kernel k is zero in the part of the square $[a, b] \times [a, b]$ where $\tau > t$ (see Figs. 55 and 56) and may not be continuous at points on the diagonal ($\tau = t$).

Problems

1. Solve by iteration, choosing $x_0 = v$:

$$x(t) - \mu \int_0^1 e^{t-\tau} x(\tau)\, d\tau = v(t) \qquad (|\mu| < 1).$$

2. (Nonlinear integral equation) If v and k are continuous on $[a, b]$ and $C = [a, b] \times [a, b] \times \mathbf{R}$, respectively, and k satisfies on G a Lipschitz condition of the form

$$|k(t, \tau, u_1) - k(t, \tau, u_2)| \leqq l\, |u_1 - u_2|,$$

show that the nonlinear integral equation

$$x(t) - \mu \int_a^b k(t, \tau, x(\tau))\, d\tau = v(t)$$

has a unique solution x for any μ such that $|\mu| < 1/l(b-a)$.

3. It is important to understand that integral equations also arise from problems in differential equations. (*a*) For example, write the initial value problem

$$\frac{dx}{dt} = f(t, x), \qquad x(t_0) = x_0$$

as an integral equation and indicate what kind of equation it is. (*b*) Show that an initial value problem

$$\frac{d^2x}{dt^2} = f(t, x), \qquad x(t_0) = x_0, \qquad x'(t_0) = x_1$$

involving a second order differential equation can be transformed into a Volterra integral equation.

4. (Neumann series) Defining an operator S by

$$Sx(t) = \int_a^b k(t, \tau) x(\tau)\, d\tau$$

and setting $z_n = x_n - x_{n-1}$, show that (6) implies

$$z_{n+1} = \mu S z_n.$$

Choosing $x_0 = v$, show that (6) yields the *Neumann series*

$$x = \lim_{n \to \infty} x_n = v + \mu S v + \mu^2 S^2 v + \mu^3 S^3 v + \cdots.$$

5. Solve the following integral equation (*a*) by a Neumann series, (*b*) by a direct approach.

$$x(t) - \mu \int_0^1 x(\tau)\, d\tau = 1.$$

6. Solve

$$x(t) - \mu \int_a^b c x(\tau)\, d\tau = \tilde{v}(t)$$

where c is a constant, and indicate how the corresponding Neumann series can be used to obtain the convergence condition (5) for the Neumann series of (1).

7. (Iterated kernel, resolvent kernel) Show that in the Neumann series in Prob. 4 we can write

$$(S^n v)(t) = \int_a^b k_{(n)}(t, \tau) v(\tau)\, d\tau \qquad n = 2, 3, \cdots,$$

where the *iterated kernel* $k_{(n)}$ is given by

$$k_{(n)}(t, \tau) = \int_a^b \cdots \int_a^b k(t, t_1) k(t_1, t_2) \cdots k(t_{n-1}, \tau)\, dt_1 \cdots dt_{n-1}$$

so that the Neumann series can be written

$$x(t) = v(t) + \mu \int_a^b k(t, \tau) v(\tau)\, d\tau + \mu^2 \int_a^b k_{(2)}(t, \tau) v(\tau)\, d\tau + \cdots$$

or, by the use of the *resolvent kernel* $\tilde{k}$ defined by

$$\tilde{k}(t, \tau, \mu) = \sum_{j=0}^{\infty} \mu^j k_{(j+1)}(t, \tau) \qquad (k_{(1)} = k)$$

it can be written

$$x(t) = v(t) + \mu \int_a^b \tilde{k}(t, \tau, \mu) v(\tau)\, d\tau.$$

8. It is of interest that the Neumann series in Prob. 4 can also be obtained by substituting a power series in μ,

$$x(t) = v_0(t) + \mu v_1(t) + \mu^2 v_2(t) + \cdots$$

into (1), integrating termwise and comparing coefficients. Show that this gives

$$v_0(t) = v(t), \qquad v_n(t) = \int_a^b k(t, \tau) v_{n-1}(\tau)\, d\tau, \qquad n = 1, 2, \cdots.$$

Assuming that $|v(t)| \leqq c_0$ and $|k(t, \tau)| \leqq c$, show that

$$|v_n(t)| \leqq c_0[c(b-a)]^n,$$

so that (5) implies convergence.

9. Using Prob. 7, solve (1), where $a = 0$, $b = 2\pi$ and

$$k(t, \tau) = \sum_{n=1}^{N} a_n \sin nt \cos n\tau.$$

10. In (1), let $a = 0$, $b = \pi$ and

$$k(t, \tau) = a_1 \sin t \sin 2\tau + a_2 \sin 2t \sin 3\tau.$$

Write the solution in terms of the resolvent kernel (cf. Prob. 7).

CHAPTER 6
FURTHER APPLICATIONS: APPROXIMATION THEORY

This chapter is optional; the material included in it will not be used in the remaining chapters.

The theory of approximation is a very extensive field which has various applications. In this chapter we give an introduction to fundamental ideas and aspects of approximation theory in normed and Hilbert spaces.

Important concepts, brief orientation about main content

In Sec. 6.1 we define *best approximations.* The existence of best approximations is discussed in the same section and the uniqueness in Sec. 6.2. If a normed space is strictly convex (cf. 6.2-2), we have uniqueness of best approximations. For Hilbert spaces this holds (cf. 6.2-4 and Sec. 6.5). For general normed spaces one may need additional conditions to guarantee uniqueness of best approximations, for instance a *Haar condition* in $C[a, b]$; cf. 6.3-2 and 6.3-4. Depending on the choice of a norm, one gets different types of approximations. Standard types include

(*i*) *uniform approximation* in $C[a, b]$ (Sec. 6.3),

(*ii*) *approximation in Hilbert space* (Sec. 6.5).

Practical aspects of uniform approximation lead to the famous Chebyshev polynomials (Sec. 6.4). Approximation in Hilbert space includes least squares approximation in $L^2[a, b]$ as a particular case. We shall also give a brief discussion of cubic splines (Sec. 6.6).

6.1 Approximation in Normed Spaces

Approximation theory is concerned with the approximation of functions of a certain kind (for instance, continuous functions on some interval) by other (probably simpler) functions (for example, polynomials). Such a situation already arises in calculus: if a function has a Taylor series, we may regard and use the partial sums of the series as approximations. To get information about the quality of such approximations, we would have to estimate the corresponding remainders.

More generally, one may want to set up practically useful criteria for the quality of approximations. Given a set X of functions to be approximated and a set Y of functions by which the elements of X are to be approximated, one may consider the problems of existence, uniqueness, and construction of a "best approximation" in the sense of such a criterion. A natural setting for the problem of approximation is as follows.

Let $X=(X, \|\cdot\|)$ be a normed space and suppose that any given $x \in X$ is to be approximated by a $y \in Y$, where Y is a fixed subspace of X. We let δ denote the distance from x to Y. By definition,

$$\delta=\delta(x, Y)=\inf_{y \in Y}\|x-y\| \tag{1}$$

(cf. Sec. 3.3). Clearly, δ depends on both x and Y, which we keep fixed, so that the simple notation δ is in order.

If there exists a $y_0 \in Y$ such that

$$\|x-y_0\|=\delta \tag{2}$$

then y_0 is called a **best approximation** *to x out of Y*.

We see that a best approximation y_0 is an element of minimum distance from the given x. Such a $y_0 \in Y$ may or may not exist; this raises the *problem of existence*. The *problem of uniqueness* is of practical interest, too, since for given x and Y there may be more than one best approximation, as we shall see.

In many applications, Y will be finite dimensional. Then we have the following

6.1-1 Existence Theorem (Best approximations). *If Y is a finite dimensional subspace of a normed space $X=(X, \|\cdot\|)$, then for each $x \in X$ there exists a best approximation to x out of Y.*

Proof. Let $x \in X$ be given. Consider the closed ball

$$\tilde{B}=\{y \in Y \mid \|y\| \leqq 2\|x\|\}.$$

Then $0 \in \tilde{B}$, so that for the distance from x to $\tilde{B}$ we obtain

$$\delta(x, \tilde{B})=\inf_{\tilde{y} \in \tilde{B}}\|x-\tilde{y}\| \leqq\|x-0\|=\|x\|.$$

Now if $y \notin \tilde{B}$, then $\|y\| > 2\|x\|$ and

$$\|x-y\| \geqq \|y\| - \|x\| > \|x\| \geqq \delta(x, \tilde{B}). \tag{3}$$

This shows that $\delta(x, \tilde{B}) = \delta(x, Y) = \delta$, and this value cannot be assumed by a $y \in Y - \tilde{B}$, because of $>$ in (3). Hence if a best approximation to x exists, it must lie in $\tilde{B}$. We now see the reason for the use of $\tilde{B}$. Instead of the whole subspace Y we may now consider the compact subset $\tilde{B}$, where compactness follows from 2.5-3 since $\tilde{B}$ is closed and bounded and Y is finite dimensional. The norm is continuous by (2), Sec. 2.2. Corollary 2.5-7 thus implies that there is a $y_0 \in \tilde{B}$ such that $\|x-y\|$ assumes a minimum at $y = y_0$. By definition, y_0 is a best approximation to x out of Y. ∎

Examples

6.1-2 Space $C[a, b]$. A finite dimensional subspace of the space $C[a, b]$ is

$$Y = \operatorname{span}\{x_0, \cdots, x_n\}, \qquad x_j(t) = t^j \qquad (n \text{ fixed}).$$

This is the set of all polynomials of degree at most n, together with $x = 0$ (for which no degree is defined in the usual discussion of degree). Theorem 6.1-1 implies that for a given continuous function x on $[a, b]$ there exists a polynomial p_n of degree at most n such that for every $y \in Y$,

$$\max_{t \in J} |x(t) - p_n(t)| \leqq \max_{t \in J} |x(t) - y(t)|$$

where $J = [a, b]$. Approximation in $C[a, b]$ is called *uniform approximation* and will be considered in detail in the next sections.

6.1-3 Polynomials. Finite dimensionality of Y in Theorem 6.1-1 is essential. In fact, let Y be the set of all polynomials on $[0, \frac{1}{2}]$ of any degree, considered as a subspace of $C[0, \frac{1}{2}]$. Then $\dim Y = \infty$. Let $x(t) = (1-t)^{-1}$. Then for every $\varepsilon > 0$ there is an N such that, setting

$$y_n(t) = 1 + t + \cdots + t^n,$$

we have $\|x - y_n\| < \varepsilon$ for all $n > N$. Hence $\delta(x, Y) = 0$. However, since x

is not a polynomial, we see that there is no $y_0 \in Y$ satisfying $\delta = \delta(x, Y) = \|x - y_0\| = 0$. ∎

Problems will be included at the end of the next section.

6.2 Uniqueness, Strict Convexity

In this section we consider the problem of uniqueness of best approximations. To understand what is going on, let us start with two simple examples.

If $X = \mathbf{R}^3$ and Y is the $\xi_1\xi_2$-plane ($\xi_3 = 0$), then we know that for a given point $x_0 = (\xi_{10}, \xi_{20}, \xi_{30})$ a best approximation out of Y is the point $y_0 = (\xi_{10}, \xi_{20}, 0)$, the distance from x_0 to Y is $\delta = |\xi_{30}|$ and that best approximation y_0 is unique. These simple facts are well known from elementary geometry.

In other spaces, uniqueness of best approximations may not hold, even if the spaces are relatively simple.

For instance, let $X = (X, \|\cdot\|_1)$ be the vector space of ordered pairs $x = (\xi_1, \xi_2), \cdots$ of real numbers with norm defined by

$$\|x\|_1 = |\xi_1| + |\xi_2|. \tag{1}$$

Let us take the point $x = (1, -1)$ and the subspace Y shown in Fig. 57,

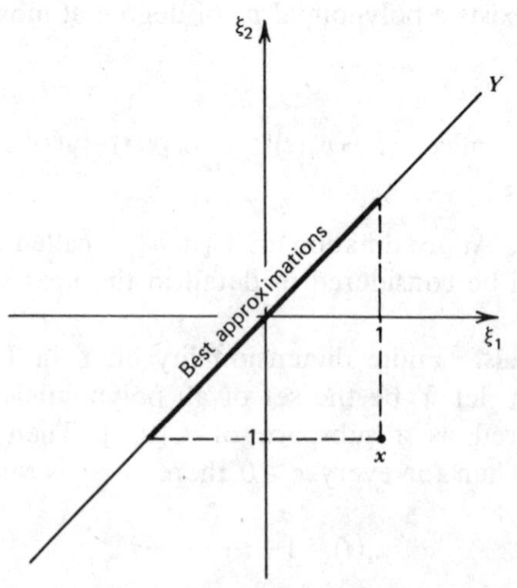

Fig. 57. Best approximations to x out of Y in the norm defined by (1)

that is, $Y=\{y=(\eta, \eta) \mid \eta \text{ real}\}$. Then for all $y \in Y$ we clearly have

$$\|x-y\|_1=|1-\eta|+|-1-\eta| \geqq 2.$$

The distance from x to Y is $\delta(x, Y)=2$, and all $y=(\eta, \eta)$ with $|\eta| \leqq 1$ are best approximations to x out of Y. This illustrates that even in such a simple space, for given x and Y we may have several best approximations, even infinitely many of them. We observe that in the present case the set of best approximations is convex, and we claim that this is typical. We also assert that convexity will be helpful in connection with our present uniqueness problem. So let us first state the definition and then find out how we can apply the concept.

A subset M of a vector space X is said to be **convex** if $y, z \in M$ implies that the set

$$W=\{v=\alpha y+(1-\alpha) z \mid 0 \leqq \alpha \leqq 1\}$$

is a subset of M. This set W is called a *closed segment.* (Why?) y and z are called the *boundary points* of the segment W. Any other point of W is called an *interior point* of W. See Fig. 58.

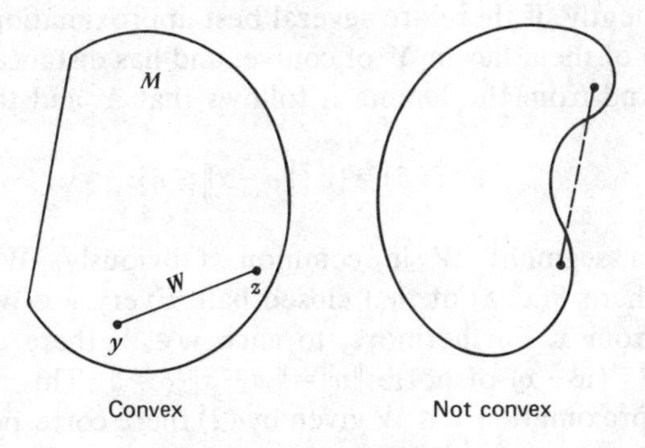

Fig. 58. Convex and not convex sets

6.2-1 Lemma (Convexity). *In a normed space $(X, \|\cdot\|)$ the set M of best approximations to a given point x out of a subspace Y of X is convex.*

Proof. Let δ denote the distance from x to Y, as before. The statement holds if M is empty or has just one point. Suppose that M

has more than one point. Then for $y, z \in M$ we have, by definition,

$$\|x-y\| = \|x-z\| = \delta.$$

We show that this implies

$$w = \alpha y + (1-\alpha) z \in M \qquad (0 \leqq \alpha \leqq 1). \tag{2}$$

Indeed, $\|x-w\| \geqq \delta$ since $w \in Y$, and $\|x-w\| \leqq \delta$ since

$$\begin{aligned} \|x-w\| &= \|\alpha(x-y) + (1-\alpha)(x-z)\| \\ &\leqq \alpha\, \|x-y\| + (1-\alpha)\, \|x-z\| \\ &= \alpha\delta + (1-\alpha)\, \delta \\ &= \delta; \end{aligned}$$

here we used that $\alpha \geqq 0$ as well as $1-\alpha \geqq 0$. Together, $\|x-w\| = \delta$. Hence $w \in M$. Since $y, z \in M$ were arbitrary, this proves that M is convex. ∎

Consequently, if there are several best approximations to x out of Y, then each of them lies in Y, of course, and has distance δ from x, by definition. And from the lemma it follows that Y and the closed ball

$$\tilde{B}(x; \delta) = \{v \mid \|v-x\| \leqq \delta\}$$

must have a segment W in common. Obviously, W lies on the boundary sphere $S(x; \delta)$ of that closed ball. Every $w \in W$ has distance $\|w-x\| = \delta$ from x. Furthermore, to each $w \in W$ there corresponds a unique $v = \delta^{-1}(w-x)$ of norm $\|v\| = \|w-x\|/\delta = 1$. This means that to each best approximation $w \in W$ given by (2) there corresponds a unique v on the unit sphere $\{x \mid \|x\| = 1\}$.

From this we see that for obtaining uniqueness of best approximations, we must exclude norms for which the unit sphere can contain segments of straight lines. This suggests the following

6.2-2 Definition (Strict convexity). A *strictly convex norm* is a norm such that for all x, y of norm 1,

$$\|x+y\| < 2 \qquad (x \neq y).$$

A normed space with such a norm is called a *strictly convex normed space.* ∎

Note that for $\|x\|=\|y\|=1$ the triangle inequality gives

$$\|x+y\| \leqq \|x\|+\|y\|=2$$

and strict convexity excludes the equality sign, except when $x=y$. We may summarize our result as follows.

6.2-3 Uniqueness Theorem (Best approximation). *In a strictly convex normed space X there is at most one best approximation to an $x \in X$ out of a given subspace Y.*

This theorem may or may not be of help in practical problems, depending on what space we use. We list two very important cases.

6.2-4 Lemma (Strict convexity). *We have:*

(a) *Hilbert space is strictly convex.*

(b) *The space $C[a, b]$ is not strictly convex.*

Proof. **(a)** For all x and $y \neq x$ of norm one we have, say, $\|x-y\|=\alpha$, where $\alpha>0$, and the parallelogram equality (Sec. 3.1) gives

$$\begin{aligned}\|x+y\|^2 &= -\|x-y\|^2+2(\|x\|^2+\|y\|^2) \\ &= -\alpha^2+2(1+1)<4,\end{aligned}$$

hence $\|x+y\|<2$.

(b) We consider x_1 and x_2 defined by

$$x_1(t)=1, \qquad x_2(t)=\frac{t-a}{b-a}$$

where $t \in [a, b]$. Clearly, $x_1, x_2 \in C[a, b]$ and $x_1 \neq x_2$. We also have $\|x_1\|=\|x_2\|=1$, and

$$\|x_1+x_2\|=\max_{t \in J}\left|1+\frac{t-a}{b-a}\right|=2$$

where $J=[a, b]$. This shows that $C[a, b]$ is not strictly convex. ∎

The first statement of this lemma had to be expected, because Theorem 3.3-1 and Lemma 3.3-2 together imply

6.2-5 Theorem (Hilbert space). *For every given x in a Hilbert space H and every given closed subspace Y of H there is a unique best approximation to x out of Y (namely, $y = Px$, where P is the projection of H onto Y).*

From the second statement in Lemma 6.2-4 we see that in uniform approximation, additional effort will be necessary to guarantee uniqueness.

Problems

1. Let Y in (1) and (2), Sec. 6.1, be finite dimensional. Under what conditions is $\|x - y_0\| = 0$ in (2)?

2. We shall confine our attention to normed spaces, but want to mention that certain discussions could be extended to general metric spaces. For instance, show that if (X, d) is a metric space and Y a compact subset, then every $x \in X$ has a best approximation y out of Y.

3. If Y is a finite dimensional subspace of a normed space X and we want to approximate an $x \in X$ out of Y, it is natural to choose a basis $\{e_1, \cdots, e_n\}$ for Y and approximate x by a linear combination $\sum \alpha_j e_j$. Show that the corresponding function f defined by

$$f(\alpha) = \left\| x - \sum_{j=1}^{n} \alpha_j e_j \right\|, \qquad \alpha = (\alpha_1, \cdots, \alpha_n)$$

depends continuously on the α_j's.

4. (Convex function) Show that f in Prob. 3 has the interesting property of being convex. A function $f\colon \mathbf{R}^n \longrightarrow \mathbf{R}$ is said to be *convex* if its domain $\mathscr{D}(f)$ is a convex set and for every $u, v \in \mathscr{D}(f)$,

$$f(\lambda u + (1-\lambda)v) \leqq \lambda f(u) + (1-\lambda) f(v),$$

where $0 \leqq \lambda \leqq 1$. (An example for $n = 1$ is shown in Fig. 59. Convex functions are useful in various minimization problems.)

5. The norm defined by (1) is not strictly convex. Prove this directly, without using 6.2-3.

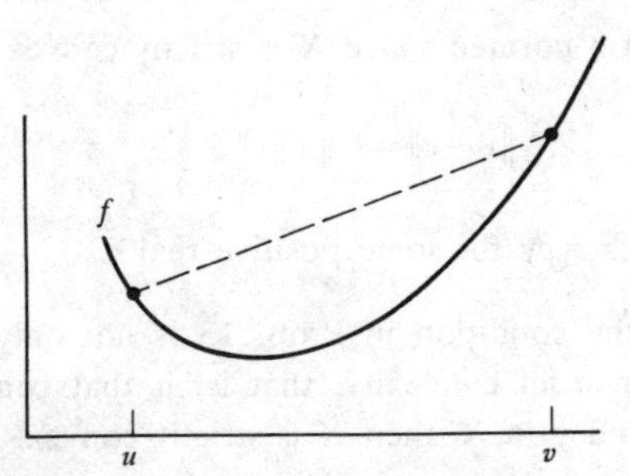

Fig. 59. Convex function f depending on a single variable t. The dashed straight line segment represents $\lambda f(u)+(1-\lambda)f(v)$, where $0 \leqq \lambda \leqq 1$.

6. Consider (1). Determine all points y of the closed unit ball $\tilde{B}$ whose distance from $x=(2, 0)$ is minimum. Determine that minimum value δ.

7. Show that the vector space of ordered pairs of real numbers with norm defined by

$$\|(\xi_1, \xi_2)\| = \max(|\xi_1|, |\xi_2|)$$

is not strictly convex. Graph the unit sphere.

8. Consider all ordered pairs of real numbers $x=(\xi_1, \xi_2)$. Find all the points which have distance $\sqrt{2}$ from $(0, 0)$ as well as from $(2, 0)$ if the distance is (*a*) Euclidean, (*b*) obtained from the norm in Prob. 7.

9. Consider the vector space of all ordered pairs of real numbers. Let $x_1=(-1, 0)$ and $x_2=(1, 0)$. Determine the intersection of the two spheres defined by $\|x-x_1\|=1$ and $\|x-x_2\|=1$ if the norm is (*a*) the Euclidean norm, (*b*) the norm defined in Prob. 7, (*c*) the norm given by (1).

10. It can be shown that l^p with $p>1$ is strictly convex whereas l^1 is not. Prove that l^1 is not strictly convex.

11. If in a normed space, the best approximation to an x out of a subspace Y is not unique, show that x has infinitely many such best approximations.

12. Show that if a norm is strictly convex, then $\|x\|=\|y\|=1$ and $x \neq y$ together imply that for all α such that $0<\alpha<1$ one has

$$\|\alpha x+(1-\alpha)y\|<1.$$

Show that this condition is also sufficient for strict convexity.

13. Show that if a normed space X is strictly convex, then

$$\|x+y\|=\|x\|+\|y\| \qquad (x\neq 0,\ y\neq 0)$$

implies that $x=cy$ for some positive real c.

14. Show that the condition in Prob. 13 is not only necessary but also sufficient for strict convexity, that is, if that condition holds for all nonzero x and y in X, then X is strictly convex.

15. An *extreme point* of a convex set M in a vector space X is a point $x\in M$ which is not an interior point of a segment $W\subset M$. Show that if X is a strictly convex normed space, then every point of the unit sphere in X is an extreme point of the closed unit ball in X.

6.3 Uniform Approximation

Depending on the choice of a norm, we get different types of approximations. The choice depends on the purpose. Two common types are as follows.

(A) *Uniform approximation* uses the norm on $C[a, b]$ defined by

$$\|x\|=\max_{t\in J}|x(t)| \qquad J=[a, b].$$

(B) *Least squares approximation* uses the norm on $L^2[a, b]$ defined by (cf. 3.1-5)

$$\|x\|=\langle x, x\rangle^{1/2}=\left(\int_a^b |x(t)|^2\, dt\right)^{1/2}.$$

This section is devoted to uniform approximation (also known as *Chebyshev approximation*). We consider the real space $X=C[a, b]$ and an n-dimensional subspace $Y\subset C[a, b]$. Then the occurring functions are real-valued and continuous on $[a, b]$. For every function $x\in X$, Theorem 6.1-1 guarantees the existence of a best approximation to x out of Y. However, since $C[a, b]$ is not strictly convex (cf. 6.2-4), the problem of uniqueness requires a special investigation. For this purpose the following concept will be of interest and importance.

6.3-1 Definition (Extremal point). An *extremal point* of an x in $C[a, b]$ is a $t_0 \in [a, b]$ such that $|x(t_0)| = \|x\|$. ∎

Hence at an extremal point t_0 of x we have either $x(t_0) = +\|x\|$ or $x(t_0) = -\|x\|$, and the definition of the norm on $C[a, b]$ shows that such a point is a $t_0 \in [a, b]$ at which $|x(t)|$ has a maximum.

The central concept of our present discussion is the following condition by A. Haar (1918) which will turn out to be necessary and sufficient for the uniqueness of approximations in $C[a, b]$.

6.3-2 Definition (Haar condition). A finite dimensional subspace Y of the real space $C[a, b]$ is said to satisfy the *Haar condition* if every $y \in Y$, $y \neq 0$, has at most $n-1$ zeros in $[a, b]$, where $n = \dim Y$. ∎

For instance, an n-dimensional subspace Y of $C[a, b]$ is given by the polynomial $y = 0$ (for which one does not define a degree in the usual discussion of degree) and all polynomials of degree not exceeding $n-1$ and with real coefficients. Since any such polynomial $y \neq 0$ has at most $n-1$ zeros, Y satisfies the Haar condition. Actually, this is the model case which suggested Def. 6.3-2. We shall return to this case after we proved that the Haar condition is necessary and sufficient for the uniqueness of best approximations in $C[a, b]$.

To gain flexibility in our further work, let us first prove the following.

The Haar condition is equivalent to the condition that for every basis $\{y_1, \cdots, y_n\} \subset Y$ *and every n-tuple of distinct points* $t_1, \cdots, t_n$ *in the interval* $J = [a, b]$,

$$\text{(1)} \qquad \begin{vmatrix} y_1(t_1) & y_1(t_2) & \cdots & y_1(t_n) \\ y_2(t_1) & y_2(t_2) & \cdots & y_2(t_n) \\ \cdot & \cdot & \cdots & \cdot \\ y_n(t_1) & y_n(t_2) & \cdots & y_n(t_n) \end{vmatrix} \neq 0.$$

Proof. Every $y \in Y$ has a representation $y = \sum \alpha_k y_k$. The subspace Y satisfies the Haar condition if and only if every $y = \sum \alpha_k y_k \in Y$ with n or more zeros $t_1, t_2, \cdots, t_n, \cdots$ in $J = [a, b]$ is identically zero. This means that the n conditions

$$\text{(2)} \qquad y(t_j) = \sum_{k=1}^{n} \alpha_k y_k(t_j) = 0 \qquad j = 1, \cdots, n$$

should imply $\alpha_1 = \cdots = \alpha_n = 0$. But this happens if and only if the determinant in (1) of the system (2) is not zero. ∎

The fact that the Haar condition is sufficient for the uniqueness of best approximations will be proved by the use of the following lemma.

6.3-3 Lemma (Extremal points). *Suppose a subspace Y of the real space $C[a, b]$ satisfies the Haar condition. If for a given $x \in C[a, b]$ and a $y \in Y$ the function $x - y$ has less than $n+1$ extremal points, then y is not a best approximation to x out of Y; here, $n = \dim Y$.*

Proof. By assumption the function $v = x - y$ has $m \leqq n$ extremal points $t_1, \cdots, t_m$. If $m < n$, we choose any additional points t_j in $J = [a, b]$ until we have n distinct points $t_1, \cdots, t_n$. Using these points and a basis $\{y_1, \cdots, y_n\}$ for Y, we consider the nonhomogeneous system of linear equations

$$\sum_{k=1}^{n} \beta_k y_k(t_j) = v(t_j) \qquad j = 1, \cdots, n \tag{3}$$

in the unknowns $\beta_1, \cdots, \beta_n$. Since Y satisfies the Haar condition, (1) holds. Hence (3) has a unique solution. We use this solution to define

$$y_0 = \beta_1 y_1 + \cdots + \beta_n y_n$$

as well as

$$\tilde{y} = y + \varepsilon y_0 \qquad (\varepsilon > 0).$$

We show that for a sufficiently small ε the function $\tilde{v} = x - \tilde{y}$ satisfies

$$\|\tilde{v}\| < \|v\|, \tag{4}$$

so that y cannot be a best approximation to x out of Y.

To obtain (4), we estimate $\tilde{v}$, breaking $J = [a, b]$ up into two sets N and $K = J - N$ where N contains the extremal points $t_1, \cdots, t_m$ of v.

At the extremal points, $|v(t_i)| = \|v\|$, and $\|v\| > 0$ since $v = x - y \neq 0$. Also $y_0(t_i) = v(t_i)$ by (3) and the definition of y_0. Hence, by continuity, for each t_i there is an open neighborhood N_i such that in the union $N = N_1 \cup \cdots \cup N_m$ we have

$$\mu = \inf_{t \in N} |v(t)| > 0, \qquad \inf_{t \in N} |y_0(t)| \geqq \tfrac{1}{2}\|v\|. \tag{5}$$

Since $y_0(t_i) = v(t_i) \neq 0$, for all $t \in N$ we have $y_0(t)/v(t) > 0$ by (5), and (5) also yields

$$\frac{y_0(t)}{v(t)} = \frac{|y_0(t)|}{|v(t)|} \geqq \frac{\inf |y_0(t)|}{\|v\|} \geqq \frac{1}{2}.$$

Let $M_0 = \sup_{t \in N} |y_0(t)|$. Then for every positive $\varepsilon < \mu/M_0$ and every $t \in N$ we obtain

$$\frac{\varepsilon y_0(t)}{v(t)} = \frac{\varepsilon\, |y_0(t)|}{|v(t)|} \leqq \frac{\varepsilon M_0}{\mu} < 1.$$

Since $\tilde{v} = x - \tilde{y} = x - y - \varepsilon y_0 = v - \varepsilon y_0$, using these inequalities, we see that for all $t \in N$ and $0 < \varepsilon < \mu/M_0$,

$$\begin{aligned} |\tilde{v}(t)| &= |v(t) - \varepsilon\, y_0(t)| \\ &= |v(t)| \left(1 - \frac{\varepsilon y_0(t)}{v(t)}\right) \\ &\leqq \|v\| \left(1 - \frac{\varepsilon}{2}\right) \\ &< \|v\|. \end{aligned} \tag{6}$$

We turn to the complement $K = J - N$, which is closed, and define

$$M_1 = \max_{t \in K} |y_0(t)|, \qquad M_2 = \max_{t \in K} |v(t)|.$$

Since N contains all the extremal points of v, we have $M_2 < \|v\|$ and may write

$$\|v\| = M_2 + \eta \qquad \text{where} \qquad \eta > 0.$$

Choosing a positive $\varepsilon < \eta/M_1$, we have $\varepsilon M_1 < \eta$ and thus obtain for all $t \in K$

$$\begin{aligned} |\tilde{v}(t)| &\leqq |v(t)| + \varepsilon\, |y_0(t)| \\ &\leqq M_2 + \varepsilon M_1 \\ &< \|v\|. \end{aligned}$$

We see that $|\tilde{v}(t)|$ does not exceed a bound which is independent of $t \in K$ and strictly less than $\|v\|$; similarly in (6), where $t \in N$ and $\varepsilon > 0$ is sufficiently small. Choosing $\varepsilon < \min\{\mu/M_0, \eta/M_1\}$ and taking the supremum, we thus have $\|\tilde{v}\| < \|v\|$. This is (4) and completes the proof.

Using this lemma, we shall now obtain the basic

6.3-4 Haar Uniqueness Theorem (Best approximation). *Let Y be a finite dimensional subspace of the real space $C[a, b]$. Then the best approximation out of Y is unique for every $x \in C[a, b]$ if and only if Y satisfies the Haar condition.*

Proof. **(a)** *Sufficiency.* Suppose Y satisfies the Haar condition, but both $y_1 \in Y$ and $y_2 \in Y$ are best approximations to some fixed $x \in C[a, b]$. Then, setting

$$v_1 = x - y_1, \qquad v_2 = x - y_2,$$

we have $\|v_1\| = \|v_2\| = \delta$, where δ is the distance from x to Y, as before. Lemma 6.2-1 implies that $y = \frac{1}{2}(y_1 + y_2)$ is also a best approximation to x. By Lemma 6.3-3 the function

$$v = x - y = x - \tfrac{1}{2}(y_1 + y_2) = \tfrac{1}{2}(v_1 + v_2) \tag{7}$$

has at least $n+1$ extremal points $t_1, \cdots, t_{n+1}$. At such a point we have $|v(t_j)| = \|v\| = \delta$. From this and (7) we obtain

$$2v(t_j) = v_1(t_j) + v_2(t_j) = +2\delta \text{ or } -2\delta.$$

Now $|v_1(t_j)| \leqq \|v_1\| = \delta$ (see before) and similarly for v_2. Hence there is only one way in which the equation can hold, namely, both terms must have the same sign and the maximum possible absolute value, that is,

$$v_1(t_j) = v_2(t_j) = +\delta \text{ or } -\delta$$

where $j = 1, \cdots, n+1$. But this implies that $y_1 - y_2 = v_2 - v_1$ has $n+1$ zeros in $[a, b]$. Hence $y_1 - y_2 = 0$ by the Haar condition. Thus $y_1 = y_2$, the uniqueness.

(b) *Necessity.* We assume that Y does not satisfy the Haar condition and show that then we do not have uniqueness of best approximations for all $x \in C[a, b]$. As we have shown in connection with 6.3-2, under our present assumption there is a basis for Y and n values t_j in $[a, b]$ such that the determinant in (1) is zero. Hence the homogeneous system

$$\gamma_1 y_k(t_1) + \gamma_2 y_k(t_2) + \cdots + \gamma_n y_k(t_n) = 0$$

$(k = 1, \cdots, n)$ has a nontrivial solution $\gamma_1, \cdots, \gamma_n$. Using this solution and any $y = \sum \alpha_k y_k \in Y$, we have

$$\sum_{j=1}^{n} \gamma_j y(t_j) = \sum_{k=1}^{n} \alpha_k \left[\sum_{j=1}^{n} \gamma_j y_k(t_j) \right] = 0. \tag{8}$$

Furthermore, the transposed system

$$\beta_1 y_1(t_j) + \beta_2 y_2(t_j) + \cdots + \beta_n y_n(t_j) = 0$$

$(j = 1, \cdots, n)$ also has a nontrivial solution $\beta_1, \cdots, \beta_n$. Using this solution, we define $y_0 = \sum \beta_k y_k$. Then $y_0 \neq 0$, and y_0 is zero at $t_1, \cdots, t_n$. Let λ be such that $\|\lambda y_0\| \leqq 1$. Let $z \in C[a, b]$ be such that $\|z\| = 1$ and

$$z(t_j) = \operatorname{sgn} \gamma_j = \begin{cases} -1 & \text{if } \gamma_j < 0 \\ 1 & \text{if } \gamma_j \geqq 0. \end{cases}$$

Define $x \in C[a, b]$ by

$$x(t) = z(t)(1 - |\lambda y_0(t)|).$$

Then $x(t_j) = z(t_j) = \operatorname{sgn} \gamma_j$ since $y_0(t_j) = 0$. Also $\|x\| = 1$. We show that this function x has infinitely many best approximations out of Y.

Using $|z(t)| \leqq \|z\| = 1$ and $|\lambda y_0(t)| \leqq \|\lambda y_0\| \leqq 1$, for every $\varepsilon \in [-1, 1]$ we obtain

$$\begin{aligned} |x(t) - \varepsilon \lambda y_0(t)| &\leqq |x(t)| + |\varepsilon \lambda y_0(t)| \\ &= |z(t)|\,(1 - |\lambda y_0(t)|) + |\varepsilon \lambda y_0(t)| \\ &\leqq 1 - |\lambda y_0(t)| + |\varepsilon \lambda y_0(t)| \\ &= 1 - (1 - |\varepsilon|)\,|\lambda y_0(t)| \\ &\leqq 1. \end{aligned}$$

Hence every $\varepsilon\lambda y_0$, $-1 \leqq \varepsilon \leqq 1$, is a best approximation to x, provided

$$\|x-y\| \geqq 1 \qquad \text{for all } y \in Y. \tag{9}$$

We prove (9) for an arbitrary $y = \sum \alpha_k y_k \in Y$. The proof is indirect. Suppose that $\|x-\tilde{y}\| < 1$ for a $\tilde{y} \in Y$. Then the conditions

$$x(t_j) = \operatorname{sgn} \gamma_j = \pm 1,$$

$$|x(t_j) - \tilde{y}(t_j)| \leqq \|x - \tilde{y}\| < 1$$

together imply that for all $\gamma_j \neq 0$,

$$\operatorname{sgn} \tilde{y}(t_j) = \operatorname{sgn} x(t_j) = \operatorname{sgn} \gamma_j.$$

But this contradicts (8) with $y = \tilde{y}$ because $\gamma_j \neq 0$ for some j, so that

$$\sum_{j=1}^{n} \gamma_j \tilde{y}(t_j) = \sum_{j=1}^{n} \gamma_j \operatorname{sgn} \gamma_j = \sum_{j=1}^{n} |\gamma_j| \neq 0.$$

Hence (9) must hold. ∎

Note that if Y is the set of all real polynomials of degree not exceeding n, together with the polynomial $y = 0$ (for which a degree is not defined in the usual discussion of degree), then $\dim Y = n+1$ and Y satisfies the Haar condition. (Why?) This yields

6.3-5 Theorem (Polynomials). *The best approximation to an x in the real space $C[a, b]$ out of Y_n is unique; here Y_n is the subspace consisting of $y = 0$ and all polynomials of degree not exceeding a fixed given n.*

In this theorem, it is worthwhile to compare the approximations for various n and see what happens as $n \longrightarrow \infty$. Let $\delta_n = \|x - p_n\|$, where $p_n \in Y_n$ is the best approximation to a fixed given x. Since $Y_0 \subset Y_1 \subset \cdots$, we have the monotonicity

$$\delta_0 \geqq \delta_1 \geqq \delta_2 \geqq \cdots \tag{10}$$

and the Weierstrass approximation theorem 4.11-5 implies that

$$\lim_{n \to \infty} \delta_n = 0. \tag{11}$$

It goes almost without saying that Theorem 6.3-5 characterizes the prototype of problems that initiated Haar's work. In fact, one may wonder why in general one cannot expect uniqueness of best approximations but still has uniqueness if one approximates by polynomials. Hence one may ask for the property which causes polynomials to be so "exceptionally well-behaved." The answer is that they satisfy the Haar condition defined in 6.3-2.

Problems

1. If $Y \subset C[a, b]$ is a subspace of dimension n and satisfies the Haar condition, show that the restrictions of the elements of Y to any subset of $[a, b]$ consisting of n points constitute a vector space which still has dimension n (whereas ordinarily the dimension would decrease under such a restriction).

2. Let $x_1(t) = 1$ and $x_2(t) = t^2$. Does $Y = \operatorname{span}\{x_1, x_2\}$ satisfy the Haar condition if Y is regarded as a subspace (*a*) of $C[0, 1]$, (*b*) of $C[-1, 1]$? (To understand what is going on, approximate x defined by $x(t) = t^3$ in both cases.)

3. Show that $Y = \operatorname{span}\{y_1, \cdots, y_n\} \subset C[a, b]$ satisfies the Haar condition if and only if, for every n-tuple $\{t_1 \cdots, t_n\} \subset [a, b]$ consisting of n different points, the n vectors $v_j = (y_1(t_j), \cdots, y_n(t_j))$, $j = 1, \cdots, n$, form a linearly independent set.

4. (Vandermonde's determinant) Write the determinant in (1) for

$$y_1(t) = 1, \qquad y_2(t) = t, \qquad y_3(t) = t^2, \qquad \cdots, \qquad y_n(t) = t^{n-1}.$$

This determinant is called *Vandermonde's determinant* (or *Cauchy's determinant*). It can be shown that it equals the product of *all* factors $(t_k - t_j)$, where j and k satisfy $0 \leqq j < k \leqq n$. Prove that this implies the existence of a unique polynomial of degree not exceeding $n-1$ which assumes prescribed values at n distinct points.

5. (De la Vallée-Poussin theorem) Let $Y \subset C[a, b]$ satisfy the Haar condition and consider any $x \in C[a, b]$. If $y \in Y$ is such that $x - y$ has alternately positive and negative values at $n+1$ consecutive points in $[a, b]$, where $n = \dim Y$, show that the distance δ of the best approximation to x out of Y is at least equal to the smallest of the absolute values of those $n+1$ values of $x - y$.

6. In $C[0, 1]$ find the best approximation to x defined by $x(t) = e^t$ out of $Y = \operatorname{span}\{y_1, y_2\}$, where $y_1(t) = 1$, $y_2(t) = t$. Compare the approximation with that by the linear Taylor polynomial given by $1 + t$.

7. Same task as in Prob. 6, when $x(t) = \sin(\pi t/2)$.

8. Probs. 6 and 7 are concerned with the approximation of a function x on $[a, b]$ whose second derivative does not change sign on $[a, b]$. Show that in this case the best approximating linear function y is given by $y(t) = \alpha_1 + \alpha_2 t$, where

$$\alpha_1 = \frac{x(a) + x(c)}{2} - \alpha_2 \frac{a + c}{2},$$

$$\alpha_2 = \frac{x(b) - x(a)}{b - a}$$

and c is the solution of $x'(t) - y'(t) = 0$. Interpret the formula geometrically.

9. (Inconsistent linear equations) If a system of $r > n$ linear equations in n unknowns

$$\gamma_{j1}\omega_1 + \gamma_{j2}\omega_2 + \cdots + \gamma_{jn}\omega_n = \beta_j \qquad (j = 1, \cdots, r)$$

is inconsistent, it does not have a solution $w = (\omega_1, \cdots, \omega_n)$, but we can look for an 'approximate solution' $z = (\zeta_1, \cdots, \zeta_n)$ such that

$$\max \left| \beta_j - \sum_{k=1}^{n} \gamma_{jk}\zeta_k \right|$$

is as small as possible. How does this problem fit into our present consideration and what form does the Haar condition take in this case?

10. To get a better feeling for what is going on in Prob. 9, the reader may wish to consider a system which is so simple that he can graph $\beta_j - \sum \gamma_{jk}\zeta_k$ and find the approximate solution, say, the system

$$\omega = 1$$

$$4\omega = 2.$$

Graph $f(\zeta)=\max_j |\beta_j-\gamma_j\zeta|$. Notice that f is convex (cf. Prob. 4, Sec. 6.2). Find the approximate solution ζ as defined by the condition in Prob. 9.

6.4 Chebyshev Polynomials

The preceding section was devoted to the theoretical aspects of uniform approximation. The remaining practical problem is the determination of best approximations in terms of explicit formulas which we can use for calculations and other purposes. This is not easy, in general, and explicit solutions of the problem are known for relatively few functions $x \in C[a, b]$. In this connection, a useful tool is the following.

6.4-1 Definition (Alternating set). Let $x \in C[a, b]$ and $y \in Y$, where Y is a subspace of the real space $C[a, b]$. A set of points $t_0, \cdots, t_k$ in $[a, b]$, where $t_0 < t_1 < \cdots < t_k$, is called an *alternating set* for $x-y$ if $x(t_j)-y(t_j)$ has alternately the values $+\|x-y\|$ and $-\|x-y\|$ at consecutive points t_j. ∎

We see that these $k+1$ points in the definition are extremal points of $x-y$ as defined in 6.3-1 and the values of $x-y$ at these points are alternating positive and negative.

The importance of alternating sets is shown to some extent by the following lemma, which states that the existence of a sufficiently large alternating set for $x-y$ implies that y is the best approximation to x. Actually, this condition is also necessary for y to be the best approximation to x; but since we shall not need this fact, we do not prove it. [The proof would be somewhat more difficult than our next proof; cf. e.g., E. W. Cheney (1966), p. 75.]

6.4-2 Lemma (Best approximation). *Let Y be a subspace of the real space $C[a, b]$ satisfying the Haar condition* 6.3-2. *Given $x \in C[a, b]$, let $y \in Y$ be such that for $x-y$ there exists an alternating set of $n+1$ points, where $n = \dim Y$. Then y is the best uniform approximation to x out of Y.*

Proof. By 6.1-1 and 6.3-4 there is a unique best approximation to x out of Y. If this is not y, it is some other $y_0 \in Y$ and then

$$\|x-y\| > \|x-y_0\|.$$

This inequality implies that at those $n+1$ extremal points the function

$$y_0 - y = (x - y) - (x - y_0)$$

has the same sign as $x - y$, because $x - y$ equals $\pm\|x - y\|$ at such a point whereas the other term on the right, $x - y_0$, can never exceed $\|x - y_0\|$ in absolute value, which is strictly less than $\|x - y\|$. This shows that $y_0 - y$ is alternating positive and negative at those $n+1$ points, so that it must have at least n zeros in $[a, b]$. But this is impossible unless $y_0 - y = 0$, since $y_0 - y \in Y$ and Y satisfies the Haar condition. Hence y must be the best approximation to x out of Y. ∎

A very important classical problem and application of this lemma is the approximation of $x \in C[-1, 1]$ defined by

$$x(t) = t^n \qquad n \in \mathbf{N} \text{ fixed} \tag{1}$$

out of $Y = \operatorname{span}\{y_0, \cdots, y_{n-1}\}$, where

$$y_j(t) = t^j \qquad j = 0, \cdots, n-1. \tag{2}$$

Obviously, this means that we want to approximate x on $[-1, 1]$ by a real polynomial y of degree less than n. Such a polynomial is of the form

$$y(t) = \alpha_{n-1}t^{n-1} + \alpha_{n-2}t^{n-2} + \cdots + \alpha_0.$$

Hence for $z = x - y$ we have

$$z(t) = t^n - (\alpha_{n-1}t^{n-1} + \alpha_{n-2}t^{n-2} + \cdots + \alpha_0)$$

and we want to find y such that $\|z\|$ becomes as small as possible. Note that $\|z\| = \|x - y\|$ is the distance from x to y. We see from the last formula that z is a polynomial of degree n with leading coefficient 1. Hence our original problem is equivalent to the following one.

Find the polynomial z which, among all polynomials of degree n and with leading coefficient 1, has the smallest maximum deviation from 0 on the interval $[-1, 1]$ under consideration.

If we set

$$t = \cos\theta \tag{3}$$

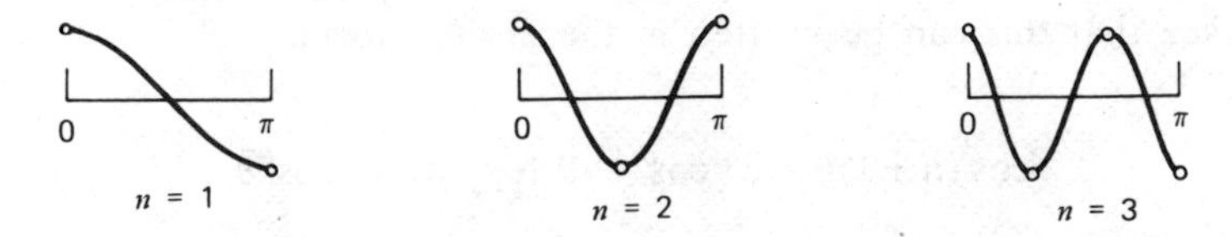

Fig. 60. The $n+1$ extremal points of $\cos n\theta$ on the interval $[0, \pi]$

and let θ vary from 0 to π, then t varies on our interval $[-1, 1]$. On $[0, \pi]$ the function defined by $\cos n\theta$ has $n+1$ extremal points, the values being ± 1 in alternating order (see Fig. 60). Because of Lemma 6.4-2 we may hope that $\cos n\theta$ will help to solve our problem, provided we are able to write $\cos n\theta$ as a polynomial in $t = \cos\theta$. Indeed, this can be done: we prove by induction that there is a representation of the form

$$\cos n\theta = 2^{n-1}\cos^n\theta + \sum_{j=0}^{n-1}\beta_{nj}\cos^j\theta \qquad (n = 1, 2, \cdots), \tag{4}$$

where the β_{nj}'s are constants.

Proof. This is true for $n = 1$ (take $\beta_{10} = 0$). Assuming it to be true for any n, we show that it holds for $n+1$. The addition formula for the cosine gives

$$\cos(n+1)\theta = \cos n\theta\cos\theta - \sin n\theta\sin\theta,$$

$$\cos(n-1)\theta = \cos n\theta\cos\theta + \sin n\theta\sin\theta.$$

Adding on both sides, we have

$$\cos(n+1)\theta + \cos(n-1)\theta = 2\cos n\theta\cos\theta. \tag{5}$$

Consequently, by the induction hypothesis,

$$\begin{aligned}\cos(n+1)\theta &= 2\cos\theta\cos n\theta - \cos(n-1)\theta \\ &= 2\cos\theta\left(2^{n-1}\cos^n\theta + \sum_{j=0}^{n-1}\beta_{nj}\cos^j\theta\right) \\ &\quad - 2^{n-2}\cos^{n-1}\theta - \sum_{j=0}^{n-2}\beta_{n-1,j}\cos^j\theta.\end{aligned}$$

We see that this can be written in the desired form

$$\cos(n+1)\theta = 2^n \cos^{n+1}\theta + \sum_{j=0}^{n} \beta_{n+1,j}\cos^j\theta,$$

and the proof is complete. ∎

Our problem is practically solved, but before we summarize our result, let us introduce a standard notation and terminology.

The functions defined by

$$\textbf{(6)} \qquad T_n(t) = \cos n\theta, \qquad \theta = \arccos t \qquad (n = 0, 1, \cdots)$$

are called **Chebyshev polynomials** *of the first kind*[1] *of order n.* They have various interesting properties, some of which are mentioned in the problem set at the end of the section. For more details, see G. Szegö (1967).

The leading coefficient in (4) is not 1, as we want it, but 2^{n-1}. Keeping this in mind, we obtain the following formulation of our result, which expresses the famous minimum property of the Chebyshev polynomials.

6.4-3 Theorem (Chebyshev polynomials). *The polynomial defined by*

$$(7) \qquad \tilde{T}_n(t) = \frac{1}{2^{n-1}} T_n(t) = \frac{1}{2^{n-1}} \cos(n \arccos t) \qquad (n \geqq 1)$$

has the smallest maximum deviation from 0 *on the interval* $[-1, 1]$, *among all real polynomials considered on* $[-1, 1]$ *which have degree n and leading coefficient* 1.

Remembering the approximation problem from which we started in this section, we can also formulate this result as follows.

The best uniform approximation to the function $x \in C[-1, 1]$ defined by $x(t) = t^n$, out of $Y = \operatorname{span}\{y_0, \cdots, y_{n-1}\}$ with y_j given by (2)

[1] T is suggested by Tchebichef, another transliteration of Чебышев. *Chebyshev polynomials of the second kind* are defined by

$$U_n(t) = [\sin(n+1)\theta]/\sin\theta \qquad (n = 1, 2, \cdots).$$

(that is, the approximation by a real polynomial of degree less than n) is y defined by

$$y(t) = x(t) - \frac{1}{2^{n-1}} T_n(t) \qquad (n \geqq 1). \tag{8}$$

Note that in (8) the highest power t^n drops out, so that the degree of y in t does not exceed $n-1$, as required.

Theorem 6.4-3 also helps in more general problems if a real polynomial $\tilde{x}$ of degree n with leading term $\beta_n t^n$ is given and we are looking for the best approximation $\tilde{y}$ to $\tilde{x}$ on $[-1, 1]$, where $\tilde{y}$ is a polynomial of lower degree, at most $n-1$. Then we may write

$$\tilde{x} = \beta_n x$$

and see that x has the leading term t^n. From Theorem 6.4-3 we conclude that $\tilde{y}$ must satisfy

$$\frac{1}{\beta_n}(\tilde{x} - \tilde{y}) = \tilde{T}_n.$$

The solution is

$$\tilde{y}(t) = \tilde{x}(t) - \frac{\beta_n}{2^{n-1}} T_n(t) \qquad (n \geqq 1). \tag{9}$$

This generalizes (8).

Explicit expressions of the first few Chebyshev polynomials can be readily obtained as follows. We see that $T_0(t) = \cos 0 = 1$ and, furthermore, $T_1(t) = \cos \theta = t$. Formula (6) shows that (5) can be written

$$T_{n+1}(t) + T_{n-1}(t) = 2tT_n(t).$$

This recursion formula

$$T_{n+1}(t) = 2tT_n(t) - T_{n-1}(t) \qquad (n = 1, 2, \cdots) \tag{10}$$

yields successively (Fig. 61)

$$\begin{aligned}
&T_0(t) = 1, && T_1(t) = t \\
(11^*) \quad &T_2(t) = 2t^2 - 1, && T_3(t) = 4t^3 - 3t \\
&T_4(t) = 8t^4 - 8t^2 + 1, && T_5(t) = 16t^5 - 20t^3 + 5t,
\end{aligned}$$

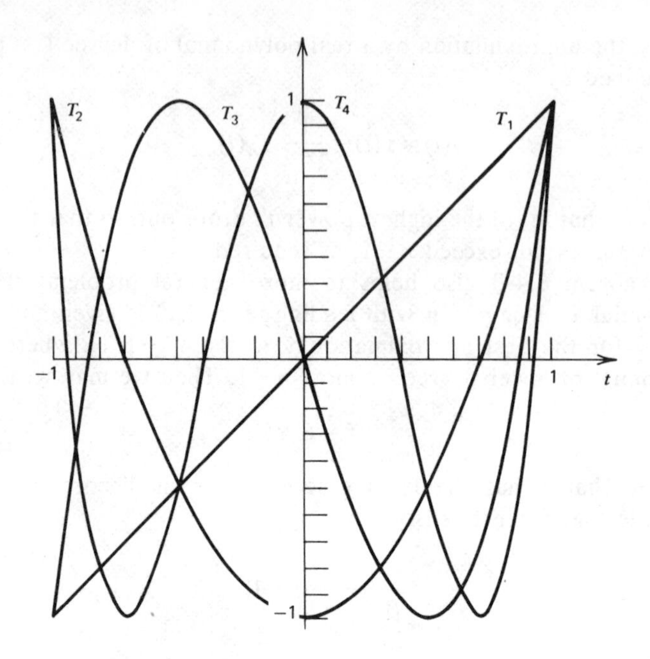

Fig. 61. Chebyshev polynomials T_1, T_2, T_3, T_4

etc. The general formula is

$$T_n(t) = \frac{n}{2} \sum_{j=0}^{[n/2]} (-1)^j \frac{(n-j-1)!}{j!(n-2j)!} (2t)^{n-2j} \qquad (n = 1, 2, \cdots), \tag{11}$$

where $[n/2] = n/2$ for even n and $[n/2] = (n-1)/2$ for odd n.

Problems

1. Verify (11*) by the use of (*a*) formula (11), (*b*) formula (10). Obtain T_6.

2. Find the best approximation to x defined by $x(t) = t^3 + t^2$, $t \in [-1, 1]$ by a quadratic polynomial y. Graph the result. What is the maximum deviation?

3. In certain applications, the zeros of the Chebyshev polynomials are of interest. Show that all the zeros of T_n are real and simple and lie in the interval $[-1, 1]$.

4. Between any two neighboring zeros of T_n there is precisely one zero of T_{n-1}. Prove this property. (This is called *interlacing of zeros* and also occurs in connection with other functions, for example, Bessel functions.)

5. Show that T_n and T_{n-1} have no common zeros.

6. Show that every real polynomial $x \in C[a, b]$ of degree $n \geqq 1$ with leading term $\beta_n t^n$ satisfies

$$\|x\| \geqq |\beta_n| \frac{(b-a)^n}{2^{2n-1}}.$$

7. Show that T_n is a solution of the differential equation

$$(1-t^2)T_n'' - tT_n' + n^2 T_n = 0.$$

8. The *hypergeometric differential equation* is

$$\tau(1-\tau)\frac{d^2w}{d\tau^2} + [c-(a+b+1)\tau]\frac{dw}{d\tau} - abw = 0,$$

where a, b, c are constants. Application of the Frobenius method (extended power series method) shows that a solution is given by

$$\begin{aligned} w(\tau) &= F(a, b, c; \tau) \\ &= 1 + \sum_{m=1}^{\infty} \frac{a(a+1)\cdots(a+m-1)b(b+1)\cdots(b+m-1)}{m!\,c(c+1)\cdots(c+m-1)} \tau^m \end{aligned}$$

where $c \neq 0, -1, -2, \cdots$. The series on the right is called the *hypergeometric series.* Under what conditions does the series reduce to a finite sum? $F(a, b, c; \tau)$ is called the *hypergeometric function.* It has been investigated in great detail. Many functions can be expressed in terms of this function. This includes the Chebyshev polynomials. In fact, show that

$$T_n(t) = F\left(-n, n, \frac{1}{2}; \frac{1}{2} - \frac{t}{2}\right).$$

9. (Orthogonality) Show that in the space $L^2[-1, 1]$ (cf. 2.2-7 and 3.1-5) the functions defined by $(1-t^2)^{-1/4}T_n(t)$ are orthogonal, that is,

$$\int_{-1}^{1} (1-t^2)^{-1/2} T_n(t) T_m(t)\, dt = 0 \qquad (m \neq n).$$

Show that the integral has the value π if $m = n = 0$ and the value $\pi/2$ if $m = n = 1, 2, \cdots$.

10. We want to mention that (3) suggests a relation between developments in terms of Chebyshev polynomials and *Fourier series.* As an example, represent $\tilde{x}$ defined by $\tilde{x}(\theta) = |\theta|$, $-\pi \leqq \theta \leqq \pi$, by a Fourier cosine series and write the result in terms of Chebyshev polynomials. Graph the function and the first few partial sums.

6.5 Approximation in Hilbert Space

For any given x in a Hilbert space H and a closed subspace $Y \subset H$ there exists a unique best approximation to x out of Y. Cf. 6.2-5.

In fact, Theorem 3.3-4 gives

$$H = Y \oplus Z \qquad (Z = Y^{\perp}), \tag{1a}$$

so that for each $x \in H$,

$$x = y + z, \tag{1b}$$

where $z = x - y \perp y$, hence $\langle x - y, y \rangle = 0$.

If Y is finite dimensional, say, $\dim Y = n$, we can determine y in terms of a basis $\{y_1, \cdots, y_n\}$ for Y as follows. We have a unique representation

$$y = \alpha_1 y_1 + \cdots + \alpha_n y_n \tag{2}$$

and $x - y \perp Y$ gives the n conditions

$$\langle y_j, x - y \rangle = \langle y_j, x - \sum \alpha_k y_k \rangle = 0$$

that is,

$$\langle y_j, x \rangle - \bar{\alpha}_1 \langle y_j, y_1 \rangle - \cdots - \bar{\alpha}_n \langle y_j, y_n \rangle = 0 \tag{3}$$

where $j=1, \cdots, n$. This is a nonhomogeneous system of n linear equations in n unknowns $\bar{\alpha}_1, \cdots, \bar{\alpha}_n$. The determinant of the coefficients is

$$\text{(4)}\quad G(y_1, \cdots, y_n) = \begin{vmatrix} \langle y_1, y_1\rangle & \langle y_1, y_2\rangle & \cdots & \langle y_1, y_n\rangle \\ \langle y_2, y_1\rangle & \langle y_2, y_2\rangle & \cdots & \langle y_2, y_n\rangle \\ \cdot & \cdot & \cdots & \cdot \\ \langle y_n, y_1\rangle & \langle y_n, y_2\rangle & \cdots & \langle y_n, y_n\rangle \end{vmatrix}.$$

Since y exists and is unique, that system has a unique solution. Hence $G(y_1, \cdots, y_n)$ must be different from 0. The determinant $G(y_1, \cdots, y_n)$ is called the **Gram determinant** of $y_1, \cdots, y_n$. It was introduced by J. P. Gram (1883). For simplicity we shall also write G for $G(y_1, \cdots, y_n)$ whenever it is clear to what functions we refer.

Cramer's rule now yields $\alpha_j = \bar{G}_j/\bar{G}$, where bars denote complex conjugates, G is given by (4) and G_j is obtained from G if we replace the jth column of G by the column with elements $\langle y_1, x\rangle, \cdots, \langle y_n, x\rangle$.

We also note a useful criterion involving G:

6.5-1 Theorem (Linear independence). *Elements $y_1, \cdots, y_n$ of a Hilbert space H constitute a linearly independent set in H if and only if*

$$G(y_1, \cdots, y_n) \neq 0.$$

Proof. Our preceding discussion shows that in the case of linear independence, $G \neq 0$. On the other hand, if $\{y_1, \cdots, y_n\}$ is linearly dependent, one of the vectors, call it y_j, is a linear combination of the others. Then the jth column of G is a linear combination of the other columns, so that $G=0$. ∎

It is interesting that the distance $\|z\| = \|x-y\|$ between x and the best approximation y to x can also be expressed by Gram determinants:

6.5-2 Theorem (Distance). *If* $\dim Y < \infty$ *in* (1) *and* $\{y_1, \cdots, y_n\}$ *is any basis for* Y, *then*

$$\text{(5)}\qquad \|z\|^2 = \frac{G(x, y_1, \cdots, y_n)}{G(y_1, \cdots, y_n)}.$$

Here, by the definition,

$$G(x, y_1, \cdots, y_n) = \begin{vmatrix} \langle x, x\rangle & \langle x, y_1\rangle & \cdots & \langle x, y_n\rangle \\ \langle y_1, x\rangle & \langle y_1, y_1\rangle & \cdots & \langle y_1, y_n\rangle \\ \cdot & \cdot & \cdots & \cdot \\ \langle y_n, x\rangle & \langle y_n, y_1\rangle & \cdots & \langle y_n, y_n\rangle \end{vmatrix}.$$

Proof. We have $\langle y, z\rangle = 0$, where $z = x - y$, so that, by (2), we obtain

$$\|z\|^2 = \langle z, z\rangle + \langle y, z\rangle = \langle x, x - y\rangle = \langle x, x\rangle - \langle x, \textstyle\sum \alpha_k y_k\rangle.$$

This can be written

$$-\|z\|^2 + \langle x, x\rangle - \bar{\alpha}_1 \langle x, y_1\rangle - \cdots - \bar{\alpha}_n \langle x, y_n\rangle = 0. \tag{6}$$

We now remember the n equations (3):

$$\langle y_j, x\rangle - \bar{\alpha}_1 \langle y_j, y_1\rangle - \cdots - \bar{\alpha}_n \langle y_j, y_n\rangle = 0,$$

where $j = 1, \cdots, n$. Equations (6) and (3) together can be regarded as a homogeneous system of $n+1$ linear equations in the $n+1$ "unknowns" $1, -\bar{\alpha}_1, \cdots, -\bar{\alpha}_n$. Since the system has a nontrivial solution, the determinant of its coefficients must be zero, that is,

$$\begin{vmatrix} \langle x, x\rangle - \|z\|^2 & \langle x, y_1\rangle & \cdots & \langle x, y_n\rangle \\ \langle y_1, x\rangle + 0 & \langle y_1, y_1\rangle & \cdots & \langle y_1, y_n\rangle \\ \cdot & \cdot & \cdots & \cdot \\ \langle y_n, x\rangle + 0 & \langle y_n, y_1\rangle & \cdots & \langle y_n, y_n\rangle \end{vmatrix} = 0. \tag{7}$$

We can write this determinant as the sum of two determinants. The first determinant is $G(x, y_1, \cdots, y_n)$. The second determinant has the elements $-\|z\|^2, 0, \cdots, 0$ in its first column; developing it by that column, we see that we can write (7) in the form

$$G(x, y_1, \cdots, y_n) - \|z\|^2 G(y_1, \cdots, y_n) = 0.$$

This gives (5) since $G(y_1, \cdots, y_n) \neq 0$ by 6.5-1. ∎

If the basis $\{y_1, \cdots, y_n\}$ in (5) is orthonormal, then $G(y_1, \cdots, y_n) = 1$ (why?), and by developing $G(x, y_1, \cdots, y_n)$ by its first row and noting that $\langle x, y_1\rangle\langle y_1, x\rangle = |\langle x, y_1\rangle|^2$, etc., we obtain from (5)

$$\|z\|^2 = \|x\|^2 - \sum_{k=1}^{n} |\langle x, y_k\rangle|^2, \tag{8}$$

in agreement with (11), Sec. 3.4, where y_k is denoted by e_k.

Problems

1. Show that permuting $y_1, \cdots, y_n$ leaves the value of $G(y_1, \cdots, y_n)$ unchanged.

2. Show that

$$G(\cdots, \alpha y_j, \cdots) = |\alpha|^2 G(\cdots, y_j, \cdots)$$

where the y_k's indicated by dots are the same on both sides.

3. If $G(y_1, \cdots, y_n) \neq 0$, show that $G(y_1, \cdots, y_j) \neq 0$ for $j = 1, \cdots, n-1$. Find a similar relation if $G(y_1, \cdots, y_n) = 0$.

4. Express the Schwarz inequality in terms of a Gram determinant. Use Theorem 6.5-1 to obtain the condition under which the equality sign holds. (Cf. 3.2-1.)

5. Show that $G(y_1, \cdots, y_n) \geqq 0$. Conclude from this that a finite subset of a Hilbert space is linearly independent if and only if the Gram determinant of its elements is positive.

6. Show that

$$G(y_1, \cdots, y_{n-1}, y_n + \alpha y_j) = G(y_1, \cdots, y_n) \qquad (j < n)$$

and indicate how this relation can be used to obtain Theorem 6.5-2.

7. Let $M = \{y_1, \cdots, y_n\}$ be a linearly independent set in a Hilbert space H. Show that for any subset $\{y_k, \cdots, y_m\}$ (where $k < m < n$),

$$\frac{G(y_k, \cdots, y_n)}{G(y_{k+1}, \cdots, y_n)} \leqq \frac{G(y_k, \cdots, y_m)}{G(y_{k+1}, \cdots, y_m)}.$$

Why is this geometrically plausible? Show that

$$\frac{G(y_m, \cdots, y_n)}{G(y_{m+1}, \cdots, y_n)} \leqq G(y_m).$$

8. Let $\{y_1, \cdots, y_n\}$ be a linearly independent set in a Hilbert space H. Show that for $m = 1, \cdots, n-1$ we have

$$G(y_1, \cdots, y_n) \leqq G(y_1, \cdots, y_m) G(y_{m+1}, \cdots, y_n)$$

and the equality sign holds if and only if each element of the set $M_1 = \{y_1, \cdots, y_m\}$ is orthogonal to each element of the set $M_2 = \{y_{m+1}, \cdots y_n\}$. (Use Prob. 7.)

9. (Hadamard's determinant theorem) Show that in Prob. 8,

$$G(y_1, \cdots, y_n) \leqq \langle y_1, y_1 \rangle \cdots \langle y_n, y_n \rangle$$

and the equality sign holds if and only if the y_j's are mutually orthogonal. Using this, show that the determinant of an n-rowed real square matrix $A = (\alpha_{jk})$ satisfies

$$(\det A)^2 \leqq a_1 \cdots a_n \qquad \text{where} \qquad a_j = \sum_{k=1}^{n} |\alpha_{jk}|^2.$$

10. Show that a linearly independent set $\{x_1, x_2, \cdots\}$ is dense in a Hilbert space H if and only if for every $x \in H$,

$$\frac{G(x, x_1, \cdots, x_n)}{G(x_1, \cdots, x_n)} \longrightarrow 0 \qquad \text{as } n \longrightarrow \infty.$$

6.6 Splines

Spline approximation is piecewise polynomial approximation. This means that we approximate a given function x on an interval $J = [a, b]$ by a function y which in the subintervals of a partition of $[a, b]$ is given by polynomials, one polynomial per subinterval, and it is required that y be several times differentiable at the common endpoints of those subintervals. Hence, instead of approximating x by a single polynomial on the entire interval $[a, b]$, we now approximate x by n polynomials, where n is the number of subintervals of the partition. In this way we lose analyticity, but may obtain approximating functions y which are more suitable in many problems of approximation and interpolation. For instance, they may not be as oscillatory between nodes as a single polynomial on $[a, b]$ oftentimes is. Since splines are of increasing practical importance, we want to give a short introduction to this field.

The simplest continuous piecewise polynomial approximation would be by piecewise linear functions. But such functions are not differentiable at certain points (at the endpoints of those subintervals), and it is preferable to use functions which have a certain number of derivatives everywhere on $[a, b]$.

We shall consider **cubic splines** on $J=[a, b]$. By definition, these are real-valued functions y which are twice continuously differentiable on $[a, b]$; this is written

$$y \in C^2[a, b];$$

and in each subinterval of a given partition P_n:

$$a = t_0 < t_1 < \cdots < t_n = b \tag{1}$$

of J, such a function y agrees with a polynomial of degree not exceeding 3. We call the t_j's the *nodes* of P_n and denote the corresponding vector space of all these cubic splines by

$$Y(P_n).$$

Let us now explain how a given real-valued function x on $[a, b]$ can be approximated by splines. We first choose a partition P_n of $J=[a, b]$ of the form (1). The desired approximation to x will be obtained by interpolation, which is one of the most important methods of effectively determining approximating functions. **Interpolation** of x by y means the construction of y such that at each of the nodes $t_0, \cdots, t_n$ the function y has the same value as x. Classical interpolation would mean that we use one of the interpolation formulas (for instance, by Lagrange, Newton or Everett[2]) to obtain a single polynomial on $[a, b]$ of degree n whose value at each node is the same as that of x. This polynomial approximates x quite well near the nodes but

[2] Most books on numerical analysis contain a chapter on interpolation. For a short introduction, see E. Kreyszig (1972), pp. 648–653.

may deviate considerably at points farther away from the nodes. In spline interpolation by cubic splines we take a spline y as just defined, whose value at each node agrees with the corresponding value of x. We prove that such a y exists and that we get a unique y if we prescribe the values of the derivative y' at the endpoints a and b of our interval. This is the content of the following theorem.

6.6-1 Theorem (Spline interpolation). *Let x be defined on $J=[a, b]$ and real-valued. Let P_n be any partition of J of the form (1), and let k_0' and k_n' be any two real numbers. Then there exists a unique cubic spline $y \in Y(P_n)$ satisfying the $n+3$ conditions*

$$\text{(2)} \qquad \begin{aligned} &\text{(a)} \qquad && y(t_j)=x(t_j) \\ &\text{(b)} \qquad && y'(t_0)=k_0' \qquad y'(t_n)=k_n' \end{aligned}$$

where $j=0, \cdots, n$.

Proof. In every subinterval $I_j=[t_j, t_{j+1}] \subset J$, $j=0, \cdots, n-1$, the spline y must agree with a cubic polynomial p_j such that

$$p_j(t_j)=x(t_j), \qquad p_j(t_{j+1})=x(t_{j+1}).$$

We write $1/(t_{j+1}-t_j)=\tau_j$ and

$$p_j'(t_j)=k_j' \qquad p_j'(t_{j+1})=k_{j+1}'$$

with given constants k_0' and k_n', and $k_1', \cdots, k_{n-1}'$ to be determined later. By direct calculation we can verify that the unique cubic polynomial p_j satisfying those four conditions is given by

$$\begin{aligned} p_j(t)=x(t_j)\tau_j^2(t-t_{j+1})^2[1+2\tau_j(t-t_j)] \\ +x(t_{j+1})\tau_j^2(t-t_j)^2[1-2\tau_j(t-t_{j+1})] \\ +k_j'\tau_j^2(t-t_j)(t-t_{j+1})^2 \\ +k_{j+1}'\tau_j^2(t-t_j)^2(t-t_{j+1}). \end{aligned}$$

Differentiating twice, we obtain

$$p_j''(t_j)=-6\tau_j^2x(t_j)+6\tau_j^2x(t_{j+1})-4\tau_jk_j'-2\tau_jk_{j+1}' \tag{3}$$

$$p_j''(t_{j+1})=6\tau_j^2x(t_j)-6\tau_j^2x(t_{j+1})+2\tau_jk_j'+4\tau_jk_{j+1}'. \tag{4}$$

Since $y \in C^2[a, b]$, at the nodes the second derivatives of the two corresponding polynomials must agree:

$$p''_{j-1}(t_j) = p_j''(t_j) \qquad j = 1, \cdots, n-1.$$

Using (4), with $j-1$ instead of j, and (3), we see that these $n-1$ equations take the form

$$\tau_{j-1}k'_{j-1} + 2(\tau_{j-1} + \tau_j)k_j' + \tau_j k'_{j+1} = 3[\tau_{j-1}^2 \, \Delta x_j + \tau_j^2 \, \Delta x_{j+1}],$$

where $\Delta x_j = x(t_j) - x(t_{j-1})$ and $\Delta x_{j+1} = x(t_{j+1}) - x(t_j)$ and $j = 1, \cdots, n-1$, as before. This system of $n-1$ linear equations has a unique solution $k_1', \cdots, k'_{n-1}$. In fact, this follows from Theorem 5.2-1 because all the elements of the coefficient matrix are nonnegative, and each element in the principal diagonal is greater than the sum of the other elements in the same row. Hence we are able to determine unique values $k_1', \cdots, k'_{n-1}$ of the first derivative of y at the nodes. This completes the proof. ∎

We conclude this introduction to splines by deriving an interesting minimum property. Suppose in Theorem 6.6-1 we have $x \in C^2[a, b]$ and (2b) is of the form

$$y'(a) = x'(a), \qquad y'(b) = x'(b). \tag{5}$$

Then $x' - y'$ is zero at a and b. Integrating by parts, we thus obtain

$$\int_a^b y''(t)[x''(t) - y''(t)] \, dt = -\int_a^b y'''(t)[x'(t) - y'(t)] \, dt.$$

Since y''' is constant on each subinterval of the partition, the integral on the right is zero by (2a). This proves

$$\int_a^b [x''(t) - y''(t)]^2 \, dt = \int_a^b x''(t)^2 \, dt - \int_a^b y''(t)^2 \, dt.$$

The integrand on the left is nonnegative and so is the integral. Hence if $x \in C^2[a, b]$ and y is the cubic spline corresponding to x and a partition P_n of $[a, b]$ and satisfying (2a) and (5), we have

$$\int_a^b x''(t)^2 \, dt \geqq \int_a^b y''(t)^2 \, dt. \tag{6}$$

And equality holds if and only if x is the cubic spline y. This is a minimum property of splines and explains the name, as follows. Engineers have for long times used thin rods called *splines* to fit curves through given points, and the strain energy minimized by such splines is proportional, approximately, to the integral of the square of the second derivative of the spline.

For splines of higher order, splines in several variables, convergence problems, applications and other topics, see the references given by A. Sard and S. Weintraub (1971), pp. 107–119.

Problems

1. Show that all the cubic splines corresponding to a given partition P_n of an interval $[a, b]$ form a vector space $Y(P_n)$. What is the dimension of this space?

2. Show that for a given partition P_n of the form (1) there exist $n+1$ unique cubic splines $y_0, \cdots, y_n$ such that

$$y_j(t_k) = \delta_{jk}$$

$$y_j'(a) = y_j'(b) = 0.$$

How can these be used to obtain a basis for $Y(P_n)$?

3. Approximate x defined by $x(t) = t^4$ on $[-1, 1]$ by a cubic spline corresponding to the partition $P_2 = \{-1, 0, 1\}$ and satisfying (2a) and (5). First guess what y may look like, then calculate.

4. Let x be defined on $[-1, 1]$ by $x(t) = t^4$. Find the Chebyshev approximation $\tilde{y}$ to x out of the space of all polynomials of degree not exceeding 3. Does $\tilde{y}$ satisfy (2a) and (5)? Graph and compare $\tilde{y}$ with the spline approximation given in the answer to Prob. 3.

5. Show that the Chebyshev approximation in Prob. 4 has a larger maximum deviation from x than the spline approximation in Prob. 3. Comment.

6. If a cubic spline y on $[a, b]$ is three times continuously differentiable, show that y must be a polynomial.

7. It may sometimes happen that a spline function is represented by the same polynomial in adjacent subintervals of $[a, b]$. To illustrate this, find the cubic spline y for the partition $\{-\pi/2, 0, \pi/2\}$ corresponding to x, where $x(t) = \sin t$, and satisfying (2a) and (5).

8. A possible geometric interpretation of (6) is that a cubic spline function minimizes the integral of the square of the curvature, at least approximately. Explain.

9. For $x, y \in C^2[a, b]$ define

$$\langle x, y\rangle_2 = \int_a^b x''(t)y''(t)\,dt, \qquad p(x) = \langle x, x\rangle_2^{1/2},$$

where the subscript 2 indicates that we use second derivatives. Show that p is a seminorm (cf. Sec. 2.3, Prob. 12), but not a norm. Write the derivation of (6), as given in the text, in terms of $\langle x, y\rangle_2$ and p.

10. Show that for any $x \in C^2[a, b]$ and its spline function y satisfying (2a) and (5) we can estimate the deviation in terms of p (cf. Prob. 9), independent of the particular choice of a partition:

$$\|x - y\|_2 \leqq p(x).$$

CHAPTER 7

SPECTRAL THEORY OF LINEAR OPERATORS IN NORMED SPACES

Spectral theory is one of the main branches of modern functional analysis and its applications. Roughly speaking, it is concerned with certain inverse operators, their general properties and their relations to the original operators. Such inverse operators arise quite naturally in connection with the problem of solving equations (systems of linear algebraic equations, differential equations, integral equations). For instance, the investigations of boundary value problems by Sturm and Liouville and Fredholm's famous theory of integral equations were important to the development of the field.

The spectral theory of operators is very important for an understanding of the operators themselves, as we shall see.

In Chaps. 7 to 9 we give an introduction to the spectral theory of bounded linear operators $T: X \longrightarrow X$ on normed and inner product spaces. This includes the consideration of classes of operators which are of great practical interest, in particular, compact operators (Chap. 8) and self-adjoint operators (Chap. 9). Spectral theory of unitary operators follows a little later (in Sec. 10.5, which can be read without reference to other sections in Chap. 10).

Unbounded linear operators in Hilbert spaces will be considered in Chap. 10 and their application in quantum mechanics in Chap. 11.

Brief orientation about main content of Chap. 7

We begin with finite dimensional vector spaces. Spectral theory in this case is essentially matrix eigenvalue theory (Sec. 7.1) and is much simpler than that of operators in infinite dimensional spaces. Nevertheless it is of great practical importance, and the number of research papers in the field is enormous, a good deal of them being in numerical analysis. Matrix eigenvalue problems also suggest part of the general setting and some of the concepts of spectral theory in infinite dimensional normed spaces as defined in Sec. 7.2, although the infinite dimensional case is much more complicated than the finite dimensional one.

Important properties of the spectrum of bounded linear operators on normed and Banach spaces are discussed in Secs. 7.3 and 7.4.

Complex analysis is a valuable tool in spectral theory, but to maintain an elementary level, we shall give merely an introduction to some basic facts in that direction. The corresponding section (Sec. 7.5) can be omitted if students do not have the background.

In Secs. 7.6 and 7.7 we show that some of the considerations can be generalized to Banach algebras.

General assumption

We exclude the trivial vector space $\{0\}$ and assume all spaces to be complex unless otherwise stated, in order to obtain a satisfactory theory.

7.1 Spectral Theory in Finite Dimensional Normed Spaces

Let X be a finite dimensional normed space and $T: X \longrightarrow X$ a linear operator. Spectral theory of such operators is simpler than that of operators defined on infinite dimensional spaces. In fact, from Sec. 2.9 we know that we can represent T by matrices (which depend on the choice of bases for X), and we shall see that spectral theory of T is essentially matrix eigenvalue theory. So let us begin with matrices.

We note that the present section is algebraic, but we shall soon make use of the norm, starting in the next section.

For a given (real or complex) n-rowed square matrix $A = (\alpha_{jk})$ the concepts of eigenvalues and eigenvectors are defined in terms of the equation

(1) $$Ax = \lambda x$$

as follows.

7.1-1 Definition (Eigenvalues, eigenvectors, eigenspaces, spectrum, resolvent set of a matrix). An *eigenvalue* of a square matrix $A = (\alpha_{jk})$ is a number λ such that (1) has a solution $x \neq 0$. This x is called an *eigenvector* of A corresponding to that eigenvalue λ. The eigenvectors corresponding to that eigenvalue λ and the zero vector form a vector subspace of X which is called the *eigenspace* of A corresponding to that eigenvalue λ. The set $\sigma(A)$ of all eigenvalues of A is called the *spectrum* of A. Its complement $\rho(A) = \mathbf{C} - \sigma(A)$ in the complex plane is called the *resolvent set* of A. ∎

For example, by direct calculation we can verify that

$$x_1 = \begin{bmatrix} 4 \\ 1 \end{bmatrix} \text{ and } x_2 = \begin{bmatrix} 1 \\ -1 \end{bmatrix} \qquad \text{are eigenvectors of} \qquad A = \begin{bmatrix} 5 & 4 \\ 1 & 2 \end{bmatrix}$$

corresponding to the eigenvalues $\lambda_1 = 6$ and $\lambda_2 = 1$ of A, respectively. How did we obtain this result, and what can we say about the existence of eigenvalues of a matrix in general?

To answer this question, let us first note that (1) can be written

$$(A - \lambda I)x = 0 \tag{2}$$

where I is the n-rowed unit matrix. This is a homogeneous system of n linear equations in n unknowns $\xi_1, \cdots, \xi_n$, the components of x. The determinant of the coefficients is $\det(A - \lambda I)$ and must be zero in order that (2) have a solution $x \neq 0$. This gives the **characteristic equation** of A:

$$\det(A - \lambda I) = \begin{vmatrix} \alpha_{11} - \lambda & \alpha_{12} & \cdots & \alpha_{1n} \\ \alpha_{21} & \alpha_{22} - \lambda & \cdots & \alpha_{2n} \\ \cdot & \cdot & \cdots & \cdot \\ \alpha_{n1} & \alpha_{n2} & \cdots & \alpha_{nn} - \lambda \end{vmatrix} = 0. \tag{3}$$

$\det(A - \lambda I)$ is called the **characteristic determinant** of A. By developing it we obtain a polynomial in λ of degree n, the **characteristic polynomial** of A. Equation (3) is called the *characteristic equation* of A.

Our result is the basic

7.1-2 Theorem (Eigenvalues of a matrix). *The eigenvalues of an n-rowed square matrix $A = (\alpha_{jk})$ are given by the solutions of the characteristic equation* (3) *of A. Hence A has at least one eigenvalue (and at most n numerically different eigenvalues).*

The second statement holds since, by the so-called fundamental theorem of algebra and the factorization theorem, a polynomial of positive degree n and with coefficients in $\mathbf{C}$ has a root in $\mathbf{C}$ (and at most n numerically different roots). Note that roots may be complex even if A is real.

In our above example,

$$\det(A-\lambda I)=\begin{vmatrix} 5-\lambda & 4 \\ 1 & 2-\lambda \end{vmatrix}=\lambda^2-7\lambda+6=0,$$

the spectrum is $\{6, 1\}$, and eigenvectors of A corresponding to 6 and 1 are obtained from

$$\begin{aligned} -\xi_1+4\xi_2&=0 \\ \xi_1-4\xi_2&=0 \end{aligned} \qquad \text{and} \qquad \begin{aligned} 4\xi_1+4\xi_2&=0 \\ \xi_1+\ \xi_2&=0 \end{aligned}$$

respectively. Observe that in each case we need only one of the two equations. (Why?)

How can we apply our result to a linear operator $T: X \longrightarrow X$ on a normed space X of dimension n? Let $e=\{e_1, \cdots, e_n\}$ be any basis for X and $T_e=(\alpha_{jk})$ the matrix representing T with respect to that basis (whose elements are kept in the given order). Then the eigenvalues of the matrix T_e are called the **eigenvalues** *of the operator* T, and similarly for the *spectrum* and the *resolvent set.* This is justified by

7.1-3 Theorem (Eigenvalues of an operator). *All matrices representing a given linear operator $T: X \longrightarrow X$ on a finite dimensional normed space X relative to various bases for X have the same eigenvalues.*

Proof. We must see what happens in the transition from one basis for X to another. Let $e=(e_1, \cdots, e_n)$ and $\tilde{e}=(\tilde{e}_1, \cdots, \tilde{e}_n)$ be any bases for X, written as row vectors. By the definition of a basis, each e_j is a linear combination of the $\tilde{e}_k$'s and conversely. We can write this

$$\tilde{e}=eC \qquad \text{or} \qquad \tilde{e}^\mathsf{T}=C^\mathsf{T}e^\mathsf{T} \tag{4}$$

where C is a nonsingular n-rowed square matrix. Every $x \in X$ has a unique representation with respect to each of the two bases, say,

$$x=ex_1=\sum \xi_j e_j=\tilde{e}x_2=\sum \tilde{\xi}_k \tilde{e}_k$$

where $x_1 = (\xi_j)$ and $x_2 = (\tilde{\xi}_k)$ are column vectors. From this and (4) we have $ex_1 = \tilde{e}x_2 = eCx_2$. Hence

$$x_1 = Cx_2. \tag{5}$$

Similarly, for $Tx = y = ey_1 = \tilde{e}y_2$ we have

$$y_1 = Cy_2. \tag{6}$$

Consequently, if T_1 and T_2 denote the matrices which represent T with respect to e and $\tilde{e}$, respectively, then

$$y_1 = T_1x_1 \qquad \text{and} \qquad y_2 = T_2x_2,$$

and from this and (5) and (6),

$$CT_2x_2 = Cy_2 = y_1 = T_1x_1 = T_1Cx_2.$$

Premultiplying by C^{-1} we obtain the transformation law

$$T_2 = C^{-1}T_1C \tag{7}$$

with C determined by the bases according to (4) (and independent of T). Using (7) and $\det(C^{-1}) \det C = 1$, we can now show that the characteristic determinants of T_2 and T_1 are equal:

$$\begin{aligned} \det(T_2 - \lambda I) &= \det(C^{-1}T_1C - \lambda C^{-1}IC) \\ &= \det(C^{-1}(T_1 - \lambda I)C) \\ &= \det(C^{-1}) \det(T_1 - \lambda I) \det C \\ &= \det(T_1 - \lambda I). \end{aligned} \tag{8}$$

Equality of the eigenvalues of T_1 and T_2 now follows from Theorem 7.1-2. ▮

We mention in passing that we can also express our result in terms of the following concept, which is of general interest. An $n \times n$ matrix T_2 is said to be *similar* to an $n \times n$ matrix T_1, if there exists a nonsingular matrix C such that (7) holds. T_1 and T_2 are then called *similar matrices*. In terms of this concept, our proof shows the following.

(i) Two matrices representing the same linear operator T on a finite dimensional normed space X relative to any two bases for X are similar.

(ii) Similar matrices have the same eigenvalues.

Furthermore, Theorems 7.1-2 and 7.1-3 imply

7.1-4 Existence Theorem (Eigenvalues). *A linear operator on a finite dimensional complex normed space $X \neq \{0\}$ has at least one eigenvalue.*

In general we cannot say more (cf. Prob. 13).

Furthermore, (8) with $\lambda = 0$ gives $\det T_2 = \det T_1$. Hence the value of the determinant represents an intrinsic property of the operator T, so that we can speak unambiguously of the quantity $\det T$.

Problems

1. Find the eigenvalues and eigenvectors of the following matrices, where a and b are real and $b \neq 0$.

$$A = \begin{bmatrix} 1 & 2 \\ -8 & 11 \end{bmatrix}, \qquad B = \begin{bmatrix} a & b \\ -b & a \end{bmatrix}.$$

2. (Hermitian matrix) Show that the eigenvalues of a Hermitian matrix $A = (\alpha_{jk})$ are real. (Definition in Sec. 3.10.)

3. (Skew-Hermitian matrix) Show that the eigenvalues of a skew-Hermitian matrix $A = (\alpha_{jk})$ are pure imaginary or zero. (Definition in Sec. 3.10.)

4. (Unitary matrix) Show that the eigenvalues of a unitary matrix have absolute value 1. (Definition in Sec. 3.10.)

5. Let X be a finite dimensional inner product space and $T: X \longrightarrow X$ a linear operator. If T is self-adjoint, show that its spectrum is real. If T is unitary, show that its eigenvalues have absolute value 1.

6. (Trace) Let $\lambda_1, \cdots, \lambda_n$ be the n eigenvalues of an n-rowed square matrix $A = (\alpha_{jk})$, where some or all of the λ_j's may be equal. Show that the product of the eigenvalues equals $\det A$ and their sum equals the *trace* of A, that is, the sum of the elements of the *principal diagonal*:

$$\text{trace } A = \alpha_{11} + \alpha_{22} + \cdots + \alpha_{nn}.$$

7. (Inverse) Show that the inverse A^{-1} of a square matrix A exists if and only if all the eigenvalues $\lambda_1, \cdots, \lambda_n$ of A are different from zero. If A^{-1} exists, show that it has the eigenvalues $1/\lambda_1, \cdots, 1/\lambda_n$.

8. Show that a two-rowed nonsingular matrix

$$A = \begin{bmatrix} a_{11} & a_{12} \\ a_{21} & a_{22} \end{bmatrix} \qquad \text{has the inverse} \qquad A^{-1} = \frac{1}{\det A} \begin{bmatrix} a_{22} & -a_{12} \\ -a_{21} & a_{11} \end{bmatrix}.$$

How does it follow from this formula that A^{-1} has the eigenvalues $1/\lambda_1$, $1/\lambda_2$, where λ_1, λ_2 are the eigenvalues of A?

9. If a square matrix $A = (\alpha_{jk})$ has eigenvalues λ_j, $j = 1, \cdots, n$, show that kA has the eigenvalues $k\lambda_j$ and A^m $(m \in \mathbf{N})$ has the eigenvalues $\lambda_j^{\,m}$.

10. If A is a square matrix with eigenvalues $\lambda_1, \cdots, \lambda_n$ and p is any polynomial, show that the matrix $p(A)$ has the eigenvalues $p(\lambda_j)$, $j = 1, \cdots, n$.

11. If x_j is an eigenvector of an n-rowed square matrix A corresponding to an eigenvalue λ_j and C is any nonsingular n-rowed square matrix, show that λ_j is an eigenvalue of $\tilde{A} = C^{-1}AC$ and a corresponding eigenvector is $y_j = C^{-1}x_j$.

12. Illustrate with a simple example that an n-rowed square matrix may not have eigenvectors which constitute a basis for $\mathbf{R}^n$ (or $\mathbf{C}^n$). For instance, consider

$$A = \begin{bmatrix} 1 & 1 \\ 0 & 1 \end{bmatrix}.$$

13. (Multiplicity) The *algebraic multiplicity* of an eigenvalue λ of a matrix A is the multiplicity of λ as a root of the characteristic polynomial, and the dimension of the eigenspace of A corresponding to λ may be called the *geometric multiplicity* of λ. Find the eigenvalues and their multiplicities of the matrix corresponding to the following transformation and comment:

$$\eta_j = \xi_j + \xi_{j+1} \qquad (j = 1, 2, \cdots, n-1), \qquad \eta_n = \xi_n.$$

14. Show that the geometric multiplicity of an eigenvalue cannot exceed the algebraic multiplicity (cf. Prob. 13).

15. Let T be the differential operator on the space X consisting of all polynomials of degree not exceeding $n-1$ and the polynomial $x=0$ (for which the degree is not defined in the usual discussion of degree). Find all eigenvalues and eigenvectors of T and their algebraic and geometric multiplicities.

7.2 Basic Concepts

In the preceding section the spaces were finite dimensional. In this section we consider normed spaces of any dimension, and we shall see that in infinite dimensional spaces, spectral theory becomes more complicated.

Let $X \neq \{0\}$ be a complex normed space and $T: \mathscr{D}(T) \longrightarrow X$ a linear operator with domain $\mathscr{D}(T) \subset X$. With T we associate the operator

$$T_\lambda = T - \lambda I \tag{1}$$

where λ is a complex number and I is the identity operator on $\mathscr{D}(T)$. If T_λ has an inverse, we denote it by $R_\lambda(T)$, that is,

$$R_\lambda(T) = T_\lambda^{-1} = (T - \lambda I)^{-1} \tag{2}$$

and call it the *resolvent operator* of T or, simply, the **resolvent**[1] of T. Instead of $R_\lambda(T)$ we also write simply R_λ if it is clear to what operator T we refer in a specific discussion.

The name "resolvent" is appropriate, since $R_\lambda(T)$ helps to *solve* the equation $T_\lambda x = y$. Thus, $x = T_\lambda^{-1} y = R_\lambda(T) y$ provided $R_\lambda(T)$ exists.

More important, the investigation of properties of R_λ will be basic for an understanding of the operator T itself. Naturally, many properties of T_λ and R_λ depend on λ. And *spectral theory* is concerned with those properties. For instance, we shall be interested in the set of all λ in the complex plane such that R_λ exists. Boundedness of R_λ is another property that will be essential. We shall also ask for what λ's the domain of R_λ is dense in X, to name just a few aspects.

[1] Some authors define the resolvent by $(\lambda I - T)^{-1}$. In the older literature on integral equations the resolvent is defined by $(I - \mu T)^{-1}$. Of course, the transition to (2) is elementary, but it is annoying, and the reader is advised to check on this point before making comparisons between different publications on spectral theory.

We further note that $R_\lambda(T)$ is a linear operator by Theorem 2.6-10(b).

For our investigation of T, T_λ and R_λ we shall need the following concepts which are basic in spectral theory.

7.2-1 Definition (Regular value, resolvent set, spectrum). Let $X \neq \{0\}$ be a complex normed space and $T\colon \mathfrak{D}(T) \longrightarrow X$ a linear operator with domain $\mathfrak{D}(T) \subset X$. A *regular value* λ of T is a complex number such that

(R1) $R_\lambda(T)$ exists,

(R2) $R_\lambda(T)$ is bounded,

(R3) $R_\lambda(T)$ is defined on a set which is dense in X.

The *resolvent set* $\rho(T)$ of T is the set of all regular values λ of T. Its complement $\sigma(T) = \mathbf{C} - \rho(T)$ in the complex plane $\mathbf{C}$ is called the *spectrum* of T, and a $\lambda \in \sigma(T)$ is called a *spectral value* of T. Furthermore, the spectrum $\sigma(T)$ is partitioned into three disjoint sets as follows.

The **point spectrum** or *discrete spectrum* $\sigma_p(T)$ is the set such that $R_\lambda(T)$ does not exist. A $\lambda \in \sigma_p(T)$ is called an *eigenvalue* of T.

The **continuous spectrum** $\sigma_c(T)$ is the set such that $R_\lambda(T)$ exists and satisfies (R3) but not (R2), that is, $R_\lambda(T)$ is unbounded.

The **residual spectrum** $\sigma_r(T)$ is the set such that $R_\lambda(T)$ exists (and may be bounded or not) but does not satisfy (R3), that is, the domain of $R_\lambda(T)$ is not dense in X. ∎

To avoid trivial misunderstandings, let us say that some of the sets in this definition may be empty. This is an existence problem which we shall have to discuss. For instance, $\sigma_c(T) = \sigma_r(T) = \varnothing$ in the finite dimensional case, as we know from Sec. 7.1.

The conditions stated in Def. 7.2-1 can be summarized in the following table.

Satisfied	Not satisfied	λ belongs to:
(R1), (R2), (R3)		$\rho(T)$
	(R1)	$\sigma_p(T)$
(R1) (R3)	(R2)	$\sigma_c(T)$
(R1)	(R3)	$\sigma_r(T)$

To gain an understanding of these concepts, we begin with some general remarks as follows.

We first note that the four sets in the table are disjoint and their union is the whole complex plane:

$$\begin{aligned}\mathbf{C} &= \rho(T) \cup \sigma(T) \\ &= \rho(T) \cup \sigma_p(T) \cup \sigma_c(T) \cup \sigma_r(T)\end{aligned}$$

Furthermore, if the resolvent $R_\lambda(T)$ exists, it is linear by Theorem 2.6-10, as was mentioned before. That theorem also shows that $R_\lambda(T)$: $\mathscr{R}(T_\lambda) \longrightarrow \mathscr{D}(T_\lambda)$ exists if and only if $T_\lambda x = 0$ implies $x = 0$, that is, the null space of T_λ is $\{0\}$; here, $\mathscr{R}(T_\lambda)$ denotes the range of T_λ (cf. Sec. 2.6).

Hence if $T_\lambda x = (T - \lambda I)x = 0$ for some $x \neq 0$, then $\lambda \in \sigma_p(T)$, by definition, that is, λ is an eigenvalue of T. The vector x is then called an **eigenvector** of T (or **eigenfunction** of T if X is a function space) corresponding to the eigenvalue λ. The subspace of $\mathscr{D}(T)$ consisting of 0 and all eigenvectors of T corresponding to an eigenvalue λ of T is called the *eigenspace* of T corresponding to that eigenvalue λ.

We see that our present definition of an eigenvalue is in harmony with that in the preceding section. We also see that the spectrum of a linear operator on a finite dimensional space is a *pure point spectrum*, that is, both the continuous spectrum and the residual spectrum are empty, as was mentioned before, so that every spectral value is an eigenvalue.

A similar motivation for the partition of $\sigma(T) - \sigma_p(T)$ into $\sigma_c(T)$ and $\sigma_r(T)$ is given by the fact that $\sigma_r(T) = \varnothing$ for the important class of self-adjoint linear operators on Hilbert spaces (cf. Def. 3.10-1); this will be proved in 9.2-4.

If X is infinite dimensional, then T can have spectral values which are not eigenvalues:

7.2-2 Example (Operator with a spectral value which is not an eigenvalue). On the Hilbert sequence space $X = l^2$ (cf. 3.1-6) we define a linear operator T: $l^2 \longrightarrow l^2$ by

$$(\xi_1, \xi_2, \cdots) \longmapsto (0, \xi_1, \xi_2, \cdots), \tag{3}$$

where $x = (\xi_j) \in l^2$. The operator T is called the *right-shift operator*. T is bounded (and $\|T\|=1$) because

$$\|Tx\|^2=\sum_{j=1}^{\infty}|\xi_j|^2=\|x\|^2.$$

The operator $R_0(T)=T^{-1}\colon T(X)\longrightarrow X$ exists; in fact, it is the *left-shift operator* given by

$$(\xi_1, \xi_2, \cdots)\longmapsto(\xi_2, \xi_3, \cdots).$$

But $R_0(T)$ does not satisfy (R3), because (3) shows that $T(X)$ is not dense in X; indeed, $T(X)$ is the subspace Y consisting of all $y=(\eta_j)$ with $\eta_1=0$. Hence, by definition, $\lambda=0$ is a spectral value of T. Furthermore, $\lambda=0$ is not an eigenvalue. We can see this directly from (3) since $Tx=0$ implies $x=0$ and the zero vector is not an eigenvector. ∎

In our present discussion, the bounded inverse theorem 4.12-2 contributes the following. If $T\colon X\longrightarrow X$ is bounded and linear and X is complete, and if for some λ the resolvent $R_\lambda(T)$ exists and is defined on the whole space X, then for that λ the resolvent is bounded.

Furthermore, the following facts (to be needed later) may also be helpful for a better understanding of the present concepts.

7.2-3 Lemma (Domain of R_λ). *Let X be a complex Banach space, $T\colon X\longrightarrow X$ a linear operator, and $\lambda\in\rho(T)$. Assume that (a) T is closed or (b) T is bounded. Then $R_\lambda(T)$ is defined on the whole space X and is bounded.*

Proof. (*a*) Since T is closed, so is T_λ by 4.13-3. Hence R_λ is closed. R_λ is bounded by (R2). Hence its domain $\mathfrak{D}(R_\lambda)$ is closed by 4.13-5(*b*) applied to R_λ, so that (R3) implies $\overline{\mathfrak{D}(R_\lambda)}=\mathfrak{D}(R_\lambda)=X$.

(*b*) Since $\mathfrak{D}(T)=X$ is closed, T is closed by 4.13-5(*a*) and the statement follows from part (*a*) of this proof. ∎

Problems

1. **(Identity operator)** For the identity operator I on a normed space X, find the eigenvalues and eigenspaces as well as $\sigma(I)$ and $R_\lambda(I)$.

2. Show that for a given linear operator T, the sets $\rho(T)$, $\sigma_p(T)$, $\sigma_c(T)$ and $\sigma_r(T)$ are mutually disjoint and their union is the complex plane.

3. (Invariant subspace) A subspace Y of a normed space X is said to be *invariant* under a linear operator $T\colon X \longrightarrow X$ if $T(Y) \subset Y$. Show that an eigenspace of T is invariant under T. Give examples.

4. If Y is an invariant subspace under a linear operator T on an n-dimensional normed space X, what can be said about a matrix representing T with respect to a basis $\{e_1, \cdots, e_n\}$ for X such that $Y = \operatorname{span}\{e_1, \cdots, e_m\}$?

5. Let (e_k) be a total orthonormal sequence in a separable Hilbert space H and let $T\colon H \longrightarrow H$ be defined at e_k by

$$Te_k = e_{k+1} \qquad (k = 1, 2, \cdots)$$

and then linearly and continuously extended to H. Find invariant subspaces. Show that T has no eigenvalues.

6. (Extension) The behavior of the various parts of the spectrum under extension of an operator is of practical interest. If T is a bounded linear operator and T_1 is a linear extension of T, show that we have $\sigma_p(T_1) \supset \sigma_p(T)$ and for any $\lambda \in \sigma_p(T)$ the eigenspace of T is contained in the eigenspace of T_1.

7. Show that $\sigma_r(T_1) \subset \sigma_r(T)$ in Prob. 6.

8. Show that $\sigma_c(T) \subset \sigma_c(T_1) \cup \sigma_p(T_1)$ in Prob. 6.

9. Show directly (without using Probs. 6 and 8) that $\rho(T_1) \subset \rho(T) \cup \sigma_r(T)$ in Prob. 6.

10. How does the statement in Prob. 9 follow from Probs. 6 and 8?

7.3 Spectral Properties of Bounded Linear Operators

What general properties will the spectrum of a given operator have? This will depend on the kind of space on which the operator is defined (as a comparison of Secs. 7.1 and 7.2 illustrates) and on the kind of operator we consider. This situation suggests separate investigations of large classes of operators with common spectral properties, and in this section we begin with bounded linear operators T on a complex Banach space X. Thus $T \in B(X, X)$, where X is complete; cf. Sec. 2.10.

Our first theorem is a key to various parts of the theory, as we shall see.

7.3-1 Theorem (Inverse). *Let $T \in B(X, X)$, where X is a Banach space. If $\|T\| < 1$, then $(I-T)^{-1}$ exists as a bounded linear operator on the whole space X and*

$$(I-T)^{-1} = \sum_{j=0}^{\infty} T^j = I + T + T^2 + \cdots \tag{1}$$

[*where the series on the right is convergent in the norm on $B(X, X)$*].

Proof. We have $\|T^j\| \leqq \|T\|^j$ by (7), Sec. 2.7. We also remember that the geometric series $\sum \|T\|^j$ converges for $\|T\| < 1$. Hence the series in (1) is absolutely convergent for $\|T\| < 1$. Since X is complete, so is $B(X, X)$ by Theorem 2.10-2. Absolute convergence thus implies convergence, as we know from Sec. 2.3.

We denote the sum of the series in (1) by S. It remains to show that $S = (I-T)^{-1}$. For this purpose we calculate

$$\begin{aligned} &(I-T)(I+T+\cdots+T^n) \\ &\qquad = (I+T+\cdots+T^n)(I-T) \\ &\qquad = I - T^{n+1}. \end{aligned} \tag{2}$$

We now let $n \longrightarrow \infty$. Then $T^{n+1} \longrightarrow 0$ because $\|T\| < 1$. We thus obtain

$$(I-T)S = S(I-T) = I. \tag{3}$$

This shows that $S = (I-T)^{-1}$. ∎

As a first application of this theorem, let us prove the important fact that the spectrum of a bounded linear operator is a closed set in the complex plane. ($\sigma \neq \varnothing$ will be shown in 7.5-4.)

7.3-2 Theorem (Spectrum closed). *The resolvent set $\rho(T)$ of a bounded linear operator T on a complex Banach space X is open; hence the spectrum $\sigma(T)$ is closed.*

Proof. If $\rho(T)=\varnothing$, it is open. (Actually, $\rho(T)\neq\varnothing$ as we shall see in Theorem 7.3-4.) Let $\rho(T)\neq\varnothing$. For a fixed $\lambda_0\in\rho(T)$ and any $\lambda\in\mathbf{C}$ we have

$$\begin{aligned} T-\lambda I &= T-\lambda_0 I-(\lambda-\lambda_0)I \\ &= (T-\lambda_0 I)[I-(\lambda-\lambda_0)(T-\lambda_0 I)^{-1}]. \end{aligned}$$

Denoting the operator in the brackets $[\cdots]$ by V, we can write this in the form

$$T_\lambda = T_{\lambda_0}V \qquad \text{where} \qquad V=I-(\lambda-\lambda_0)R_{\lambda_0}. \tag{4}$$

Since $\lambda_0\in\rho(T)$ and T is bounded, Lemma 7.2-3(*b*) implies that $R_{\lambda_0}=T_{\lambda_0}^{-1}\in B(X,X)$. Furthermore, Theorem 7.3-1 shows that V has an inverse

$$V^{-1}=\sum_{j=0}^{\infty}[(\lambda-\lambda_0)R_{\lambda_0}]^j=\sum_{j=0}^{\infty}(\lambda-\lambda_0)^jR_{\lambda_0}{}^j \tag{5a}$$

in $B(X,X)$ for all λ such that $\|(\lambda-\lambda_0)R_{\lambda_0}\|<1$, that is,

$$|\lambda-\lambda_0|<\frac{1}{\|R_{\lambda_0}\|}. \tag{5b}$$

Since $T_{\lambda_0}{}^{-1}=R_{\lambda_0}\in B(X,X)$, we see from this and (4) that for every λ satisfying (5b) the operator T_λ has an inverse

$$R_\lambda=T_\lambda{}^{-1}=(T_{\lambda_0}V)^{-1}=V^{-1}R_{\lambda_0}. \tag{6}$$

Hence (5b) represents a neighborhood of λ_0 consisting of regular values λ of T. Since $\lambda_0\in\rho(T)$ was arbitrary, $\rho(T)$ is open, so that its complement $\sigma(T)=\mathbf{C}-\rho(T)$ is closed. ∎

It is of great interest to note that in this proof we have also obtained a basic representation of the resolvent by a power series in powers of λ. In fact, from (5) and (6) we immediately have the following

7.3-3 Representation Theorem (Resolvent). *For X and T as in Theorem 7.3-2 and every $\lambda_0 \in \rho(T)$ the resolvent $R_\lambda(T)$ has the representation*

$$R_\lambda = \sum_{j=0}^{\infty} (\lambda - \lambda_0)^j R_{\lambda_0}^{j+1}, \tag{7}$$

the series being absolutely convergent for every λ in the open disk given by [cf. (5b)]

$$|\lambda - \lambda_0| < \frac{1}{\|R_{\lambda_0}\|}$$

in the complex plane. This disk is a subset of $\rho(T)$.

This theorem will also provide a way of applying complex analysis to spectral theory, as we shall see in Sec. 7.5.

As another consequence of Theorem 7.3-1, let us prove the important fact that for a bounded linear operator the spectrum is a bounded set in the complex plane. The precise statement is as follows.

7.3-4 Theorem (Spectrum). *The spectrum $\sigma(T)$ of a bounded linear operator $T\colon X \longrightarrow X$ on a complex Banach space X is compact and lies in the disk given by*

$$|\lambda| \leqq \|T\|. \tag{8}$$

Hence the resolvent set $\rho(T)$ of T is not empty. [$\sigma(T) \neq \varnothing$ *will be shown in* 7.5-4.]

Proof. Let $\lambda \neq 0$ and $\kappa = 1/\lambda$. From Theorem 7.3-1 we obtain the representation

$$R_\lambda = (T - \lambda I)^{-1} = -\frac{1}{\lambda}(I - \kappa T)^{-1} = -\frac{1}{\lambda}\sum_{j=0}^{\infty} (\kappa T)^j = -\frac{1}{\lambda}\sum_{j=0}^{\infty}\left(\frac{1}{\lambda}T\right)^j \tag{9}$$

where, by Theorem 7.3-1, the series converges for all λ such that

$$\left\|\frac{1}{\lambda}T\right\| = \frac{\|T\|}{|\lambda|} < 1 \qquad \text{that is,} \qquad |\lambda| > \|T\|.$$

The same theorem also shows that any such λ is in $\rho(T)$. Hence the spectrum $\sigma(T)=\mathbf{C}-\rho(T)$ must lie in the disk (8), so that $\sigma(T)$ is bounded. Furthermore, $\sigma(T)$ is closed by Theorem 7.3-2. Hence $\sigma(T)$ is compact. ∎

Since from the theorem just proved we know that for a bounded linear operator T on a complex Banach space the spectrum is bounded, it seems natural to ask for the smallest disk about the origin which contains the whole spectrum. This question suggests the following concept.

7.3-5 Definition (Spectral radius). The *spectral radius* $r_\sigma(T)$ of an operator $T \in B(X, X)$ on a complex Banach space X is the radius

$$r_\sigma(T) = \sup_{\lambda \in \sigma(T)} |\lambda|$$

of the smallest closed disk centered at the origin of the complex λ-plane and containing $\sigma(T)$. ∎

From (8) it is obvious that for the spectral radius of a bounded linear operator T on a complex Banach space we have

$$r_\sigma(T) \leqq \|T\|, \tag{10}$$

and in Sec. 7.5 we shall prove that

$$r_\sigma(T) = \lim_{n \to \infty} \sqrt[n]{\|T^n\|}. \tag{11}$$

Problems

1. Let $X = C[0, 1]$ and define $T: X \longrightarrow X$ by $Tx = vx$, where $v \in X$ is fixed. Find $\sigma(T)$. Note that $\sigma(T)$ is closed.

2. Find a linear operator $T: C[0, 1] \longrightarrow C[0, 1]$ whose spectrum is a given interval $[a, b]$.

3. If Y is the eigenspace corresponding to an eigenvalue λ of an operator T, what is the spectrum of $T|_Y$?

4. Let $T\colon l^2 \longrightarrow l^2$ be defined by $y = Tx$, $x = (\xi_j)$, $y = (\eta_j)$, $\eta_j = \alpha_j \xi_j$, where (α_j) is dense in $[0, 1]$. Find $\sigma_p(T)$ and $\sigma(T)$.

5. *If* $\lambda \in \sigma(T) - \sigma_p(T)$ in Prob. 4, show that $R_\lambda(T)$ is unbounded.

6. Extending Prob. 4, find a linear operator $T\colon l^2 \longrightarrow l^2$ whose eigenvalues are dense in a given compact set $K \subset \mathbf{C}$ and $\sigma(T) = K$.

7. Let $T \in B(X, X)$. Show that $\|R_\lambda(T)\| \longrightarrow 0$ as $\lambda \longrightarrow \infty$.

8. Let $X = C[0, \pi]$ and define $T\colon \mathcal{D}(T) \longrightarrow X$ by $x \longmapsto x''$, where

$$\mathcal{D}(T) = \{x \in X \mid x', x'' \in X, x(0) = x(\pi) = 0\}.$$

Show that $\sigma(T)$ is not compact.

9. Let $T\colon l^\infty \longrightarrow l^\infty$ be defined by $x \longmapsto (\xi_2, \xi_3, \cdots)$, where x is given by $x = (\xi_1, \xi_2, \cdots)$. (*a*) If $|\lambda| > 1$, show that $\lambda \in \rho(T)$. (*b*) If $|\lambda| \leqq 1$, show that λ is an eigenvalue and find the eigenspace Y.

10. Let $T\colon l^p \longrightarrow l^p$ be defined by $x \longmapsto (\xi_2, \xi_3, \cdots)$, where x is given by $x = (\xi_1, \xi_2, \cdots)$ and $1 \leqq p < +\infty$. If $|\lambda| = 1$, is λ an eigenvalue of T (as in Prob. 9)?

7.4 Further Properties of Resolvent and Spectrum

Some further interesting and basic properties of the resolvent are expressed in the following

7.4-1 Theorem (Resolvent equation, commutativity). *Let X be a complex Banach space, $T \in B(X, X)$ and $\lambda, \mu \in \rho(T)$* [cf. 7.2-1]. *Then:*

(a) *The resolvent R_λ of T satisfies the Hilbert relation or resolvent equation*

$$R_\mu - R_\lambda = (\mu - \lambda) R_\mu R_\lambda \qquad [\lambda, \mu \in \rho(T)]. \tag{1}$$

(b) *R_λ commutes with any $S \in B(X, X)$ which commutes with T.*

(c) *We have*

$$R_\lambda R_\mu = R_\mu R_\lambda \qquad [\lambda, \mu \in \rho(T)]. \tag{2}$$

Proof. **(a)** By 7.2-3 the range of T_λ is all of X. Hence $I = T_\lambda R_\lambda$, where I is the identity operator on X. Also $I = R_\mu T_\mu$. Consequently,

$$\begin{aligned} R_\mu - R_\lambda &= R_\mu(T_\lambda R_\lambda) - (R_\mu T_\mu)R_\lambda \\ &= R_\mu(T_\lambda - T_\mu)R_\lambda \\ &= R_\mu[T - \lambda I - (T - \mu I)]R_\lambda \\ &= (\mu - \lambda)R_\mu R_\lambda. \end{aligned}$$

(b) By assumption, $ST = TS$. Hence $ST_\lambda = T_\lambda S$. Using $I = T_\lambda R_\lambda = R_\lambda T_\lambda$, we thus obtain

$$R_\lambda S = R_\lambda S T_\lambda R_\lambda = R_\lambda T_\lambda S R_\lambda = S R_\lambda.$$

(c) R_μ commutes with T by (*b*). Hence R_λ commutes with R_μ by (*b*). ∎

Our next result will be the important spectral mapping theorem, and we start with a motivation suggested by matrix eigenvalue theory.

If λ is an eigenvalue of a square matrix A, then $Ax = \lambda x$ for some $x \neq 0$. Application of A gives

$$A^2 x = A\lambda x = \lambda A x = \lambda^2 x.$$

Continuing in this way, we have for every positive integer m

$$A^m x = \lambda^m x;$$

that is, if λ is an eigenvalue of A, then λ^m is an eigenvalue of A^m. More generally, then

$$p(\lambda) = \alpha_n \lambda^n + \alpha_{n-1}\lambda^{n-1} + \cdots + \alpha_0$$

is an eigenvalue of the matrix

$$p(A) = \alpha_n A^n + \alpha_{n-1} A^{n-1} + \cdots + \alpha_0 I.$$

It is quite remarkable that this property extends to complex Banach spaces of any dimension, as we shall prove. In the proof we shall use the fact that a bounded linear operator has a nonempty spectrum. This will be shown later (in 7.5-4), by methods of complex analysis.

A convenient notation for formulating the desired theorem is

$$p(\sigma(T)) = \{\mu \in \mathbf{C} \mid \mu = p(\lambda),\ \lambda \in \sigma(T)\}, \tag{3}$$

that is, $p(\sigma(T))$ is the set of all complex numbers μ such that $\mu = p(\lambda)$ for some $\lambda \in \sigma(T)$. We shall also use $p(\rho(T))$ in a similar sense.

7.4-2 Spectral Mapping Theorem for Polynomials. *Let X be a complex Banach space, $T \in B(X, X)$ and*

$$p(\lambda) = \alpha_n\lambda^n + \alpha_{n-1}\lambda^{n-1} + \cdots + \alpha_0 \qquad (\alpha_n \neq 0).$$

Then

$$\sigma(p(T)) = p(\sigma(T)); \tag{4}$$

that is, the spectrum $\sigma(p(T))$ of the operator

$$p(T) = \alpha_n T^n + \alpha_{n-1} T^{n-1} + \cdots + \alpha_0 I$$

consists precisely of all those values which the polynomial p assumes on the spectrum $\sigma(T)$ of T.

Proof. We assume that $\sigma(T) \neq \varnothing$; this will be proved in 7.5-4. The case $n = 0$ is trivial; then $p(\sigma(T)) = \{\alpha_0\} = \sigma(p(T))$. Let $n > 0$. In part (a) we prove

$$\sigma(p(T)) \subset p(\sigma(T)) \tag{4a}$$

and in part (b) we prove

$$p(\sigma(T)) \subset \sigma(p(T)), \tag{4b}$$

so that we obtain (4). The details are as follows.

(a) For simplicity we also write $S = p(T)$ and

$$S_\mu = p(T) - \mu I \qquad (\mu \in \mathbf{C}).$$

Then if S_μ^{-1} exists, the formula for S_μ shows that S_μ^{-1} is the resolvent operator of $p(T)$. We keep μ fixed. Since X is complex, the polynomial given by $s_\mu(\lambda)=p(\lambda)-\mu$ must factor completely into linear terms, say,

$$s_\mu(\lambda)=p(\lambda)-\mu=\alpha_n(\lambda-\gamma_1)(\lambda-\gamma_2)\cdots(\lambda-\gamma_n), \tag{5}$$

where $\gamma_1,\cdots,\gamma_n$ are the zeros of s_μ (which depend on μ, of course). Corresponding to (5) we have

$$S_\mu=p(T)-\mu I=\alpha_n(T-\gamma_1 I)(T-\gamma_2 I)\cdots(T-\gamma_n I).$$

If each γ_j is in $\rho(T)$, then each $T-\gamma_j I$ has a bounded inverse which, by 7.2-3, is defined on all of X, and the same holds for S_μ; in fact, by (6) in Sec. 2.6,

$$S_\mu^{-1}=\frac{1}{\alpha_n}(T-\gamma_n I)^{-1}\cdots(T-\gamma_1 I)^{-1}.$$

Hence in this case, $\mu\in\rho(p(T))$. From this we conclude that

$$\mu\in\sigma(p(T)) \quad\Longrightarrow\quad \gamma_j\in\sigma(T) \text{ for some } j.$$

Now (5) gives

$$s_\mu(\gamma_j)=p(\gamma_j)-\mu=0,$$

thus

$$\mu=p(\gamma_j)\in p(\sigma(T)).$$

Since $\mu\in\sigma(p(T))$ was arbitrary, this proves (4a):

$$\sigma(p(T))\subset p(\sigma(T)).$$

(b) We prove (4b):

$$p(\sigma(T))\subset\sigma(p(T)).$$

We do this by showing that

$$\kappa\in p(\sigma(T)) \quad\Longrightarrow\quad \kappa\in\sigma(p(T)). \tag{6}$$

Let $\kappa \in p(\sigma(T))$. By definition this means that

$$\kappa = p(\beta) \qquad \text{for some} \qquad \beta \in \sigma(T).$$

There are now two possibilities:

(A) $T - \beta I$ has no inverse.

(B) $T - \beta I$ has an inverse.

We consider these cases one after the other.

(A) From $\kappa = p(\beta)$ we have $p(\beta) - \kappa = 0$. Hence β is a zero of the polynomial given by

$$s_\kappa(\lambda) = p(\lambda) - \kappa.$$

It follows that we can write

$$s_\kappa(\lambda) = p(\lambda) - \kappa = (\lambda - \beta) g(\lambda),$$

where $g(\lambda)$ denotes the product of the other $n-1$ linear factors and α_n. Corresponding to this representation we have

$$S_\kappa = p(T) - \kappa I = (T - \beta I) g(T). \tag{7}$$

Since the factors of $g(T)$ all commute with $T - \beta I$, we also have

$$S_\kappa = g(T)(T - \beta I). \tag{8}$$

If S_κ had an inverse, (7) and (8) would yield

$$I = (T - \beta I) g(T) S_\kappa^{-1} = S_\kappa^{-1} g(T)(T - \beta I)$$

which shows that $T - \beta I$ has an inverse, contradictory to our assumption. Hence for our present κ the resolvent S_κ^{-1} of $p(T)$ does not exist, so that $\kappa \in \sigma(p(T))$. Since $\kappa \in p(\sigma(T))$ was arbitrary, this proves (6) under the assumption that $T - \beta I$ has no inverse.

(B) Suppose that $\kappa = p(\beta)$ for some $\beta \in \sigma(T)$, as before, but now assume that the inverse $(T - \beta I)^{-1}$ exists. Then for the range of $T - \beta I$ we must have

$$\mathcal{R}(T - \beta I) \neq X \tag{9}$$

because otherwise $(T - \beta I)^{-1}$ would be bounded by the bounded inverse theorem 4.12-2, applied to $T - \beta I$, so that $\beta \in \rho(T)$, which would contradict $\beta \in \sigma(T)$. From (7) and (9) we obtain

$$\mathcal{R}(S_\kappa) \neq X.$$

This shows that $\kappa \in \sigma(p(T))$, since $\kappa \in \rho(p(T))$ would imply that $\mathcal{R}(S_\kappa) = X$ by Lemma 7.2-3(*b*), applied to $p(T)$. This proves (6) under the assumption that $T - \beta I$ has an inverse. Theorem 7.4-2 is proved. ∎

We finally consider a basic property of the eigenvectors.

7.4-3 Theorem (Linear independence). *Eigenvectors $x_1, \cdots, x_n$ corresponding to different eigenvalues $\lambda_1, \cdots, \lambda_n$ of a linear operator T on a vector space X constitute a linearly independent set.*

Proof. We assume that $\{x_1, \cdots, x_n\}$ is linearly dependent and derive a contradiction. Let x_m be the first of the vectors which is a linear combination of its predecessors, say,

$$x_m = \alpha_1 x_1 + \cdots + \alpha_{m-1} x_{m-1}. \tag{10}$$

Then $\{x_1, \cdots, x_{m-1}\}$ is linearly independent. Applying $T - \lambda_m I$ on both sides of (10), we obtain

$$\begin{aligned}(T-\lambda_m I)x_m &= \sum_{j=1}^{m-1} \alpha_j (T - \lambda_m I) x_j \\ &= \sum_{j=1}^{m-1} \alpha_j (\lambda_j - \lambda_m) x_j.\end{aligned}$$

Since x_m is an eigenvector corresponding to λ_m, the left side is zero. Since the vectors on the right form a linearly independent set, we must have

$$\alpha_j(\lambda_j - \lambda_m) = 0, \qquad \text{hence} \qquad \alpha_j = 0 \qquad (j = 1, \cdots, m-1)$$

since $\lambda_j - \lambda_m \neq 0$. But then $x_m = 0$ by (10). This contradicts the fact that $x_m \neq 0$ since x_m is an eigenvector and completes the proof. ∎

Problems

1. Prove (2) directly, without using (1) or 7.4-1(*b*).

2. Obtain (2) from (1).

3. If $S, T \in B(X, X)$, show that for any $\lambda \in \rho(S) \cap \rho(T)$,

$$R_\lambda(S) - R_\lambda(T) = R_\lambda(S)(T-S)R_\lambda(T).$$

4. Let X be a complex Banach space, $T \in B(X, X)$ and p a polynomial. Show that the equation

$$p(T)x = y \qquad (x, y \in X)$$

has a unique solution x for every $y \in X$ if and only if $p(\lambda) \neq 0$ for all $\lambda \in \sigma(T)$.

5. Why is it necessary in Theorem 7.4-2 that X be complex?

6. Using Theorem 7.4-3, find a sufficient condition in order that an n-rowed square matrix have n eigenvectors that span the whole space $\mathbf{C}^n$ (or $\mathbf{R}^n$).

7. Show that for any operator $T \in B(X, X)$ on a complex Banach space X,

$$r_\sigma(\alpha T) = |\alpha|\, r_\sigma(T), \qquad r_\sigma(T^k) = [r_\sigma(T)]^k \qquad (k \in \mathbf{N})$$

where r_σ denotes the spectral radius (cf. 7.3-5).

8. Determine the eigenvalues of the following matrix (i) by direct calculation, (ii) by showing that $A^2 = I$.

$$A = \begin{bmatrix} 0 & 1 & 0 \\ 1 & 0 & 0 \\ 0 & 0 & 1 \end{bmatrix}.$$

9. (Idempotent operator). Let T be a bounded linear operator on a Banach space. T is said to be *idempotent* if $T^2 = T$. (Cf. also Sec. 3.3.) Give examples. Show that if $T \neq 0, \neq I$, then its spectrum is $\sigma(T) = \{0, 1\}$; prove this (a) by means of (9) in Sec. 7.3, (b) by Theorem 7.4-2.

10. Show that the matrix

$$T_B = \begin{bmatrix} 1/2 & 1/2 & 0 \\ 1/2 & 1/2 & 0 \\ 0 & 0 & 1 \end{bmatrix}$$

represents an idempotent operator $T: \mathbf{R}^3 \longrightarrow \mathbf{R}^3$ with respect to an orthonormal basis B. Determine the spectrum by the use of (*i*) Prob. 9, (*ii*) direct calculation. Find eigenvectors and the eigenspaces.

7.5 Use of Complex Analysis in Spectral Theory

An important tool in spectral theory is complex analysis. Connections between the two areas can be obtained by means of complex line integrals or power series. We shall use power series only. In this way we shall be able to keep the discussion on a more elementary level and need only a few basic concepts and facts, as follows.[2]

A metric space is said to be *connected* if it is not the union of two disjoint nonempty open subsets. A subset of a metric space is said to be *connected* if it is connected regarded as a subspace.

By a **domain** G in the complex plane $\mathbf{C}$ we mean an open connected subset G of $\mathbf{C}$.

It can be shown that an open subset G of $\mathbf{C}$ is connected if and only if every pair of points of G can be joined by a broken line consisting of finitely many straight line segments all points of which belong to G. (In most books on complex analysis this is used as a definition of connectedness.)

A complex valued function h of a complex variable λ is said to be **holomorphic** (or *analytic*) on a domain G of the complex λ-plane if h is defined and differentiable on G, that is, the derivative h' of h, defined by

$$h'(\lambda) = \lim_{\Delta\lambda \to 0} \frac{h(\lambda + \Delta\lambda) - h(\lambda)}{\Delta\lambda}$$

exists for every $\lambda \in G$. The function h is said to be *holomorphic at a point* $\lambda_0 \in \mathbf{C}$ if h is holomorphic on some ε-neighborhood of λ_0.

h is holomorphic on G if and only if at every $\lambda_0 \in G$ it has a power

[2] Readers unfamiliar with the elements of complex analysis may just look at the main results (7.5-3, 7.5-4, 7.5-5), skipping the proofs, and then proceed to the next section.

"Domain" of a mapping (e.g., of an operator) means *domain of definition* of the mapping, that is, the set of all points for which the mapping is defined; hence this is a different use of the term "domain."

series representation

$$h(\lambda)=\sum_{j=0}^{\infty} c_j(\lambda-\lambda_0)^j$$

with a nonzero radius of convergence.

This and Theorem 7.3-3 will be the keys for the application of complex analysis to spectral theory.

The resolvent R_λ is an operator which depends on the complex parameter λ. This suggests the general setting of our present approach as follows.

By a *vector valued function or* **operator function** we mean a mapping

$$\begin{aligned} S\colon \Lambda &\longrightarrow B(X, X) \\ \lambda &\longmapsto S_\lambda \end{aligned} \tag{1}$$

where Λ is any subset of the complex λ-plane. (We write S_λ instead of $S(\lambda)$, to have a notation similar to R_λ, because later we shall consider $S_\lambda = R_\lambda$ and $\Lambda = \rho(T)$.)

S being given, we may choose any $x \in X$, so that we get a mapping

$$\begin{aligned} \Lambda &\longrightarrow X \\ \lambda &\longmapsto S_\lambda x. \end{aligned} \tag{2}$$

We may also choose $x \in X$ and any $f \in X'$ (cf. 2.10-3) to get a mapping of Λ into the complex plane, namely,

$$\begin{aligned} \Lambda &\longrightarrow \mathbf{C} \\ \lambda &\longmapsto f(S_\lambda x). \end{aligned} \tag{3}$$

Formula (3) suggests the following

7.5-1 Definition (Local holomorphy, holomorphy). Let Λ be an open subset of $\mathbf{C}$ and X a complex Banach space. Then S in (1) is said to be *locally holomorphic* on Λ if for every $x \in X$ and $f \in X'$ the function h defined by

$$h(\lambda)=f(S_\lambda x)$$

is holomorphic at every $\lambda_0 \in \Lambda$ in the usual sense (as stated at the beginning). S is said to be *holomorphic* on Λ if S is locally holomorphic on Λ and Λ is a domain. S is said to be *holomorphic at a point* $\lambda_0 \in \mathbf{C}$ if S is holomorphic on some ε-neighborhood of λ_0. ∎

This definition takes care of the following situation. The resolvent set $\rho(T)$ of a bounded linear operator T is open (by 7.3-2) but may not always be a domain, so that, in general, it is the union of disjoint domains (disjoint connected open sets). We shall see that the resolvent is holomorphic at every point of $\rho(T)$. Hence in any case it is locally holomorphic on $\rho(T)$ (thus defining holomorphic operator functions, one on each of those domains), and it is holomorphic on $\rho(T)$ if and only if $\rho(T)$ is connected, so that $\rho(T)$ is a single domain.

Before we discuss this matter in detail, let us include the following general

Remark. Def. 7.5-1 is suitable, a fact which is by no means trivial and deserves an explanation as follows. From Sec. 4.9 we remember that we defined three kinds of convergence in connection with bounded linear operators. Accordingly, we can define three corresponding kinds of derivative S_λ' of S_λ with respect to λ by the formulas

$$\left\|\frac{1}{\Delta\lambda}[S_{\lambda+\Delta\lambda} - S_\lambda] - S_\lambda'\right\| \longrightarrow 0$$

$$\left\|\frac{1}{\Delta\lambda}[S_{\lambda+\Delta\lambda}x - S_\lambda x] - S_\lambda' x\right\| \longrightarrow 0 \qquad (x \in X)$$

$$\left|\frac{1}{\Delta\lambda}[f(S_{\lambda+\Delta\lambda}x) - f(S_\lambda x)] - f(S_\lambda' x)\right| \longrightarrow 0 \qquad (x \in X,\ f \in X').$$

The existence of the derivative in the sense of the last formula for all λ in a domain Λ means that h defined by $h(\lambda) = f(S_\lambda x)$ is a holomorphic function on Λ in the usual sense; hence this is our definition of the derivative. It can be shown that the existence of this derivative (for every $x \in X$ and every $f \in X'$) implies the existence of the other two kinds of derivative; cf. E. Hille and R. S. Phillips (1957), p. 93 (listed in Appendix 3). This is quite remarkable and justifies Def. 7.5-1. It is also practically important, since holomorphy as defined in 7.5-1 is often much easier to check. ∎

As indicated before, our key to the application of complex analysis to spectral theory will be Theorem 7.3-3. The theorem states that for every value $\lambda_0 \in \rho(T)$ the resolvent $R_\lambda(T)$ of an operator $T \in B(X, X)$ on a complex Banach space X has a power series representation

$$\textbf{(4)} \qquad R_\lambda(T) = \sum_{j=0}^{\infty} R_{\lambda_0}(T)^{j+1}(\lambda - \lambda_0)^j$$

which converges absolutely for each λ in the disk (Fig. 62)

$$(5) \qquad |\lambda - \lambda_0| < \frac{1}{\|R_{\lambda_0}\|}.$$

Taking any $x \in X$ and $f \in X'$ and defining h by

$$h(\lambda) = f(R_\lambda(T)x),$$

we obtain from (4) the power series representation

$$h(\lambda) = \sum_{j=0}^{\infty} c_j(\lambda - \lambda_0)^j, \qquad c_j = f(R_{\lambda_0}(T)^{j+1}x)$$

which is absolutely convergent on the disk (5). This proves

7.5-2 Theorem (Holomorphy of R_λ). *The resolvent $R_\lambda(T)$ of a bounded linear operator $T\colon X \longrightarrow X$ on a complex Banach space X is*

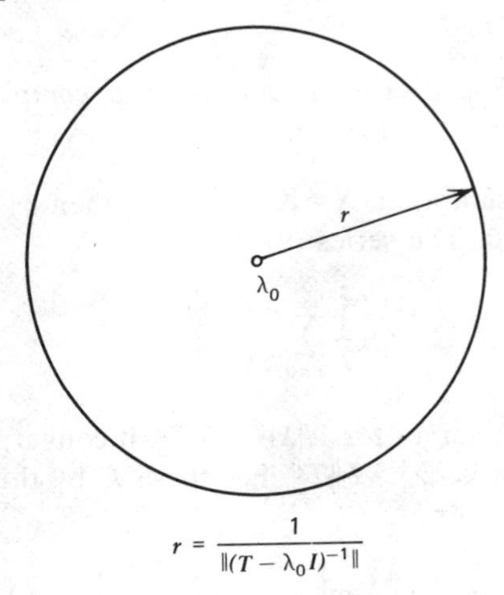

Fig. 62. Open disk of radius $r = 1/\|R_{\lambda_0}\|$ in the complex λ-plane represented by (5)

holomorphic at every point λ_0 of the resolvent set $\rho(T)$ of T. Hence it is locally holomorphic on $\rho(T)$.

Furthermore, $\rho(T)$ is the largest set on which the resolvent of T is locally holomorphic. Indeed, the resolvent cannot be continued analytically into points of the spectrum. This may be seen from (7) in the following

7.5-3 Theorem (Resolvent). *If $T \in B(X, X)$, where X is a complex Banach space, and $\lambda \in \rho(T)$, then*

$$\|R_\lambda(T)\| \geqq \frac{1}{\delta(\lambda)} \qquad \text{where} \qquad \delta(\lambda) = \inf_{s \in \sigma(T)} |\lambda - s| \tag{6}$$

is the distance from λ to the spectrum $\sigma(T)$. Hence

$$\|R_\lambda(T)\| \longrightarrow \infty \qquad \text{as} \qquad \delta(\lambda) \longrightarrow 0. \tag{7}$$

Proof. For every $\lambda_0 \in \rho(T)$ the disk (5) is a subset of $\rho(T)$; cf. 7.3-3. Hence, assuming $\sigma(T) \neq \varnothing$ (proof below), we see that the distance from λ_0 to the spectrum must at least equal the radius of the disk, that is, $\delta(\lambda_0) \geqq 1/\|R_{\lambda_0}\|$. This implies (6). ∎

It is of great theoretical and practical importance that the spectrum of a bounded linear operator T on a complex Banach space can never be the empty set:

7.5-4 Theorem (Spectrum). *If $X \neq \{0\}$ is a complex Banach space and $T \in B(X, X)$, then $\sigma(T) \neq \varnothing$.*

Proof. By assumption, $X \neq \{0\}$. If $T = 0$, then $\sigma(T) = \{0\} \neq \varnothing$. Let $T \neq 0$. Then $\|T\| \neq 0$. The series (9), Sec. 7.3, is

$$R_\lambda = -\frac{1}{\lambda} \sum_{j=0}^{\infty} \left(\frac{1}{\lambda} T\right)^j \qquad (|\lambda| > \|T\|). \tag{8}$$

Since this series converges for $1/|\lambda| < 1/\|T\|$, it converges absolutely for $1/|\lambda| < 1/2\,\|T\|$, that is, $|\lambda| > 2\,\|T\|$. For these λ, by the formula for the sum of a geometric series,

$$\|R_\lambda\| \leqq \frac{1}{|\lambda|} \sum_{j=0}^{\infty} \left\|\frac{1}{\lambda} T\right\|^j = \frac{1}{|\lambda| - \|T\|} \leqq \frac{1}{\|T\|} \qquad (|\lambda| \geqq 2\,\|T\|). \tag{9}$$

We show that the assumption $\sigma(T)=\varnothing$ leads to a contradiction. $\sigma(T)=\varnothing$ implies $\rho(T)=\mathbf{C}$. Hence R_λ is holomorphic for all λ, by 7.5-2. Consequently, for a fixed $x \in X$ and a fixed $f \in X'$ the function h defined by

$$h(\lambda)=f(R_\lambda x)$$

is holomorphic on $\mathbf{C}$, that is, h is an entire function. Since holomorphy implies continuity, h is continuous and thus bounded on the compact disk $|\lambda| \leqq 2\,\|T\|$. But h is also bounded for $|\lambda| \geqq 2\,\|T\|$ since $\|R_\lambda\| < 1/\|T\|$ by (9) and

$$\begin{aligned} |h(\lambda)| = |f(R_\lambda x)| &\leqq \|f\|\,\|R_\lambda x\| \leqq \|f\|\,\|R_\lambda\|\,\|x\| \\ &\leqq \|f\|\,\|x\|/\|T\|. \end{aligned}$$

Hence h is bounded on $\mathbf{C}$ and thus constant, by *Liouville's theorem*, which states that an entire function which is bounded on the whole complex plane is a constant. Since $x \in X$ and $f \in X'$ in h were arbitrary, $h = const$ implies that R_λ is independent of λ, and so is $R_\lambda^{-1} = T - \lambda I$. But this is impossible, and the theorem is proved. ∎

Using (8), we finally prove the following result by I. Gelfand (1941).

7.5-5 Theorem (Spectral radius). *If T is a bounded linear operator on a complex Banach space, then for the spectral radius $r_\sigma(T)$ of T we have*

$$r_\sigma(T) = \lim_{n\to\infty} \sqrt[n]{\|T^n\|}. \qquad \textbf{(10)}$$

Proof. We have $\sigma(T^n)=[\sigma(T)]^n$ by the spectral mapping theorem 7.4-2, so that

$$r_\sigma(T^n) = [r_\sigma(T)]^n. \qquad \textbf{(11)}$$

From (10) in Sec. 7.3, applied to T^n instead of T, we see that

$$r_\sigma(T^n) \leqq \|T^n\|.$$

Together,

$$r_\sigma(T) = \sqrt[n]{r_\sigma(T^n)} \leqq \sqrt[n]{\|T^n\|}$$

for every n. Hence

$$r_\sigma(T) \leqq \varliminf_{n\to\infty} \sqrt[n]{\|T^n\|} \leqq \varlimsup_{n\to\infty} \sqrt[n]{\|T^n\|}. \tag{12}$$

We show that the last expression equals $r_\sigma(T)$; then (12) implies (10).

A power series $\sum c_n \kappa^n$ converges absolutely for $|\kappa| < r$ with radius of convergence r given by the well-known *Hadamard formula*

$$\frac{1}{r} = \varlimsup_{n\to\infty} \sqrt[n]{|c_n|}. \tag{13}$$

This formula is included and proved in many books on complex analysis, for instance, in E. Hille (1973), p. 118.

Setting $\kappa = 1/\lambda$, we can write (8) in the form

$$R_\lambda = -\kappa \sum_{n=0}^{\infty} T^n \kappa^n \qquad (|\kappa| < r).$$

Then, writing $|c_n| = \|T^n\|$, we obtain

$$\left\| \sum_{n=0}^{\infty} T^n \kappa^n \right\| \leqq \sum_{n=0}^{\infty} \|T^n\| \, |\kappa|^n = \sum_{n=0}^{\infty} |c_n| \, |\kappa|^n.$$

The Hadamard formula (13) shows that we have absolute convergence for $|\kappa| < r$, hence for

$$|\lambda| = \frac{1}{|\kappa|} > \frac{1}{r} = \varlimsup_{n\to\infty} \sqrt[n]{\|T^n\|}.$$

From Theorems 7.5-2 and 7.5-3 we know that R_λ is locally holomorphic precisely on the resolvent set $\rho(T)$ in the complex λ-plane, as was mentioned before. To $\rho(T)$ there corresponds a set in the complex κ-plane, call it M. Then it is known from complex analysis that the radius of convergence r is the radius of the largest open circular disk about $\kappa = 0$ which lies entirely in M. (For instance, see E. Hille (1973), p. 197; note that our power series has center $\kappa = 0$.) Hence $1/r$ is the radius of the smallest circle about $\lambda = 0$ in the λ-plane whose exterior lies entirely in $\rho(T)$. By definition this means that $1/r$ is the spectral radius of T. Hence by (13),

$$r_\sigma(T) = \frac{1}{r} = \overline{\lim_{n\to\infty}} \sqrt[n]{\|T^n\|}.$$

From this and (12) we obtain (10). ∎

Problems

1. **(Nilpotent operator)** A linear operator T is said to be *nilpotent* if there is a positive integer m such that $T^m = 0$. Find the spectrum of a nilpotent operator $T\colon X \longrightarrow X$ on a complex Banach space $X \neq \{0\}$.

2. How does the result of Prob. 1 follow from (8)?

3. Determine $(A - \lambda I)^{-1}$ from (8) where A is the following matrix. (Use $A^2 = I$.)

$$A = \begin{bmatrix} 0 & 1 & 0 \\ 1 & 0 & 0 \\ 0 & 0 & 1 \end{bmatrix}$$

4. Clearly, Theorem 7.3-4 implies that

$$r_\sigma(T) \leqq \|T\|.$$

How does this follow from Theorem 7.5-5?

5. If X is a complex Banach space, $S, T \in B(X, X)$ and $ST = TS$, show that

$$r_\sigma(ST) \leqq r_\sigma(S) r_\sigma(T).$$

6. Show that in Prob. 5, commutativity $ST = TS$ cannot be dropped.

7. It is worthwhile noting that the sequence $(\|T^n\|^{1/n})$ in (10) need not be monotone. To illustrate this, consider $T\colon l^1 \longrightarrow l^1$ defined by

$$x = (\xi_1, \xi_2, \xi_3, \cdots) \longmapsto (0, \xi_1, 2\xi_2, \xi_3, 2\xi_4, \xi_5, \cdots).$$

8. **(Schur's inequality)** Let $A = (a_{jk})$ be an n-rowed square matrix and let $\lambda_1, \cdots, \lambda_n$ be its eigenvalues. Then it can be shown that *Schur's*

inequality

$$\sum_{m=1}^{n} |\lambda_m|^2 \leqq \sum_{j=1}^{n} \sum_{k=1}^{n} |a_{jk}|^2$$

holds, where the equality sign is valid if and only if A is normal (cf. 3.10-2). Derive an upper bound for the spectral radius of A from this inequality.

9. If T is a normal operator on a Hilbert space H, show that $r_\sigma(T) = \|T\|$. (Cf. Def. 3.10-1.)

10. Show that the existence of the limit in (10) already follows from $\|T^{m+n}\| \leqq \|T^m\| \|T^n\|$; cf. (7), Sec. 2.7. (Set $a_n = \|T^n\|$, $b_n = \ln a_n$, $\alpha = \inf (b_n/n)$ and show that $b_n/n \longrightarrow \alpha$.)

7.6 Banach Algebras

It is interesting to note that spectra also occur in connection with Banach algebras; these are Banach spaces which are at the same time algebras. We shall explain this fact, beginning with some relevant concepts.

An **algebra** A over a field K is a vector space A over K such that for each ordered pair of elements $x, y \in A$ a unique product $xy \in A$ is defined with the properties

$$(xy)z = x(yz) \tag{1}$$

$$x(y+z) = xy + xz \tag{2a}$$

$$(x+y)z = xz + yz \tag{2b}$$

$$\alpha(xy) = (\alpha x)y = x(\alpha y) \tag{3}$$

for all $x, y, z \in A$ and scalars α.

If $K = \mathbf{R}$ or $\mathbf{C}$, then A is said to be *real* or *complex*, respectively.

A is said to be **commutative** (or *abelian*) if the multiplication is commutative, that is, if for all $x, y \in A$,

$$xy = yx. \tag{4}$$

A is called an *algebra with identity* if A contains an element e such that for all $x \in A$,

$$ex = xe = x. \tag{5}$$

This element e is called an **identity** of A.

If A has an identity, the identity is unique.

In fact, if e' is another identity of A, then $e' = e$ because

$$ee' = e \qquad \text{(since } e' \text{ is an identity)},$$

$$ee' = e' \qquad \text{(since } e \text{ is an identity)}.$$

7.6-1 Definition (Normed algebra, Banach algebra). A *normed algebra* A is a normed space which is an algebra such that for all $x, y \in A$,

$$\|xy\| \leqq \|x\| \, \|y\| \tag{6}$$

and if A has an identity e,

$$\|e\| = 1. \tag{7}$$

A *Banach algebra* is a normed algebra which is complete, considered as a normed space. ∎

Note that (6) relates multiplication and norm and makes the product a jointly continuous function of its factors. We can see this from

$$\begin{aligned} \|xy - x_0 y_0\| &= \|x(y - y_0) + (x - x_0) y_0\| \\ &\leqq \|x\| \, \|y - y_0\| + \|x - x_0\| \, \|y_0\|. \end{aligned}$$

The following examples illustrate that many spaces of importance are Banach algebras.

Examples

7.6-2 Spaces R and C. The real line **R** and the complex plane **C** are commutative Banach algebras with identity $e = 1$.

7.6-3 Space $C[a, b]$. The space $C[a, b]$ is a commutative Banach algebra with identity $(e=1)$, the product xy being defined as usual:

$$(xy)(t)=x(t)y(t) \qquad t\in[a, b].$$

Relation (6) is readily verified.

The subspace of $C[a, b]$ consisting of all polynomials is a commutative normed algebra with identity $(e=1)$.

7.6-4 Matrices. The vector space X of all complex $n\times n$ matrices $(n>1$, fixed) is a noncommutative algebra with identity I (the n-rowed unit matrix), and by defining a norm on X we obtain a Banach algebra. (For such norms, see Sec. 2.7, Prob. 12.)

7.6-5 Space $B(X, X)$. The Banach space $B(X, X)$ of all bounded linear operators on a complex Banach space $X\neq\{0\}$ is a Banach algebra with identity I (the identity operator on X), the multiplication being composition of operators, by definition. Relation (6) is (cf. (7) in Sec. 2.7)

$$\|T_1T_2\|\leqq\|T_1\|\,\|T_2\|.$$

$B(X, X)$ is not commutative, unless $\dim X=1$. ∎

Let A be an algebra with identity. An $x\in A$ is said to be **invertible**[3] if it has an **inverse** in A, that is, if A contains an element, written x^{-1}, such that

$$x^{-1}x=xx^{-1}=e. \tag{8}$$

If x is invertible, the inverse is unique. Indeed, $yx=e=xz$ implies

$$y=ye=y(xz)=(yx)z=ez=z.$$

Using these concepts, we can now formulate the following definition.

7.6-6 Definition (Resolvent set, spectrum). Let A be a complex Banach algebra with identity. Then the *resolvent set* $\rho(x)$ of an $x\in A$ is the set of all λ in the complex plane such that $x-\lambda e$ is invertible. The

[3] Some authors use the terms "regular" (for invertible) and "singular" (for noninvertible).

spectrum $\sigma(x)$ of x is the complement of $\rho(x)$ in the complex plane; thus $\sigma(x) = \mathbf{C} - \rho(x)$. Any $\lambda \in \sigma(x)$ is called a *spectral value* of x. ∎

Hence the spectral values of $x \in A$ are those λ for which $x - \lambda e$ is not invertible.

If X is a complex Banach space, then $B(X, X)$ is a Banach algebra, so that Def. 7.6-6 is applicable. This immediately raises the question whether in this case, 7.6-6 agrees with our earlier Def. 7.2-1. We prove that this is so.

Let $T \in B(X, X)$ and λ in the resolvent set $\rho(T)$ defined by 7.6-6. Then by the definition, $R_\lambda(T) = (T - \lambda I)^{-1}$ exists and is an element of $B(X, X)$, that is, $R_\lambda(T)$ is a bounded linear operator defined on X. Hence $\lambda \in \rho(T)$ with $\rho(T)$ defined by 7.2-1.

Conversely, suppose that $\lambda \in \rho(T)$ with $\rho(T)$ defined by 7.2-1. Then $R_\lambda(T)$ exists and is linear (by 2.6-10), bounded and defined on a dense subset of X. But since T is bounded, Lemma 7.2-3(*b*) implies that $R_\lambda(T)$ is defined on all of X. Hence $\lambda \in \rho(T)$ with $\rho(T)$ defined by 7.6-6. This shows that 7.2-1 and 7.6-6 agree with respect to resolvent sets. Hence they also agree with respect to spectra since spectra are complements of resolvent sets. ∎

Problems

1. Why is X in Example 7.6-4 complete?

2. Show that (6) holds in Example 7.6-3.

3. How can we make the vector space of all ordered n-tuples of complex numbers into a Banach algebra?

4. What are the invertible elements (*a*) in 7.6-2, (*b*) in 7.6-3, (*c*) in 7.6-4?

5. Show that for the elements of X in 7.6-4, the definition of a spectrum in 7.6-6 agrees with that in Def. 7.1-1.

6. Find $\sigma(x)$ of $x \in c[0, 2\pi]$, where $x(t) = \sin t$. Find $\sigma(x)$ for any $x \in C[a, b]$.

7. Show that the set of all linear operators on a vector space into itself forms an algebra.

8. Let A be a complex Banach algebra with identity e. If for an $x \in A$ there are $y, z \in A$ such that $yx = e$ and $xz = e$, show that x is invertible and $y = z = x^{-1}$

9. If $x \in A$ is invertible and commutes with $y \in A$, show that x^{-1} and y also commute.

10. A subset A_1 of an algebra A is called a *subalgebra* of A if the application of the algebraic operations to elements of A_1 yields again elements of A_1. The *center* C of A is the set of those elements of A which commute with all elements of A. Give examples. Show that C is a commutative subalgebra of A.

7.7 Further Properties of Banach Algebras

We want to illustrate the interesting fact that certain results and proofs in the previous sections of this chapter can be generalized to Banach algebras.

7.7-1 Theorem (Inverse). *Let A be a complex Banach algebra with identity e. If $x \in A$ satisfies $\|x\| < 1$, then $e - x$ is invertible, and*

$$(e-x)^{-1} = e + \sum_{j=1}^{\infty} x^j. \tag{1}$$

Proof. From (6) in the last section we have $\|x^j\| \leqq \|x\|^j$, so that $\sum \|x^j\|$ converges since $\|x\| < 1$. Hence the series in (1) converges absolutely, so that it converges because A is complete (cf. Sec. 2.3). Let s denote its sum, and let us show that $s = (e-x)^{-1}$. By direct calculation,

$$\begin{aligned} &(e-x)(e+x+\cdots+x^n) \\ &= (e+x+\cdots+x^n)(e-x) \\ &= e - x^{n+1}. \end{aligned} \tag{2}$$

We now let $n \longrightarrow \infty$. Then $x^{n+1} \longrightarrow 0$ since $\|x\| < 1$, and (2) yields

$$(e-x)s = s(e-x) = e$$

since multiplication in A is continuous. Hence $s = (e-x)^{-1}$ and (1) holds. ∎

A complex Banach algebra A with identity e being given, we may consider the subset G of all invertible elements of A. We write G since G is a group. (Definition in A1.8; cf. Appendix 1.)

In fact, $e \in G$. Also if $x \in G$, then x^{-1} exists and is in G since it has an inverse $(x^{-1})^{-1} = x$. Furthermore, if $x, y \in G$, then $xy \in G$ because $y^{-1}x^{-1}$ is the inverse of xy:

$$(xy)(y^{-1}x^{-1}) = x(yy^{-1})x^{-1} = xex^{-1} = e$$

and similarly $(y^{-1}x^{-1})(xy) = e$.

We show that G is open:

7.7-2 Theorem (Invertible elements). *Let A be a complex Banach algebra with identity. Then the set G of all invertible elements of A is an open subset of A; hence the subset $M = A - G$ of all non-invertible elements of A is closed.*

Proof. Let $x_0 \in G$. We have to show that every $x \in A$ sufficiently close to x_0, say,

$$\|x - x_0\| < \frac{1}{\|x_0^{-1}\|}$$

belongs to G. Let $y = x_0^{-1}x$ and $z = e - y$. Then, using (6) in the last section, we obtain

$$\begin{aligned}\|z\| = \|-z\| &= \|y - e\| \\ &= \|x_0^{-1}x - x_0^{-1}x_0\| \\ &= \|x_0^{-1}(x - x_0)\| \\ &\leqq \|x_0^{-1}\| \, \|x - x_0\| < 1.\end{aligned}$$

Thus $\|z\| < 1$, so that $e - z$ is invertible by 7.7-1. Hence $e - z = y \in G$. Since G is a group, also

$$x_0 y = x_0 x_0^{-1} x = x \in G.$$

Since $x_0 \in G$ was arbitrary, this proves that G is open, and the complement M is closed. ∎

Referring to Def. 7.6-6, we define the **spectral radius** $r_\sigma(x)$ of an $x \in A$ by

$$r_\sigma(x) = \sup_{\lambda \in \sigma(x)} |\lambda| \tag{3}$$

and prove

7.7-3 Theorem (Spectrum). *Let A be a complex Banach algebra with identity e. Then for any $x \in A$, the spectrum $\sigma(x)$ is compact, and the spectral radius satisfies*

$$r_\sigma(x) \leqq \|x\|. \tag{4}$$

Proof. If $|\lambda| > \|x\|$, then $\|\lambda^{-1}x\| < 1$, so that $e - \lambda^{-1}x$ is invertible by 7.7-1. Hence $-\lambda(e - \lambda^{-1}x) = x - \lambda e$ is invertible, too, so that we have $\lambda \in \rho(x)$. This proves (4).

Hence $\sigma(x)$ is bounded. We show that $\sigma(x)$ is closed by proving that $\rho(x) = \mathbf{C} - \sigma(x)$ is open.

If $\lambda_0 \in \rho(x)$, then $x - \lambda_0 e$ is invertible, by definition. By 7.7-2 there is a neighborhood $N \subset A$ of $x - \lambda_0 e$ consisting wholly of invertible elements. Now for a fixed x, the mapping $\lambda \longmapsto x - \lambda e$ is continuous. Hence all $x - \lambda e$ with λ close to λ_0, say, $|\lambda - \lambda_0| < \delta$, where $\delta > 0$, lie in N, so that these $x - \lambda e$ are invertible. This means that the corresponding λ belong to $\rho(x)$. Since $\lambda_0 \in \rho(x)$ was arbitrary, we see that $\rho(x)$ is open and $\sigma(x) = \mathbf{C} - \rho(x)$ is closed. ∎

The theorem shows that $\rho(x) \neq \varnothing$. We also have

7.7-4 Theorem (Spectrum). $\sigma(x) \neq \varnothing$ *under the assumptions of the preceding theorem.*

Proof. Let λ, $\mu \in \rho(x)$ and let us write

$$v(\lambda) = (x - \lambda e)^{-1}, \qquad w = (\mu - \lambda)v(\lambda).$$

Then

$$\begin{aligned} x - \mu e &= x - \lambda e - (\mu - \lambda)e \\ &= (x - \lambda e)(e - w). \end{aligned}$$

Taking inverses, we have

$$v(\mu) = (e - w)^{-1} v(\lambda), \tag{5}$$

a formula to be used shortly. Suppose μ is so close to λ that $\|w\|<\frac{1}{2}$. Then by (1),

$$\|(e-w)^{-1}-e-w\|=\left\|\sum_{j=2}^{\infty} w^j\right\| \leqq \sum_{j=2}^{\infty} \|w\|^j=\frac{\|w\|^2}{1-\|w\|} \leqq 2\,\|w\|^2.$$

From this and (5),

$$\begin{aligned}\|v(\mu)-v(\lambda)-(\mu-\lambda)v(\lambda)^2\| &= \|(e-w)^{-1}v(\lambda)-(e+w)v(\lambda)\| \\ &\leqq \|v(\lambda)\|\,\|(e-w)^{-1}-(e+w)\| \\ &\leqq 2\,\|w\|^2\|v(\lambda)\|.\end{aligned}$$

$\|w\|^2$ contains a factor $|\mu-\lambda|^2$. Hence, dividing the inequality by $|\mu-\lambda|$ and letting $\mu \longrightarrow \lambda$, we see that $\|w\|^2/|\mu-\lambda| \longrightarrow 0$, so that on the left,

$$\text{(6)} \qquad \frac{1}{\mu-\lambda}[v(\mu)-v(\lambda)] \quad \longrightarrow \quad v(\lambda)^2.$$

This will be used as follows.

Let $f \in A'$, where A' is the dual space of A, considered as a Banach space. We define h: $\rho(x) \longrightarrow \mathbf{C}$ by $h(\lambda)=f(v(\lambda))$. Since f is continuous, so is h. Applying f to (6), we thus obtain

$$\lim_{\mu\to\lambda} \frac{h(\mu)-h(\lambda)}{\mu-\lambda}=f(v(\lambda)^2).$$

This shows that h is holomorphic at every point of $\rho(x)$.

If $\sigma(x)$ were empty, then $\rho(x)=\mathbf{C}$, so that h would be an entire function. Since $v(\lambda)=-\lambda^{-1}(e-\lambda^{-1}x)^{-1}$ and $(e-\lambda^{-1}x)^{-1} \longrightarrow e^{-1}=e$ as $|\lambda| \longrightarrow \infty$, we obtain

$$\text{(7)} \qquad |h(\lambda)|=|f(v(\lambda))| \leqq \|f\|\,\|v(\lambda)\|=\|f\|\frac{1}{|\lambda|}\left\|\left(e-\frac{1}{\lambda}x\right)^{-1}\right\| \quad \longrightarrow \quad 0$$

as $|\lambda| \longrightarrow \infty$. This shows that h would be bounded on $\mathbf{C}$, hence a constant by Liouville's theorem (cf. Sec. 7.5), which is zero by (7). Since $f \in A'$ was arbitrary, $h(\lambda)=f(v(\lambda))=0$ implies $v(\lambda)=0$ by 4.3-4. But this is impossible since it would mean that

$$\|e\|=\|(x-\lambda e)v(\lambda)\|=\|0\|=0$$

and contradict $\|e\|=1$. Hence $\sigma(x)=\varnothing$ cannot hold. ∎

The existence of an identity e was vital to our discussion. In many applications, A will have an identity, but we should say a word about what can be done if A has no identity. In this case we can supply A with an identity in the following canonical fashion.

Let $\tilde{A}$ be the set of all ordered pairs (x, α), where $x \in A$ and α is a scalar. Define

$$(x, \alpha)+(y, \beta)=(x+y, \alpha+\beta)$$

$$\beta(x, \alpha)=(\beta x, \beta\alpha)$$

$$(x, \alpha)(y, \beta)=(xy+\alpha y+\beta x, \alpha\beta)$$

$$\|(x, \alpha)\|=\|x\|+|\alpha|$$

$$\tilde{e}=(0, 1).$$

Then $\tilde{A}$ is a Banach algebra with identity $\tilde{e}$; in fact, (1) to (3) and (6), (7) of Sec. 7.6 are readily verified, and completeness follows from that of A and $\mathbf{C}$.

Furthermore, the mapping $x \longmapsto (x, 0)$ is an isomorphism of A onto a subspace of $\tilde{A}$, both regarded as normed spaces. The subspace has codimension 1. If we identify x with $(x, 0)$, then $\tilde{A}$ is simply A plus the one-dimensional vector space generated by $\tilde{e}$.

Problems

1. If $\|x-e\|<1$, show that x is invertible, and

$$x^{-1}=e+\sum_{j=1}^{\infty}(e-x)^j$$

2. Show that in Theorem 7.7-1,

$$\|(e-x)^{-1}-e-x\| \leqq \frac{\|x\|^2}{1-\|x\|}.$$

3. If x is invertible and y is such that $\|yx^{-1}\|<1$, show that $x-y$ is invertible and, writing $a^0=e$ for any $a \in A$,

$$(x-y)^{-1}=\sum_{j=0}^{\infty} x^{-1}(yx^{-1})^j.$$

4. Show that the set of all complex matrices of the form

$$x = \begin{bmatrix} \alpha & \beta \\ 0 & 0 \end{bmatrix}$$

forms a subalgebra of the algebra of all complex 2×2 matrices and find $\sigma(x)$.

5. It should be noted that the spectrum $\sigma(x)$ of an element x of a Banach algebra A depends on A. In fact, show that if B is a subalgebra of A, then $\tilde{\sigma}(x) \supset \sigma(x)$, where $\tilde{\sigma}(x)$ is the spectrum of x regarded as an element of B.

6. Let $\lambda, \mu \in \rho(x)$. Prove the *resolvent equation*

$$v(\mu) - v(\lambda) = (\mu - \lambda)v(\mu)v(\lambda),$$

where $v(\lambda) = (x - \lambda e)^{-1}$.

7. A **division algebra** is an algebra with identity such that every nonzero element is invertible. If a complex Banach algebra A is a division algebra, show that A is the set of all scalar multiples of the identity.

8. Let G be defined as in 7.7-2. Show that the mapping $G \longrightarrow G$ given by $x \longmapsto x^{-1}$ is continuous.

9. A *left inverse* of an $x \in A$ is a $y \in A$ such that $yx = e$. Similarly, if $xz = e$, then z is called a *right inverse* of x. If every element $x \neq 0$ of an algebra A has a left inverse, show that A is a division algebra.

10. If (x_n) and (y_n) are Cauchy sequences in a normed algebra A, show that $(x_n y_n)$ is Cauchy in A. Moreover, if $x_n \longrightarrow x$ and $y_n \longrightarrow y$, show that $x_n y_n \longrightarrow xy$.

CHAPTER 8

COMPACT LINEAR OPERATORS ON NORMED SPACES AND THEIR SPECTRUM

Compact linear operators are very important in applications. For instance, they play a central role in the theory of integral equations and in various problems of mathematical physics.

Their theory served as a model for the early work in functional analysis. Their properties closely resemble those of operators on finite dimensional spaces. For a compact linear operator, spectral theory can be treated fairly completely in the sense that Fredholm's famous theory of linear integral equations may be extended to linear functional equations $Tx - \lambda x = y$ with a complex parameter λ. This generalized theory is called the *Riesz-Schauder theory;* cf. F. Riesz (1918) and J. Schauder (1930).

Brief orientation about main content

Compactness of a linear operator (Def. 8.1-1) was suggested by integral equations. It was the property that was essential in Fredholm's theory. We shall discuss important general properties of compact linear operators in Secs. 8.1 and 8.2 and spectral properties in Secs. 8.3 and 8.4. The *Riesz-Schauder theory* is based on Secs. 8.3 and 8.4, and the results about operator equations are presented in Secs. 8.5 to 8.7. This includes applications to integral equations in Sec. 8.7.

8.1 Compact Linear Operators on Normed Spaces

Compact linear operators are defined as follows.

8.1-1 Definition (Compact linear operator). Let X and Y be normed spaces. An operator $T: X \longrightarrow Y$ is called a *compact linear operator* (or *completely continuous linear operator*) if T is linear and if for every bounded subset M of X, the image $T(M)$ is *relatively compact*, that is, the closure $\overline{T(M)}$ is compact. (Cf. Def. 2.5-1.) ∎

Many linear operators in analysis are compact. A systematic theory of compact linear operators emerged from the theory of *integral equations* of the form

$$(1) \qquad (T-\lambda I)x(s)=y(s) \qquad \text{where} \qquad Tx(s)=\int_a^b k(s,t)x(t)\,dt.$$

Here, $\lambda \in \mathbf{C}$ is a parameter,[1] y and the *kernel* k are given functions (subject to certain conditions), and x is the unknown function. Such equations also play a role in the theory of ordinary and partial differential equations. D. Hilbert (1912) discovered the surprising fact that the essential results about the solvability of (1) ("Fredholm's theory") do not depend on the existence of the integral representation of T in (1) but only on the assumption that T in (1) is a compact linear operator. F. Riesz (1918) put Fredholm's theory in an abstract axiomatic form, in his famous paper of 1918. (We shall consider integral equations in Sec. 8.7.)

The term "compact" is suggested by the definition. The older term "completely continuous" can be motivated by the following lemma, which shows that a compact linear operator is continuous, whereas the converse is not generally true.

8.1-2 Lemma (Continuity). *Let X and Y be normed spaces. Then:*

(a) *Every compact linear operator $T\colon X \longrightarrow Y$ is bounded, hence continuous.*

(b) *If* $\dim X=\infty$, *the identity operator $I\colon X \longrightarrow X$ (which is continuous) is not compact.*

Proof. **(a)** The unit sphere $U=\{x \in X \mid \|x\|=1\}$ is bounded. Since T is compact, $\overline{T(U)}$ is compact, and is bounded by 2.5-2, so that

$$\sup_{\|x\|=1} \|Tx\| < \infty.$$

Hence T is bounded and 2.7-9 shows that it is continuous.

[1] We assume that $\lambda \neq 0$. Then (1) is said to be *of the second kind*. With $\lambda=0$ it is said to be *of the first kind*. The two corresponding theories are quite different, for reasons which cannot be explained in a few words; see R. Courant and D. Hilbert (1953–62), vol. 1, p. 159.—The introduction of a variable parameter λ was the idea of H. Poincaré (1896).

(b) Of course, the closed unit ball $M = \{x \in X \mid \|x\| \leqq 1\}$ is bounded. If $\dim X = \infty$, then 2.5-5 implies that M cannot be compact; thus, $I(M) = M = \bar{M}$ is not relatively compact. ∎

From the definition of compactness of a set (cf. 2.5-1) we readily obtain a useful criterion for operators:

8.1-3 Theorem (Compactness criterion). *Let X and Y be normed spaces and $T: X \longrightarrow Y$ a linear operator. Then T is compact if and only if it maps every bounded sequence (x_n) in X onto a sequence (Tx_n) in Y which has a convergent subsequence.*

Proof. If T is compact and (x_n) is bounded, then the closure of (Tx_n) in Y is compact and Def. 2.5-1 shows that (Tx_n) contains a convergent subsequence.

Conversely, assume that every bounded sequence (x_n) contains a subsequence (x_{n_k}) such that (Tx_{n_k}) converges in Y. Consider any bounded subset $B \subset X$, and let (y_n) be any sequence in $T(B)$. Then $y_n = Tx_n$ for some $x_n \in B$, and (x_n) is bounded since B is bounded. By assumption, (Tx_n) contains a convergent subsequence. Hence $\overline{T(B)}$ is compact by Def. 2.5-1 because (y_n) in $T(B)$ was arbitrary. By definition, this shows that T is compact. ∎

From this theorem it is almost obvious that the sum $T_1 + T_2$ of two compact linear operators $T_j: X \longrightarrow Y$ is compact. (Cf. Prob. 2.) Similarly, αT_1 is compact, where α is any scalar. Hence we have the following result.

The compact linear operators from X into Y form a vector space.

Furthermore, Theorem 8.1-3 also implies that certain simplifications take place in the finite dimensional case:

8.1-4 Theorem (Finite dimensional domain or range). *Let X and Y be normed spaces and $T: X \longrightarrow Y$ a linear operator. Then:*

(a) *If T is bounded and* $\dim T(X) < \infty$, *the operator T is compact.*

(b) *If* $\dim X < \infty$, *the operator T is compact.*

Proof. **(a)** Let (x_n) be any bounded sequence in X. Then the inequality $\|Tx_n\| \leqq \|T\| \|x_n\|$ shows that (Tx_n) is bounded. Hence (Tx_n) is relatively compact by 2.5-3 since $\dim T(X)<\infty$. It follows that (Tx_n) has a convergent subsequence. Since (x_n) was an arbitrary bounded sequence in X, the operator T is compact by 8.1-3.

(b) follows from (a) by noting that $\dim X<\infty$ implies boundedness of T by 2.7-8 and $\dim T(X)\leqq \dim X$ by 2.6-9(b). ∎

We mention that an operator $T\in B(X, Y)$ with $\dim T(X)<\infty$ [cf. 8.1-4(a)] is often called an *operator of finite rank.*

The following theorem states conditions under which the limit of a sequence of compact linear operators is compact. The theorem is also important as a tool for proving compactness of a given operator by exhibiting it as the uniform operator limit of a sequence of compact linear operators.

8.1-5 Theorem (Sequence of compact linear operators). *Let (T_n) be a sequence of compact linear operators from a normed space X into a Banach space Y. If (T_n) is uniformly operator convergent, say, $\|T_n - T\| \longrightarrow 0$ (cf. Sec. 4.9), then the limit operator T is compact.*

Proof. Using a "diagonal method," we show that for any bounded sequence (x_m) in X the image (Tx_m) has a convergent subsequence, and then apply Theorem 8.1-3.

Since T_1 is compact, (x_m) has a subsequence $(x_{1,m})$ such that $(T_1x_{1,m})$ is Cauchy. Similarly, $(x_{1,m})$ has a subsequence $(x_{2,m})$ such that $(T_2x_{2,m})$ is Cauchy. Continuing in this way we see that the "diagonal sequence" $(y_m)=(x_{m,m})$ is a subsequence of (x_m) such that for every fixed positive integer n the sequence $(T_ny_m)_{m\in\mathbf{N}}$ is Cauchy. (x_m) is bounded, say, $\|x_m\|\leqq c$ for all m. Hence $\|y_m\|\leqq c$ for all m. Let $\varepsilon>0$. Since $T_m \longrightarrow T$, there is an $n=p$ such that $\|T-T_p\|<\varepsilon/3c$. Since $(T_py_m)_{m\in\mathbf{N}}$ is Cauchy, there is an N such that

$$\|T_py_j - T_py_k\| < \frac{\varepsilon}{3} \qquad (j, k>N).$$

Hence we obtain for $j, k>N$

$$\begin{aligned} \|Ty_j - Ty_k\| &\leqq \|Ty_j - T_py_j\| + \|T_py_j - T_py_k\| + \|T_py_k - Ty_k\| \\ &\leqq \|T-T_p\|\,\|y_j\| + \frac{\varepsilon}{3} + \|T_p - T\|\,\|y_k\| \\ &< \frac{\varepsilon}{3c}\,c + \frac{\varepsilon}{3} + \frac{\varepsilon}{3c}\,c = \varepsilon. \end{aligned}$$

This shows that (Ty_m) is Cauchy and converges since Y is complete. Remembering that (y_m) is a subsequence of the arbitrary bounded sequence (x_m), we see that Theorem 8.1-3 implies compactness of the operator T. ∎

Note that the present theorem becomes false if we replace uniform operator convergence by strong operator convergence $\|T_nx - Tx\| \longrightarrow 0$. This can be seen from $T_n\colon l^2 \longrightarrow l^2$ defined by $T_nx = (\xi_1, \cdots, \xi_n, 0, 0, \cdots)$, where $x = (\xi_j) \in l^2$. Since T_n is linear and bounded, T_n is compact by 8.1-4(*a*). Clearly, $T_nx \longrightarrow x = Ix$, but I is not compact since $\dim l^2 = \infty$; cf. 8.1-2(*b*).

The following example illustrates how the theorer can be used to prove compactness of an operator.

8.1-6 Example (Space l^2). Prove compactness of $T\colon l^2 \longrightarrow l^2$ defined by $y = (\eta_j) = Tx$ where $\eta_j = \xi_j/j$ for $j = 1, 2, \cdots$.

Solution. T is linear. If $x = (\xi_j) \in l^2$, then $y = (\eta_j) \in l^2$. Let $T_n\colon l^2 \longrightarrow l^2$ be defined by

$$T_nx = \left(\xi_1, \frac{\xi_2}{2}, \frac{\xi_3}{3}, \cdots, \frac{\xi_n}{n}, 0, 0, \cdots\right).$$

T_n is linear and bounded, and is compact by 8.1-4(*a*). Furthermore,

$$\|(T - T_n)x\|^2 = \sum_{j=n+1}^{\infty} |\eta_j|^2 = \sum_{j=n+1}^{\infty} \frac{1}{j^2} |\xi_j|^2$$

$$\leq \frac{1}{(n+1)^2} \sum_{j=n+1}^{\infty} |\xi_j|^2 \leq \frac{\|x\|^2}{(n+1)^2}.$$

Taking the supremum over all x of norm 1, we see that

$$\|T - T_n\| \leq \frac{1}{n+1}.$$

Hence $T_n \longrightarrow T$, and T is compact by Theorem 8.1-5. ∎

Another interesting and basic property of a compact linear operator is that it transforms weakly convergent sequences into strongly convergent sequences as follows.

8.1-7 Theorem (Weak convergence). *Let X and Y be normed spaces and $T\colon X \longrightarrow Y$ a compact linear operator. Suppose that (x_n) in X is weakly convergent, say, $x_n \xrightarrow{w} x$. Then (Tx_n) is strongly convergent in Y and has the limit $y = Tx$.*

Proof. We write $y_n = Tx_n$ and $y = Tx$. First we show that

$$(2) \qquad y_n \xrightarrow{w} y.$$

Then we show that

$$(3) \qquad y_n \longrightarrow y.$$

Let g be any bounded linear functional on Y. We define a functional f on X by setting

$$f(z) = g(Tz) \qquad (z \in X).$$

f is linear. f is bounded because T is compact, hence bounded, and

$$|f(z)| = |g(Tz)| \leqq \|g\| \, \|Tz\| \leqq \|g\| \, \|T\| \, \|z\|.$$

By definition, $x_n \xrightarrow{w} x$ implies $f(x_n) \longrightarrow f(x)$, hence by the definition, $g(Tx_n) \longrightarrow g(Tx)$, that is, $g(y_n) \longrightarrow g(y)$. Since g was arbitrary, this proves (2).

We prove (3). Assume that (3) does not hold. Then (y_n) has a subsequence (y_{n_k}) such that

$$(4) \qquad \|y_{n_k} - y\| \geqq \eta$$

for some $\eta > 0$. Since (x_n) is weakly convergent, (x_n) is bounded by 4.8-3(*c*), and so is (x_{n_k}). Compactness of T now implies by 8.1-3 that (Tx_{n_k}) has a convergent subsequence, say, $(\tilde{y}_j)$. Let $\tilde{y}_j \longrightarrow \tilde{y}$. A fortiori, $\tilde{y}_j \xrightarrow{w} \tilde{y}$. Hence $\tilde{y} = y$ by (2) and 4.8-3(*b*). Consequently,

$$\|\tilde{y}_j - y\| \longrightarrow 0. \qquad \text{But} \qquad \|\tilde{y}_j - y\| \geqq \eta > 0$$

by (4). This contradicts, so that (3) must hold. ∎

Problems

1. Show that the zero operator on any normed space is compact.

2. If T_1 and T_2 are compact linear operators from a normed space X into a normed space Y, show that T_1+T_2 is a compact linear operator. Show that the compact linear operators from X into Y constitute a subspace $C(X, Y)$ of $B(X, Y)$.

3. If Y is a Banach space, show that $C(X, Y)$ in Prob. 2 is a closed subset of $B(X, Y)$.

4. If Y is a Banach space, show that $C(X, Y)$ in Prob. 2 is a Banach space.

5. It was shown in the text that Theorem 8.1-5 becomes false if we replace uniform operator convergence by strong operator convergence. Prove that the operators T_n used for that purpose are bounded.

6. It is worth noting that the condition in 8.1-1 could be weakened without altering the concept of a compact linear operator. In fact, show that a linear operator $T\colon X \longrightarrow Y$ (X, Y normed spaces) is compact if and only if the image $T(M)$ of the unit ball $M \subset X$ is relatively compact in Y.

7. Show that a linear operator $T\colon X \longrightarrow X$ is compact if and only if for every sequence (x_n) of vectors of norm not exceeding 1 the sequence (Tx_n) has a convergent subsequence.

8. If z is a fixed element of a normed space X and $f \in X'$, show that the operator $T\colon X \longrightarrow X$ defined by $Tx = f(x)z$ is compact.

9. If X is an inner product space, show that $Tx = \langle x, y\rangle z$ with fixed y and z defines a compact linear operator on X.

10. Let Y be a Banach space and let $T_n\colon X \longrightarrow Y$, $n = 1, 2, \cdots$, be operators of finite rank. If (T_n) is uniformly operator convergent, show that the limit operator is compact.

11. Show that the projection of a Hilbert space H onto a finite dimensional subspace of H is compact.

12. Show that $T\colon l^2 \longrightarrow l^2$ defined by $Tx = y = (\eta_j)$, $\eta_j = \xi_j/2^j$, is compact.

13. Show that $T\colon l^p \longrightarrow l^p$, $1 \leqq p < +\infty$, with T defined by $y = (\eta_j) = Tx$, $\eta_j = \xi_j/j$, is compact.

14. Show that $T: l^\infty \longrightarrow l^\infty$ with T defined by $y=(\eta_j)=Tx$, $\eta_j=\xi_j/j$, is compact.

15. (Continuous mapping) If $T: X \longrightarrow Y$ is a continuous mapping of a metric space X into a metric space Y, show that the image of a relatively compact set $A \subset X$ is relatively compact.

8.2 Further Properties of Compact Linear Operators

In this section we prove that a compact linear operator on a normed space has a separable range and a compact adjoint operator. These properties will be needed in the study of the spectrum of compact linear operators, which starts in the next section.

We base our consideration on two related concepts which are of general interest in connection with the compactness of sets:

8.2-1 Definition (ε-net, total boundedness). Let B be a subset of a metric space X and let $\varepsilon > 0$ be given. A set $M_\varepsilon \subset X$ is called an *ε-net for B* if for every point $z \in B$ there is a point of M_ε at a distance from z less than ε. The set B is said to be *totally bounded* if for every $\varepsilon > 0$ there is a *finite* ε-net $M_\varepsilon \subset X$ for B, where "finite" means that M_ε is a finite set (that is, consists of finitely many points). ∎

Consequently, total boundedness of B means that for every given $\varepsilon > 0$ the set B is contained in the union of finitely many open balls of radius ε.

We can see the significance and usefulness of the concepts just defined from the following lemma, which will also play the key role in the proofs in this section.

8.2-2 Lemma (Total boundedness). *Let B be a subset of a metric space X. Then:*

(a) *If B is relatively compact, B is totally bounded.*

(b) *If B is totally bounded and X is complete, B is relatively compact.*

(c) *If B is totally bounded, for every $\varepsilon > 0$ it has a finite ε-net $M_\varepsilon \subset B$.*

(d) *If B is totally bounded, B is separable.*

Proof. **(a)** We assume B to be relatively compact and show that, any fixed $\varepsilon_0 > 0$ being given, there exists a finite ε_0-net for B. If $B = \emptyset$, then $\emptyset$ is an ε_0-net for B. If $B \neq \emptyset$, we pick any $x_1 \in B$. If $d(x_1, z) < \varepsilon_0$ for all $z \in B$, then $\{x_1\}$ is an ε_0-net for B. Otherwise, let $x_2 \in B$ be such that $d(x_1, x_2) \geqq \varepsilon_0$. If for all $z \in B$,

$$d(x_j, z) < \varepsilon_0 \qquad (j = 1 \text{ or } 2) \tag{1}$$

then $\{x_1, x_2\}$ is an ε_0-net for B. Otherwise let $z = x_3 \in B$ be a point not satisfying (1). If for all $z \in B$,

$$d(x_j, z) < \varepsilon_0 \qquad (j = 1, 2 \text{ or } 3)$$

then $\{x_1, x_2, x_3\}$ is an ε_0-net for B. Otherwise we continue by selecting an $x_4 \in B$, etc. We assert the existence of a positive integer n such that the set $\{x_1, \cdots, x_n\}$ obtained after n such steps is an ε_0-net for B. In fact, if there were no such n, our construction would yield a sequence (x_j) satisfying

$$d(x_j, x_k) \geqq \varepsilon_0 \qquad \text{for } j \neq k.$$

Obviously, (x_j) could not have a subsequence which is Cauchy. Hence (x_j) could not have a subsequence which converges in X. But this contradicts the relative compactness of B because (x_j) lies in B, by the construction. Hence there must be a finite ε_0-net for B. Since $\varepsilon_0 > 0$ was arbitrary, we conclude that B is totally bounded.

(b) Let B be totally bounded and X complete. We consider any sequence (x_n) in B and show that it has a subsequence which converges in X, so that B will then be relatively compact. By assumption, B has a finite ε-net for $\varepsilon = 1$. Hence B is contained in the union of finitely many open balls of radius 1. From these balls we can pick a ball B_1 which contains infinitely many terms of (x_n) (counting repetitions). Let $(x_{1,n})$ be the subsequence of (x_n) which lies in B_1. Similarly, by assumption, B is also contained in the union of finitely many balls of radius $\varepsilon = 1/2$. From these balls we can pick a ball B_2 which contains a subsequence $(x_{2,n})$ of the subsequence $(x_{1,n})$. We continue inductively, choosing $\varepsilon = 1/3, 1/4, \cdots$ and setting $y_n = x_{n,n}$. Then for every given $\varepsilon > 0$ there is an N (depending on ε) such that all y_n with $n > N$ lie in a ball of radius ε. Hence (y_n) is Cauchy. It converges in X, say, $y_n \longrightarrow y \in X$, since X is complete. Also, $y_n \in B$ implies $y \in \bar{B}$. This entails the following. By the definition of the closure, for every sequence (z_n) in $\bar{B}$ there is a sequence (x_n) in B which satisfies $d(x_n, z_n) \leqq 1/n$ for every n. Since (x_n) is in B, it has a subsequence which converges in $\bar{B}$, as we have just shown. Hence (z_n) also has a subsequence which converges in $\bar{B}$ since $d(x_n, z_n) \leqq 1/n$, so that $\bar{B}$ is compact and B is relatively compact.

(c) The case $B = \varnothing$ is obvious. Let $B \neq \varnothing$. By assumption, an $\varepsilon > 0$ being given, there is a finite ε_1-net $M_{\varepsilon_1} \subset X$ for B, where $\varepsilon_1 = \varepsilon/2$. Hence B is contained in the union of finitely many balls of radius ε_1 with the elements of M_{ε_1} as centers. Let $B_1, \cdots, B_n$ be those balls which intersect B, and let $x_1, \cdots, x_n$ be their centers. We seiect a point $z_j \in B \cap B_j$. See Fig. 63. Then $M_\varepsilon = \{z_1, \cdots, z_n\} \subset B$ is an ε-net for B, because for every $z \in B$ there is a B_j containing z, and

$$d(z, z_j) \leqq d(z, x_j) + d(x_j, z_j) < \varepsilon_1 + \varepsilon_1 = \varepsilon.$$

(d) Suppose B is totally bounded. Then by (c) the set B contains a finite ε-net $M_{1/n}$ for itself, where $\varepsilon = \varepsilon_n = 1/n$, $n = 1, 2, \cdots$. The union M of all these nets is countable. M is dense in B. In fact, for any given $\varepsilon > 0$ there is an n such that $1/n < \varepsilon$; hence for any $z \in B$ there is an $a \in M_{1/n} \subset M$ such that $d(z, a) < \varepsilon$. This shows that B is separable. ∎

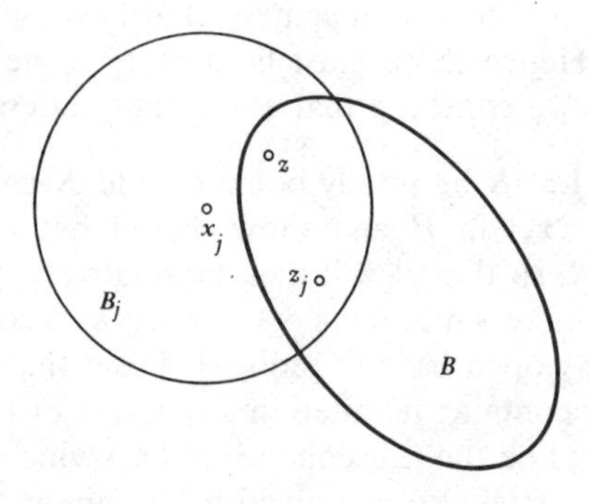

Fig. 63. Notation in the proof of Lemma 8.2-2, part (c)

Total boundedness implies boundedness. The converse does not generally hold.

Indeed, the first statement is almost obvious. The second one follows if we note that the closed unit ball $U=\{x \mid \|x\|\leqq 1\}\subset l^2$ is bounded but not totally bounded, because l^2 is infinite dimensional and complete, so that U is not compact (cf. 2.5-5), hence not totally bounded by 8.2-2(b).

Lemma 8.2-2 includes those properties which we need in our further work. Other properties which are of interest but not needed for our purpose are stated in the problem set; see, in particular, Probs. 2 to 4.

Using that lemma, we can now readily prove the following

8.2-3 Theorem (Separability of range). *The range $\mathcal{R}(T)$ of a compact linear operator $T: X \longrightarrow Y$ is separable; here, X and Y are normed spaces.*

Proof. Consider the ball $B_n = B(0; n)\subset X$. Since T is compact, the image $C_n = T(B_n)$ is relatively compact. C_n is separable by Lemma 8.2-2. The norm of any $x\in X$ is finite, so that $\|x\|<n$, hence $x\in B_n$ with n sufficiently large. Consequently,

$$\text{(2)}\quad \text{(a)}\quad X=\bigcup_{n=1}^{\infty} B_n, \qquad \text{(b)}\quad T(X)=\bigcup_{n=1}^{\infty} T(B_n)=\bigcup_{n=1}^{\infty} C_n.$$

Since C_n is separable, it has a countable dense subset D_n, and the union

$$D=\bigcup_{n=1}^{\infty} D_n$$

is countable. (2b) shows that D is dense in the range $\mathcal{R}(T)=T(X)$. ∎

In the next theorem we show that a compact linear operator on a normed space X can be extended to the completion of X, the extended operator being linear and compact.

8.2-4 Theorem (Compact extension). *A compact linear operator $T: X \longrightarrow Y$ from a normed space X into a Banach space Y has a compact linear extension $\tilde{T}: \hat{X} \longrightarrow Y$, where $\hat{X}$ is the completion of X.* (Cf. 2.3-2.)

Proof. We may regard X as a subspace of $\hat{X}$; cf. Theorem 2.3-2. Since T is bounded (cf. 8.1-2), it has a bounded linear extension $\tilde{T}: \hat{X} \longrightarrow Y$; cf. 2.7-11. We show that the compactness of T implies that $\tilde{T}$ is also compact. For this purpose we consider an arbitrary bounded sequence $(\hat{x}_n)$ in $\hat{X}$ and show that $(\tilde{T}\hat{x}_n)$ has a convergent subsequence.

Since X is dense in $\hat{X}$, there is a sequence (x_n) in X such that $\hat{x}_n - x_n \longrightarrow 0$. Clearly, (x_n) is bounded, too. Since T is compact, (Tx_n) has a convergent subsequence (Tx_{n_k}); let

$$(3) \qquad Tx_{n_k} \longrightarrow y_0 \in Y.$$

Now $\hat{x}_n - x_n \longrightarrow 0$ implies $\hat{x}_{n_k} - x_{n_k} \longrightarrow 0$. Since $\tilde{T}$ is linear and bounded, it is continuous. We thus obtain (cf. 1.4-8)

$$\tilde{T}\hat{x}_{n_k} - Tx_{n_k} = \tilde{T}(\hat{x}_{n_k} - x_{n_k}) \qquad \longrightarrow \qquad \tilde{T}0 = 0.$$

By (3) this implies that $\tilde{T}\hat{x}_{n_k} \longrightarrow y_0$. We have shown that the arbitrary bounded sequence $(\hat{x}_n)$ has a subsequence $(\hat{x}_{n_k})$ such that $(\tilde{T}\hat{x}_{n_k})$ converges. This proves compactness of $\tilde{T}$ by 8.1-3. ∎

We shall see later in this chapter that compact linear operators occur in operator equations of great practical and theoretical importance. A general theory of solvability of those equations will make essential use of adjoint operators. Most crucial in that connection will be the fact that the adjoint operator of a compact linear operator is itself compact. Let us prove this important property of compact linear operators and then, in the next section, begin with the discussion of the spectrum of those operators.

8.2-5 Theorem (Adjoint operator). *Let $T: X \longrightarrow Y$ be a linear operator. If T is compact, so is its adjoint operator $T^\times: Y' \longrightarrow X'$; here X and Y are normed spaces and X' and Y' the dual spaces of X and Y* (cf. 2.10-3).

Proof. We consider any subset B of Y' which is bounded, say

$$\|g\| \leqq c \qquad \text{for all } g \in B$$

and show that the image $T^\times(B) \subset X'$ is totally bounded, so that $T^\times(B)$ is relatively compact by 8.2-2(*b*), since X' is complete (cf. 2.10-4).

Accordingly, we must prove that for any fixed $\varepsilon_0 > 0$ the image $T^\times(B)$ has a finite ε_0-net. Since T is compact, the image $T(U)$ of the unit ball

$$U = \{x \in X \mid \|x\| \leqq 1\}$$

is relatively compact. Hence $T(U)$ is totally bounded by 8.2-2(a). From 8.2-2(c) it follows that there is a finite ε_1-net $M \subset T(U)$ for $T(U)$, where $\varepsilon_1 = \varepsilon_0/4c$. This means that U contains points $x_1, \cdots, x_n$ such that each $x \in U$ satisfies

$$\|Tx - Tx_j\| < \frac{\varepsilon_0}{4c} \qquad \text{for some } j. \tag{4}$$

We define a linear operator $A\colon\ Y' \longrightarrow \mathbf{R}^n$ by

$$Ag = (g(Tx_1), g(Tx_2), \cdots, g(Tx_n)). \tag{5}$$

g is bounded by assumption and T is bounded by 8.1-2(a). Hence A is compact by 8.1-4. Since B is bounded, $A(B)$ is relatively compact. Hence $A(B)$ is totally bounded by 8.2-2(a). By 8.2-2(c) it contains a finite ε_2-net $\{Ag_1, \cdots, Ag_m\}$ for itself, where $\varepsilon_2 = \varepsilon_0/4$. This means that each $g \in B$ satisfies

$$\|Ag - Ag_k\|_0 < \tfrac{1}{4}\varepsilon_0 \qquad \text{for some } k, \tag{6}$$

where $\|\cdot\|_0$ is the norm on $\mathbf{R}^n$. We shall show that $\{T^\times g_1, \cdots, T^\times g_m\}$ is the desired ε_0-net for $T^\times(B)$; this will then complete the proof.

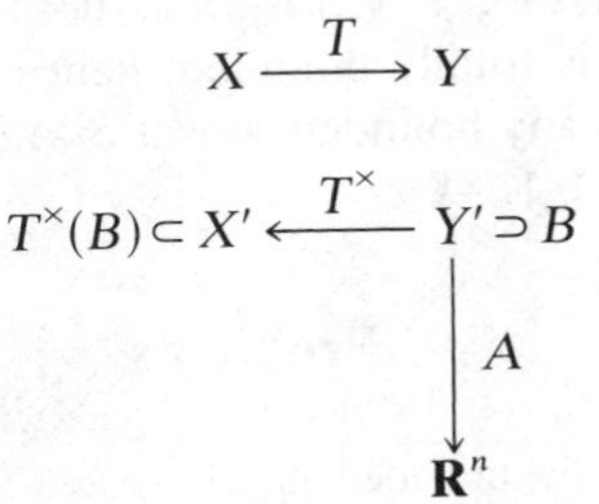

$$\begin{array}{ccc} X & \xrightarrow{\ T\ } & Y \\ & & \\ T^\times(B) \subset X' & \xleftarrow{\ T^\times\ } & Y' \supset B \\ & & \big\downarrow A \\ & & \mathbf{R}^n \end{array}$$

Fig. 64. Notations in the proof of Theorem 8.2-5

From (5) and (6) we see immediately that for every j and every $g \in B$ there is a k such that

$$
\begin{aligned}
|g(Tx_j)-g_k(Tx_j)|^2 &\leqq \sum_{j=1}^{n} |g(Tx_j)-g_k(Tx_j)|^2 \\
&= \|A(g-g_k)\|_0^{\ 2} \qquad (7)\\
&< (\tfrac{1}{4}\varepsilon_0)^2.
\end{aligned}
$$

Let $x \in U$ be arbitrary. Then there is a j for which (4) holds. Let $g \in B$ be arbitrary. Then there is a k such that (6) holds, and (7) holds for that k and every j. We thus obtain

$$
\begin{aligned}
|g(Tx)-g_k(Tx)| &\leqq |g(Tx)-g(Tx_j)| + |g(Tx_j)-g_k(Tx_j)| \\
&\qquad + |g_k(Tx_j)-g_k(Tx)| \\
&< \|g\|\,\|Tx-Tx_j\| + \frac{\varepsilon_0}{4} + \|g_k\|\,\|Tx_j-Tx\| \\
&\leqq c\frac{\varepsilon_0}{4c} + \frac{\varepsilon_0}{4} + c\frac{\varepsilon_0}{4c} < \varepsilon_0.
\end{aligned}
$$

Since this holds for every $x \in U$ and since by the definition of $T^\times$ we have $g(Tx) = (T^\times g)(x)$, etc., we finally obtain

$$
\begin{aligned}
\|T^\times g - T^\times g_k\| &= \sup_{\|x\|=1} |(T^\times(g-g_k))(x)| \\
&= \sup_{\|x\|=1} |g(Tx)-g_k(Tx)| < \varepsilon_0.
\end{aligned}
$$

This shows that $\{T^\times g_1, \cdots, T^\times g_m\}$ is an ε_0-net for $T^\times(B)$. Since $\varepsilon_0 > 0$ was arbitrary, $T^\times(B)$ is totally bounded, hence relatively compact by 8.2-2(*b*). Since B was any bounded subset of Y', this proves compactness of $T^\times$ by Def. 8.1-1. ∎

Problems

1. Let X be a totally bounded metric space. Show that every infinite subset $Y \subset X$ has an infinite subset Z of diameter less than a given $\varepsilon > 0$.

2. If a metric space X is compact, show that X is complete. Show that completeness does not imply compactness.

3. Illustrate with an example that total boundedness is necessary for compactness but not sufficient.

4. Show that a metric space X is compact if and only if it is complete and totally bounded.

5. If a metric space (X, d) is compact, show that for any $\varepsilon > 0$ the space X has a finite subset M such that every point $x \in X$ has distance $\delta(x, M) = \inf_{y \in M} d(x, y) < \varepsilon$ from M.

6. Define $T: l^2 \longrightarrow l^2$ by $Tx = y = (\eta_j)$, where $x = (\xi_j)$ and

$$\eta_j = \sum_{k=1}^{\infty} \alpha_{jk}\xi_k, \qquad \sum_{j=1}^{\infty}\sum_{k=1}^{\infty} |\alpha_{jk}|^2 < \infty.$$

Show that T is compact. (Use 8.1-5.)

7. Show that the operators of the kind defined in Prob. 6 constitute a subspace of $B(l^2, l^2)$. Illustrate with an example that the condition in Prob. 6 is sufficient for compactness but not necessary.

8. Does there exist a surjective compact linear operator $T: l^\infty \longrightarrow l^\infty$?

9. If $T \in B(X, Y)$ is not compact, can the restriction of T to an infinite dimensional subspace of X be compact?

10. Let (λ_n) be a sequence of scalars such that $\lambda_n \longrightarrow 0$ as $n \longrightarrow \infty$. Define $T: l^2 \longrightarrow l^2$ by $Tx = y = (\eta_j)$ where $x = (\xi_j)$ and $\eta_j = \lambda_j \xi_j$. Show that T is compact.

8.3 Spectral Properties of Compact Linear Operators on Normed Spaces

In this section and the next one we consider spectral properties of a compact linear operator $T: X \longrightarrow X$ on a normed space X. For this purpose we shall again use the operator

$$T_\lambda = T - \lambda I \qquad (\lambda \in \mathbf{C}) \tag{1}$$

and the basic concepts of spectral theory as defined in Sec. 7.2.

The spectral theory of compact linear operators is a relatively simple generalization of the eigenvalue theory of finite matrices (Sec. 7.1) and resembles that finite dimensional case in many ways. This can be seen from the following summary of Secs. 8.3 and 8.4 which we include here as an orientation for the reader, so that he can find his way through the details. In this summary we also give the numbers of corresponding theorems (whose order in the text is suggested by the dependence of the proofs upon one another).

Summary. *A compact linear operator $T: X \longrightarrow X$ on a normed space X has the following properties.*

The set of the eigenvalues of T is countable (perhaps finite or even empty). (Cf. 8.3-1.)

$\lambda = 0$ is the only possible point of accumulation of that set. (Cf. 8.3-1.)

Every spectral value $\lambda \neq 0$ is an eigenvalue. (Cf. 8.4-4.) *If X is infinite dimensional, then $0 \in \sigma(T)$.*

For $\lambda \neq 0$ the dimension of any eigenspace of T is finite. (Cf. 8.3-3.)

For $\lambda \neq 0$ the null spaces of $T_\lambda, T_\lambda^2, T_\lambda^3, \cdots$ are finite dimensional (cf. 8.3-3, 8.3-4) *and the ranges of these operators are closed* (cf. 8.3-5, 8.3-6).

There is a number r (depending on λ, where $\lambda \neq 0$) such that

$$X = \mathcal{N}(T_\lambda^r) \oplus T_\lambda^r(X)$$

(cf. 8.4-5); *furthermore, the null spaces satisfy*

$$\mathcal{N}(T_\lambda^r) = \mathcal{N}(T_\lambda^{r+1}) = \mathcal{N}(T_\lambda^{r+2}) = \cdots$$

and the ranges satisfy

$$T_\lambda^r(X) = T_\lambda^{r+1}(X) = T_\lambda^{r+2}(X) = \cdots$$

(cf. 8.4-3); *if $r > 0$, proper inclusions are* (cf. 8.4-3)

$$\mathcal{N}(T_\lambda^0) \subset \mathcal{N}(T_\lambda) \subset \quad \cdots \quad \subset \mathcal{N}(T_\lambda^r)$$

and

$$T_\lambda^0(X) \supset T_\lambda(X) \supset \quad \cdots \quad \supset T_\lambda^r(X). \qquad \blacksquare$$

Our first theorem is concerned with eigenvalues. It tells us that the point spectrum of a compact linear operator is not complicated. The theorem is even much more powerful as it looks at the moment. In fact, in the next section we shall see that every spectral value $\lambda \neq 0$ which a compact linear operator may have (or may not have!) is an eigenvalue. This shows that the spectrum of a compact linear operator resembles that of an operator on a finite dimensional space to a large extent.

8.3-1 Theorem (Eigenvalues). *The set of eigenvalues of a compact linear operator* $T: X \longrightarrow X$ *on a normed space* X *is countable (perhaps finite or even empty), and the only possible point of accumulation is* $\lambda = 0$.

Proof. Obviously, it suffices to show that for every real $k > 0$ the set of all $\lambda \in \sigma_p(T)$ such that $|\lambda| \geqq k$ is finite.

Suppose the contrary for some $k_0 > 0$. Then there is a sequence (λ_n) of infinitely many distinct eigenvalues such that $|\lambda_n| \geqq k_0$. Also $Tx_n = \lambda_n x_n$ for some $x_n \neq 0$. The set of all the x_n's is linearly independent, by Theorem 7.4-3. Let $M_n = \operatorname{span}\{x_1, \cdots, x_n\}$. Then every $x \in M_n$ has a unique representation

$$x = \alpha_1 x_1 + \cdots + \alpha_n x_n.$$

We apply $T - \lambda_n I$ and use $Tx_j = \lambda_j x_j$:

$$(T - \lambda_n I)x = \alpha_1(\lambda_1 - \lambda_n)x_1 + \cdots + \alpha_{n-1}(\lambda_{n-1} - \lambda_n)x_{n-1}.$$

We see that x_n no longer occurs on the right. Hence

$$(T - \lambda_n I)x \in M_{n-1} \qquad \text{for all } x \in M_n. \tag{2}$$

The M_n's are closed (cf. 2.4-3). By Riesz's lemma 2.5-4 there is a sequence (y_n) such that

$$y_n \in M_n, \qquad \|y_n\| = 1, \qquad \|y_n - x\| \geqq \tfrac{1}{2} \text{ for all } x \in M_{n-1}.$$

We show that

$$\|Ty_n - Ty_m\| \geqq \tfrac{1}{2} k_0 \qquad (n > m) \tag{3}$$

so that (Ty_n) has no convergent subsequence because $k_0>0$. This contradicts the compactness of T since (y_n) is bounded.

By adding and subtracting a term we can write

$$(4) \quad Ty_n - Ty_m = \lambda_n y_n - \tilde{x} \qquad \text{where} \qquad \tilde{x} = \lambda_n y_n - Ty_n + Ty_m.$$

Let $m<n$. We show that $\tilde{x} \in M_{n-1}$. Since $m \leqq n-1$, we see that $y_m \in M_m \subset M_{n-1} = \operatorname{span}\{x_1, \cdots, x_{n-1}\}$. Hence $Ty_m \in M_{n-1}$ since $Tx_j = \lambda_j x_j$. By (2),

$$\lambda_n y_n - Ty_n = -(T-\lambda_n I)y_n \in M_{n-1}.$$

Together, $\tilde{x} \in M_{n-1}$. Thus also $x = \lambda_n^{-1}\tilde{x} \in M_{n-1}$, so that

$$(5) \qquad \|\lambda_n y_n - \tilde{x}\| = |\lambda_n| \, \|y_n - x\| \geqq \tfrac{1}{2}|\lambda_n| \geqq \tfrac{1}{2}k_0$$

because $|\lambda_n| \geqq k_0$. From this and (4) we have (3). Hence the assumption that there are infinitely many eigenvalues satisfying $|\lambda_n| \geqq k_0$ for some $k_0>0$ must be false and the theorem is proved. ∎

This theorem shows that if a compact linear operator on a normed space has infinitely many eigenvalues, we can arrange these eigenvalues in a sequence converging to zero.

Composition of a compact linear operator and a bounded linear operator yields compact linear operators. This interesting fact is the content of the following lemma which has many applications and, at present, will be used to prove a basic property of compact linear operators (8.3-4, below).

8.3-2 Lemma (Compactness of product). *Let $T\colon X \longrightarrow X$ be a compact linear operator and $S\colon X \longrightarrow X$ a bounded linear operator on a normed space X. Then TS and ST are compact.*

Proof. Let $B \subset X$ be any bounded set. Since S is a bounded operator, $S(B)$ is a bounded set, and the set $T(S(B)) = TS(B)$ is relatively compact because T is compact. Hence TS is a compact linear operator.

We prove that ST is also compact. Let (x_n) be any bounded sequence in X. Then (Tx_n) has a convergent subsequence (Tx_{n_k}) by 8.1-3, and (STx_{n_k}) converges by 1.4-8. Hence ST is compact by 8.1-3. ∎

It was claimed at the beginning of the chapter that the spectral theory of compact linear operators is almost as simple as that of linear operators on finite dimensional spaces (which is essentially eigenvalue theory of finite matrices; cf. Sec. 7.1). An important property supporting that claim is as follows. For every nonzero eigenvalue which a compact linear operator may (or may not) have the eigenspace is finite dimensional. Indeed, this is implied by

8.3-3 Theorem (Null space). *Let $T: X \longrightarrow X$ be a compact linear operator on a normed space X. Then for every $\lambda \neq 0$ the null space $\mathcal{N}(T_\lambda)$ of $T_\lambda = T - \lambda I$ is finite dimensional.*

Proof. We show that the closed unit ball M in $\mathcal{N}(T_\lambda)$ is compact and then apply Theorem 2.5-5.

Let (x_n) be in M. Then (x_n) is bounded ($\|x_n\| \leqq 1$), and (Tx_n) has a convergent subsequence (Tx_{n_k}) by 8.1-3. Now $x_n \in M \subset \mathcal{N}(T_\lambda)$ implies $T_\lambda x_n = Tx_n - \lambda x_n = 0$, so that $x_n = \lambda^{-1} Tx_n$ because $\lambda \neq 0$. Consequently, $(x_{n_k}) = (\lambda^{-1} Tx_{n_k})$ also converges. The limit is in M since M is closed. Hence M is compact by Def. 2.5-1 because (x_n) was arbitrary in M. This proves $\dim \mathcal{N}(T_\lambda) < \infty$ by 2.5-5. ∎

8.3-4 Corollary (Null spaces). *In Theorem* 8.3-3,

$$\dim \mathcal{N}(T_\lambda{}^n) < \infty \qquad n = 1, 2, \cdots \tag{6}$$

and

$$\{0\} = \mathcal{N}(T_\lambda{}^0) \subset \mathcal{N}(T_\lambda) \subset \mathcal{N}(T_\lambda{}^2) \subset \quad \cdots. \tag{7}$$

Proof. Since T_λ is linear, it maps 0 onto 0 (cf. (3), Sec. 2.6). Hence $T_\lambda{}^n x = 0$ implies $T_\lambda^{n+1} x = 0$, and (7) follows.

We prove (6). By the binomial theorem,

$$\begin{aligned} T_\lambda{}^n = (T - \lambda I)^n &= \sum_{k=0}^{n} \binom{n}{k} T^k (-\lambda)^{n-k} \\ &= (-\lambda)^n I + T \sum_{k=1}^{n} \binom{n}{k} T^{k-1} (-\lambda)^{n-k}. \end{aligned}$$

This can be written

$$T_\lambda{}^n = W - \mu I \qquad \mu = -(-\lambda)^n,$$

where $W = TS = ST$ and S denotes the sum on the right. T is compact, and S is bounded since T is bounded by 8.1-2(a). Hence W is compact by Lemma 8.3-2, so that we obtain (6) by applying Theorem 8.3-3 to $W - \mu I$. ∎

We shall now consider the ranges of T_λ, $T_\lambda^2, \cdots$ for a compact linear operator T and any $\lambda \neq 0$. In this connection we should first remember that for a bounded linear operator the null space is always closed but the range need not be closed. [Cf. 2.7-10(b) and Prob. 6 in Sec. 2.7.] However, if T is compact, then T_λ has a closed range for every $\lambda \neq 0$, and the same holds for T_λ^2, $T_\lambda^3, \cdots$. Let us first prove this for T_λ. The extension of the result to T_λ^n for any $n \in \mathbf{N}$ will then be immediate.

8.3-5 Theorem (Range). *Let $T: X \longrightarrow X$ be a compact linear operator on a normed space X. Then for every $\lambda \neq 0$ the range of $T_\lambda = T - \lambda I$ is closed.*

Proof. The proof is indirect. Thus we assume that the range $T_\lambda(X)$ is not closed and derive from this a contradiction, proceeding according to the following idea.

(a) We consider a y in the closure of $T_\lambda(X)$ but not in $T_\lambda(X)$ and a sequence $(T_\lambda x_n)$ converging to y. We show that $x_n \notin \mathcal{N}(T_\lambda)$ but $\mathcal{N}(T_\lambda)$ contains a sequence (z_n) such that $\|x_n - z_n\| < 2\delta_n$, where δ_n is the distance from x_n to $\mathcal{N}(T_\lambda)$.

(b) We show that $a_n \longrightarrow \infty$, where $a_n = \|x_n - z_n\|$.

(c) We obtain the anticipated contradiction by considering the sequence (w_n), where $w_n = a_n^{-1}(x_n - z_n)$.

The details are as follows.

(a) Suppose that $T_\lambda(X)$ is not closed. Then there is a $y \in \overline{T_\lambda(X)}$, $y \notin T_\lambda(X)$ and a sequence (x_n) in X such that

$$y_n = T_\lambda x_n \longrightarrow y. \tag{8}$$

Since $T_\lambda(X)$ is a vector space, $0 \in T_\lambda(X)$. But $y \notin T_\lambda(X)$, so that $y \neq 0$. This implies $y_n \neq 0$ and $x_n \notin \mathcal{N}(T_\lambda)$ for all sufficiently large n. Without loss of generality we may assume that this holds for all n. Since $\mathcal{N}(T_\lambda)$ is closed, the distance δ_n from x_n to $\mathcal{N}(T_\lambda)$ is positive, that is,

$$\delta_n = \inf_{z \in \mathcal{N}(T_\lambda)} \|x_n - z\| > 0.$$

By the definition of an infimum there is a sequence (z_n) in $\mathcal{N}(T_\lambda)$ such that

$$a_n = \|x_n - z_n\| < 2\delta_n. \tag{9}$$

(b) We show that

$$a_n = \|x_n - z_n\| \longrightarrow \infty \qquad (n \longrightarrow \infty). \tag{10}$$

Suppose (10) does not hold. Then $(x_n - z_n)$ has a bounded subsequence. Since T is compact, it follows from 8.1-3 that $(T(x_n - z_n))$ has a convergent subsequence. Now from $T_\lambda = T - \lambda I$ and $\lambda \neq 0$ we have $I = \lambda^{-1}(T - T_\lambda)$. Using $T_\lambda z_n = 0$ [remember that $z_n \in \mathcal{N}(T_\lambda)$], we thus obtain

$$\begin{aligned} x_n - z_n &= \frac{1}{\lambda}(T - T_\lambda)(x_n - z_n) \\ &= \frac{1}{\lambda}[T(x_n - z_n) - T_\lambda x_n]. \end{aligned}$$

$(T(x_n - z_n))$ has a convergent subsequence and $(T_\lambda x_n)$ converges by (8); hence $(x_n - z_n)$ has a convergent subsequence, say, $x_{n_k} - z_{n_k} \longrightarrow v$. Since T is compact, T is continuous and so is T_λ. Hence Theorem 1.4-8 yields

$$T_\lambda(x_{n_k} - z_{n_k}) \longrightarrow T_\lambda v.$$

Here $T_\lambda z_{n_k} = 0$ because $z_n \in \mathcal{N}(T_\lambda)$, so that by (8) we also have

$$T_\lambda(x_{n_k} - z_{n_k}) = T_\lambda x_{n_k} \longrightarrow y,$$

hence $T_\lambda v = y$. Thus $y \in T_\lambda(X)$, which contradicts $y \notin T_\lambda(X)$, cf. at the beginning of part (a) of the proof. This contradiction resulted from our assumption that (10) does not hold, so that (10) is now proved.

(c) Using again a_n as defined in (10) and setting

$$w_n = \frac{1}{a_n}(x_n - z_n) \tag{11}$$

we have $\|w_n\|=1$. Since $a_n \longrightarrow \infty$ whereas $T_\lambda z_n = 0$ and $(T_\lambda x_n)$ converges, it follows that

$$(12) \qquad T_\lambda w_n = \frac{1}{a_n} T_\lambda x_n \longrightarrow 0.$$

Using again $I = \lambda^{-1}(T - T_\lambda)$, we obtain

$$(13) \qquad w_n = \frac{1}{\lambda}(Tw_n - T_\lambda w_n).$$

Since T is compact and (w_n) is bounded, (Tw_n) has a convergent subsequence. Furthermore, $(T_\lambda w_n)$ converges by (12). Hence (13) shows that (w_n) has a convergent subsequence, say

$$(14) \qquad w_{n_j} \longrightarrow w.$$

A comparison with (12) implies that $T_\lambda w = 0$. Hence $w \in \mathcal{N}(T_\lambda)$. Since $z_n \in \mathcal{N}(T_\lambda)$, also

$$u_n = z_n + a_n w \in \mathcal{N}(T_\lambda).$$

Hence for the distance from x_n to u_n we must have

$$\|x_n - u_n\| \geqq \delta_n.$$

Writing u_n out and using (11) and (9), we thus obtain

$$\begin{aligned} \delta_n &\leqq \|x_n - z_n - a_n w\| \\ &= \|a_n w_n - a_n w\| \\ &= a_n \|w_n - w\| \\ &< 2\delta_n \|w_n - w\|. \end{aligned}$$

Dividing by $2\delta_n > 0$, we have $\frac{1}{2} < \|w_n - w\|$. This contradicts (14) and proves the theorem. ∎

8.3-6 Corollary (Ranges). *Under the assumptions in Theorem* 8.3-5 *the range of T_λ^n is closed for every $n = 0, 1, 2, \cdots$. Furthermore,*

$$X = T_\lambda^0(X) \supset T_\lambda(X) \supset T_\lambda^2(X) \supset \cdots.$$

Proof. The first statement follows from Theorem 8.3-5 by noting that W in the proof of 8.3-4 is compact. The second statement follows by induction. Indeed, we have $T_\lambda^0(X) = I(X) = X \supset T_\lambda(X)$, and application of T_λ to $T_\lambda^{n-1}(X) \supset T_\lambda^n(X)$ gives $T_\lambda^n(X) \supset T_\lambda^{n+1}(X)$. ∎

Problems

1. Prove Theorem 8.3-1, assuming that T^p is a compact linear operator for some positive integer p.

2. Let X, Y and Z be normed spaces, and let $T_1: X \longrightarrow Y$ and $T_2: Y \longrightarrow Z$. If T_1 and T_2 are compact linear operators, show that $T_2T_1: X \longrightarrow Z$ is a compact linear operator.

3. If T is a compact linear operator, show that for any given number $k > 0$ there are at most finitely many linearly independent eigenvectors of T corresponding to eigenvalues of absolute value greater than k.

4. Let $T_j: X_j \longrightarrow X_{j+1}$, $j = 1, 2, 3$, be bounded linear operators on normed spaces. If T_2 is compact, show that $T = T_3T_2T_1: X_1 \longrightarrow X_4$ is compact.

5. Give a proof of compactness of TS in Lemma 8.3-2 based on the consideration of bounded sequences.

6. Let H be a Hilbert space, $T: H \longrightarrow H$ a bounded linear operator and T^* the Hilbert-adjoint operator of T. Show that T is compact if and only if T^*T is compact.

7. If T in Prob. 6 is compact, show that T^* is compact.

8. If a compact linear operator $T: X \longrightarrow X$ on an infinite dimensional normed space X has an inverse which is defined on all of X, show that the inverse cannot be bounded.

9. Prove Theorem 8.3-3 by the use of Riesz's lemma 2.5-4 (instead of Theorem 2.5-5).

10. Prove Theorem 8.3-3 under the weaker assumption that T^p is a compact linear operator for a $p \in \mathbf{N}$. (Use the proof in Prob. 9.)

11. Show by simple examples that in Theorem 8.3-3 the conditions that T be compact and $\lambda \neq 0$ cannot be omitted.

12. Give an independent proof of Theorem 8.3-3 if X is a Hilbert space.

13. Prove Corollary 8.3-4 under the weaker assumption that T^p is a compact linear operator for a $p \in \mathbf{N}$.

14. Let $T: X \longrightarrow X$ be a compact linear operator on a normed space. If $\dim X = \infty$, show that $0 \in \sigma(T)$.

15. Let $T: l^2 \longrightarrow l^2$ be defined by $y = (\eta_j) = Tx$, $x = (\xi_j)$, $\eta_{2k} = \xi_{2k}$ and $\eta_{2k-1} = 0$, where $k = 1, 2, \cdots$. Find $\mathcal{N}(T_\lambda^n)$. Is T compact?

8.4 Further Spectral Properties of Compact Linear Operators

From the preceding section we know that for a compact linear operator T on a normed space X and $\lambda \neq 0$ the null spaces $\mathcal{N}(T_\lambda^n)$, $n = 1, 2, \cdots$, are finite dimensional and satisfy $\mathcal{N}(T_\lambda^n) \subset \mathcal{N}(T_\lambda^{n+1})$; and the ranges $T_\lambda^n(X)$ are closed and satisfy $T_\lambda^n(X) \supset T_\lambda^{n+1}(X)$.

We can say more, as follows. From some $n = r$ on, these null spaces are all equal (Lemma 8.4-1, below); from an $n = q$ on, those ranges are equal (Lemma 8.4-2), and $q = r$ (Theorem 8.4-3; here, q and r are the smallest integers with those properties). Let us begin with

8.4-1 Lemma (Null spaces). *Let $T: X \longrightarrow X$ be a compact linear operator on a normed space X, and let $\lambda \neq 0$. Then there exists a smallest integer r (depending on λ) such that from $n = r$ on, the null spaces $\mathcal{N}(T_\lambda^n)$ are all equal, and if $r > 0$, the inclusions*

$$\mathcal{N}(T_\lambda^0) \subset \mathcal{N}(T_\lambda) \subset \cdots \subset \mathcal{N}(T_\lambda^r)$$

are all proper.

Proof. Let us write $\mathcal{N}_n = \mathcal{N}(T_\lambda^n)$, for simplicity. The idea of the proof is as follows.

(*a*) We assume that $\mathcal{N}_m = \mathcal{N}_{m+1}$ for no m and derive a contradiction. As an essential tool we use Riesz's lemma 2.5-4.

(*b*) We show that $\mathcal{N}_m = \mathcal{N}_{m+1}$ implies $\mathcal{N}_n = \mathcal{N}_{n+1}$ for all $n > m$.

The details are as follows.

(a) We know that $\mathcal{N}_m \subset \mathcal{N}_{m+1}$ by 8.3-4. Suppose that $\mathcal{N}_m = \mathcal{N}_{m+1}$ for no m. Then $\mathcal{N}_n$ is a proper subspace of $\mathcal{N}_{n+1}$ for every n. Since these null spaces are closed, Riesz's lemma 2.5-4 thus implies the existence of a sequence (y_n) such that

$$(1) \qquad y_n \in \mathcal{N}_n, \qquad \|y_n\| = 1, \qquad \|y_n - x\| \geqq \tfrac{1}{2} \text{ for all } x \in \mathcal{N}_{n-1}.$$

We show that

$$(2) \qquad \|Ty_n - Ty_m\| \geqq \tfrac{1}{2}|\lambda| \qquad (m<n),$$

so that (Ty_n) has no convergent subsequence because $|\lambda|>0$. This contradicts the compactness of T since (y_n) is bounded.

From $T_\lambda = T - \lambda I$ we have $T = T_\lambda + \lambda I$ and

$$(3) \qquad Ty_n - Ty_m = \lambda y_n - \tilde{x} \qquad \text{where} \qquad \tilde{x} = T_\lambda y_m + \lambda y_m - T_\lambda y_n.$$

Let $m<n$. We show that $\tilde{x} \in \mathcal{N}_{n-1}$. Since $m \leqq n-1$ we clearly have $\lambda y_m \in \mathcal{N}_m \subset \mathcal{N}_{n-1}$. Also $y_m \in \mathcal{N}_m$ implies

$$0 = T_\lambda{}^m y_m = T_\lambda^{m-1}(T_\lambda y_m),$$

that is, $T_\lambda y_m \in \mathcal{N}_{m-1} \subset \mathcal{N}_{n-1}$. Similarly, $y_n \in \mathcal{N}_n$ implies $T_\lambda y_n \in \mathcal{N}_{n-1}$. Together, $\tilde{x} \in \mathcal{N}_{n-1}$. Also $x = \lambda^{-1}\tilde{x} \in \mathcal{N}_{n-1}$, so that by (1),

$$\|\lambda y_n - \tilde{x}\| = |\lambda| \, \|y_n - x\| \geqq \tfrac{1}{2}|\lambda|.$$

From this and (3) we have (2). Hence our assumption that $\mathcal{N}_m = \mathcal{N}_{m+1}$ for no m is false and we must have $\mathcal{N}_m = \mathcal{N}_{m+1}$ for some m.

(b) We prove that $\mathcal{N}_m = \mathcal{N}_{m+1}$ implies $\mathcal{N}_n = \mathcal{N}_{n+1}$ for all $n>m$. Suppose this does not hold. Then $\mathcal{N}_n$ is a proper subspace of $\mathcal{N}_{n+1}$ for some $n>m$. We consider an $x \in \mathcal{N}_{n+1} - \mathcal{N}_n$. By definition,

$$T_\lambda^{n+1} x = 0 \qquad \text{but} \qquad T_\lambda{}^n x \neq 0.$$

Since $n>m$, we have $n-m>0$. We set $z = T_\lambda^{n-m} x$. Then

$$T_\lambda^{m+1} z = T_\lambda^{n+1} x = 0 \qquad \text{but} \qquad T_\lambda{}^m z = T_\lambda{}^n x \neq 0.$$

Hence $z \in \mathcal{N}_{m+1}$ but $z \notin \mathcal{N}_m$, so that $\mathcal{N}_m$ is a proper subspace of $\mathcal{N}_{m+1}$. This contradicts $\mathcal{N}_m = \mathcal{N}_{m+1}$. The first statement is proved, where r is the smallest n such that $\mathcal{N}_n = \mathcal{N}_{n+1}$. Consequently, if $r>0$, the inclusions stated in the lemma are proper. ▮

The lemma just proved concerns the null spaces of the operators T_λ, $T_\lambda^2, \cdots$, where T is a compact linear operator and $\lambda \neq 0$. Let us show that similar statements are true for the ranges of those operators:

8.4-2 Lemma (Ranges). *Under the assumptions of Lemma 8.4-1 there exists a smallest integer q (depending on λ) such that from $n = q$ on, the ranges $T_\lambda^n(X)$ are all equal; and if $q > 0$, the inclusions*

$$T_\lambda^0(X) \supset T_\lambda(X) \supset \cdots \supset T_\lambda^q(X)$$

are all proper.

Proof. The proof is again indirect and parallels the previous one. We write simply $\mathcal{R}_n = T_\lambda^n(X)$. Suppose that $\mathcal{R}_s = \mathcal{R}_{s+1}$ for no s. Then $\mathcal{R}_{n+1}$ is a proper subspace of $\mathcal{R}_n$ for every n (cf. 8.3-6). Since these ranges are closed by 8.3-6, Riesz's lemma 2.5-4 thus implies the existence of a sequence (x_n) such that

$$\text{(4)} \qquad x_n \in \mathcal{R}_n, \qquad \|x_n\| = 1, \qquad \|x_n - x\| \geqq \tfrac{1}{2} \text{ for all } x \in \mathcal{R}_{n+1}.$$

Let $m < n$. Since $T = T_\lambda + \lambda I$, we can write

$$\text{(5)} \qquad Tx_m - Tx_n = \lambda x_m - (-T_\lambda x_m + T_\lambda x_n + \lambda x_n).$$

On the right, $\lambda x_m \in \mathcal{R}_m$, $x_m \in \mathcal{R}_m$, so that $T_\lambda x_m \in \mathcal{R}_{m+1}$ and, since $n > m$, also $T_\lambda x_n + \lambda x_n \in \mathcal{R}_n \subset \mathcal{R}_{m+1}$. Hence (5) is of the form

$$Tx_m - Tx_n = \lambda(x_m - x) \qquad x \in \mathcal{R}_{m+1}.$$

Consequently, by (4),

$$\text{(6)} \qquad \|Tx_m - Tx_n\| = |\lambda| \, \|x_m - x\| \geqq \tfrac{1}{2}|\lambda| > 0.$$

(x_n) is bounded and T is compact. Hence (Tx_n) has a convergent subsequence. This contradicts (6) and proves that $\mathcal{R}_s = \mathcal{R}_{s+1}$ for some s. Let q be the smallest s such that $\mathcal{R}_s = \mathcal{R}_{s+1}$. Then, if $q > 0$, the inclusions stated in the lemma (which follow from 8.3-6) are proper.

Furthermore, $\mathcal{R}_{q+1} = \mathcal{R}_q$ means that T_λ maps $\mathcal{R}_q$ onto itself. Hence repeated application of T_λ gives $\mathcal{R}_{n+1} = \mathcal{R}_n$ for every $n > q$. ∎

Combining Lemmas 8.4-1 and 8.4-2, we now obtain the important

8.4-3 Theorem (Null spaces and ranges). *Let $T\colon X \longrightarrow X$ be a compact linear operator on a normed space X, and let $\lambda \neq 0$. Then there exists a smallest integer $n = r$ (depending on λ) such that*

$$\mathcal{N}(T_\lambda^r) = \mathcal{N}(T_\lambda^{r+1}) = \mathcal{N}(T_\lambda^{r+2}) = \cdots \tag{7}$$

and

$$T_\lambda^r(X) = T_\lambda^{r+1}(X) = T_\lambda^{r+2}(X) = \cdots; \tag{8}$$

and if $r > 0$, the following inclusions are proper:

$$\mathcal{N}(T_\lambda^0) \subset \mathcal{N}(T_\lambda) \subset \cdots \subset \mathcal{N}(T_\lambda^r) \tag{9}$$

and

$$T_\lambda^0(X) \supset T_\lambda(X) \supset \cdots \supset T_\lambda^r(X). \tag{10}$$

Proof. Lemma 8.4-1 gives (7) and (9). Lemma 8.4-2 gives (8) and (10) with q instead of r. All we have to show is that $q = r$. We prove $q \geqq r$ in part (a) and $q \leqq r$ in part (b). As before we simply write $\mathcal{N}_n = \mathcal{N}(T_\lambda^n)$ and $\mathcal{R}_n = T_\lambda^n(X)$.

(a) We have $\mathcal{R}_{q+1} = \mathcal{R}_q$ by Lemma 8.4-2. This means that $T_\lambda(\mathcal{R}_q) = \mathcal{R}_q$. Hence

$$y \in \mathcal{R}_q \qquad \Longrightarrow \qquad y = T_\lambda x \text{ for some } x \in \mathcal{R}_q. \tag{11}$$

We show:

$$T_\lambda x = 0,\ x \in \mathcal{R}_q \qquad \Longrightarrow \qquad x = 0. \tag{12}$$

Suppose (12) does not hold. Then $T_\lambda x_1 = 0$ for some nonzero $x_1 \in \mathcal{R}_q$. Now (11) with $y = x_1$ gives $x_1 = T_\lambda x_2$ for some $x_2 \in \mathcal{R}_q$. Similarly, $x_2 = T_\lambda x_3$ for some $x_3 \in \mathcal{R}_q$, etc. For every n we thus obtain by substitution

$$0 \neq x_1 = T_\lambda x_2 = \cdots = T_\lambda^{n-1} x_n \qquad \text{but} \qquad 0 = T_\lambda x_1 = T_\lambda^n x_n.$$

Hence $x_n \notin \mathcal{N}_{n-1}$ but $x_n \in \mathcal{N}_n$. We have $\mathcal{N}_{n-1} \subset \mathcal{N}_n$ by 8.4-1, and our present result shows that this inclusion is proper for every n since n was arbitrary. This contradicts 8.4-1 and proves (12).

Remembering that $\mathcal{R}_{q+1}=\mathcal{R}_q$ by 8.4-2, we prove that $\mathcal{N}_{q+1}=\mathcal{N}_q$; this then implies $q \geqq r$ by 8.4-1 since r is the smallest integer for which we have equality.

We have $\mathcal{N}_{q+1} \supset \mathcal{N}_q$ by 8.3-4. We prove that $\mathcal{N}_{q+1} \subset \mathcal{N}_q$, that is, $T_\lambda^{q+1}x=0$ implies $T_\lambda{}^q x=0$. Suppose this is false. Then for some x_0,

$$y=T_\lambda{}^q x_0 \neq 0 \qquad \text{but} \qquad T_\lambda y = T_\lambda^{q+1}x_0=0.$$

Hence $y \in \mathcal{R}_q$, $y \neq 0$ and $T_\lambda y=0$. But this contradicts (12) with $x=y$ and proves $\mathcal{N}_{q+1} \subset \mathcal{N}_q$. Hence $\mathcal{N}_{q+1}=\mathcal{N}_q$, and $q \geqq r$.

(b) We prove that $q \leqq r$. If $q=0$, this holds. Let $q \geqq 1$. We prove $q \leqq r$ by showing that $\mathcal{N}_{q-1}$ is a proper subspace of $\mathcal{N}_q$. This implies $q \leqq r$ since r is the smallest integer n such that $\mathcal{N}_n=\mathcal{N}_{n+1}$; cf. 8.4-1.

By the definition of q in 8.4-2 the inclusion $\mathcal{R}_q \subset \mathcal{R}_{q-1}$ is proper. Let $y \in \mathcal{R}_{q-1}-\mathcal{R}_q$. Then $y \in \mathcal{R}_{q-1}$, so that $y=T_\lambda^{q-1}x$ for some x. Also $T_\lambda y \in \mathcal{R}_q=\mathcal{R}_{q+1}$ implies that $T_\lambda y = T_\lambda^{q+1}z$ for some z. Since $T_\lambda{}^q z \in \mathcal{R}_q$ but $y \notin \mathcal{R}_q$, we have

$$T_\lambda^{q-1}(x-T_\lambda z)=y-T_\lambda{}^q z \neq 0.$$

Hence $x-T_\lambda z \notin \mathcal{N}_{q-1}$. But $x-T_\lambda z \in \mathcal{N}_q$ because

$$T_\lambda{}^q(x-T_\lambda z)=T_\lambda y - T_\lambda y=0.$$

This proves that $\mathcal{N}_{q-1} \neq \mathcal{N}_q$, so that $\mathcal{N}_{q-1}$ is a proper subspace of $\mathcal{N}_q$. Hence $q \leqq r$, and $q=r$ since $q \geqq r$ was shown in part (*a*) of the proof. ∎

An almost immediate consequence of this theorem is the following important characterization of the spectrum of a compact linear operator on a Banach space. (In 8.6-4 we shall see that the conclusion continues to hold even if the space is not complete.)

8.4-4 Theorem (Eigenvalues). *Let $T: X \longrightarrow X$ be a compact linear operator on a Banach space X. Then every spectral value $\lambda \neq 0$ of T* (if it exists[2]) *is an eigenvalue of T.* (This holds even for general normed spaces. Proof in 8.6-4.)

[2] Prob. 5 shows that T may not have eigenvalues. A *self-adjoint* compact linear operator T on a complex Hilbert space $H \neq \{0\}$ always has at least one eigenvalue, as we shall see in Sec. 9.2.

Proof. If $\mathcal{N}(T_\lambda) \neq \{0\}$, then λ is an eigenvalue of T. Suppose that $\mathcal{N}(T_\lambda) = \{0\}$, where $\lambda \neq 0$. Then $T_\lambda x = 0$ implies that $x = 0$ and $T_\lambda^{-1}\colon T_\lambda(X) \longrightarrow X$ exists by 2.6-10. Since

$$\{0\} = \mathcal{N}(I) = \mathcal{N}(T_\lambda^{\,0}) = \mathcal{N}(T_\lambda),$$

we have $r = 0$ by 8.4-3. Hence $X = T_\lambda^{\,0}(X) = T_\lambda(X)$, also by 8.4-3. It follows that T_λ is bijective, T_λ^{-1} is bounded by the bounded inverse theorem 4.12-2 since X is complete, and $\lambda \in \rho(T)$ by definition. ∎

The value $\boldsymbol{\lambda = 0}$ was excluded in many theorems of this chapter, so that it is natural to ask what we can say about $\lambda = 0$ in the case of a compact operator $T\colon X \longrightarrow X$ on a complex normed space X. If X is finite dimensional, then T has representations by matrices and it is clear that 0 may or may not belong to $\sigma(T) = \sigma_p(T)$; that is, if $\dim X < \infty$, we may have $0 \notin \sigma(T)$; then $0 \in \rho(T)$. However, if $\dim X = \infty$, then we must have $0 \in \sigma(T)$; cf. Prob. 14 in the previous section. And all three cases

$$0 \in \sigma_p(T), \qquad 0 \in \sigma_c(T), \qquad 0 \in \sigma_r(T)$$

are possible, as is illustrated by Probs. 4 and 5 of this section and Prob. 7 of Sec. 9.2. ∎

As another interesting and important application of Theorem 8.4-3, let us establish a representation of X as the direct sum (Sec. 3.3) of two *closed* subspaces, namely, the null space and the range of $T_\lambda^{\,r}$:

8.4-5 Theorem (Direct sum). *Let X, T, λ and r be as in Theorem* 8.4-3. *Then*[3] *X can be represented in the form*

$$X = \mathcal{N}(T_\lambda^{\,r}) \oplus T_\lambda^{\,r}(X). \tag{13}$$

Proof. We consider any $x \in X$. We must show that x has a unique representation of the form

$$x = y + z \qquad (y \in \mathcal{N}_r,\ z \in \mathcal{R}_r),$$

[3] If X is a vector space, then for any subspace $Y \subset X$ there exists a subspace $Z \subset X$ such that $X = Y \oplus Z$; cf. Sec. 3.3. If X is a normed space (even a Banach space) and $Y \subset X$ is a *closed* subspace, there may not exist a *closed* subspace $Z \subset X$ such that $X = Y \oplus Z$. (For examples, see F. J. Murray (1937) and A. Sobczyk (1941); cf. Appendix 3.) If X is a Hilbert space, then for every closed subspace Y one always has $X = Y \oplus Z$, where $Z = Y^\perp$ is closed (cf. 3.3-3 and 3.3-4). Note that the subspaces in (13) are closed.

where $\mathcal{N}_n = \mathcal{N}(T_\lambda^n)$ and $\mathcal{R}_n = T_\lambda^n(X)$, as before. Let $z = T_\lambda^r x$. Then $z \in \mathcal{R}_r$. Now $\mathcal{R}_r = \mathcal{R}_{2r}$ by Theorem 8.4-3. Hence $z \in \mathcal{R}_{2r}$, so that $z = T_\lambda^{2r} x_1$ for some $x_1 \in X$. Let $x_0 = T_\lambda^r x_1$. Then $x_0 \in \mathcal{R}_r$, and

$$T_\lambda^r x_0 = T_\lambda^{2r} x_1 = z = T_\lambda^r x.$$

This shows that $T_\lambda^r(x - x_0) = 0$. Hence $x - x_0 \in \mathcal{N}_r$, and

$$x = (x - x_0) + x_0 \qquad (x - x_0 \in \mathcal{N}_r,\ x_0 \in \mathcal{R}_r). \tag{14}$$

This proves (13), provided (14) is unique.

We show uniqueness. In addition to (14), let

$$x = (x - \tilde{x}_0) + \tilde{x}_0 \qquad (x - \tilde{x}_0 \in \mathcal{N}_r,\ \tilde{x}_0 \in \mathcal{R}_r).$$

Let $v_0 = x_0 - \tilde{x}_0$. Then $v_0 \in \mathcal{R}_r$ since $\mathcal{R}_r$ is a vector space. Hence $v_0 = T_\lambda^r v$ for some $v \in X$. Also

$$v_0 = x_0 - \tilde{x}_0 = (x - \tilde{x}_0) - (x - x_0);$$

hence $v_0 \in \mathcal{N}_r$, and $T_\lambda^r v_0 = 0$. Together,

$$T_\lambda^{2r} v = T_\lambda^r v_0 = 0,$$

and $v \in \mathcal{N}_{2r} = \mathcal{N}_r$ (cf. 8.4-3). This implies that

$$v_0 = T_\lambda^r v = 0,$$

that is, $v_0 = x_0 - \tilde{x}_0 = 0$, $x_0 = \tilde{x}_0$, the representation (14) is unique, and the sum $\mathcal{N}_r + \mathcal{R}_r$ is direct. ∎

Problems

1. Prove Lemma 8.4-1 under the weaker assumption that T^p is compact for a $p \in \mathbf{N}$.

2. In the proof of Lemma 8.4-1 it was shown that $\mathcal{N}_m = \mathcal{N}_{m+1}$ implies $\mathcal{N}_n = \mathcal{N}_{n+1}$ for all $n > m$. The proof was indirect. Give a direct proof.

3. To obtain Theorem 8.4-4 for a general normed space, we could try to use the present proof for $\tilde{T}$ in 8.2-4 and then make conclusions about T. What would be the difficulty?

4. Show that $T: l^2 \longrightarrow l^2$ defined by

$$Tx = \left(\frac{\xi_2}{1}, \frac{\xi_3}{2}, \frac{\xi_4}{3}, \cdots\right)$$

is compact and $\sigma_p(T) = \{0\}$; here, $x = (\xi_1, \xi_2, \cdots)$.

5. In Theorem 8.4-4 we had to include the phrase "if it exists" since a compact linear operator may not have eigenvalues. Show that an operator of that kind is $T: l^2 \longrightarrow l^2$ defined by

$$Tx = \left(0, \frac{\xi_1}{1}, \frac{\xi_2}{2}, \frac{\xi_3}{3}, \cdots\right)$$

where $x = (\xi_1, \xi_2, \cdots)$. Show that $\sigma(T) = \sigma_r(T) = \{0\}$. (Note that Prob. 4 shows that 0 may belong to the point spectrum. 0 may also belong to the continuous spectrum, as we shall see in Sec. 9.2, Prob. 7.)

6. Find the eigenvalues of $T_n: \mathbf{R}^n \longrightarrow \mathbf{R}^n$ defined by

$$T_n x = \left(0, \frac{\xi_1}{1}, \frac{\xi_2}{2}, \cdots, \frac{\xi_{n-1}}{n-1}\right)$$

where $x = (\xi_1, \cdots, \xi_n)$. Compare with Prob. 5 and explain what happens as $n \longrightarrow \infty$.

7. Let $T: l^2 \longrightarrow l^2$ be defined by $y = Tx$, $x = (\xi_j)$, $y = (\eta_j)$, $\eta_j = \alpha_j \xi_j$, where (α_j) is dense on $[0, 1]$. Show that T is not compact.

8. Let $T: l^2 \longrightarrow l^2$ be defined by

$$x = (\xi_1, \xi_2, \cdots) \qquad \longmapsto \qquad Tx = (\xi_2, \xi_3, \cdots).$$

Let $m = m_0$ and $n = n_0$ be the smallest numbers such that we have $\mathcal{N}(T^m) = \mathcal{N}(T^{m+1})$ and $T^{n+1}(X) = T^n(X)$. Find $\mathcal{N}(T^m)$. Does there exist a finite m_0? Find n_0.

9. Let $T: C[0, 1] \longrightarrow C[0, 1]$ be defined by $Tx = vx$, where $v(t) = t$. Show that T is not compact.

10. Derive the representation (13) in the case of the linear operator $T: \mathbf{R}^2 \longrightarrow \mathbf{R}^2$ represented by the matrix

$$\begin{bmatrix} 1 & -1 \\ -1 & 1 \end{bmatrix}.$$

8.5 Operator Equations Involving Compact Linear Operators

I. Fredholm (1903) investigated linear integral equations, and his famous work suggested a theory of solvability of certain equations involving a compact linear operator. We shall introduce the reader to this theory which was developed mainly by F. Riesz (1918) with an important contribution by J. Schauder (1930).

We shall consider a compact linear operator $T: X \longrightarrow X$ on a normed space X, the adjoint operator $T^\times: X' \longrightarrow X'$ as defined in 4.5-1, the equation

(1) $$Tx - \lambda x = y \qquad (y \in X \text{ given}, \lambda \neq 0),$$

the corresponding homogeneous equation

(2) $$Tx - \lambda x = 0 \qquad (\lambda \neq 0),$$

and two similar equations involving the adjoint operator, namely

(3) $$T^\times f - \lambda f = g \qquad (g \in X' \text{ given}, \lambda \neq 0)$$

and the corresponding homogeneous equation

(4) $$T^\times f - \lambda f = 0 \qquad (\lambda \neq 0).$$

Here $\lambda \in \mathbf{C}$ is arbitrary and fixed, not zero, and we shall study the existence of solutions x and f, respectively.

Why do we consider these four equations at the same time? The answer can be seen from the following summary of results, which shows the interrelation of the equations with respect to solvability. (Numbers in parentheses refer to corresponding theorems to be considered.)

Summary. *Let $T: X \longrightarrow X$ be a compact linear operator on a normed space X and $T^\times: X' \longrightarrow X'$ the adjoint operator of T. Let $\lambda \neq 0$. Then:*

(1) is normally solvable, *that is,* (1) *has a solution x if and only if $f(y) = 0$ for all solutions f of* (4). *Hence if $f = 0$ is the only solution of* (4), *then for every y the equation* (1) *is solvable.* (Cf. 8.5-1.)

(3) has a solution f if and only if $g(x)=0$ *for all solutions x of* (2). *Hence if* $x=0$ *is the only solution of* (2), *then for every g the equation* (3) *is solvable.* (Cf. 8.5-3.)

(1) *has a solution x for every* $y\in X$ *if and only if* $x=0$ *is the only solution of* (2). (Cf. 8.6-1*a*.)

(3) *has a solution f for every* $g\in X'$ *if and only if* $f=0$ *is the only solution of* (4). (Cf. 8.6-1*b*.)

(2) *and* (4) *have the same number of linearly independent solutions.* (Cf. 8.6-3.)

T_λ *satisfies the* Fredholm alternative. (Cf. 8.7-2.) ∎

Our first theorem gives a necessary and sufficient condition for the solvability of (1):

8.5-1 Theorem (Solutions of (1)). *Let* $T\colon X\longrightarrow X$ *be a compact linear operator on a normed space X and let* $\lambda\neq 0$. *Then* (1) *has a solution x if and only if y is such that*

$$f(y)=0 \tag{5}$$

for all $f\in X'$ *satisfying* (4).

Hence if (4) *has only the trivial solution* $f=0$, *then* (1) *with any given* $y\in X$ *is solvable.*

Proof. (*a*) Suppose (1) has a solution $x=x_0$, that is,

$$y=Tx_0-\lambda x_0=T_\lambda x_0.$$

Let f be any solution of (4). Then we first have

$$f(y)=f(Tx_0-\lambda x_0)=f(Tx_0)-\lambda f(x_0).$$

Now $f(Tx_0)=(T^\times f)(x_0)$ by the definition of the adjoint operator (cf. 4.5-1, where g plays the role of our present f). Hence by (4),

$$f(y)=(T^\times f)(x_0)-\lambda f(x_0)=0.$$

(*b*) Conversely, we assume that y in (1) satisfies (5) for every solution of (4) and show that then (1) has a solution.

Suppose (1) has no solution. Then $y=T_\lambda x$ for no x. Hence $y\notin T_\lambda(X)$. Since $T_\lambda(X)$ is closed by 8.3-5, the distance δ from y to

$T_\lambda(X)$ is positive. By Lemma 4.6-7 there exists an $\tilde{f} \in X'$ such that $\tilde{f}(y) = \delta$ and $\tilde{f}(z) = 0$ for every $z \in T_\lambda(X)$. Since $z \in T_\lambda(X)$, we have $z = T_\lambda x$ for some $x \in X$, so that $\tilde{f}(z) = 0$ becomes

$$\begin{aligned}\tilde{f}(T_\lambda x) &= \tilde{f}(Tx) - \lambda \tilde{f}(x) \\ &= (T^\times \tilde{f})(x) - \lambda \tilde{f}(x) = 0.\end{aligned}$$

This holds for every $x \in X$ since $z \in T_\lambda(X)$ was arbitrary. Hence $\tilde{f}$ is a solution of (4). By assumption it satisfies (5), that is, $\tilde{f}(y) = 0$. But this contradicts $\tilde{f}(y) = \delta > 0$. Consequently, (1) must have a solution. This proves the first statement of the theorem, which immediately implies the second one. ∎

The situation characterized by this theorem suggests the following concept. Let

$$Ax = y \qquad (y \text{ given}), \tag{6}$$

where $A: X \longrightarrow X$ is a bounded linear operator on a normed space X. Suppose that (6) has a solution $x \in X$ if and only if y satisfies $f(y) = 0$ for every solution $f \in X'$ of the equation

$$A^\times f = 0, \tag{7}$$

where $A^\times$ is the adjoint operator of A. Then (6) is said to be **normally solvable.**

Theorem 8.5-1 *shows that* (1) *with a compact linear operator* T *and* $\lambda \neq 0$ *is normally solvable.*

For equation (3) there is an analogue of Theorem 8.5-1 which we shall obtain from the following lemma. The positive real number c in the lemma may depend on λ, which is given. Note well that (8) holds for *some* solution—call it a *solution of minimum norm*—but not necessarily for *all* solutions. Hence the lemma does *not* imply the existence of $R_\lambda = T_\lambda^{-1}$ (by Prob. 7 in Sec. 2.7).

8.5-2 Lemma (Bound for certain solutions of (1)). *Let* $T: X \longrightarrow X$ *be a compact linear operator on a normed space and let* $\lambda \neq 0$ *be given. Then there exists a real number* $c > 0$ *which is independent of* y *in* (1) *and such that for every* y *for which* (1) *has a solution, at least one of* *these solutions, call it* $x = \tilde{x}$, *satisfies*

$$\|\tilde{x}\| \leqq c\|y\|, \tag{8}$$

where $y = T_\lambda \tilde{x}$.

Proof. We subdivide the proof into two steps:

(a) We show that if (1) with a given y has solutions at all, the set of these solutions contains a solution of minimum norm, call it $\tilde{x}$.

(b) We show that there is a $c > 0$ such that (8) holds for a solution $\tilde{x}$ of minimum norm corresponding to any $y = T_\lambda \tilde{x}$ for which (1) has solutions.

The details are as follows.

(a) Let x_0 be a solution of (1). If x is any other solution of (1), the difference $z = x - x_0$ satisfies (2). Hence every solution of (1) can be written

$$x = x_0 + z \qquad \text{where} \qquad z \in \mathcal{N}(T_\lambda)$$

and, conversely, for every $z \in \mathcal{N}(T_\lambda)$ the sum $x_0 + z$ is a solution of (1). For a fixed x_0 the norm of x depends on z; let us write

$$p(z) = \|x_0 + z\| \qquad \text{and} \qquad k = \inf_{z \in \mathcal{N}(T_\lambda)} p(z).$$

By the definition of an infimum, $\mathcal{N}(T_\lambda)$ contains a sequence (z_n) such that

$$p(z_n) = \|x_0 + z_n\| \longrightarrow k \qquad (n \longrightarrow \infty). \tag{9}$$

Since $(p(z_n))$ converges, it is bounded. Also (z_n) is bounded because

$$\|z_n\| = \|(x_0 + z_n) - x_0\| \leqq \|x_0 + z_n\| + \|x_0\| = p(z_n) + \|x_0\|.$$

Since T is compact, (Tz_n) has a convergent subsequence. But $z_n \in \mathcal{N}(T_\lambda)$ means that $T_\lambda z_n = 0$, that is, $Tz_n = \lambda z_n$, where $\lambda \neq 0$. Hence (z_n) has a convergent subsequence, say,

$$z_{n_j} \longrightarrow z_0$$

where $z_0 \in \mathcal{N}(T_\lambda)$ since $\mathcal{N}(T_\lambda)$ is closed by 2.7-10. Also

$$p(z_{n_j}) \longrightarrow p(z_0)$$

since p is continuous. We thus obtain from (9)

$$p(z_0) = \|x_0 + z_0\| = k.$$

This shows that if (1) with a given y has solutions, the set of these solutions contains a solution $\tilde{x} = x_0 + z_0$ of minimum norm.

(b) We prove that there is a $c > 0$ (independent of y) such that (8) holds for a solution $\tilde{x}$ of minimum norm corresponding to any $y = T_\lambda \tilde{x}$ for which (1) is solvable.

Suppose our assertion does not hold. Then there is a sequence (y_n) such that

$$\frac{\|\tilde{x}_n\|}{\|y_n\|} \longrightarrow \infty \qquad (n \longrightarrow \infty), \tag{10}$$

where $\tilde{x}_n$ is of minimum norm and satisfies $T_\lambda \tilde{x}_n = y_n$. Multiplication by an α shows that to αy_n there corresponds $\alpha \tilde{x}_n$ as a solution of minimum norm. Hence we may assume that $\|\tilde{x}_n\| = 1$, without loss of generality. Then (10) implies $\|y_n\| \longrightarrow 0$. Since T is compact and $(\tilde{x}_n)$ is bounded, $(T\tilde{x}_n)$ has a convergent subsequence, say, $T\tilde{x}_{n_j} \longrightarrow v_0$ or, if we write $v_0 = \lambda \tilde{x}_0$ for convenience,

$$T\tilde{x}_{n_j} \longrightarrow \lambda \tilde{x}_0 \qquad (j \longrightarrow \infty). \tag{11}$$

Since $y_n = T_\lambda \tilde{x}_n = T\tilde{x}_n - \lambda \tilde{x}_n$, we have $\lambda \tilde{x}_n = T\tilde{x}_n - y_n$. Using (11) and $\|y_n\| \longrightarrow 0$, we thus obtain, noting that $\lambda \neq 0$,

$$\tilde{x}_{n_j} = \frac{1}{\lambda}(T\tilde{x}_{n_j} - y_{n_j}) \longrightarrow \tilde{x}_0. \tag{12}$$

From this, since T is continuous, we have

$$T\tilde{x}_{n_j} \longrightarrow T\tilde{x}_0.$$

Hence $T\tilde{x}_0 = \lambda \tilde{x}_0$ by (11). Since $T_\lambda \tilde{x}_n = y_n$, we see that $x = \tilde{x}_n - \tilde{x}_0$ satisfies $T_\lambda x = y_n$. Since $\tilde{x}_n$ is of minimum norm,

$$\|x\| = \|\tilde{x}_n - \tilde{x}_0\| \geqq \|\tilde{x}_n\| = 1.$$

But this contradicts the convergence in (12). Hence (10) cannot hold but the sequence of quotients in (10) must be bounded; that is, we must have

$$c = \sup_{y \in T_\lambda(X)} \frac{\|\tilde{x}\|}{\|y\|} < \infty,$$

where $y = T_\lambda \tilde{x}$. This implies (8). ∎

Using this lemma, we can now give a characterization of the solvability of (3) similar to that for (1) given in Theorem 8.5-1:

8.5-3 Theorem (Solutions of (3)). *Let $T: X \longrightarrow X$ be a compact linear operator on a normed space X and let $\lambda \neq 0$. Then (3) has a solution f if and only if g is such that*

$$g(x) = 0 \tag{13}$$

for all $x \in X$ which satisfy (2).

Hence if (2) has only the trivial solution $x = 0$, then (3) with any given $g \in X'$ is solvable.

Proof. (*a*) If (3) has a solution f and x satisfies (2), then (13) holds because

$$g(x) = (T^\times f)(x) - \lambda f(x) = f(Tx - \lambda x) = f(0) = 0.$$

(*b*) Conversely, assume that g satisfies (13) for every solution x of (2). We show that then (3) has a solution f. We consider any $x \in X$ and set $y = T_\lambda x$. Then $y \in T_\lambda(X)$. We may define a functional f_0 on $T_\lambda(X)$ by

$$f_0(y) = f_0(T_\lambda x) = g(x).$$

This definition is unambiguous because if $T_\lambda x_1 = T_\lambda x_2$, then $T_\lambda(x_1 - x_2) = 0$, so that $x_1 - x_2$ is a solution of (2); hence $g(x_1 - x_2) = 0$ by assumption, that is, $g(x_1) = g(x_2)$.

f_0 is linear since T_λ and g are linear. We show that f_0 is bounded. Lemma 8.5-2 implies that for every $y \in T_\lambda(X)$ at least one of the corresponding x's satisfies

$$\|x\| \leqq c\|y\| \qquad (y = T_\lambda x)$$

where c does not depend on y. Boundedness of f_0 can now be seen from

$$|f_0(y)| = |g(x)| \leqq \|g\| \, \|x\| \leqq c\|g\| \, \|y\| = \tilde{c}\|y\|,$$

where $\tilde{c} = c\|g\|$. By the Hahn-Banach theorem 4.3-2 the functional f_0 has an extension f on X, which is a bounded linear functional defined on all of X. By the definition of f_0,

$$f(Tx - \lambda x) = f(T_\lambda x) = f_0(T_\lambda x) = g(x).$$

On the left, by the definition of the adjoint operator we have for all $x \in X$,

$$f(Tx - \lambda x) = f(Tx) - \lambda f(x) = (T^\times f)(x) - \lambda f(x).$$

Together with the preceding formula this shows that f is a solution of (3) and proves the first statement of the theorem. The second statement follows readily from the first one. ∎

The next section is an immediate continuation of the present one. A joint problem set for both sections is included at the end of the next section.

8.6 Further Theorems of Fredholm Type

In this section we present further results about the solvability of the operator equations

(1) $$Tx - \lambda x = y \qquad (y \text{ given})$$

(2) $$Tx - \lambda x = 0$$

(3) $$T^\times f - \lambda f = g \qquad (g \text{ given})$$

(4) $$T^\times f - \lambda f = 0.$$

The assumptions are literally the same as in the previous section, namely, $T\colon X \longrightarrow X$ is a compact linear operator on a normed space X, the operator $T^\times$ is the adjoint operator of T and $\lambda \neq 0$ is fixed.

The theory in the last section and in the present one generalizes Fredholm's famous theory of integral equations, as was mentioned before.

The main results of the last section characterize the solvability of (1) in terms of (4) (Theorem 8.5-1) and that of (3) in terms of (2) (Theorem 8.5-3). It is natural to look for similar relations between (1) and (2) and for relations between (3) and (4):

8.6-1 Theorem (Solutions of (1)). *Let $T: X \longrightarrow X$ be a compact linear operator on a normed space X and let $\lambda \neq 0$. Then:*

(a) *Equation* (1) *has a solution x for every $y \in X$ if and only if the homogeneous equation* (2) *has only the trivial solution $x = 0$. In this case the solution of* (1) *is unique, and T_λ has a bounded inverse.*

(b) *Equation* (3) *has a solution f for every $g \in X'$ if and only if* (4) *has only the trivial solution $f = 0$. In this case the solution of* (3) *is unique.*

Proof. **(a)** We prove that if for every $y \in X$ the equation (1) is solvable, then $x = 0$ is the only solution of (2).

Otherwise (2) would have a solution $x_1 \neq 0$. Since (1) with any y is solvable, $T_\lambda x = y = x_1$ has a solution $x = x_2$, that is, $T_\lambda x_2 = x_1$. For the same reason there is an x_3 such that $T_\lambda x_3 = x_2$, etc. By substitution we thus have for every $k = 2, 3, \cdots$

$$0 \neq x_1 = T_\lambda x_2 = T_\lambda^{\ 2} x_3 = \cdots = T_\lambda^{k-1} x_k$$

and

$$0 = T_\lambda x_1 = T_\lambda^{\ k} x_k.$$

Hence $x_k \in \mathcal{N}(T_\lambda^{\ k})$ but $x_k \notin \mathcal{N}(T_\lambda^{k-1})$. This means that the null space $\mathcal{N}(T_\lambda^{k-1})$ is a proper subspace of $\mathcal{N}(T_\lambda^{\ k})$ for all k. But this contradicts Theorem 8.4-3. Hence $x = 0$ must be the only solution of (2).

Conversely, suppose that $x = 0$ is the only solution of (2). Then (3) with any g is solvable, by Theorem 8.5-3. Now $T^\times$ is compact (cf. 8.2-5), so that we can apply the first part of the present proof to $T^\times$ and conclude that $f = 0$ must be the only solution of (4). Solvability of (1) with any y now follows from Theorem 8.5-1.

Uniqueness results from the fact that the difference of two solutions of (1) is a solution of (2). Clearly, such a unique solution

$x = T_\lambda^{-1}y$ is the solution of minimum norm, and boundedness of T_λ^{-1} follows from Lemma 8.5-2:

$$\|x\| = \|T_\lambda^{-1}y\| \leq c\|y\|.$$

(b) is a consequence of (*a*) and the fact that $T^\times$ is compact (cf. 8.2-5). ∎

The homogeneous equations (2) and (4) are also related: we shall see that they have the same number of linearly independent solutions. For the proof of this fact we shall need the existence of certain sets in X and X' which are related by (5), below, and are often called a *biorthogonal system.*

8.6-2 Lemma (Biorthogonal system). *Given a linearly independent set $\{f_1, \cdots, f_m\}$ in the dual space X' of a normed space X, there are elements $z_1, \cdots, z_m$ in X such that*

$$f_j(z_k) = \delta_{jk} = \begin{cases} 0 & (j \neq k) \\ 1 & (j = k) \end{cases} \qquad (j, k = 1, \cdots, m). \tag{5}$$

Proof. Since it does not matter how we order the f_j's, it suffices to prove that there exists a z_m such that

$$f_m(z_m) = 1, \qquad f_j(z_m) = 0 \qquad (j = 1, \cdots, m-1). \tag{6}$$

When $m = 1$ this holds since $f_1 \neq 0$ by the linear independence, so that $f_1(x_0) \neq 0$ for some x_0, and $z_1 = \alpha x_0$ with $\alpha = 1/f_1(x_0)$ gives $f_1(z_1) = 1$.

We now let $m > 1$ and make the induction hypothesis that the lemma holds for $m-1$, that is, X contains elements $z_1, \cdots, z_{m-1}$ such that

$$f_k(z_k) = 1, \qquad f_n(z_k) = 0, \quad n \neq k \qquad (k, n = 1, \cdots, m-1). \tag{7}$$

We consider the set

$$M = \{x \in X \mid f_1(x) = 0, \cdots, f_{m-1}(x) = 0\}$$

and show that M contains a $\tilde{z}_m$ such that $f_m(\tilde{z}_m) = \beta \neq 0$, which clearly yields (6), where $z_m = \beta^{-1}\tilde{z}_m$.

Otherwise $f_m(x)=0$ for all $x \in M$. We now take any $x \in X$ and set

$$\tilde{x} = x - \sum_{j=1}^{m-1} f_j(x) z_j. \tag{8}$$

Then, by (7), for $k \leqq m-1$,

$$f_k(\tilde{x}) = f_k(x) - \sum_{j=1}^{m-1} f_j(x) f_k(z_j) = f_k(x) - f_k(x) = 0.$$

This shows that $\tilde{x} \in M$, so that $f_m(\tilde{x})=0$ by our assumption. From (8),

$$\begin{aligned} f_m(x) &= f_m\left(\tilde{x} + \sum f_j(x) z_j\right) \\ &= f_m(\tilde{x}) + \sum f_j(x) f_m(z_j) \\ &= \sum \alpha_j f_j(x) \qquad\qquad [\alpha_j = f_m(z_j)] \end{aligned}$$

(sum over j from 1 to $m-1$). Since $x \in X$ was arbitrary, this is a representation of f_m as a linear combination of $f_1, \cdots, f_{m-1}$ and contradicts the linear independence of $\{f_1, \cdots, f_m\}$. Hence $f_m(x)=0$ for all $x \in M$ is impossible, so that M must contain a z_m such that (6) holds, and the lemma is proved. ∎

Using this lemma, we can now show that $\dim \mathcal{N}(T_\lambda) = \dim \mathcal{N}(T_\lambda^\times)$, where $T_\lambda^\times = (T-\lambda I)^\times = T^\times - \lambda I$. In terms of the operator equations under consideration, this equality of the dimensions means the following.

8.6-3 Theorem (Null spaces of T_λ and $T_\lambda^\times$). *Let $T: X \longrightarrow X$ be a compact linear operator on a normed space X, and let $\lambda \neq 0$. Then equations (2) and (4) have the same number of linearly independent solutions.*

Proof. T and $T^\times$ are compact (cf. 8.2-5), so that $\mathcal{N}(T_\lambda)$ and $\mathcal{N}(T_\lambda^\times)$ are finite dimensional by 8.3-3, say

$$\dim \mathcal{N}(T_\lambda) = n, \qquad \dim \mathcal{N}(T_\lambda^\times) = m.$$

We subdivide the proof into three parts (*a*), (*b*), (*c*) which are devoted to

(*a*) the case $m=n=0$ and a preparation for $m>0$, $n>0$,
(*b*) the proof that $n<m$ is impossible,
(*c*) the proof that $n>m$ is impossible.

The details are as follows.

(a) If $n=0$, the only solution of (2) is $x=0$. Then (3) with any given g is solvable; cf. 8.5-3. By 8.6-1(*b*) this implies that $f=0$ is the only solution of (4). Hence $m=0$. Similarly, from $m=0$ it follows that $n=0$.

Suppose $m>0$ and $n>0$. Let $\{x_1, \cdots, x_n\}$ be a basis for $\mathcal{N}(T_\lambda)$. Clearly, $x_1 \notin Y_1 = \operatorname{span}\{x_2, \cdots, x_n\}$. By Lemma 4.6-7 there is a $\tilde{g}_1 \in X'$ which is zero everywhere on Y_1 and $\tilde{g}_1(x_1)=\delta$, where $\delta>0$ is the distance from x_1 to Y_1. Hence $g_1=\delta^{-1}\tilde{g}_1$ satisfies $g_1(x_1)=1$ and $g_1(x_2)=0, \cdots, g_1(x_n)=0$. Similarly, there is a g_2 such that $g_2(x_2)=1$ and $g_2(x_j)=0$ for $j\neq 2$, etc. Hence X' contains $g_1, \cdots, g_n$ such that

$$g_k(x_j)=\delta_{jk}=\begin{cases}0 & \text{if} \quad j\neq k\\ 1 & \text{if} \quad j=k\end{cases} \qquad (j, k=1, \cdots, n). \tag{9}$$

Similarly, if $\{f_1, \cdots, f_m\}$ is a basis for $\mathcal{N}(T_\lambda^{\times})$, then by Lemma 8.6-2 there are elements $z_1, \cdots, z_m$ of X such that

$$f_j(z_k)=\delta_{jk} \qquad (j, k=1, \cdots, m). \tag{10}$$

(b) We show that $n<m$ is impossible. Let $n<m$ and define $S\colon X \longrightarrow X$ by

$$Sx = Tx + \sum_{j=1}^{n} g_j(x) z_j. \tag{11}$$

S is compact since $g_j(x)z_j$ represents a compact linear operator by 8.1-4(*a*) and a sum of compact operators is compact. Let us prove that

$$\text{(12)} \qquad \text{(a)} \quad S_\lambda x_0 = Sx_0 - \lambda x_0 = 0 \quad \Longrightarrow \quad \text{(b)} \quad x_0=0.$$

By (12a) we have $f_k(S_\lambda x_0)=f_k(0)=0$ for $k=1, \cdots, m$. Hence by (11) and (10) we obtain

$$0 = f_k(S_\lambda x_0) = f_k\left(T_\lambda x_0 + \sum_{j=1}^{n} g_j(x_0) z_j\right)$$

$$(13) \qquad = f_k(T_\lambda x_0) + \sum_{j=1}^{n} g_j(x_0) f_k(z_j)$$

$$= (T_\lambda^\times f_k)(x_0) + g_k(x_0).$$

Since $f_k \in \mathcal{N}(T_\lambda^\times)$, we have $T_\lambda^\times f_k = 0$. Hence (13) yields

$$(14) \qquad g_k(x_0) = 0, \qquad k = 1, \cdots, m.$$

This implies $Sx_0 = Tx_0$ by (11) and $T_\lambda x_0 = S_\lambda x_0 = 0$ by (12a). Hence $x_0 \in \mathcal{N}(T_\lambda)$. Since $\{x_1, \cdots, x_n\}$ is a basis for $\mathcal{N}(T_\lambda)$,

$$x_0 = \sum_{j=1}^{n} \alpha_j x_j,$$

where the α_j's are suitable scalars. Applying g_k and using (14) and (9), we have

$$0 = g_k(x_0) = \sum_{j=1}^{n} \alpha_j g_k(x_j) = \alpha_k \qquad (k = 1, \cdots, n).$$

Hence $x_0 = 0$. This proves (12). Theorem 8.6-1(*a*) now implies that $S_\lambda x = y$ with any y is solvable. We choose $y = z_{n+1}$. Let $x = v$ be a corresponding solution, that is, $S_\lambda v = z_{n+1}$. As in (13) we calculate, using (10) and (11):

$$1 = f_{n+1}(z_{n+1}) = f_{n+1}(S_\lambda v)$$

$$= f_{n+1}\left(T_\lambda v + \sum_{j=1}^{n} g_j(v) z_j\right)$$

$$= (T_\lambda^\times f_{n+1})(v) + \sum_{j=1}^{n} g_j(v) f_{n+1}(z_j)$$

$$= (T_\lambda^\times f_{n+1})(v).$$

Since we assumed $n<m$, we have $n+1 \leqq m$ and $f_{n+1} \in \mathcal{N}(T_\lambda{}^\times)$. Hence $T_\lambda{}^\times f_{n+1}=0$. This gives a contradiction in the previous equation and shows that $n<m$ is impossible.

(c) We show that $n>m$ is also impossible. The reasoning is similar to that in part (*b*). Let $n>m$ and define $\tilde{S}\colon X' \longrightarrow X'$ by

$$\tilde{S}f = T^\times f + \sum_{j=1}^{m} f(z_j) g_j. \tag{15}$$

$T^\times$ is compact by 8.2-5, and $\tilde{S}$ is compact since $f(z_j)g_j$ represents a compact linear operator by 8.1-4(*a*). Instead of (12) we now prove:

$$\text{(a)} \quad \tilde{S}_\lambda f_0 = \tilde{S} f_0 - \lambda f_0 = 0 \qquad \Longrightarrow \qquad \text{(b)} \quad f_0 = 0. \tag{16}$$

Using (16a), then (15) with $f=f_0$, then the definition of the adjoint operator and finally (9), we obtain for each $k=1, \cdots, m$

$$\begin{aligned} 0 = (\tilde{S}_\lambda f_0)(x_k) &= (T_\lambda{}^\times f_0)(x_k) + \sum_{j=1}^{m} f_0(z_j) g_j(x_k) \\ &= f_0(T_\lambda x_k) + f_0(z_k). \end{aligned} \tag{17}$$

Our assumption $m<n$ implies that $x_k \in \mathcal{N}(T_\lambda)$ for $k=1, \cdots, m$. [Remember that $\{x_1, \cdots, x_n\}$ is a basis for $\mathcal{N}(T_\lambda)$.] Hence $f_0(T_\lambda x_k) = f_0(0)=0$, so that (17) yields

$$f_0(z_k)=0 \qquad (k=1, \cdots, m). \tag{18}$$

Consequently, $\tilde{S}f_0 = T^\times f_0$ by (15). From this and (16a) it follows that $T_\lambda{}^\times f_0 = \tilde{S}_\lambda f_0 = 0$. Hence $f_0 \in \mathcal{N}(T_\lambda{}^\times)$. Since $\{f_1, \cdots, f_m\}$ is a basis for $\mathcal{N}(T_\lambda{}^\times)$,

$$f_0 = \sum_{j=1}^{m} \beta_j f_j,$$

where the β_j's are suitable scalars. Using (18) and (10), we thus obtain for each $k=1, \cdots, m$

$$0 = f_0(z_k) = \sum_{j=1}^{m} \beta_j f_j(z_k) = \beta_k.$$

Hence $f_0=0$. This proves (16). Theorem 8.6-1(*b*) now implies that $\tilde{S}_\lambda f=g$ with any g is solvable. We choose $g=g_{m+1}$. Let $f=h$ be a corresponding solution, that is, $\tilde{S}_\lambda h=g_{m+1}$. Using (9), (15) and again (9), we obtain

$$\begin{aligned} 1=g_{m+1}(x_{m+1}) &= (\tilde{S}_\lambda h)(x_{m+1}) \\ &= (T_\lambda^\times h)(x_{m+1})+\sum_{j=1}^{m} h(z_j)g_j(z_{m+1}) \\ &= (T_\lambda^\times h)(x_{m+1}) \\ &= h(T_\lambda x_{m+1}). \end{aligned}$$

Our assumption $m<n$ implies $m+1\leqq n$, so that $x_{m+1}\in \mathcal{N}(T_\lambda)$. Hence $h(T_\lambda x_{m+1})=h(0)=0$. This gives a contradiction in the previous equation and shows that $m<n$ is impossible. Since $n<m$ is impossible, too, by part (*b*), we conclude that we must have $n=m$. ∎

Theorem 8.6-1(*a*) can also be used to show that one of our earlier results for Banach spaces (Theorem 8.4-4) even holds for general normed spaces:

8.6-4 Theorem (Eigenvalues). *Let $T: X \longrightarrow X$ be a compact linear operator on a normed space X. Then if T has nonzero spectral values, every one of them must be an eigenvalue of T.*

Proof. If the resolvent $R_\lambda=T_\lambda^{-1}$ does not exist, $\lambda\in\sigma_p(T)$ by definition. Let $\lambda\neq 0$ and assume that $R_\lambda=T_\lambda^{-1}$ exists. Then $T_\lambda x=0$ implies $x=0$ by 2.6-10. This means that (2) has only the trivial solution. Theorem 8.6-1(*a*) now shows that (1) with any y is solvable, that is, R_λ is defined on all of X and is bounded. Hence $\lambda\in\rho(T)$. ∎

Problems

1. Show that the functional f_0 in the proof of Theorem 8.5-3 is linear.

2. What does Theorem 8.5-1 imply in the case of a system of n linear algebraic equations in n unknowns?

3. Consider a system $Ax = y$ consisting of n linear equations in n unknowns. Assuming that the system has a solution x, show that y must satisfy a condition of the form (5), Sec. 8.5.

4. A system $Ax = y$ of n linear equations in n unknowns has a (unique) solution for any given y if and only if $Ax = 0$ has only the trivial solution $x = 0$. How does this follow from one of our present theorems?

5. A system $Ax = y$ consisting of n linear equations in n unknowns has a solution x if and only if the *augmented matrix*

$$\begin{bmatrix} \alpha_{11} & \alpha_{12} & \cdots & \alpha_{1n} & \eta_1 \\ \alpha_{21} & \alpha_{22} & \cdots & \alpha_{2n} & \eta_2 \\ \cdot & \cdot & \cdots & \cdot & \cdot \\ \alpha_{n1} & \alpha_{n2} & \cdots & \alpha_{nn} & \eta_n \end{bmatrix}$$

has the same rank as the coefficient matrix $A = (\alpha_{jk})$; here $y = (\eta_j)$. Obtain this familiar criterion from Theorem 8.5-1.

6. If (2) has a solution $x \neq 0$ and (1) is solvable, show that the solution of (1) cannot be unique; similarly, if (4) has a solution $f \neq 0$ and (3) is solvable, show that the solution of (3) cannot be unique.

7. Show that the first statement in Theorem 8.6-1 may also be formulated as follows. $R_\lambda(T)\colon X \longrightarrow X$ with $\lambda \neq 0$ exists if and only if $Tx = \lambda x$ implies $x = 0$.

8. Two sequences $(z_1, z_2, \cdots)$ in a normed space X and $(f_1, f_2, \cdots)$ in the dual space X' are called a *biorthogonal system* if they satisfy $f_j(z_k) = \delta_{jk}$, where $j, k = 1, 2, \cdots$; cf. (5). Given (z_k), show that there is a sequence (f_j) in X' such that (z_k), (f_j) is a biorthogonal system if and only if $z_m \notin \bar{A}_m$ for all $m \in \mathbf{N}$, where

$$A_m = \text{span}\,\{z_k \mid k = 1, 2, \cdots; k \neq m\}.$$

9. Show that for a finite biorthogonal system, as defined in the text, the condition stated in Prob. 8 is automatically satisfied.

10. If two sets $\{z_1, \cdots, z_n\}$ and $\{y_1, \cdots, y_n\}$ in an inner product space satisfy $\langle z_k, y_j \rangle = \delta_{kj}$, show that each one of them is linearly independent.

11. What form does a biorthogonal system assume in a Hilbert space?

12. State and prove Lemma 8.6-2 if X is a Hilbert space H.

13. What does Theorem 8.6-3 imply in the case of a system of n linear equation in n unknowns?

14. If a linear operator $T: X \longrightarrow Y$ on a normed space X has a finite dimensional range $\mathcal{R}(T) = T(X)$, show that T has a representation of the form

$$Tx = f_1(x)y_1 + \cdots + f_n(x)y_n$$

where $\{y_1, \cdots, y_n\}$ and $\{f_1, \cdots, f_n\}$ are linearly independent sets in Y and X' (the dual space of X), respectively.

15. We may wonder what would happen to our present theorems if $\lambda = 0$, so that (1) and (2) would be

$$Tx = y \qquad \text{and} \qquad Tx = 0,$$

respectively. For these equations, Theorem 8.6-1 may no longer hold. To see this, consider $T: C[0, \pi] \longrightarrow C[0, \pi]$ defined by

$$Tx(s) = \int_0^\pi k(s, t)x(t)\, dt, \qquad k(s, t) = \sum_{n=1}^\infty \frac{1}{n^2} \sin ns \sin nt.$$

8.7 Fredholm Alternative

The preceding two sections were devoted to the study of the behavior of compact linear operators with respect to solvability of operator equations. The results obtained suggest the following concept.

8.7-1 Definition (Fredholm alternative). A bounded linear operator $A: X \longrightarrow X$ on a normed space X is said to satisfy the *Fredholm alternative* if A is such that either (I) or (II) holds:

(I) The nonhomogeneous equations

$$Ax = y, \qquad A^\times f = g$$

($A^\times: X' \longrightarrow X'$ the adjoint operator of A) have solutions x and f, respectively, for every given $y \in X$ and $g \in X'$, the solutions being unique. The corresponding homogeneous equations

$$Ax = 0, \qquad A^{\times}f = 0$$

have only the trivial solutions $x = 0$ and $f = 0$, respectively.

(II) The homogeneous equations

$$Ax = 0, \qquad A^{\times}f = 0$$

have the same number of linearly independent solutions

$$x_1, \cdots, x_n \quad \text{and} \quad f_1, \cdots, f_n \qquad (n \geqq 1),$$

respectively. The nonhomogeneous equations

$$Ax = y, \qquad A^{\times}f = g$$

are not solvable for all y and g, respectively; they have a solution if and only if y and g are such that

$$f_k(y) = 0, \qquad g(x_k) = 0$$

$(k = 1, \cdots, n)$, respectively. ∎

We see that this concept can be used for summarizing the results of the last two sections:

8.7-2 Theorem (Fredholm alternative). *Let $T: X \longrightarrow X$ be a compact linear operator on a normed space X, and let $\lambda \neq 0$. Then $T_\lambda = T - \lambda I$ satisfies the Fredholm alternative.*

This statement of exclusive alternatives is particularly important for applications because, instead of showing the existence of a solution directly, it is often simpler to prove that the homogeneous equation has only the trivial solution.

We have already mentioned (in Sec. 8.5) that Riesz's theory of compact linear operators was suggested by Fredholm's theory of integral equations of the second kind

$$x(s) - \mu \int_a^b k(s, t)x(t)\, dt = \tilde{y}(s) \tag{1}$$

and generalizes Fredholm's famous results, which antedate the development of the theory of Hilbert and Banach spaces. We shall give a brief introduction to the application of the theory of compact linear operators to equations of the form (1).

Setting $\mu = 1/\lambda$ and $\tilde{y}(s) = -y(s)/\lambda$, where $\lambda \neq 0$, we have

$$Tx - \lambda x = y \qquad (\lambda \neq 0); \tag{2}$$

with T defined by

$$(Tx)(s) = \int_a^b k(s, t)x(t)\, dt. \tag{3}$$

Consequences of the general theory can now be interpreted for (2). In fact, we have

8.7-3 Theorem (Fredholm alternative for integral equations). *If k in (1) is such that $T\colon X \longrightarrow X$ in (2) and (3) is a compact linear operator on a normed space X, then the Fredholm alternative holds for T_λ; thus either (1) has a unique solution for all $\tilde{y} \in X$ or the homogeneous equation corresponding to (1) has finitely many linearly independent nontrivial solutions (that is, solutions $x \neq 0$).*

Suppose that T in (2) is compact (conditions for this will be given below). Then if λ is in the resolvent set $\rho(T)$ of T, the resolvent $R_\lambda(T) = (T - \lambda I)^{-1}$ exists, is defined on all of X, is bounded [cf. 8.6-1(*a*)] and gives the unique solution

$$x = R_\lambda(T)y$$

of (2) for every $y \in X$. Since $R_\lambda(T)$ is linear, $R_\lambda(T)0 = 0$, which implies that the homogeneous equation $Tx - \lambda x = 0$ has only the trivial solution $x = 0$. Hence $\lambda \in \rho(T)$ yields case (I) of the Fredholm alternative.

Let $|\lambda| > \|T\|$. Assuming that X is a complex Banach space, we have $\lambda \in \rho(T)$ by Theorem 7.3-4. Furthermore, (9) in Sec. 7.3 yields

$$R_\lambda(T) = -\lambda^{-1}(I + \lambda^{-1}T + \lambda^{-2}T^2 + \cdots). \tag{4}$$

Consequently, for the solution $x = R_\lambda(T)y$ we have the representation

$$x = -\frac{1}{\lambda}\left(y + \frac{1}{\lambda}\, Ty + \frac{1}{\lambda^2}\, T^2 y + \cdots\right), \tag{5}$$

which is called a **Neumann series.**

Case (II) of the Fredholm alternative is obtained if we take a nonzero $\lambda \in \sigma(T)$ (if such a λ exists), where $\sigma(T)$ is the spectrum of T. Theorem 8.6-4 implies that λ is an eigenvalue. The dimension of the corresponding eigenspace is finite, by Theorem 8.3-3, and equal to the dimension of the corresponding eigenspace of $T_\lambda^\times$, by Theorem 8.6-3.

In connection with Theorem 8.7-3, two spaces of particular interest are

$$X = L^2[a, b] \qquad \text{and} \qquad X = C[a, b].$$

To apply the theorem, one needs conditions for the kernel k in (1) which are sufficient for T to be compact.

If $X = L^2[a, b]$, such a condition is that k be in $L^2(J \times J)$, where $J = [a, b]$. The proof would require measure theory and lies outside the scope of this book.

In the case $X = C[a, b]$, where $[a, b]$ is compact, continuity of k will imply compactness of T.

We shall obtain this result by the use of a standard theorem (8.7-4, below) as follows.

A sequence (x_n) in $C[a, b]$ is said to be **equicontinuous** if for every $\varepsilon > 0$ there is a $\delta > 0$, depending only on ε, such that for all x_n and all $s_1, s_2 \in [a, b]$ satisfying $|s_1 - s_2| < \delta$ we have

$$|x_n(s_1) - x_n(s_2)| < \varepsilon.$$

We see from this definition that each x_n is uniformly continuous on $[a, b]$ and δ does not depend on n either.

8.7-4 Ascoli's Theorem (Equicontinuous sequence). *A bounded equicontinuous sequence (x_n) in $C[a, b]$ has a subsequence which converges (in the norm on $C[a, b]$).*

For a proof, see e.g. E. J. McShane (1944), p. 336. Using this theorem, we shall obtain the desired result in the case of $X = C[a, b]$, as follows.

8.7-5 Theorem (Compact integral operator). *Let $J = [a, b]$ be any compact interval and suppose that k is continuous on $J \times J$. Then the operator $T\colon X \longrightarrow X$ defined by (3), where $X = C[a, b]$, is a compact linear operator.*

Proof. T is linear. Boundedness of T follows from

$$\|Tx\| = \max_{s \in J} \left| \int_a^b k(s, t)x(t)\, dt \right| \leqq \|x\| \max_{s \in J} \int_a^b |k(s, t)|\, dt,$$

which is of the form $\|Tx\| \leqq \tilde{c}\, \|x\|$. Let (x_n) be any bounded sequence in X, say, $\|x_n\| \leqq c$ for all n. Let $y_n = Tx_n$. Then $\|y_n\| \leqq \|T\|\, \|x_n\|$. Hence (y_n) is also bounded. We show that (y_n) is equicontinuous. Since the kernel k is continuous on $J \times J$ by assumption and $J \times J$ is compact, k is uniformly continuous on $J \times J$. Hence, given any $\varepsilon > 0$, there is a $\delta > 0$ such that for all $t \in J$ and all $s_1, s_2 \in J$ satisfying $|s_1 - s_2| < \delta$ we have

$$|k(s_1, t) - k(s_2, t)| < \frac{\varepsilon}{(b-a)c}.$$

Consequently, for s_1, s_2 as before and every n we obtain

$$\begin{aligned} |y_n(s_1) - y_n(s_2)| &= \left| \int_a^b [k(s_1, t) - k(s_2, t)] x_n(t)\, dt \right| \\ &< (b-a) \frac{\varepsilon}{(b-a)c}\, c = \varepsilon. \end{aligned}$$

This proves equicontinuity of (y_n). Ascoli's theorem implies that (y_n) has a convergent subsequence. Since (x_n) was an arbitrary bounded sequence and $y_n = Tx_n$, compactness of T follows from Theorem 8.1-3. ∎

Problems

1. Formulate the Fredholm alternative for a system of n linear algebraic equations in n unknowns.

2. Show directly that (1) may not always have a solution.

3. Give an example of a discontinuous kernel k in (3) such that for a continuous x, the image Tx is discontinuous. Comment.

4. **(Neumann series)** Show that in terms of $\mu = 1/\lambda$ and $\tilde{y}$ in (1) the Neumann series (5) takes the form

$$x = \tilde{y} + \mu T \tilde{y} + \mu^2 T^2 \tilde{y} + \cdots.$$

Consider (1) in $C[a, b]$. If k is continuous on $[a, b]\times[a, b]$, so that, say, $|k(s, t)|<M$, and if $|\mu|<1/M(b-a)$, show that the Neumann series converges.

5. Solve the following integral equation. Compare the result with the Neumann series in Prob. 4.

$$x(s)-\mu\int_0^1 x(t)\,dt=1.$$

Find all solutions of the corresponding homogeneous equation. Comment.

6. Solve the following equation and show that if $|\mu|<1/k_0(b-a)$, the corresponding Neumann series (cf. Prob. 4) converges.

$$x(s)-\mu\int_a^b k_0 x(t)\,dt=\tilde{y}(s).$$

Here, k_0 is a constant.

7. (Iterated kernel, resolvent kernel) Show that in the Neumann series in Prob. 4 we can write

$$(T^n\tilde{y})(s)=\int_a^b k_{(n)}(s, t)\tilde{y}(t)\,dt$$

where $n=2, 3, \cdots$ and the *iterated kernel* k_n is given by

$$k_{(n)}(s, t)=\int_a^b\cdots\int_a^b k(s, t_1)k(t_1, t_2)\cdots k(t_{n-1}, t)\,dt_1\cdots dt_{n-1},$$

so that the Neumann series in Prob. 4 can be written

$$x(s)=\tilde{y}(s)+\mu\int_a^b k(s, t)\tilde{y}(t)\,dt+\mu^2\int_a^b k_{(2)}(s, t)\tilde{y}(t)\,dt+\cdots.$$

Show that this representation can be written as an integral equation

$$x(s)=\tilde{y}(s)+\mu\int_a^b \tilde{k}(s, t, \mu)\tilde{y}(t)\,dt$$

where the *resolvent kernel*[4] $\tilde{k}$ is given by

$$\tilde{k}(s, t, \mu) = \sum_{j=0}^{\infty} \mu^j k_{(j+1)}(s, t) \qquad [k_{(1)} = k].$$

Show that the iterated kernels satisfy

$$k_n(s, t) = \int_a^b k_{(n-1)}(s, u) k(u, t)\, du.$$

8. Determine the resolvent kernel for (1) with $a = 0$, $b = \pi$ and

$$k(s, t) = a_1 \sin s \sin 2t + a_2 \sin 2s \sin 3t.$$

9. Using the Neumann series in Prob. 4, solve (1), where $a = 0$, $b = 2\pi$ and

$$k(s, t) = \sum_{n=1}^{\infty} a_n \sin ns \cos nt \qquad \left(\sum_{n=1}^{\infty} |a_n| < \infty\right).$$

10. In (1), let $k(s, t) = s(1+t)$ and $a = 0$, $b = 1$. Determine the eigenvalues and eigenfunctions. Solve the equation when $\lambda = 1/\mu$ is not an eigenvalue.

11. In (1), let $k(s, t) = 2e^{s+t}$ and $\tilde{y}(s) = e^s$, and $a = 0$, $b = 1$. Find the eigenvalues and eigenfunctions.

12. Solve

$$x(s) - \mu \int_0^{2\pi} \sin s \cos t\, x(t)\, dt = \tilde{y}(s).$$

13. Ascoli's theorem 8.7-4 is concerned with subsequences which converge in the norm of $C[a, b]$. We know that this is uniform convergence on $[a, b]$; cf. 1.5-6. Illustrate with an example that a sequence of continuous functions may be convergent at each point of $[a, b]$ but may fail to contain any subsequence which converges uniformly on $[a, b]$.

14. (Degenerate kernel) A kernel k of the form

$$k(s, t) = \sum_{j=1}^{n} a_j(s) b_j(t)$$

[4] The resolvent kernel must not be confused with the resolvent of an operator (cf. Sec. 7.2).

is called a *degenerate kernel.* Here we may assume each of the two sets $\{a_1, \cdots, a_n\}$ and $\{b_1, \cdots, b_n\}$ to be linearly independent on $[a, b]$, since otherwise the number of terms of the sum could be reduced. If an equation (1) with such a kernel has a solution x, show that it must be of the form

$$x(s) = \tilde{y}(s) + \mu \sum_{j=1}^{n} c_j a_j(s), \qquad c_j = \int_a^b b_j(t)x(t)\,dt$$

and the unknown constants must satisfy

$$c_j - \mu \sum_{k=1}^{n} \alpha_{jk} c_k = y_j, \qquad \alpha_{jk} = \int_a^b b_j(t)a_k(t)\,dt$$

where

$$y_j = \int_a^b b_j(t)\tilde{y}(t)\,dt, \qquad j = 1, \cdots, n.$$

15. Consider

$$x(s) - \mu \int_0^1 (s+t)x(t)\,dt = \tilde{y}(s).$$

(*a*) Assuming $\mu^2 + 12\mu - 12 \neq 0$ and using Prob. 14, solve the equation. (*b*) Find the eigenvalues and eigenfunctions.

CHAPTER 9

SPECTRAL THEORY OF BOUNDED SELF-ADJOINT LINEAR OPERATORS

Bounded self-adjoint linear operators on Hilbert spaces were defined and considered in Sec. 3.10. This chapter is devoted to their spectral theory, which is very highly developed since these operators are particularly important in applications.

Important concepts, brief orientation about main content

In Secs. 9.1 and 9.2 we discuss spectral properties of bounded self-adjoint linear operators. In Secs. 9.3 to 9.8 we develop material which is of interest in itself and will be needed for establishing "spectral representations" of those operators in Secs. 9.9 and 9.10.

The spectrum of a bounded self-adjoint linear operator T is real (cf. 9.1-3) and lies in the interval $[m, M]$, where m and M are the infimum and supremum of $\langle Tx, x\rangle$, taken over all x of norm 1 (cf. 9.2-1), and eigenvectors corresponding to different eigenvalues are orthogonal (cf. 9.1-1).

Such an operator T can be represented by an integral ("spectral theorems" 9.9-1 and 9.10-1) which involves a spectral family $\mathscr{E}$ associated with T (cf. 9.8-3), where a *spectral family* or *decomposition of unity* (cf. 9.7-1) is a family of projection operators having certain properties. We remember that projection operators were used in Sec. 3.3. However, for the present purpose we need various general properties of these operators (Secs. 9.5, 9.6) as well as the concepts of a *positive operator* (Sec. 9.3) and its *square roots* (Sec. 9.4).

In Sec. 9.11 we characterize the behavior of the spectral family of a bounded self-adjoint linear operator at points of the resolvent set, at the eigenvalues and at points of the continuous spectrum. (The residual spectrum of that operator is empty; cf. 9.2-4.)

9.1 Spectral Properties of Bounded Self-Adjoint Linear Operators

Throughout this chapter we shall consider bounded linear operators which are defined on a complex Hilbert space H and map H into itself. Furthermore, these operators will be self-adjoint. It will take us only a minute to recall two relevant definitions from Chap. 3:

Let $T: H \longrightarrow H$ be a bounded linear operator on a complex Hilbert space H. Then the **Hilbert-adjoint operator** $T^*: H \longrightarrow H$ is defined to be the operator satisfying

$$\langle Tx, y\rangle = \langle x, T^*y\rangle \qquad \text{for all } x, y \in H.$$

This is Def. 3.9-1 (with $H_1 = H_2 = H$), and we know from 3.9-2 that T^* exists as a bounded linear operator of norm $\|T^*\| = \|T\|$ on H and is unique.

Furthermore, T is said to be **self-adjoint** or **Hermitian**[1] if

$$T = T^*.$$

This is Def. 3.10-1. Then $\langle Tx, y\rangle = \langle x, T^*y\rangle$ becomes

$$\langle Tx, y\rangle = \langle x, Ty\rangle. \tag{1}$$

If T is self-adjoint, then $\langle Tx, x\rangle$ is real for all $x \in H$. Conversely, this condition implies self-adjointness of T since H is complex. Cf. 3.10-3.

This was our brief review, and we now begin our investigation of the spectrum of a bounded self-adjoint linear operator. We shall see that such a spectrum has several general properties of practical importance.

A bounded self-adjoint linear operator T may not have eigenvalues (cf. Prob. 9), but if T has eigenvalues, the following basic facts can readily be established.

[1] A distinction is made between these two terms in the theory of *unbounded* operators.—We mention that boundedness of T follows automatically from (1) and our assumption that T is defined on all of H. (Cf. Prob. 10.)

9.1-1 Theorem (Eigenvalues, eigenvectors). *Let $T\colon H \longrightarrow H$ be a bounded self-adjoint linear operator on a complex Hilbert space H. Then:*

(a) *All the eigenvalues of T (if they exist) are real.*

(b) *Eigenvectors corresponding to (numerically) different eigenvalues of T are orthogonal.*

Proof. **(a)** Let λ be any eigenvalue of T and x a corresponding eigenvector. Then $x \neq 0$ and $Tx = \lambda x$. Using the self-adjointness of T, we obtain

$$\lambda\langle x, x\rangle = \langle \lambda x, x\rangle = \langle Tx, x\rangle$$

$$= \langle x, Tx\rangle = \langle x, \lambda x\rangle = \bar{\lambda}\langle x, x\rangle.$$

Here $\langle x, x\rangle = \|x\|^2 \neq 0$ since $x \neq 0$, and division by $\langle x, x\rangle$ gives $\lambda = \bar{\lambda}$. Hence λ is real.

(b) Let λ and μ be eigenvalues of T, and let x and y be corresponding eigenvectors. Then $Tx = \lambda x$ and $Ty = \mu y$. Since T is self-adjoint and μ is real,

$$\lambda\langle x, y\rangle = \langle \lambda x, y\rangle = \langle Tx, y\rangle$$

$$= \langle x, Ty\rangle = \langle x, \mu y\rangle = \mu\langle x, y\rangle.$$

Since $\lambda \neq \mu$, we must have $\langle x, y\rangle = 0$, which means orthogonality of x and y. ∎

Even the whole spectrum of a bounded self-adjoint operator T is real. This remarkable result (Theorem 9.1-3, below) will be obtained from the following characterization of the resolvent set $\rho(T)$ of T.

9.1-2 Theorem (Resolvent set). *Let $T\colon H \longrightarrow H$ be a bounded self-adjoint linear operator on a complex Hilbert space H. Then a number λ belongs to the resolvent set $\rho(T)$ of T if and only if there exists a $c > 0$ such that for every $x \in H$,*

$$\|T_\lambda x\| \geqq c\,\|x\| \qquad (T_\lambda = T - \lambda I). \tag{2}$$

Proof. **(a)** If $\lambda \in \rho(T)$, then $R_\lambda = T_\lambda^{-1}\colon H \longrightarrow H$ exists and is bounded (cf. 7.2-3), say, $\|R_\lambda\| = k$, where $k > 0$ since $R_\lambda \neq 0$. Now $I = R_\lambda T_\lambda$, so that for every $x \in H$ we have

$$\|x\| = \|R_\lambda T_\lambda x\| \leqq \|R_\lambda\| \, \|T_\lambda x\| = k \, \|T_\lambda x\|.$$

This gives $\|T_\lambda x\| \geqq c \, \|x\|$, where $c = 1/k$.

(b) Conversely, suppose (2) with a $c > 0$ holds for all $x \in H$. Let us prove that then:

(α) $T_\lambda\colon H \longrightarrow T_\lambda(H)$ is bijective;

(β) $T_\lambda(H)$ is dense in H;

(γ) $T_\lambda(H)$ is closed in H,

so that $T_\lambda(H) = H$ and $R_\lambda = T_\lambda^{-1}$ is bounded by the bounded inverse theorem 4.12-2.

(α) We must show that $T_\lambda x_1 = T_\lambda x_2$ implies $x_1 = x_2$. But this follows from (2) since T_λ is linear and

$$0 = \|T_\lambda x_1 - T_\lambda x_2\| = \|T_\lambda(x_1 - x_2)\| \geqq c \, \|x_1 - x_2\|;$$

hence $\|x_1 - x_2\| = 0$ because $c > 0$, and $x_1 = x_2$. Since x_1, x_2 were arbitrary, this shows that $T_\lambda\colon H \longrightarrow T_\lambda(H)$ is bijective.

(β) We show that $x_0 \perp \overline{T_\lambda(H)}$ implies $x_0 = 0$, so that $\overline{T_\lambda(H)} = H$ by the projection theorem 3.3-4. Let $x_0 \perp \overline{T_\lambda(H)}$. Then $x_0 \perp T_\lambda(H)$. Hence for all $x \in H$ we have

$$0 = \langle T_\lambda x, x_0 \rangle = \langle Tx, x_0 \rangle - \lambda \langle x, x_0 \rangle.$$

Since T is self-adjoint, we thus obtain

$$\langle x, Tx_0 \rangle = \langle Tx, x_0 \rangle = \langle x, \bar{\lambda} x_0 \rangle,$$

so that $Tx_0 = \bar{\lambda} x_0$ by 3.8-2. A solution is $x_0 = 0$, and $x_0 \neq 0$ is impossible since it would mean that $\bar{\lambda}$ is an eigenvalue of T, so that $\bar{\lambda} = \lambda$ by 9.1-1 and $Tx_0 - \lambda x_0 = T_\lambda x_0 = 0$, and (2) would yield the contradiction

$$0 = \|T_\lambda x_0\| \geqq c \, \|x_0\| > 0$$

since $c > 0$. The result is $x_0 = 0$. Thus $\overline{T_\lambda(H)}^\perp = \{0\}$ because x_0 was any vector orthogonal to $T_\lambda(H)$. Hence $\overline{T_\lambda(H)} = H$ by 3.3-4, that is, $T_\lambda(H)$ is dense in H.

(γ) We finally prove that $y \in \overline{T_\lambda(H)}$ implies $y \in T_\lambda(H)$, so that $T_\lambda(H)$ is closed, and $T_\lambda(H) = H$ by (β). Let $y \in \overline{T_\lambda(H)}$. By 1.4-6($a$) there is a sequence (y_n) in $T_\lambda(H)$ which converges to y. Since $y_n \in T_\lambda(H)$, we have $y_n = T_\lambda x_n$ for some $x_n \in H$. By (2),

$$\|x_n - x_m\| \leqq \frac{1}{c} \|T_\lambda(x_n - x_m)\| = \frac{1}{c} \|y_n - y_m\|.$$

Hence (x_n) is Cauchy since (y_n) converges. H is complete, so that (x_n) converges, say, $x_n \longrightarrow x$. Since T is continuous, so is T_λ, and $y_n = T_\lambda x_n \longrightarrow T_\lambda x$ by 1.4-8. By definition, $T_\lambda x \in T_\lambda(H)$. Since the limit is unique, $T_\lambda x = y$, so that $y \in T_\lambda(H)$. Hence $T_\lambda(H)$ is closed because $y \in \overline{T_\lambda(H)}$ was arbitrary. We thus have $T_\lambda(H) = H$ by (β). This means that $R_\lambda = T_\lambda^{-1}$ is defined on all of H, and is bounded, as follows from the bounded inverse theorem 4.12-2 or directly from (2). Hence $\lambda \in \rho(T)$. ∎

From this theorem we shall now immediately obtain the basic

9.1-3 Theorem (Spectrum). *The spectrum $\sigma(T)$ of a bounded self-adjoint linear operator $T\colon H \longrightarrow H$ on a complex Hilbert space H is real.*

Proof. Using Theorem 9.1-2, we show that a $\lambda = \alpha + i\beta$ (α, β real) with $\beta \neq 0$ must belong to $\rho(T)$, so that $\sigma(T) \subset \mathbf{R}$.

For every $x \neq 0$ in H we have

$$\langle T_\lambda x, x\rangle = \langle Tx, x\rangle - \lambda\langle x, x\rangle$$

and, since $\langle x, x\rangle$ and $\langle Tx, x\rangle$ are real (cf. 3.10-3),

$$\overline{\langle T_\lambda x, x\rangle} = \langle Tx, x\rangle - \bar{\lambda}\langle x, x\rangle.$$

Here, $\bar{\lambda} = \alpha - i\beta$. By subtraction,

$$\overline{\langle T_\lambda x, x\rangle} - \langle T_\lambda x, x\rangle = (\lambda - \bar{\lambda})\langle x, x\rangle = 2i\beta \|x\|^2.$$

The left side is $-2i \operatorname{Im} \langle T_\lambda x, x\rangle$, where Im denotes the imaginary part. The latter cannot exceed the absolute value, so that, dividing by 2, taking absolute values and applying the Schwarz inequality, we obtain

$$|\beta| \, \|x\|^2 = |\mathrm{Im} \langle T_\lambda x, x\rangle| \leqq |\langle T_\lambda x, x\rangle| \leqq \|T_\lambda x\| \, \|x\|.$$

Division by $\|x\| \neq 0$ gives $|\beta| \, \|x\| \leqq \|T_\lambda x\|$. If $\beta \neq 0$, then $\lambda \in \rho(T)$ by Theorem 9.1-2. Hence for $\lambda \in \sigma(T)$ we must have $\beta = 0$, that is, λ is real. ∎

Problems

1. It was mentioned in the text that for a self-adjoint linear operator T the inner product $\langle Tx, x\rangle$ is real. What does this imply for matrices? What familiar theorem on matrices does Theorem 9.1-1 include as a special case?

2. If in the finite dimensional case, a self-adjoint linear operator T is represented by a diagonal matrix, show that the matrix must be real. What is the spectrum of T?

3. Show that in Theorem 9.1-2, boundedness of R_λ also follows from (2).

4. Illustrate Theorem 9.1-2 by an operator whose spectrum consists of a single given value λ_0. What is the largest c in this case?

5. Let $T: H \longrightarrow H$ and $W: H \longrightarrow H$ be bounded linear operators on a complex Hilbert space H. If T is self-adjoint, show that $S = W^*TW$ is self-adjoint.

6. Let $T: l^2 \longrightarrow l^2$ be defined by $(\xi_1, \xi_2, \cdots) \longmapsto (0, 0, \xi_1, \xi_2, \cdots)$. Is T bounded? Self-adjoint? Find $S: l^2 \longrightarrow l^2$ such that $T = S^2$.

7. Let $T: l^2 \longrightarrow l^2$ be defined by $y = (\eta_j) = Tx$, $x = (\xi_j)$, $\eta_j = \lambda_j \xi_j$, where (λ_j) is a bounded sequence on $\mathbf{R}$, and $a = \inf \lambda_j$, $b = \sup \lambda_j$. Show that each λ_j is an eigenvalue of T. Under what condition will $\sigma(T) \supset [a, b]$?

8. Using Theorem 9.1-2, prove that the spectrum of the operator T in Prob. 7 is the closure of the set of eigenvalues.

9. From 2.2-7 and 3.1-5 we know that the Hilbert space $L^2[0, 1]$ is the completion of the inner product space X of all continuous functions on $[0, 1]$ with inner product defined by

$$\langle x, y\rangle = \int_0^1 x(t)\overline{y(t)} \, dt.$$

Show that $T: L^2[0, 1] \longrightarrow L^2[0, 1]$ defined by

$$y(t) = Tx(t) = tx(t)$$

is a bounded self-adjoint linear operator without eigenvalues.

10. It is interesting to note that a linear operator T which is defined on all of a complex Hilbert space H and satisfies (1) for all $x, y \in H$ must be bounded (so that our assumption of boundedness at the beginning of the text would not be necessary). Prove this fact.

9.2 Further Spectral Properties of Bounded Self-Adjoint Linear Operators

The spectrum $\sigma(T)$ of a bounded self-adjoint linear operator T is real. This important fact was proved in the last section. We shall now see that the spectrum of such an operator can be characterized in more detail since it has a number of general properties which are mathematically interesting and practically important. It is clear by 7.3-4 that $\sigma(T)$ must be compact, but in the present case we can say more:

9.2-1 Theorem (Spectrum). *The spectrum $\sigma(T)$ of a bounded self-adjoint linear operator $T: H \longrightarrow H$ on a complex Hilbert space H lies in the closed interval $[m, M]$ on the real axis, where*

(1) $$m = \inf_{\|x\|=1} \langle Tx, x\rangle, \qquad M = \sup_{\|x\|=1} \langle Tx, x\rangle.$$

Proof. $\sigma(T)$ lies on the real axis (by 9.1-3). We show that any real $\lambda = M + c$ with $c > 0$ belongs to the resolvent set $\rho(T)$. For every $x \neq 0$ and $v = \|x\|^{-1}x$ we have $x = \|x\|\, v$ and

$$\langle Tx, x\rangle = \|x\|^2\langle Tv, v\rangle \leqq \|x\|^2 \sup_{\|\tilde{v}\|=1} \langle T\tilde{v}, \tilde{v}\rangle = \langle x, x\rangle M.$$

Hence $-\langle Tx, x\rangle \geqq -\langle x, x\rangle M$, and by the Schwarz inequality we obtain

$$\begin{aligned}\|T_\lambda x\|\,\|x\| &\geqq -\langle T_\lambda x, x\rangle = -\langle Tx, x\rangle + \lambda\langle x, x\rangle \\ &\geqq (-M+\lambda)\langle x, x\rangle \\ &= c\,\|x\|^2,\end{aligned}$$

where $c = \lambda - M > 0$ by assumption. Division by $\|x\|$ yields the inequality $\|T_\lambda x\| \geqq c\|x\|$. Hence $\lambda \in \rho(T)$ by 9.1-2. For a real $\lambda < m$ the idea of proof is the same. ∎

m and M in (1) are related to the norm of T in an interesting fashion:

9.2-2 Theorem (Norm). *For any bounded self-adjoint linear operator T on a complex Hilbert space H we have* [cf. (1)]

$$\|T\| = \max(|m|, |M|) = \sup_{\|x\|=1} |\langle Tx, x\rangle|. \tag{2}$$

Proof. By the Schwarz inequality,

$$\sup_{\|x\|=1} |\langle Tx, x\rangle| \leqq \sup_{\|x\|=1} \|Tx\| \, \|x\| = \|T\|,$$

that is, $K \leqq \|T\|$, where K denotes the expression on the left. We show that $\|T\| \leqq K$. If $Tz = 0$ for all z of norm 1, then $T = 0$ (why?) and we are done. Otherwise for any z of norm 1 such that $Tz \neq 0$, we set $v = \|Tz\|^{1/2} z$ and $w = \|Tz\|^{-1/2} Tz$. Then $\|v\|^2 = \|w\|^2 = \|Tz\|$. We now set $y_1 = v + w$ and $y_2 = v - w$. Then by straightforward calculation, since a number of terms drop out and T is self-adjoint,

$$\begin{aligned} \langle Ty_1, y_1\rangle - \langle Ty_2, y_2\rangle &= 2(\langle Tv, w\rangle + \langle Tw, v\rangle) \\ &= 2(\langle Tz, Tz\rangle + \langle T^2 z, z\rangle) \\ &= 4\,\|Tz\|^2. \end{aligned} \tag{3}$$

Now for every $y \neq 0$ and $x = \|y\|^{-1} y$ we have $y = \|y\|\, x$ and

$$|\langle Ty, y\rangle| = \|y\|^2\, |\langle Tx, x\rangle| \leqq \|y\|^2 \sup_{\|\tilde{x}\|=1} |\langle T\tilde{x}, \tilde{x}\rangle| = K\,\|y\|^2,$$

so that by the triangle inequality and straightforward calculation we obtain

$$\begin{aligned} |\langle Ty_1, y_1\rangle - \langle Ty_2, y_2\rangle| &\leqq |\langle Ty_1, y_1\rangle| + |\langle Ty_2, y_2\rangle| \\ &\leqq K(\|y_1\|^2 + \|y_2\|^2) \\ &= 2K(\|v\|^2 + \|w\|^2) \\ &= 4K\,\|Tz\|. \end{aligned}$$

From this and (3) we see that $4\|Tz\|^2 \leqq 4K\|Tz\|$. Hence $\|Tz\| \leqq K$. Taking the supremum over all z of norm 1, we obtain $\|T\| \leqq K$. Together with $K \leqq \|T\|$ this yields (2). ∎

Actually, the bounds for $\sigma(T)$ in Theorem 9.2-1 cannot be tightened. This may be seen from

9.2-3 Theorem (m and M as spectral values). *Let H and T be as in Theorem* 9.2-1 *and $H \neq \{0\}$. Then m and M defined in* (1) *are spectral values of T.*

Proof. We show that $M \in \sigma(T)$. By the spectral mapping theorem 7.4-2 the spectrum of $T + kI$ (k a real constant) is obtained from that of T by a translation, and

$$M \in \sigma(T) \qquad \Longleftrightarrow \qquad M + k \in \sigma(T + kI).$$

Hence we may assume $0 \leqq m \leqq M$ without loss of generality. Then by the previous theorem we have

$$M = \sup_{\|x\|=1} \langle Tx, x \rangle = \|T\|.$$

By the definition of a supremum there is a sequence (x_n) such that

$$\|x_n\| = 1, \qquad \langle Tx_n, x_n \rangle = M - \delta_n, \qquad \delta_n \geqq 0, \qquad \delta_n \longrightarrow 0.$$

Then $\|Tx_n\| \leqq \|T\| \|x_n\| = \|T\| = M$, and since T is self-adjoint,

$$\begin{aligned} \|Tx_n - Mx_n\|^2 &= \langle Tx_n - Mx_n, Tx_n - Mx_n \rangle \\ &= \|Tx_n\|^2 - 2M\langle Tx_n, x_n \rangle + M^2 \|x_n\|^2 \\ &\leqq M^2 - 2M(M - \delta_n) + M^2 = 2M\delta_n \qquad \longrightarrow \qquad 0. \end{aligned}$$

Hence there is no positive c such that

$$\|T_M x_n\| = \|Tx_n - Mx_n\| \geqq c = c\,\|x_n\| \qquad (\|x_n\| = 1).$$

Theorem 9.1-2 now shows that $\lambda = M$ cannot belong to the resolvent set of T. Hence $M \in \sigma(T)$. For $\lambda = m$ the proof is similar. ∎

The subdivision of the spectrum of a linear operator into the point spectrum and another part seems natural since that "other part" is absent in finite dimensional spaces, as is well known from matrix theory (cf. Sec. 7.1). A similar justification can now be given for the subdivision of that "other part" into the continuous and residual spectrum since the latter is absent for the large and important class of self-adjoint linear operators:

9.2-4 Theorem (Residual spectrum). *The residual spectrum $\sigma_r(T)$ of a bounded self-adjoint linear operator $T\colon H \longrightarrow H$ on a complex Hilbert space H is empty.*

Proof. We show that the assumption $\sigma_r(T) \neq \varnothing$ leads to a contradiction. Let $\lambda \in \sigma_r(T)$. By the definition of $\sigma_r(T)$, the inverse of T_λ exists but its domain $\mathfrak{D}(T_\lambda^{-1})$ is not dense in H. Hence, by the projection theorem 3.3-4 there is a $y \neq 0$ in H which is orthogonal to $\mathfrak{D}(T_\lambda^{-1})$. But $\mathfrak{D}(T_\lambda^{-1})$ is the range of T_λ, hence

$$\langle T_\lambda x, y\rangle = 0 \qquad \text{for all } x \in H.$$

Since λ is real (cf. 9.1-3) and T is self-adjoint, we thus obtain $\langle x, T_\lambda y\rangle = 0$ for all x. Taking $x = T_\lambda y$, we get $\|T_\lambda y\|^2 = 0$, so that

$$T_\lambda y = Ty - \lambda y = 0.$$

Since $y \neq 0$, this shows that λ is an eigenvalue of T. But this contradicts $\lambda \in \sigma_r(T)$. Hence $\sigma_r(T) \neq \varnothing$ is impossible and $\sigma_r(T) = \varnothing$ follows. ∎

Problems

1. Give a proof of Theorem 9.2-1 for a $\lambda < m$.

2. What theorem about the eigenvalues of a Hermitian matrix $A = (\alpha_{jk})$ do we obtain from Theorem 9.2-1?

3. Find m and M (cf. Theorem 9.2-1) if T is the projection operator of a Hilbert space H onto a proper subspace $Y \neq \{0\}$ of H.

4. Prove that $m \in \sigma(T)$ in Theorem 9.2-3.

5. Show that the spectrum of a bounded self-adjoint linear operator on a complex Hilbert space $H \neq \{0\}$ is not empty, using one of the theorems of the present section.

6. Show that a compact self-adjoint linear operator $T: H \longrightarrow H$ on a complex Hilbert space $H \neq \{0\}$ has at least one eigenvalue.

7. Consider the operator $T: l^2 \longrightarrow l^2$ defined by $y = Tx$, where $x = (\xi_j)$, $y = (\eta_j)$ and $\eta_j = \xi_j / j$, $j = 1, 2, \cdots$. It was shown in 8.1-6 that T is compact. Find the spectrum of T. Show that $0 \in \sigma_c(T)$ and, actually, $\sigma_c(T) = \{0\}$. (For compact operators with $0 \in \sigma_p(T)$ or $0 \in \sigma_r(T)$, see Probs. 4, 5, Sec. 8.4.)

8. (Rayleigh quotient) Show that (1) can be written

$$\sigma(T) \subset \left[\inf_{x \neq 0} q(x), \sup_{x \neq 0} q(x) \right], \qquad \text{where} \qquad q(x) = \frac{\langle Tx, x \rangle}{\langle x, x \rangle}$$

is called the *Rayleigh quotient.*

9. If $\lambda_1 \geqq \lambda_2 \geqq \cdots \geqq \lambda_n$ are the eigenvalues of a Hermitian matrix A, show that

$$\lambda_1 = \max_{x \neq 0} q(x), \qquad \lambda_n = \min_{x \neq 0} q(x) \qquad \text{where} \qquad q(x) = \frac{\bar{x}^{\mathsf{T}} A x}{\bar{x}^{\mathsf{T}} x}.$$

Show that, furthermore,

$$\lambda_j = \max_{\substack{x \in Y_j \\ x \neq 0}} q(x) \qquad\qquad j = 2, 3, \cdots, n$$

where Y_j is the subspace of $\mathbf{C}^n$ consisting of all vectors that are orthogonal to the eigenvectors corresponding to $\lambda_1, \cdots, \lambda_{j-1}$.

10. Show that a real symmetric square matrix $A = (\alpha_{jk})$ with positive elements has a positive eigenvalue. (It can be shown that the statement holds even without the assumption of symmetry; this is part of a famous theorem by Perron and Frobenius; cf. F. R. Gantmacher (1960), vol. II, p. 53.)

9.3 Positive Operators

If T is self-adjoint, $\langle Tx, x \rangle$ is real, as we know from Sec. 9.1. Hence we may consider the set of all bounded self-adjoint linear operators on a complex Hilbert space H and introduce on this set a *partial ordering* $\leqq$

(cf. Sec. 4.1) by defining

$$(1)\qquad T_1 \leqq T_2 \qquad \text{if and only if} \qquad \langle T_1 x, x\rangle \leqq \langle T_2 x, x\rangle$$

for all $x \in H$. Instead of $T_1 \leqq T_2$ we also write $T_2 \geqq T_1$.

An important particular case is the following one. A bounded self-adjoint linear operator $T\colon H \longrightarrow H$ is said to be **positive,** written

$$(2)\qquad T \geqq 0, \qquad \text{if and only if} \qquad \langle Tx, x\rangle \geqq 0$$

for all $x \in H$. Instead of $T \geqq 0$ we also write $0 \leqq T$. Actually, such an operator should be called "nonnegative", but "positive" is the usual term.

Note the simple relation between (1) and (2), namely,

$$T_1 \leqq T_2 \qquad \Longleftrightarrow \qquad 0 \leqq T_2 - T_1,$$

that is, (1) holds if and only if $T_2 - T_1$ is positive.

We devote this section and the next one to positive operators and their square roots, a topic which is interesting in itself and, moreover, will serve as a tool in the derivation of a spectral representation for bounded self-adjoint linear operators later in this chapter.

The sum of positive operators is positive.

This is obvious from the definition. Let us turn to products. From 3.10-4 we know that a product (composite) of bounded self-adjoint linear operators is self-adjoint if and only if the operators commute, and we shall now see that in this case, positivity is preserved, too. This fact will be used quite often in our further work.

9.3-1 Theorem (Product of positive operators). *If two bounded self-adjoint linear operators S and T on a Hilbert space H are positive and commute ($ST = TS$), then their product ST is positive.*

Proof. We must show that $\langle STx, x\rangle \geqq 0$ for all $x \in H$. If $S = 0$, this holds. Let $S \neq 0$. We proceed in two steps (a) and (b):

(a) We consider

$$(3)\qquad S_1 = \frac{1}{\|S\|}\, S, \qquad S_{n+1} = S_n - S_n{}^2 \qquad (n = 1, 2, \cdots)$$

and prove by induction that

$$0 \leqq S_n \leqq I. \tag{4}$$

(b) We prove that $\langle STx, x\rangle \geqq 0$ for all $x \in H$.
The details are as follows.

(a) For $n=1$ the inequality (4) holds. Indeed, the assumption $0 \leqq S$ implies $0 \leqq S_1$, and $S_1 \leqq I$ is obtained by an application of the Schwarz inequality and the inequality $\|Sx\| \leqq \|S\| \|x\|$:

$$\langle S_1 x, x\rangle = \frac{1}{\|S\|}\langle Sx, x\rangle \leqq \frac{1}{\|S\|}\|Sx\| \|x\| \leqq \|x\|^2 = \langle Ix, x\rangle.$$

Suppose (4) holds for an $n = k$, that is,

$$0 \leqq S_k \leqq I, \qquad \text{thus} \qquad 0 \leqq I - S_k \leqq I.$$

Then, since S_k is self-adjoint, for every $x \in H$ and $y = S_k x$ we obtain

$$\begin{aligned}\langle S_k^{\,2}(I-S_k)x, x\rangle &= \langle (I-S_k)S_k x, S_k x\rangle \\ &= \langle (I-S_k)y, y\rangle \geqq 0.\end{aligned}$$

By definition this proves

$$S_k^{\,2}(I-S_k) \geqq 0.$$

Similarly,

$$S_k(I-S_k)^2 \geqq 0.$$

By addition and simplification,

$$0 \leqq S_k^{\,2}(I-S_k) + S_k(I-S_k)^2 = S_k - S_k^{\,2} = S_{k+1}.$$

Hence $0 \leqq S_{k+1}$. And $S_{k+1} \leqq I$ follows from $S_k^{\,2} \geqq 0$ and $I - S_k \geqq 0$ by addition; indeed,

$$0 \leqq I - S_k + S_k^{\,2} = I - S_{k+1}.$$

This completes the inductive proof of (4).

(b) We now show that $\langle STx, x\rangle \geqq 0$ for all $x \in H$. From (3) we obtain successively

$$\begin{aligned} S_1 &= S_1^2 + S_2 \\ &= S_1^2 + S_2^2 + S_3 \\ &\cdots\cdots \\ &= S_1^2 + S_2^2 + \cdots + S_n^2 + S_{n+1}. \end{aligned}$$

Since $S_{n+1} \geqq 0$, this implies

$$S_1^2 + \cdots + S_n^2 = S_1 - S_{n+1} \leqq S_1. \tag{5}$$

By the definition of $\leqq$ and the self-adjointness of S_j this means that

$$\sum_{j=1}^{n} \|S_j x\|^2 = \sum_{j=1}^{n} \langle S_j x, S_j x\rangle = \sum_{j=1}^{n} \langle S_j^2 x, x\rangle \leqq \langle S_1 x, x\rangle.$$

Since n is arbitrary, the infinite series $\|S_1 x\|^2 + \|S_2 x\|^2 + \cdots$ converges. Hence $\|S_n x\| \longrightarrow 0$ and $S_n x \longrightarrow 0$. By (5),

$$\left(\sum_{j=1}^{n} S_j^2\right) x = (S_1 - S_{n+1}) x \qquad \longrightarrow \qquad S_1 x \qquad (n \longrightarrow \infty). \tag{6}$$

All the S_j's commute with T since they are sums and products of $S_1 = \|S\|^{-1} S$, and S and T commute. Using $S = \|S\|\, S_1$, formula (6), $T \geqq 0$ and the continuity of the inner product, we thus obtain for every $x \in H$ and $y_j = S_j x$,

$$\begin{aligned} \langle STx, x\rangle &= \|S\| \langle TS_1 x, x\rangle \\ &= \|S\| \lim_{n\to\infty} \sum_{j=1}^{n} \langle TS_j^2 x, x\rangle \\ &= \|S\| \lim_{n\to\infty} \sum_{j=1}^{n} \langle Ty_j, y_j\rangle \geqq 0, \end{aligned}$$

that is, $\langle STx, x\rangle \geqq 0$. ∎

The partial order relation defined by (2) also suggests the following concept.

9.3-2 Definition (Monotone sequence). A *monotone sequence* (T_n) of self-adjoint linear operators T_n on a Hilbert space H is a sequence (T_n) which is either *monotone increasing*, that is,

$$T_1 \leqq T_2 \leqq T_3 \leqq \cdots$$

or *monotone decreasing*, that is,

$$T_1 \geqq T_2 \geqq T_3 \geqq \cdots. \qquad \blacksquare$$

A monotone increasing sequence has the following remarkable property. (A similar theorem holds for a monotone decreasing sequence.)

9.3-3 Theorem (Monotone sequence). *Let (T_n) be a sequence of bounded self-adjoint linear operators on a complex Hilbert space H such that*

$$T_1 \leqq T_2 \leqq \cdots \leqq T_n \leqq \cdots \leqq K \tag{7}$$

where K is a bounded self-adjoint linear operator on H. Suppose that any T_j commutes with K and with every T_m. Then (T_n) is strongly operator convergent ($T_nx \longrightarrow Tx$ for all $x \in H$) and the limit operator T is linear, bounded and self-adjoint and satisfies $T \leqq K$.

Proof. We consider $S_n = K - T_n$ and prove:

(*a*) The sequence $(\langle S_n^2 x, x\rangle)$ converges for every $x \in H$.

(*b*) $T_n x \longrightarrow Tx$, where T is linear and self-adjoint, and is bounded by the uniform boundedness theorem.

The details are as follows.

(a) Clearly, S_n is self-adjoint. We have

$$S_m^2 - S_n S_m = (S_m - S_n)S_m = (T_n - T_m)(K - T_m).$$

Let $m < n$. Then $T_n - T_m$ and $K - T_m$ are positive by (7). Since these operators commute, their product is positive by Theorem 9.3-1. Hence on the left, $S_m^2 - S_n S_m \geqq 0$, that is, $S_m^2 \geqq S_n S_m$ for $m < n$. Similarly,

$$S_n S_m - S_n^2 = S_n(S_m - S_n) = (K - T_n)(T_n - T_m) \geqq 0,$$

so that $S_nS_m \geqq S_n^2$. Together,

$$S_m^2 \geqq S_nS_m \geqq S_n^2 \qquad (m<n).$$

By definition, using the self-adjointness of S_n, we thus have

$$(8) \qquad \langle S_m^2x, x\rangle \geqq \langle S_nS_mx, x\rangle \geqq \langle S_n^2x, x\rangle = \langle S_nx, S_nx\rangle = \|S_nx\|^2 \geqq 0.$$

This shows that $(\langle S_n^2x, x\rangle)$ with fixed x is a monotone decreasing sequence of nonnegative numbers. Hence it converges.

(b) We show that (T_nx) converges. By assumption, every T_n commutes with every T_m and with K. Hence the S_j's all commute. These operators are self-adjoint. Since $-2\langle S_mS_nx, x\rangle \leqq -2\langle S_n^2x, x\rangle$ by (8), where $m<n$, we thus obtain

$$\begin{aligned} \|S_mx - S_nx\|^2 &= \langle (S_m - S_n)x, (S_m - S_n)x\rangle \\ &= \langle (S_m - S_n)^2x, x\rangle \\ &= \langle S_m^2x, x\rangle - 2\langle S_mS_nx, x\rangle + \langle S_n^2x, x\rangle \\ &\leqq \langle S_m^2x, x\rangle - \langle S_n^2x, x\rangle. \end{aligned}$$

From this and the convergence proved in part (*a*) we see that (S_nx) is Cauchy. It converges since H is complete. Now $T_n = K - S_n$. Hence (T_nx) also converges. Clearly the limit depends on x, so that we can write $T_nx \longrightarrow Tx$ for every $x \in H$. Hence this defines an operator $T: H \longrightarrow H$ which is linear. T is self-adjoint because T_n is self-adjoint and the inner product is continuous. Since (T_nx) converges, it is bounded for every $x \in H$. The uniform boundedness theorem 4.7-3 now implies that T is bounded. Finally, $T \leqq K$ follows from $T_n \leqq K$. ∎

Problems

1. Let S and T be bounded self-adjoint linear operators on a complex Hilbert space. If $S \leqq T$ and $S \geqq T$, show that $S = T$.

2. Show that (1) defines a partial order relation (cf. Def. 4.1-1) on the set of all bounded self-adjoint linear operators on a complex Hilbert space

H, and for any such operator T,

$$T_1 \leqq T_2 \implies T_1 + T \leqq T_2 + T$$

$$T_1 \leqq T_2 \implies \alpha T_1 \leqq \alpha T_2 \qquad (\alpha \geqq 0).$$

3. Let A, B, T be bounded self-adjoint linear operators on a complex Hilbert space H. If $T \geqq 0$ and commutes with A and B, show that

$$A \leqq B \qquad \text{implies} \qquad AT \leqq BT.$$

4. If $T\colon H \longrightarrow H$ is a bounded linear operator on a complex Hilbert space H, show that TT^* and T^*T are self-adjoint and positive. Show that the spectra of TT^* and T^*T are real and cannot contain negative values. What are the consequences of the second statement for a square matrix A?

5. Show that a bounded self-adjoint linear operator T on a complex Hilbert space H is positive if and only if its spectrum consists of nonnegative real values only. What does this imply for a matrix?

6. Let $T\colon H \longrightarrow H$ and $W\colon H \longrightarrow H$ be bounded linear operators on a complex Hilbert space H and $S = W^*TW$. Show that if T is self-adjoint and positive, so is S.

7. Let T_1 and T_2 be bounded self-adjoint linear operators on a complex Hilbert space H and suppose that $T_1T_2 = T_2T_1$ and $T_2 \geqq 0$. Show that then $T_1^{\,2}T_2$ is self-adjoint and positive.

8. Let S and T be bounded self-adjoint linear operators on a Hilbert space H. If $S \geqq 0$, show that $TST \geqq 0$.

9. Show that if $T \geqq 0$, then $(I+T)^{-1}$ exists.

10. Let T be any bounded linear operator on a complex Hilbert space. Show that the inverse of $I + T^*T$ exists.

11. Show that an illustrative example for Theorem 9.3-3 is given by the sequence (P_n), where P_n is the projection of l^2 onto the subspace consisting of all sequences $x = (\xi_j) \in l^2$ such that $\xi_j = 0$ for all $j > n$.

12. If T is a bounded self-adjoint linear operator on a complex Hilbert space H, show that T^2 is positive. What does this imply for a matrix?

13. If T is a bounded self-adjoint linear operator on a complex Hilbert space H, show that the spectrum of T^2 cannot contain a negative value. What theorem on matrices does this generalize?

14. If $T: H \longrightarrow H$ and $S: H \longrightarrow H$ are bounded linear operators and T is compact and $S^*S \leqq T^*T$, show that S is compact.

15. Let $T: H \longrightarrow H$ be a bounded linear operator on an infinite dimensional complex Hilbert space H. If there is a $c>0$ such that we have $\|Tx\| \geqq c\,\|x\|$ for all $x \in H$, show that T is not compact.

9.4 Square Roots of a Positive Operator

If T is self-adjoint, then T^2 is positive since $\langle T^2x, x\rangle = \langle Tx, Tx\rangle \geqq 0$. We consider the converse problem: given a positive operator T, find a self-adjoint A such that $A^2 = T$. This suggests the following concept, which will be basic in connection with spectral representations.

9.4-1 Definition (Positive square root). Let $T: H \longrightarrow H$ be a positive bounded self-adjoint linear operator on a complex Hilbert space H. Then a bounded self-adjoint linear operator A is called a *square root* of T if

$$A^2 = T. \qquad (1)$$

If, in addition, $A \geqq 0$, then A is called a *positive square root* of T and is denoted by

$$A = T^{1/2}.$$ ∎

$T^{1/2}$ exists and is unique:

9.4-2 Theorem (Positive square root). *Every positive bounded self-adjoint linear operator $T: H \longrightarrow H$ on a complex Hilbert space H has a positive square root A, which is unique. This operator A commutes with every bounded linear operator on H which commutes with T.*

Proof. We proceed in three steps:

(a) We show that if the theorem holds under the additional assumption $T \leqq I$, it also holds without that assumption.

(b) We obtain the existence of the operator $A = T^{1/2}$ from $A_nx \longrightarrow Ax$, where $A_0 = 0$ and

$$A_{n+1} = A_n + \tfrac{1}{2}(T - A_n^2), \qquad n = 0, 1, \cdots, \qquad (2)$$

and we also prove the commutativity stated in the theorem.

(*c*) We prove uniqueness of the positive square root.

The details are as follows.

(a) If $T=0$, we can take $A=T^{1/2}=0$. Let $T\neq 0$. By the Schwarz inequality,

$$\langle Tx, x\rangle \leqq \|Tx\|\,\|x\| \leqq \|T\|\,\|x\|^2.$$

Dividing by $\|T\|\neq 0$ and setting $Q=(1/\|T\|)T$, we obtain

$$\langle Qx, x\rangle \leqq \|x\|^2 = \langle Ix, x\rangle;$$

that is, $Q \leqq I$. Assuming that Q has a unique positive square root $B=Q^{1/2}$, we have $B^2=Q$ and we see that a square root of $T=\|T\|\,Q$ is $\|T\|^{1/2}B$ because

$$(\|T\|^{1/2}B)^2 = \|T\|\,B^2 = \|T\|\,Q = T.$$

Also, it is not difficult to see that the uniqueness of $Q^{1/2}$ implies uniqueness of the *positive* square root of T.

Hence if we can prove the theorem under the additional assumption $T \leqq I$, we are done.

(b) *Existence.* We consider (2). Since $A_0=0$, we have $A_1=\frac{1}{2}T$, $A_2=T-\frac{1}{8}T^2$, etc. Each A_n is a polynomial in T. Hence the A_n's are self-adjoint and all commute, and they also commute with every operator that T commutes with. We now prove

$$A_n \leqq I \qquad n=0,1,\cdots; \tag{3}$$

$$A_n \leqq A_{n+1} \qquad n=0,1,\cdots; \tag{4}$$

$$A_n x \longrightarrow Ax, \qquad A=T^{1/2}; \tag{5}$$

$$ST=TS \quad \Longrightarrow \quad AS=SA \tag{6}$$

where S is a bounded linear operator on H.

Proof of (3):

We have $A_0 \leqq I$. Let $n>0$. Since $I-A_{n-1}$ is self-adjoint, $(I-A_{n-1})^2 \geqq 0$. Also $T \leqq I$ implies $I-T \geqq 0$. From this and (2) we obtain (3):

$$\begin{aligned} 0 &\leqq \tfrac{1}{2}(I-A_{n-1})^2+\tfrac{1}{2}(I-T) \\ &= I-A_{n-1}-\tfrac{1}{2}(T-A_{n-1}^2) \\ &= I-A_n. \end{aligned}$$

Proof of (4):

We use induction. (2) gives $0=A_0 \leqq A_1=\frac{1}{2}T$. We show that $A_{n-1} \leqq A_n$ for any fixed n implies $A_n \leqq A_{n+1}$. From (2) we calculate directly

$$\begin{aligned} A_{n+1}-A_n &= A_n+\tfrac{1}{2}(T-A_n{}^2)-A_{n-1}-\tfrac{1}{2}(T-A_{n-1}^2) \\ &= (A_n-A_{n-1})[I-\tfrac{1}{2}(A_n+A_{n-1})]. \end{aligned}$$

Here $A_n-A_{n-1} \geqq 0$ by hypothesis and $[\cdots] \geqq 0$ by (3). Hence $A_{n+1}-A_n \geqq 0$ by 9.3-1.

Proof of (5):

(A_n) is monotone by (4) and $A_n \leqq I$ by (3). Hence Theorem 9.3-3 implies the existence of a bounded self-adjoint linear operator A such that $A_n x \longrightarrow Ax$ for all $x \in H$. Since $(A_n x)$ converges, (2) gives

$$A_{n+1}x-A_n x=\tfrac{1}{2}(Tx-A_n{}^2x) \qquad \longrightarrow \qquad 0$$

as $n \longrightarrow \infty$. Hence $Tx-A^2x=0$ for all x, that is, $T=A^2$. Also $A \geqq 0$ because $0=A_0 \leqq A_n$ by (4), that is, $\langle A_n x, x\rangle \geqq 0$ for every $x \in H$, which implies $\langle Ax, x\rangle \geqq 0$ for every $x \in H$, by the continuity of the inner product (cf. 3.2-2).

Proof of (6):

From the line of text before (3) we know that $ST=TS$ implies $A_nS=SA_n$, that is, $A_nSx=SA_nx$ for all $x \in H$. Letting $n \longrightarrow \infty$, we obtain (6).

(c) *Uniqueness.* Let both A and B be positive square roots of T. Then $A^2=B^2=T$. Also $BT=BB^2=B^2B=TB$, so that $AB=BA$ by (6). Let $x \in H$ be arbitrary and $y=(A-B)x$. Then $\langle Ay, y\rangle \geqq 0$ and $\langle By, y\rangle \geqq 0$ because $A \geqq 0$ and $B \geqq 0$. Using $AB=BA$

and $A^2 = B^2$, we obtain

$$\langle Ay, y\rangle + \langle By, y\rangle = \langle (A+B)y, y\rangle = \langle (A^2 - B^2)x, y\rangle = 0.$$

Hence $\langle Ay, y\rangle = \langle By, y\rangle = 0$. Since $A \geqq 0$ and A is self-adjoint, it has itself a positive square root C, that is, $C^2 = A$ and C is self-adjoint. We thus obtain

$$0 = \langle Ay, y\rangle = \langle C^2 y, y\rangle = \langle Cy, Cy\rangle = \|Cy\|^2$$

and $Cy = 0$. Also $Ay = C^2 y = C(Cy) = 0$. Similarly, $By = 0$. Hence $(A - B)y = 0$. Using $y = (A - B)x$, we thus have for all $x \in H$

$$\|Ax - Bx\|^2 = \langle (A-B)^2 x, x\rangle = \langle (A-B)y, x\rangle = 0.$$

This shows that $Ax - Bx = 0$ for all $x \in H$ and proves that $A = B$. ∎

Applications of square roots will be considered in Sec. 9.8. Indeed, square roots will play a basic role in connection with the spectral representation of bounded self-adjoint linear operators.

Problems

1. Find operators $T: \mathbf{R}^2 \longrightarrow \mathbf{R}^2$ such that $T^2 = I$, the identity operator. Indicate which of the square roots is the positive square root of I.

2. Let $T: L^2[0, 1] \longrightarrow L^2[0, 1]$ be defined by $(Tx)(t) = tx(t)$. (Cf. 3.1-5.) Show that T is self-adjoint and positive and find its positive square root.

3. Let $T: l^2 \longrightarrow l^2$ be defined by $(\xi_1, \xi_2, \xi_3, \cdots) \longmapsto (0, 0, \xi_3, \xi_4, \cdots)$. Is T bounded? Self-adjoint? Positive? Find a square root of T

4. Show that for the square root in Theorem 9.4-2 we have

$$\|T^{1/2}\| = \|T\|^{1/2}.$$

5. Let $T: H \longrightarrow H$ be a bounded positive self-adjoint linear operator on a complex Hilbert space. Using the positive square root of T, show that for all $x, y \in H$,

$$|\langle Tx, y\rangle| \leqq \langle Tx, x\rangle^{1/2} \langle Ty, y\rangle^{1/2}.$$

6. It is interesting to note that the statement in Prob. 5 can also be proved without the use of $T^{1/2}$. Give such a proof (which is similar to that of the Schwarz inequality).

7. Show that in Prob. 5, for all $x \in H$,

$$\|Tx\| \leqq \|T\|^{1/2} \langle Tx, x \rangle^{1/2},$$

so that $\langle Tx, x \rangle = 0$ if and only if $Tx = 0$.

8. Let B be a nonsingular n-rowed real square matrix and $C = BB^{\mathsf{T}}$. Show that C has a nonsingular positive square root A.

9. Show that $D = A^{-1}B$ with A and B given in Prob. 8 is an orthogonal matrix. (Cf. 3.10-2.)

10. If S and T are positive bounded self-adjoint linear operators on a complex Hilbert space H and $S^2 = T^2$, show that $S = T$.

9.5 Projection Operators

The concept of a *projection operator* P or, briefly, *projection* P, was defined in Sec. 3.3, where a Hilbert space H was represented as the direct sum of a closed subspace Y and its orthogonal complement $Y^{\perp}$; thus

$$\begin{aligned} &H = Y \oplus Y^{\perp} \\ &x = y + z \qquad (y \in Y, z \in Y^{\perp}). \end{aligned} \tag{1}$$

Since the sum is direct, y is unique for any given $x \in H$. Hence (1) defines a linear operator

$$\begin{aligned} P\colon\ & H \longrightarrow H \\ & x \longmapsto y = Px. \end{aligned} \tag{2}$$

P is called an *orthogonal projection* or **projection** on H. More specifically, P is called the *projection of H onto Y*. Hence a linear operator $P\colon H \longrightarrow H$ is a projection on H if there is a closed subspace Y of H such that Y is the range of P and $Y^{\perp}$ is the null space of P and $P|_Y$ is the identity operator on Y.

Note in passing that in (1) we can now write

$$x = y + z = Px + (I - P)x.$$

This shows that the projection of H onto $Y^\perp$ is $I - P$.

There is another characterization of a projection on H, which is sometimes used as a definition:

9.5-1 Theorem (Projection). *A bounded linear operator $P: H \longrightarrow H$ on a Hilbert space H is a projection if and only if P is self-adjoint and idempotent (that is, $P^2 = P$).*

Proof. (*a*) Suppose that P is a projection on H and denote $P(H)$ by Y. Then $P^2 = P$ because for every $x \in H$ and $Px = y \in Y$ we have

$$P^2x = Py = y = Px.$$

Furthermore, let $x_1 = y_1 + z_1$ and $x_2 = y_2 + z_2$, where $y_1, y_2 \in Y$ and $z_1, z_2 \in Y^\perp$. Then $\langle y_1, z_2 \rangle = \langle y_2, z_1 \rangle = 0$ because $Y \perp Y^\perp$, and self-adjointness of P is seen from

$$\langle Px_1, x_2 \rangle = \langle y_1, y_2 + z_2 \rangle = \langle y_1, y_2 \rangle = \langle y_1 + z_1, y_2 \rangle = \langle x_1, Px_2 \rangle.$$

(*b*) Conversely, suppose that $P^2 = P = P^*$ and denote $P(H)$ by Y. Then for every $x \in H$,

$$x = Px + (I - P)x.$$

Orthogonality $Y = P(H) \perp (I - P)(H)$ follows from

$$\langle Px, (I - P)v \rangle = \langle x, P(I - P)v \rangle = \langle x, Pv - P^2v \rangle = \langle x, 0 \rangle = 0.$$

Y is the null space $\mathcal{N}(I - P)$ of $I - P$, because $Y \subset \mathcal{N}(I - P)$ can be seen from

$$(I - P)Px = Px - P^2x = 0,$$

and $Y \supset \mathcal{N}(I - P)$ follows if we note that $(I - P)x = 0$ implies $x = Px$. Hence Y is closed by 2.7-10(*b*). Finally, $P|_Y$ is the identity operator on Y since, writing $y = Px$, we have $Py = P^2x = Px = y$. ∎

Projections have relatively simple and perspicuous properties, as we shall see. This suggests the attempt of representing more complicated linear operators on Hilbert spaces in terms of such simple operators. The resulting representation is called a *spectral representation* of the operator because we shall see that the projections employed for that purpose are related to the spectrum of the operator. Spectral representations account for the great importance of projections.

For bounded self-adjoint linear operators a spectral representation will be obtained in Sec. 9.9. The first step required to reach this goal is a thorough investigation of general properties of projections. This will be done in this section and the next one. The second step is the definition of projections suitable for that purpose. These are one-parameter families of projections, called *spectral families* (Sec. 9.7). In the third step we associate with a given bounded self-adjoint linear operator T a spectral family in a unique way (Sec. 9.8). This is called the *spectral family associated with* T. It is used in Sec. 9.9 for obtaining the desired spectral representation of T. A generalization of that representation is discussed in Sec. 9.10. The behavior of a spectral family at different points of the spectrum is considered in Sec. 9.11. This is our program for the remaining sections of this chapter.

Let us start, as indicated, by discussing basic properties of projections. In the first place we show that projections are always positive operators:

9.5-2 Theorem (Positivity, norm). *For any projection P on a Hilbert space H,*

$$\langle Px, x\rangle = \|Px\|^2 \tag{3}$$

$$P \geqq 0 \tag{4}$$

$$\|P\| \leqq 1; \qquad \|P\| = 1 \text{ if } P(H) \neq \{0\}. \tag{5}$$

Proof. (3) and (4) follow from

$$\langle Px, x\rangle = \langle P^2x, x\rangle = \langle Px, Px\rangle = \|Px\|^2 \geqq 0.$$

By the Schwarz inequality,

$$\|Px\|^2 = \langle Px, x\rangle \leqq \|Px\| \, \|x\|,$$

so that $\|Px\|/\|x\| \leqq 1$ for every $x \neq 0$, and $\|P\| \leqq 1$. Also $\|Px\|/\|x\| = 1$ if $x \in P(H)$ and $x \neq 0$. This proves (5). ∎

The product of projections need not be a projection, but we have the basic

9.5-3 Theorem (Product of projections). *In connection with products (composites) of projections on a Hilbert space H, the following two statements hold.*

(a) *$P = P_1P_2$ is a projection on H if and only if the projections P_1 and P_2 commute, that is, $P_1P_2 = P_2P_1$. Then P projects H onto $Y = Y_1 \cap Y_2$, where $Y_j = P_j(H)$.*

(b) *Two closed subspaces Y and V of H are orthogonal if and only if the corresponding projections satisfy $P_YP_V = 0$.*

Proof. **(a)** Suppose that $P_1P_2 = P_2P_1$. Then P is self-adjoint, by Theorem 3.10-4. P is idempotent since

$$P^2 = (P_1P_2)(P_1P_2) = P_1{}^2P_2{}^2 = P_1P_2 = P.$$

Hence P is a projection by 9.5-1, and for every $x \in H$ we have

$$Px = P_1(P_2x) = P_2(P_1x).$$

Since P_1 projects H onto Y_1, we must have $P_1(P_2x) \in Y_1$. Similarly, $P_2(P_1x) \in Y_2$. Together, $Px \in Y_1 \cap Y_2$. Since $x \in H$ was arbitrary, this shows that P projects H into $Y = Y_1 \cap Y_2$. Actually, P projects H *onto* Y. Indeed, if $y \in Y$, then $y \in Y_1$, $y \in Y_2$, and

$$Py = P_1P_2y = P_1y = y.$$

Conversely, if $P = P_1P_2$ is a projection defined on H, then P is self-adjoint by 9.5-1, and $P_1P_2 = P_2P_1$ follows from Theorem 3.10-4.

(b) If $Y \perp V$, then $Y \cap V = \{0\}$ and $P_YP_Vx = 0$ for all $x \in H$ by part (*a*), so that $P_YP_V = 0$.

Conversely, if $P_YP_V = 0$, then for every $y \in Y$ and $v \in V$ we obtain

$$\langle y, v \rangle = \langle P_Yy, P_Vv \rangle = \langle y, P_YP_Vv \rangle = \langle y, 0 \rangle = 0.$$

Hence $Y \perp V$. ∎

Similarly, a sum of projections need not be a projection, but we have

9.5-4 Theorem (Sum of projections). *Let P_1 and P_2 be projections on a Hilbert space H. Then:*

(a) *The sum $P = P_1 + P_2$ is a projection on H if and only if $Y_1 = P_1(H)$ and $Y_2 = P_2(H)$ are orthogonal.*

(b) *If $P = P_1 + P_2$ is a projection, P projects H onto $Y = Y_1 \oplus Y_2$.*

Proof. **(a)** If $P = P_1 + P_2$ is a projection, $P = P^2$ by 9.5-1, written out,

$$P_1 + P_2 = (P_1 + P_2)^2 = P_1{}^2 + P_1P_2 + P_2P_1 + P_2{}^2.$$

On the right, $P_1{}^2 = P_1$ and $P_2{}^2 = P_2$ by 9.5-1. There remains

$$P_1P_2 + P_2P_1 = 0. \tag{6}$$

Multiplying by P_2 from the left, we obtain

$$P_2P_1P_2 + P_2P_1 = 0. \tag{7}$$

Multiplying this by P_2 from the right, we have $2P_2P_1P_2 = 0$, so that $P_2P_1 = 0$ by (7) and $Y_1 \perp Y_2$ by 9.5-3(*b*).

Conversely, if $Y_1 \perp Y_2$, then $P_1P_2 = P_2P_1 = 0$, again by 9.5-3(*b*). This yields (6), which implies $P^2 = P$. Since P_1 and P_2 are self-adjoint, so is $P = P_1 + P_2$. Hence P is a projection by 9.5-1.

(b) We determine the closed subspace $Y \subset H$ onto which P projects. Since $P = P_1 + P_2$, for every $x \in H$ we have

$$y = Px = P_1x + P_2x.$$

Here $P_1x \in Y_1$ and $P_2x \in Y_2$. Hence $y \in Y_1 \oplus Y_2$, so that $Y \subset Y_1 \oplus Y_2$.

We show that $Y \supset Y_1 \oplus Y_2$. Let $v \in Y_1 \oplus Y_2$ be arbitrary. Then $v = y_1 + y_2$. Here, $y_1 \in Y_1$ and $y_2 \in Y_2$. Applying P and using $Y_1 \perp Y_2$, we thus obtain

$$Pv = P_1(y_1 + y_2) + P_2(y_1 + y_2) = P_1y_1 + P_2y_2 = y_1 + y_2 = v.$$

Hence $v \in Y$ and $Y \supset Y_1 \oplus Y_2$. Together, $Y = Y_1 \oplus Y_2$. ∎

Problems

1. Show that a projection P on a Hilbert space H satisfies

$$0 \leqq P \leqq I.$$

Under what conditions will (*i*) $P=0$, (*ii*) $P=I$?

2. Let $Q=S^{-1}PS\colon H \longrightarrow H$, where S and P are bounded and linear. If P is a projection and S is unitary, show that Q is a projection.

3. Find linear operators $T\colon \mathbf{R}^2 \longrightarrow \mathbf{R}^2$ which are idempotent but not self-adjoint (so that they are not projections; cf. 9.5-1).

4. Illustrate Theorem 9.5-3 with an example of projections P_1, P_2 in $\mathbf{R}^3$ such that P_1P_2 is neither P_1 nor P_2.

5. Extend Theorem 9.5-4 to sums $P=P_1+\cdots+P_m$.

6. In Prob. 5, let $Y_j=P_j(H)$, $j=1,\cdots,m$, and $Y=P(H)$. Show that every $x \in Y$ has a representation

$$x=x_1+\cdots+x_m, \qquad x_j=P_jx \in Y_j,$$

and, conversely, if $x \in H$ can be represented in this form, then $x \in Y$ and the representation is unique.

7. Give a simple example which illustrates that the sum of two projections need not be a projection.

8. If a sum $P_1+\cdots+P_k$ of projections $P_j\colon H \longrightarrow H$ (H a Hilbert space) is a projection, show that

$$\|P_1x\|^2+\cdots+\|P_kx\|^2 \leqq \|x\|^2.$$

9. How could we obtain the Bessel inequality (Sec. 3.4) from theorems of the present section?

10. Let P_1 and P_2 be projections of a Hilbert space H onto Y_1 and Y_2, respectively, and $P_1P_2=P_2P_1$. Show that

$$P_1+P_2-P_1P_2$$

is a projection, namely, the projection of H onto Y_1+Y_2.

9.6 Further Properties of Projections

Let us consider some further properties of projections which we shall need later, for reasons explained at the beginning of the previous section.

Our first theorem refers to the partial order relation which is defined by $P_1 \leqq P_2$ (cf. Sec. 9.3) on the set of all projections on a given Hilbert space. This theorem will be a basic tool in the next three sections.

9.6-1 Theorem (Partial order). *Let P_1 and P_2 be projections defined on a Hilbert space H. Denote by $Y_1 = P_1(H)$ and $Y_2 = P_2(H)$ the subspaces onto which H is projected by P_1 and P_2, and let $\mathcal{N}(P_1)$ and $\mathcal{N}(P_2)$ be the null spaces of these projections. Then the following conditions are equivalent.*

$$P_2P_1 = P_1P_2 = P_1 \tag{1}$$

$$Y_1 \subset Y_2 \tag{2}$$

$$\mathcal{N}(P_1) \supset \mathcal{N}(P_2) \tag{3}$$

$$\|P_1x\| \leqq \|P_2x\| \qquad \text{for all } x \in H \tag{4}$$

$$P_1 \leqq P_2. \tag{5}$$

Proof. (1) $\Longrightarrow$ (4):

We have $\|P_1\| \leqq 1$ by 9.5-2. Hence (1) yields for all $x \in H$

$$\|P_1x\| = \|P_1P_2x\| \leqq \|P_1\| \, \|P_2x\| \leqq \|P_2x\|.$$

(4) $\Longrightarrow$ (5):

From (3) in Sec. 9.5 and (4) in the present theorem we have for all $x \in H$

$$\langle P_1x, x\rangle = \|P_1x\|^2 \leqq \|P_2x\|^2 = \langle P_2x, x\rangle,$$

which shows that $P_1 \leqq P_2$ by definition.

(5) $\Longrightarrow$ (3):

Let $x \in \mathcal{N}(P_2)$. Then $P_2x = 0$. By (3) in Sec. 9.5 and (5) in the present theorem,

$$\|P_1x\|^2 = \langle P_1x, x\rangle \leqq \langle P_2x, x\rangle = 0.$$

Hence $P_1x=0$, $x\in\mathcal{N}(P_1)$ and $\mathcal{N}(P_1)\supset\mathcal{N}(P_2)$ since $x\in\mathcal{N}(P_2)$ was arbitrary.

$(3)\Longrightarrow(2)$:

This is clear since $\mathcal{N}(P_j)$ is the orthogonal complement of Y_j in H, by Lemma 3.3-5.

$(2)\Longrightarrow(1)$:

For every $x\in H$ we have $P_1x\in Y_1$. Hence $P_1x\in Y_2$ by (2), so that $P_2(P_1x)=P_1x$, that is, $P_2P_1=P_1$. Since P_1 is self-adjoint by 9.5-1, Theorem 3.10-4 implies that $P_1=P_2P_1=P_1P_2$. ▮

Sums of projections were considered in the previous section, and we can now turn to differences of projections. This will be a first application of the theorem just proved.

9.6-2 Theorem (Difference of projections). *Let P_1 and P_2 be projections on a Hilbert space H. Then:*

(a) *The difference $P=P_2-P_1$ is a projection on H if and only if $Y_1\subset Y_2$, where $Y_j=P_j(H)$.*

(b) *If $P=P_2-P_1$ is a projection, P projects H onto Y, where Y is the orthogonal complement of Y_1 in Y_2.*

Proof. **(a)** If $P=P_2-P_1$ is a projection, $P=P^2$ by 9.5-1, written out,

$$P_2-P_1=(P_2-P_1)^2=P_2^2-P_2P_1-P_1P_2+P_1^2.$$

On the right, $P_2^2=P_2$ and $P_1^2=P_1$ by 9.5-1. Hence

$$P_1P_2+P_2P_1=2P_1. \tag{6}$$

Multiplication by P_2 from the left and from the right gives

$$P_2P_1P_2+P_2P_1=2P_2P_1$$
$$P_1P_2+P_2P_1P_2=2P_1P_2.$$

Hence $P_2P_1P_2=P_2P_1$, $P_2P_1P_2=P_1P_2$, and by (6),

$$P_2P_1=P_1P_2=P_1. \tag{7}$$

$Y_1\subset Y_2$ now follows from Theorem 9.6-1.

Conversely, if $Y_1 \subset Y_2$, Theorem 9.6-1 yields (7), which implies (6) and shows that P is idempotent. Since P_1 and P_2 are self-adjoint, $P = P_2 - P_1$ is self-adjoint, and P is a projection by 9.5-1.

(b) $Y = P(H)$ consists of all vectors of the form

$$y = Px = P_2x - P_1x \qquad (x \in H). \tag{8}$$

Since $Y_1 \subset Y_2$ by part (*a*), we have $P_2P_1 = P_1$ by (1) and thus obtain from (8)

$$P_2y = P_2^2x - P_2P_1x = P_2x - P_1x = y.$$

This shows that $y \in Y_2$. Also, from (8) and (1),

$$P_1y = P_1P_2x - P_1^2x = P_1x - P_1x = 0.$$

This shows that $y \in \mathcal{N}(P_1) = Y_1^\perp$; cf. 3.3-5. Together, $y \in V$ where $V = Y_2 \cap Y_1^\perp$. Since $y \in Y$ was arbitrary, $Y \subset V$.

We show that $Y \supset V$. Since the projection of H onto $Y_1^\perp$ is $I - P_1$ (cf. Sec. 9.5), every $v \in V$ is of the form

$$v = (I - P_1)y_2 \qquad (y_2 \in Y_2). \tag{9}$$

Using again $P_2P_1 = P_1$, we obtain from (9), since $P_2y_2 = y_2$,

$$\begin{aligned} Pv &= (P_2 - P_1)(I - P_1)y_2 \\ &= (P_2 - P_2P_1 - P_1 + P_1^2)y_2 \\ &= y_2 - P_1y_2 = v. \end{aligned}$$

This shows that $v \in Y$. Since $v \in V$ was arbitrary, $Y \supset V$. Together, $Y = P(H) = V = Y_2 \cap Y_1^\perp$. ∎

From this theorem and the preceding one we can now derive a basic result about the convergence of a monotone increasing sequence of projections. (A similar theorem holds for a monotone *decreasing* sequence of projections.)

9.6-3 Theorem (Monotone increasing sequence). *Let (P_n) be a monotone increasing sequence of projections P_n defined on a Hilbert space H. Then:*

(a) *(P_n) is strongly operator convergent, say, $P_n x \longrightarrow Px$ for every $x \in H$, and the limit operator P is a projection defined on H.*

(b) *P projects H onto*

$$P(H) = \overline{\bigcup_{n=1}^{\infty} P_n(H)}.$$

(c) *P has the null space*

$$\mathcal{N}(P) = \bigcap_{n=1}^{\infty} \mathcal{N}(P_n).$$

Proof. **(a)** Let $m < n$. By assumption, $P_m \leqq P_n$, so that we have $P_m(H) \subset P_n(H)$ by 9.6-1 and $P_n - P_m$ is a projection by 9.6-2. Hence for every fixed $x \in H$ we obtain by 9.5-2

$$\begin{aligned} \|P_n x - P_m x\|^2 &= \|(P_n - P_m)x\|^2 \\ &= \langle (P_n - P_m)x, x \rangle \\ &= \langle P_n x, x \rangle - \langle P_m x, x \rangle \\ &= \|P_n x\|^2 - \|P_m x\|^2. \end{aligned} \tag{10}$$

Now $\|P_n\| \leqq 1$ by 9.5-2, so that $\|P_n x\| \leqq \|x\|$ for every n. Hence $(\|P_n x\|)$ is a bounded sequence of numbers. $(\|P_n x\|)$ is also monotone by 9.6-1 since (P_n) is monotone. Hence $(\|P_n x\|)$ converges. From this and (10) we see that $(P_n x)$ is Cauchy. Since H is complete, $(P_n x)$ converges. The limit depends on x, say, $P_n x \longrightarrow Px$. This defines an operator P on H. Linearity of P is obvious. Since $P_n x \longrightarrow Px$ and the P_n's are bounded, self-adjoint and idempotent, P has the same properties. Hence P is a projection by 9.5-1.

(b) We determine $P(H)$. Let $m < n$. Then $P_m \leqq P_n$, that is, $P_n - P_m \geqq 0$ and $\langle (P_n - P_m)x, x \rangle \geqq 0$ by definition. Letting $n \longrightarrow \infty$, we obtain $\langle (P - P_m)x, x \rangle \geqq 0$ by the continuity of the inner product (cf. 3.2-2), that is, $P_m \leqq P$ and 9.6-1 yields $P_m(H) \subset P(H)$ for every m. Hence

$$\bigcup P_m(H) \subset P(H).$$

Furthermore, for every m and for every $x \in H$ we have

$$P_m x \in P_m(H) \subset \bigcup P_m(H).$$

Since $P_m x \longrightarrow Px$, we see from 1.4-6(a) that $Px \in \overline{\bigcup P_m(H)}$. Hence $P(H) \subset \overline{\bigcup P_m(H)}$. Together,

$$\bigcup P_m(H) \subset P(H) \subset \overline{\bigcup P_m(H)}.$$

From 3.3-5 we have $P(H) = \mathcal{N}(I-P)$, so that $P(H)$ is closed by 2.7-10(b). This proves (b).

(c) We determine $\mathcal{N}(P)$. Using Lemma 3.3-5, we have $\mathcal{N}(P) = P(H)^{\perp} \subset P_n(H)^{\perp}$ for every n, since $P(H) \supset P_n(H)$ by part (b) of the proof. Hence

$$\mathcal{N}(P) \subset \bigcap P_n(H)^{\perp} = \bigcap \mathcal{N}(P_n).$$

On the other hand, if $x \in \bigcap \mathcal{N}(P_n)$, then $x \in \mathcal{N}(P_n)$ for every n, so that $P_n x = 0$, and $P_n x \longrightarrow Px$ implies $Px = 0$, that is, $x \in \mathcal{N}(P)$. Since $x \in \bigcap \mathcal{N}(P_n)$ was arbitrary, $\bigcap \mathcal{N}(P_n) \subset \mathcal{N}(P)$. Together we thus obtain $\mathcal{N}(P) = \bigcap \mathcal{N}(P_n)$. ∎

Problems

1. Illustrate the various equivalent statements in Theorem 9.6-1 by simple examples of projections in Euclidean space $\mathbf{R}^3$.

2. Show that the difference $P = P_2 - P_1$ of two projections on a Hilbert space H is a projection on H if and only if $P_1 \leqq P_2$.

3. For a better understanding of Theorem 9.6-2, consider $H = \mathbf{R}^3$, and let P_2 be the projection onto the $\xi_1\xi_2$-plane and P_1 the projection onto the straight line $\xi_2 = \xi_1$ in the $\xi_1\xi_2$-plane. Sketch Y_1, Y_2, $Y_1^{\perp}$, $Y_2^{\perp}$ and the orthogonal complement of Y_1 in Y_2. Determine the coordinates of $(P_2 - P_1)x$, where $x = (\xi_1, \xi_2, \xi_3)$. Is $P_1 + P_2$ a projection?

4. **(Limit of projections)** If (P_n) is a sequence of projections defined on a Hilbert space H and $P_n \longrightarrow P$, show that P is a projection defined on H.

5. Let $P_n(H)$ in Theorem 9.6-3 be finite dimensional for every n. Show that, nevertheless, $P(H)$ may be infinite dimensional.

6. Let (P_n) be strongly operator convergent with limit P, where the P_n's are projections on a Hilbert space H. Suppose that $P_n(H)$ is infinite dimensional. Show by an example that, nevertheless, $P(H)$ may be finite dimensional. (We mention that this irregularity as well as that in Prob. 5 cannot happen in the case of *uniform* operator convergence.)

7. What is $P(H)$ in Theorem 9.6-3 in the case of a monotone decreasing sequence (P_n)?

8. If $Q_1, Q_2, \cdots$ are projections on a Hilbert space H such that we have $Q_j(H) \perp Q_k(H)$ $(j \neq k)$, show that for every $x \in H$ the series

$$Qx = \sum_{j=1}^{\infty} Q_j x$$

converges (in the norm on H) and Q is a projection. Onto what subspace of H does Q project?

9. (Invariant subspace) Let $T: H \longrightarrow H$ be a bounded linear operator. Then a subspace $Y \subset H$ is said to be *invariant* under T if $T(Y) \subset Y$. Show that a closed subspace Y of H is invariant under T if and only if $Y^{\perp}$ is invariant under T^*.

10. (Reduction of an operator) A closed subspace Y of a Hilbert space H is said to *reduce* a linear operator $T: H \longrightarrow H$ if $T(Y) \subset Y$ and $T(Y^{\perp}) \subset Y^{\perp}$, that is, if both Y and $Y^{\perp}$ are invariant under T. (The point is that then the investigation of T can be facilitated by considering $T|_Y$ and $T|_{Y^{\perp}}$ separately.) If P_1 is the projection of H onto Y and $P_1T = TP_1$, show that Y reduces T.

11. If $\dim H < \infty$ in Prob. 10 and Y reduces T, what can we say about a matrix representing T?

12. Prove the converse of the statement in Prob. 10, that is, if Y reduces T, then $P_1T = TP_1$.

13. If Y in Prob. 10 reduces T, show that $TP_2 = P_2T$, where P_2 is the projection of H onto $Y^{\perp}$.

14. Let (e_k) be a total orthonormal sequence in a separable Hilbert space H. Let $T: H \longrightarrow H$ be a linear operator defined at e_k by $Te_k = e_{k+1}$, $k = 1, 2, \cdots$, and then continuously extended to H. Let Y_n be the closure of span $\{e_n, e_{n+1}, \cdots\}$, where $n>1$. Show that T is not self-adjoint. Show that Y_n does not reduce T; (*a*) use Prob. 12, (*b*) give a direct proof.

15. Let $T: H \longrightarrow H$ be a bounded linear operator on a Hilbert space H and Y a closed subspace of H such that $T(Y) \subset Y$. If T is self-adjoint, show that $T(Y^{\perp}) \subset Y^{\perp}$, so that in this case, Y reduces T. (Note that T in Prob. 14 is not self-adjoint.)

9.7 Spectral Family

We recall from Sec. 9.5 that our present aim is a representation of bounded self-adjoint linear operators on a Hilbert space in terms of very simple operators (projections) whose properties we can readily investigate in order to obtain information about those more complicated operators. Such a representation will be called a *spectral representation* of the operator concerned. A bounded self-adjoint linear operator $T: H \longrightarrow H$ being given, we shall obtain a spectral representation of T by the use of a suitable family of projections which is called the *spectral family associated with* T. In this section we motivate and define the concept of a *spectral family* in general, that is, without reference to a given operator T. The association of a suitable spectral family with a given operator T will be considered separately, in the next section, and the resulting spectral representation of T in Sec. 9.9.

That motivation of a spectral family can be obtained from the finite dimensional case as follows. Let $T: H \longrightarrow H$ be a self-adjoint linear operator on the unitary space $H = \mathbf{C}^n$ (cf. 3.10-2). Then T is bounded (by 2.7-8), and we may choose a basis for H and represent T by a Hermitian matrix which we denote simply by T. The spectrum of the operator consists of the eigenvalues of that matrix (cf. Secs. 7.1, 7.2) which are real by 9.1-1. For simplicity let us assume that the matrix T has n different eigenvalues $\lambda_1 < \lambda_2 < \cdots < \lambda_n$. Then Theorem 9.1-1(*b*) implies that T has an orthonormal set of n eigenvectors

$$x_1, \qquad x_2, \qquad \cdots, \qquad x_n$$

where x_j corresponds to λ_j and we write these vectors as column vectors, for convenience. This is a basis for H, so that every $x \in H$ has a unique representation

$$x = \sum_{j=1}^{n} \gamma_j x_j,$$

(1)

$$\gamma_j = \langle x, x_j \rangle = x^{\mathsf{T}} \bar{x}_j.$$

In (1) we obtain the second formula from the first one by taking the inner product $\langle x, x_k \rangle$, where x_k is fixed, and using the orthonormality. The essential fact in (1) is that x_j is an eigenvector of T, so that we have $Tx_j = \lambda_j x_j$. Consequently, if we apply T to (1) we simply obtain

$$Tx = \sum_{j=1}^{n} \lambda_j \gamma_j x_j. \tag{2}$$

Thus, whereas T may act on x in a complicated way, it acts on each term of the sum in (1) in a very simple fashion. This demonstrates the great advantage of the use of eigenvectors in connection with the investigation of a linear operator on $H = \mathbf{C}^n$.

Looking at (1) more closely, we see that we can define an operator

$$P_j : H \longrightarrow H$$

(3)

$$x \longmapsto \gamma_j x_j.$$

Obviously, P_j is the projection (orthogonal projection) of H onto the eigenspace of T corresponding to λ_j. Formula (1) can now be written

$$x = \sum_{j=1}^{n} P_j x \qquad \text{hence} \qquad I = \sum_{j=1}^{n} P_j \tag{4}$$

where I is the identity operator on H. Formula (2) becomes

$$Tx = \sum_{j=1}^{n} \lambda_j P_j x \qquad \text{hence} \qquad T = \sum_{j=1}^{n} \lambda_j P_j. \tag{5}$$

This is a representation of T in terms of projections. It shows that the spectrum of T is employed to get a representation of T [namely, (5)] in terms of very simple operators.

The use of the projections P_j seems quite natural and geometrically perspicuous. Unfortunately, our present formulas would not be suitable for immediate generalization to infinite dimensional Hilbert spaces H since in that case, spectra of bounded self-adjoint linear operators may be more complicated, as we know. We shall now describe another approach, which is somewhat less perspicuous but has the advantage that it can be generalized to the infinite dimensional case.

Instead of the projections $P_1, \cdots, P_n$ themselves we take sums of such projections. More precisely, for any real λ we define

$$E_\lambda = \sum_{\lambda_j \leqq \lambda} P_j \qquad (\lambda \in \mathbf{R}). \tag{6}$$

This is a one-parameter family of projections, λ being the parameter. From (6) we see that for any λ, the operator E_λ is the projection of H onto the subspace V_λ spanned by all those x_j for which $\lambda_j \leqq \lambda$. It follows that

$$V_\lambda \subset V_\mu \qquad (\lambda \leqq \mu).$$

Roughly speaking, as λ traverses $\mathbf{R}$ in the positive sense, E_λ grows from 0 to I, the growth occurring at the eigenvalues of T and E_λ remaining unchanged for λ in any interval that is free of eigenvalues. Hence we see that E_λ has the properties

$$E_\lambda E_\mu = E_\mu E_\lambda = E_\lambda \qquad \text{if } \lambda < \mu$$

$$E_\lambda = 0 \qquad \text{if } \lambda < \lambda_1$$

$$E_\lambda = I \qquad \text{if } \lambda \geqq \lambda_n$$

$$E_{\lambda+0} = \lim_{\mu \to \lambda+0} E_\mu = E_\lambda,$$

where $\mu \longrightarrow \lambda + 0$ means that we let μ approach λ from the right. This suggests the following definition.

9.7-1 Definition (Spectral family or decomposition of unity). A real *spectral family* (or real *decomposition of unity*) is a one-parameter family $\mathscr{E} = (E_\lambda)_{\lambda \in \mathbf{R}}$ of projections E_λ defined on a Hilbert space H (of any dimension) which depends on a real parameter λ and is such that

$$E_\lambda \leqq E_\mu, \qquad \text{hence} \qquad E_\lambda E_\mu = E_\mu E_\lambda = E_\lambda \qquad (\lambda < \mu) \tag{7}$$

$$\lim_{\lambda \to -\infty} E_\lambda x = 0 \tag{8a}$$

$$\lim_{\lambda \to +\infty} E_\lambda x = x \tag{8b}$$

$$E_{\lambda+0}x = \lim_{\mu \to \lambda+0} E_\mu x = E_\lambda x \qquad (x \in H). \quad \blacksquare \tag{9}$$

We see from this definition that a real spectral family can be regarded as a mapping

$$\mathbf{R} \longrightarrow B(H, H)$$

$$\lambda \longmapsto E_\lambda;$$

to each $\lambda \in \mathbf{R}$ there corresponds a projection $E_\lambda \in B(H, H)$, where $B(H, H)$ is the space of all bounded linear operators from H into H.

Note that the two conditions in (7) are equivalent, by 9.6-1.

$\mathscr{E}$ is called a **spectral family on an interval $[a, b]$** if

$$E_\lambda = 0 \quad \text{for} \quad \lambda < a, \qquad E_\lambda = I \quad \text{for} \quad \lambda \geqq b. \tag{8*}$$

Such families will be of particular interest to us since the spectrum of a bounded self-adjoint linear operator lies in a finite interval on the real line. Observe that (8*) implies (8).

$\mu \longrightarrow \lambda + 0$ in (9) indicates that in this limit process we consider only values $\mu > \lambda$, and (9) means that $\lambda \longmapsto E_\lambda$ is strongly operator continuous from the right. As a matter of fact, continuity from the left would do equally well. We could even impose no such condition at all, but then we would have to work with limits $E_{\lambda+0}$ and $E_{\lambda-0}$, which would be an unnecessary inconvenience.

We shall see later (in the next two sections) that with any given bounded self-adjoint linear operator T on any Hilbert space we can associate a spectral family which may be used for representing T by a Riemann-Stieltjes integral. This is known as a *spectral representation*, as was mentioned before.

Then we shall also see that in the finite dimensional case considered at the beginning of this section, that integral representation reduces to a finite sum, namely, (5) written in terms of the spectral family (6). For the time being, let us show how we can write (5) in terms of (6). As before we assume, for simplicity, that the eigenvalues $\lambda_1, \cdots, \lambda_n$ of T are all different, and $\lambda_1 < \lambda_2 < \cdots < \lambda_n$. Then we have

$$
\begin{aligned}
E_{\lambda_1} &= P_1 \\
E_{\lambda_2} &= P_1 + P_2 \\
&\cdots\cdots \\
E_{\lambda_n} &= P_1 + \cdots + P_n.
\end{aligned}
$$

Hence, conversely,

$$
\begin{aligned}
P_1 &= E_{\lambda_1} \\
P_j &= E_{\lambda_j} - E_{\lambda_{j-1}} \qquad j = 2, \cdots, n.
\end{aligned}
$$

Since E_λ remains the same for λ in the interval $[\lambda_{j-1}, \lambda_j)$, this can be written

$$P_j = E_{\lambda_j} - E_{\lambda_j - 0}.$$

(4) now becomes

$$x = \sum_{j=1}^{n} P_j x = \sum_{j=1}^{n} (E_{\lambda_j} - E_{\lambda_j - 0}) x$$

and (5) becomes

$$Tx = \sum_{j=1}^{n} \lambda_j P_j x = \sum_{j=1}^{n} \lambda_j (E_{\lambda_j} - E_{\lambda_j - 0}) x.$$

If we drop the x and write

$$\delta E_\lambda = E_\lambda - E_{\lambda - 0}$$

we arrive at

$$T = \sum_{j=1}^{n} \lambda_j \delta E_{\lambda_j}. \tag{10}$$

This is the *spectral representation* of the self-adjoint linear operator T with eigenvalues $\lambda_1 < \lambda_2 < \cdots < \lambda_n$ on that n-dimensional Hilbert space H. The representation shows that for any $x, y \in H$,

$$\langle Tx, y\rangle = \sum_{j=1}^{n} \lambda_j \langle \delta E_{\lambda_j} x, y\rangle. \tag{11}$$

We note that this may be written as a Riemann-Stieltjes integral

$$\langle Tx, y\rangle = \int_{-\infty}^{+\infty} \lambda \, dw(\lambda) \tag{12}$$

where $w(\lambda) = \langle E_\lambda x, y\rangle$.

Our present discussion takes care of self-adjoint linear operators on spaces that are finite dimensional. It also paves the way for the case of arbitrary Hilbert spaces to be considered in the next section. A joint problem set will be included at the end of the next section.

9.8 Spectral Family of a Bounded Self-Adjoint Linear Operator

With a given bounded self-adjoint linear operator $T: H \longrightarrow H$ on a complex Hilbert space H we can associate a spectral family $\mathscr{E}$ such that $\mathscr{E}$ may be used for a spectral representation of T (to be obtained in the next section).

To define $\mathscr{E}$ we need the operator

$$T_\lambda = T - \lambda I, \tag{1}$$

the positive square root of $T_\lambda{}^2$, which we denote[2] by B_λ; thus

$$B_\lambda = (T_\lambda{}^2)^{1/2} \tag{2}$$

and the operator

$$T_\lambda{}^+ = \tfrac{1}{2}(B_\lambda + T_\lambda), \tag{3}$$

which is called the **positive part** of T_λ.

[2] Another notation for B_λ also used in the literature is $|T_\lambda|$.

The **spectral family** $\mathscr{E}$ of T is then defined by $\mathscr{E}=(E_\lambda)_{\lambda\in\mathbf{R}}$, where E_λ is the projection of H onto the null space $\mathcal{N}(T_\lambda^+)$ of T_λ^+.

Our task in the remainder of this section is to prove that $\mathscr{E}$ is indeed a spectral family, that is, has all the properties by which a spectral family is characterized in Def. 9.7-1. This will require some patience, but we shall be rewarded by the fact that the proofs will produce a basic tool (inequality (18), below) for the derivation of the spectral representation in the next section.

We proceed stepwise and consider at first the operators

$$B=(T^2)^{1/2} \qquad \textit{(positive square root of } T^2\textit{)}$$

$$T^+=\tfrac{1}{2}(B+T) \qquad \textit{(positive part of } T\textit{)}$$

$$T^-=\tfrac{1}{2}(B-T) \qquad \textit{(negative part of } T\textit{)}$$

and the projection of H onto the null space of T^+ which we denote by E, that is,

$$E\colon H \longrightarrow Y=\mathcal{N}(T^+).$$

By subtraction and addition we see that

$$T=T^+-T^- \tag{4}$$

$$B=T^++T^-. \tag{5}$$

Furthermore we have

9.8-1 Lemma (Operators related to T). *The operators just defined have the following properties.*

(a) *B, T^+ and T^- are bounded and self-adjoint.*

(b) *B, T^+ and T^- commute with every bounded linear operator that T commutes with; in particular,*

$$BT=TB \qquad T^+T=TT^+ \qquad T^-T=TT^- \qquad T^+T^-=T^-T^+. \tag{6}$$

(c) *E commutes with every bounded self-adjoint linear operator that T commutes with; in particular,*

$$ET=TE \qquad EB=BE. \tag{7}$$

(d) *Furthermore,*

$$(8) \qquad T^+T^- = 0 \qquad\qquad T^-T^+ = 0$$

$$(9) \qquad T^+E = ET^+ = 0 \qquad\qquad T^-E = ET^- = T^-$$

$$(10) \qquad TE = -T^- \qquad\qquad T(I-E) = T^+$$

$$(11) \qquad T^+ \geqq 0 \qquad\qquad T^- \geqq 0.$$

Proof. **(a)** is clear since T and B are bounded and self-adjoint.

(b) Suppose that $TS = ST$. Then $T^2S = TST = ST^2$, and $BS = SB$ follows from Theorem 9.4-2 applied to T^2. Hence

$$T^+S = \tfrac{1}{2}(BS + TS) = \tfrac{1}{2}(SB + ST) = ST^+.$$

The proof of $T^-S = ST^-$ is similar.

(c) For every $x \in H$ we have $y = Ex \in Y = \mathcal{N}(T^+)$. Hence $T^+y = 0$ and $ST^+y = S0 = 0$. From $TS = ST$ and (b) we have $ST^+ = T^+S$, and

$$T^+SEx = T^+Sy = ST^+y = 0.$$

Hence $SEx \in Y$. Since E projects H onto Y, we thus have $ESEx = SEx$ for every $x \in H$, that is, $ESE = SE$. A projection is self-adjoint by 9.5-1 and so is S by assumption. Using (6g), Sec. 3.9, we thus obtain

$$ES = E^*S^* = (SE)^* = (ESE)^* = E^*S^*E^* = ESE = SE.$$

(d) We prove (8)-(11).

Proof of (8):

From $B = (T^2)^{1/2}$ we have $B^2 = T^2$. Also $BT = TB$ by (6). Hence, again by (6),

$$T^+T^- = T^-T^+ = \tfrac{1}{2}(B-T)\tfrac{1}{2}(B+T) = \tfrac{1}{4}(B^2 + BT - TB - T^2) = 0.$$

Proof of (9):

By definition, $Ex \in \mathcal{N}(T^+)$, so that $T^+Ex = 0$ for all $x \in H$. Since T^+ is self-adjoint, we have $ET^+x = T^+Ex = 0$ by (6) and (c), that is $ET^+ = T^+E = 0$.

Furthermore, $T^+T^-x=0$ by (8), so that $T^-x \in \mathcal{N}(T^+)$. Hence $ET^-x = T^-x$. Since T^- is self-adjoint, (*c*) yields $T^-Ex = ET^-x = T^-x$ for all $x \in H$, that is, $T^-E = ET^- = T^-$.

Proof of (10):

From (4) and (9) we have $TE = (T^+ - T^-)E = -T^-$ and from this, again by (4),

$$T(I-E) = T - TE = T + T^- = T^+.$$

Proof of (11):

By (9) and (5) and Theorem 9.3-1,

$$T^- = ET^- + ET^+ = E(T^- + T^+) = EB \geqq 0$$

since E and B are self-adjoint and commute, $E \geqq 0$ by 9.5-2 and $B \geqq 0$ by definition. Similarly, again by Theorem 9.3-1,

$$T^+ = B - T^- = B - EB = (I-E)B \geqq 0$$

because $I - E \geqq 0$ by 9.5-2. ∎

This was the first step. In the second step, instead of T we consider $T_\lambda = T - \lambda I$. Instead of B, T^+, T^- and E we now have to take $B_\lambda = (T_\lambda^2)^{1/2}$ [cf. (2)], the positive part and negative part of T_λ, defined by

$$T_\lambda^+ = \tfrac{1}{2}(B_\lambda + T_\lambda)$$

$$T_\lambda^- = \tfrac{1}{2}(B_\lambda - T_\lambda)$$

[cf. (3)] and the projection

$$E_\lambda\colon H \longrightarrow Y_\lambda = \mathcal{N}(T_\lambda^+)$$

of H onto the null space $Y_\lambda = \mathcal{N}(T_\lambda^+)$ of T_λ^+. We now have

9.8-2 Lemma (Operators related to T_λ). *The previous lemma remains true if we replace*

$$T, B, T^+, T^-, E \qquad \text{by} \qquad T_\lambda, B_\lambda, T_\lambda^+, T_\lambda^-, E_\lambda,$$

respectively, where λ is real. Moreover, for any real κ, λ, μ, ν, τ, the following operators all commute:

$$T_\kappa, \qquad B_\lambda, \qquad T_\mu^+, \qquad T_\nu^-, \qquad E_\tau.$$

Proof. The first statement is obvious. To obtain the second statement, we note that $IS = SI$ and

$$T_\lambda = T - \lambda I = T - \mu I + (\mu - \lambda)I = T_\mu + (\mu - \lambda)I. \tag{12}$$

Hence

$$ST = TS \implies ST_\mu = T_\mu S \implies ST_\lambda = T_\lambda S \implies SB_\lambda = B_\lambda S,\ SB_\mu = B_\mu S$$

and so on. For $S = T_\kappa$ this gives $T_\kappa B_\lambda = B_\lambda T_\kappa$, etc. ∎

With this preparation we can now prove that for a given bounded self-adjoint linear operator T we may define a spectral family $\mathscr{E} = (E_\lambda)$ in a unique fashion as explained in the subsequent theorem. This $\mathscr{E}$ is called the **spectral family associated with the operator *T*.** And in the next section we shall then see that we can use $\mathscr{E}$ to obtain the desired spectral representation of T, thereby reaching our actual goal.

9.8-3 Theorem (Spectral family associated with an operator). *Let $T: H \longrightarrow H$ be a bounded self-adjoint linear operator on a complex Hilbert space H. Furthermore, let E_λ (λ real) be the projection of H onto the null space $Y_\lambda = \mathcal{N}(T_\lambda^+)$ of the positive part T_λ^+ of $T_\lambda = T - \lambda I$. Then $\mathscr{E} = (E_\lambda)_{\lambda \in \mathbf{R}}$ is a spectral family on the interval $[m, M] \subset \mathbf{R}$, where m and M are given by (1) in Sec. 9.2.*

Proof. We shall prove

$$\lambda < \mu \implies E_\lambda \leqq E_\mu \tag{13}$$

$$\lambda < m \implies E_\lambda = 0 \tag{14}$$

$$\lambda \geqq M \implies E_\lambda = I \tag{15}$$

$$\mu \longrightarrow \lambda + 0 \implies E_\mu x \longrightarrow E_\lambda x. \tag{16}$$

In the proof we use part of Lemma 9.8-1 formulated for T_λ, T_μ, T_λ^+

etc., instead of T, T^+, etc., namely,

$$(8^*) \qquad T_\mu{}^+T_\mu{}^- = 0$$

$$(10^*) \qquad T_\lambda E_\lambda = -T_\lambda{}^- \qquad T_\lambda(I - E_\lambda) = T_\lambda{}^+ \qquad T_\mu E_\mu = -T_\mu{}^-$$

$$(11^*) \qquad T_\lambda{}^+ \geqq 0 \qquad T_\lambda{}^- \geqq 0 \qquad T_\mu{}^+ \geqq 0 \qquad T_\mu{}^- \geqq 0.$$

Proof of (13):

Let $\lambda < \mu$. We have $T_\lambda = T_\lambda{}^+ - T_\lambda{}^- \leqq T_\lambda{}^+$ because $-T^- \leqq 0$ by (11*). Hence

$$T_\lambda{}^+ - T_\mu \geqq T_\lambda - T_\mu = (\mu - \lambda)I \geqq 0.$$

$T_\lambda{}^+ - T_\mu$ is self-adjoint and commutes with $T_\mu{}^+$ by 9.8-2, and $T_\mu{}^+ \geqq 0$ by (11*). Theorem 9.3-1 thus implies

$$T_\mu{}^+(T_\lambda{}^+ - T_\mu) = T_\mu{}^+(T_\lambda{}^+ - T_\mu{}^+ + T_\mu{}^-) \geqq 0.$$

Here $T_\mu{}^+T_\mu{}^- = 0$ by (8*). Hence $T_\mu{}^+T_\lambda{}^+ \geqq T_\mu{}^{+2}$, that is, for all $x \in H$,

$$\langle T_\mu{}^+T_\lambda{}^+x, x\rangle \geqq \langle T_\mu{}^{+2}x, x\rangle = \|T_\mu{}^+x\|^2 \geqq 0.$$

This shows that $T_\lambda{}^+x = 0$ implies $T_\mu{}^+x = 0$. Hence $\mathcal{N}(T_\lambda{}^+) \subset \mathcal{N}(T_\mu{}^+)$, so that $E_\lambda \leqq E_\mu$ by 9.6-1. Here, $\lambda < \mu$.

Proof of (14):

Let $\lambda < m$ but suppose that, nevertheless, $E_\lambda \neq 0$. Then $E_\lambda z \neq 0$ for some z. We set $x = E_\lambda z$. Then $E_\lambda x = E_\lambda{}^2 z = E_\lambda z = x$, and we may assume $\|x\| = 1$ without loss of generality. It follows that

$$\begin{aligned} \langle T_\lambda E_\lambda x, x\rangle &= \langle T_\lambda x, x\rangle \\ &= \langle Tx, x\rangle - \lambda \\ &\geqq \inf_{\|\tilde{x}\|=1} \langle T\tilde{x}, \tilde{x}\rangle - \lambda \\ &= m - \lambda > 0. \end{aligned}$$

But this contradicts $T_\lambda E_\lambda = -T_\lambda{}^- \leqq 0$, which is obtained from (10*) and (11*).

Proof of (15):

Suppose that $\lambda > M$ but $E_\lambda \neq I$, so that $I - E_\lambda \neq 0$. Then $(I - E_\lambda)x = x$ for some x of norm $\|x\| = 1$. Hence

$$\begin{aligned}\langle T_\lambda(I-E_\lambda)x, x\rangle &= \langle T_\lambda x, x\rangle \\ &= \langle Tx, x\rangle - \lambda \\ &\leqq \sup_{\|\tilde{x}\|=1} \langle T\tilde{x}, \tilde{x}\rangle - \lambda \\ &= M - \lambda < 0.\end{aligned}$$

But this contradicts $T_\lambda(I - E_\lambda) = T_\lambda^+ \geqq 0$ which is obtained from (10*) and (11*). Also $E_M = I$ by the continuity from the right to be proved now.

Proof of (16):

With an interval $\Delta = (\lambda, \mu]$ we associate the operator

$$E(\Delta) = E_\mu - E_\lambda.$$

Since $\lambda < \mu$ we have $E_\lambda \leqq E_\mu$ by (13), hence $E_\lambda(H) \subset E_\mu(H)$ by 9.6-1, and $E(\Delta)$ is a projection by 9.6-2. Also $E(\Delta) \geqq 0$ by 9.5-2. Again by 9.6-1,

$$\text{(17)} \qquad \begin{aligned} E_\mu E(\Delta) &= E_\mu{}^2 - E_\mu E_\lambda = E_\mu - E_\lambda = E(\Delta) \\ (I - E_\lambda)E(\Delta) &= E(\Delta) - E_\lambda(E_\mu - E_\lambda) = E(\Delta). \end{aligned}$$

Since $E(\Delta)$, T_μ^- and T_λ^+ are positive [cf. (11*)] and commute by 9.8-2, the products $T_\mu^- E(\Delta)$ and $T_\lambda^+ E(\Delta)$ are positive by 9.3-1. Hence by (17) and (10*),

$$\begin{aligned} T_\mu E(\Delta) &= T_\mu E_\mu E(\Delta) = -T_\mu^- E(\Delta) \leqq 0 \\ T_\lambda E(\Delta) &= T_\lambda(I - E_\lambda)E(\Delta) = T_\lambda^+ E(\Delta) \geqq 0. \end{aligned}$$

This implies $TE(\Delta) \leqq \mu E(\Delta)$ and $TE(\Delta) \geqq \lambda E(\Delta)$, respectively. Together,

$$\textbf{(18)} \qquad \lambda E(\Delta) \leqq TE(\Delta) \leqq \mu E(\Delta) \qquad\qquad E(\Delta) = E_\mu - E_\lambda.$$

This is an important inequality which we shall also need in the next section and in Sec. 9.11.

We keep λ fixed and let $\mu \longrightarrow \lambda$ from the right in a monotone fashion. Then $E(\Delta)x \longrightarrow P(\lambda)x$ by the analogue of Theorem 9.3-3 for a decreasing sequence. Here $P(\lambda)$ is bounded and self-adjoint. Since $E(\Delta)$ is idempotent, so is $P(\lambda)$. Hence $P(\lambda)$ is a projection. Also $\lambda P(\lambda) = TP(\lambda)$ by (18), that is, $T_\lambda P(\lambda) = 0$. From this, (10*) and 9.8-2,

$$T_\lambda^+ P(\lambda) = T_\lambda (I - E_\lambda) P(\lambda) = (I - E_\lambda) T_\lambda P(\lambda) = 0.$$

Hence $T_\lambda^+ P(\lambda)x = 0$ for all $x \in H$. This shows that $P(\lambda)x \in \mathcal{N}(T_\lambda^+)$. By definition, E_λ projects H onto $\mathcal{N}(T_\lambda^+)$. Consequently, we have $E_\lambda P(\lambda)x = P(\lambda)x$, that is, $E_\lambda P(\lambda) = P(\lambda)$. On the other hand, if we let $\mu \longrightarrow \lambda + 0$ in (17) then

$$(I - E_\lambda) P(\lambda) = P(\lambda).$$

Together, $P(\lambda) = 0$. Remembering that $E(\Delta)x \longrightarrow P(\lambda)x$, we see that this proves (16); that is, $\mathscr{E}$ is continuous from the right.

This completes the proof that $\mathscr{E} = (E_\lambda)$ given in the theorem has all the properties by which a spectral family on $[m, M]$ is defined.

Problems

1. In certain cases, left continuity of a spectral family is more convenient than right continuity (and some books even formulate the definition accordingly). To see that there is not much difference, obtain from E_λ in Def. 9.7-1 an F_λ which is continuous from the left.

2. Suppose that E_λ satisfies all conditions Def. 9.7-1 except (9). Find $\tilde{E}_\lambda$ satisfying all those conditions including (9).

3. Prove that $T^- T = TT^-$ [cf. (6)].

4. Find T^+, T^-, $(T^2)^{1/2}$ and the other square roots of T^2 if

$$T = \begin{bmatrix} 2 & 0 \\ 0 & -3 \end{bmatrix}.$$

5. If in the finite dimensional case a linear operator T is represented by a real diagonal matrix $\tilde{T}$, what is the spectrum of T? How do we obtain from $\tilde{T}$ the matrix (*a*) $\tilde{T}^+$ (representing T^+), (*b*) $\tilde{T}^-$ (representing T^-), (*c*) $\tilde{B}$ (representing B)?

6. In Prob. 5, how can we obtain the matrix representing the projection of H (*a*) onto $\mathcal{N}(T^+)$, (*b*) onto $\mathcal{N}(T_\lambda{}^+)$?

7. Obtain from $\tilde{T}$ in Prob. 5 the matrices representing (*a*) T_λ, (*b*) $T_\lambda{}^+$, (*c*) $T_\lambda{}^-$, (*d*) B_λ.

8. Show that if a bounded self-adjoint linear operator $T\colon H \longrightarrow H$ is positive, then $T = T^+$ and $T^- = 0$.

9. Find the spectral family of the zero operator $T = 0\colon H \longrightarrow H$, where $H \neq \{0\}$.

10. Let $T = I\colon H \longrightarrow H$. Find $B_\lambda = (T_\lambda{}^2)^{1/2}$, $T_\lambda{}^+$, $\mathcal{N}(T_\lambda{}^+)$ and E_λ.

9.9 Spectral Representation of Bounded Self-Adjoint Linear Operators

From the preceding section we know that with a bounded self-adjoint linear operator T on a complex Hilbert space H we can associate a spectral family $\mathscr{E} = (E_\lambda)$. We want to show that $\mathscr{E}$ may be used to obtain a spectral representation of T; this is an integral representation (1) (below) which involves $\mathscr{E}$ and is such that $\langle Tx, y\rangle$ is represented by an ordinary Riemann-Stieltjes integral (cf. Sec. 4.4).

The notation $m - 0$ occurring in the theorem will be explained at the end of the theorem, before the proof.

9.9-1 Spectral Theorem for Bounded Self-Adjoint Linear Operators. *Let $T\colon H \longrightarrow H$ be a bounded self-adjoint linear operator on a complex Hilbert space H. Then:*

(a) *T has the spectral representation*

$$T = \int_{m-0}^{M} \lambda \, dE_\lambda \tag{1}$$

where $\mathscr{E} = (E_\lambda)$ is the spectral family associated with T (cf. 9.8-3); *the integral is to be understood in the sense of uniform operator convergence*

[*convergence in the norm on* $B(H, H)$], *and for all* $x, y \in H$,

$$\text{(1*)} \qquad \langle Tx, y\rangle = \int_{m-0}^{M} \lambda \, dw(\lambda), \qquad w(\lambda) = \langle E_\lambda x, y\rangle,$$

where the integral is an ordinary Riemann-Stieltjes integral (Sec. 4.4).

(b) *More generally, if p is a polynomial in* λ *with real coefficients, say,*

$$p(\lambda) = \alpha_n \lambda^n + \alpha_{n-1}\lambda^{n-1} + \cdots + \alpha_0,$$

then the operator $p(T)$ *defined by*

$$p(T) = \alpha_n T^n + \alpha_{n-1} T^{n-1} + \cdots + \alpha_0 I$$

has the spectral representation

$$\text{(2)} \qquad p(T) = \int_{m-0}^{M} p(\lambda) \, dE_\lambda$$

and for all $x, y \in H$,

$$\text{(2*)} \qquad \langle p(T)x, y\rangle = \int_{m-0}^{M} p(\lambda)\, dw(\lambda), \qquad w(\lambda) = \langle E_\lambda x, y\rangle.$$

(Extension to continuous functions see later, in Theorem 9.10-1.)

Remark. $m-0$ is written to indicate that one must take into account a contribution at $\lambda = m$ which occurs if $E_m \neq 0$ (and $m \neq 0$); thus, using any $a < m$, we can write

$$\int_a^M \lambda \, dE_\lambda = \int_{m-0}^{M} \lambda \, dE_\lambda = mE_m + \int_m^M \lambda \, dE_\lambda.$$

Similarly,

$$\int_a^M p(\lambda)\, dE_\lambda = \int_{m-0}^{M} p(\lambda)\, dE_\lambda = p(m)E_m + \int_m^M p(\lambda)\, dE_\lambda.$$

Proof of Theorem 9.9-1.

(a) We choose a sequence $(\mathscr{P}_n)$ of partitions of $(a, b]$ where $a < m$ and $M < b$. Here every $\mathscr{P}_n$ is a partition of $(a, b]$ into intervals

$$\Delta_{nj} = (\lambda_{nj}, \mu_{nj}] \qquad j = 1, \cdots, n$$

of length $l(\Delta_{nj}) = \mu_{nj} - \lambda_{nj}$. Note that $\mu_{nj} = \lambda_{n,j+1}$ for $j = 1, \cdots, n-1$. We assume the sequence $(\mathscr{P}_n)$ to be such that

$$\text{(3)} \qquad \eta(\mathscr{P}_n) = \max_j l(\Delta_{nj}) \longrightarrow 0 \qquad \text{as } n \longrightarrow \infty.$$

We use (18) in Sec. 9.8 with $\Delta = \Delta_{nj}$, that is,

$$\lambda_{nj} E(\Delta_{nj}) \leqq TE(\Delta_{nj}) \leqq \mu_{nj} E(\Delta_{nj}).$$

By summation over j from 1 to n we obtain for every n

$$\text{(4)} \qquad \sum_{j=1}^{n} \lambda_{nj} E(\Delta_{nj}) \leqq \sum_{j=1}^{n} TE(\Delta_{nj}) \leqq \sum_{j=1}^{n} \mu_{nj} E(\Delta_{nj}).$$

Since $\mu_{nj} = \lambda_{n,j+1}$ for $j = 1, \cdots, n-1$, using (14) and (15) in Sec. 9.8, we simply have

$$T \sum_{j=1}^{n} E(\Delta_{nj}) = T \sum_{j=1}^{n} (E_{\mu_{nj}} - E_{\lambda_{nj}}) = T(I - 0) = T.$$

Formula (3) implies that for every $\varepsilon > 0$ there is an n such that $\eta(\mathscr{P}_n) < \varepsilon$; hence in (4) we then have

$$\sum_{j=1}^{n} \mu_{nj} E(\Delta_{nj}) - \sum_{j=1}^{n} \lambda_{nj} E(\Delta_{nj}) = \sum_{j=1}^{n} (\mu_{nj} - \lambda_{nj}) E(\Delta_{nj}) < \varepsilon I.$$

From this and (4) it follows that, given any $\varepsilon > 0$, there is an N such that for every $n > N$ and every choice of $\hat{\lambda}_{nj} \in \Delta_{nj}$ we have

$$\text{(5)} \qquad \left\| T - \sum_{j=1}^{n} \hat{\lambda}_{nj} E(\Delta_{nj}) \right\| < \varepsilon.$$

Since E_λ is constant for $\lambda < m$ and for $\lambda \geqq M$, the particular choice of an $a < m$ and a $b > M$ is immaterial. This proves (1), where (5) shows that the integral is to be understood in the sense of uniform operator convergence. The latter implies strong operator convergence (cf. Sec. 4.9), the inner product is continuous and the sum in (5) is of Stieltjes type. Hence (1) implies (1*) for every choice of x and y in H.

(b) We prove the theorem for polynomials, starting with $p(\lambda) = \lambda^r$, where $r \in \mathbf{N}$. For any $\kappa < \lambda \leqq \mu < \nu$ we have from (7) in Sec. 9.7,

$$\begin{aligned}(E_\lambda - E_\kappa)(E_\mu - E_\nu) &= E_\lambda E_\mu - E_\lambda E_\nu - E_\kappa E_\mu + E_\kappa E_\nu \\ &= E_\lambda - E_\lambda - E_\kappa + E_\kappa = 0.\end{aligned}$$

This shows that $E(\Delta_{nj})E(\Delta_{nk}) = 0$ for $j \neq k$. Also, since $E(\Delta_{nj})$ is a projection, $E(\Delta_{nj})^s = E(\Delta_{nj})$ for every $s = 1, 2, \cdots$. Consequently, we obtain

$$\left[\sum_{j=1}^{n} \hat{\lambda}_{nj} E(\Delta_{nj})\right]^r = \sum_{j=1}^{n} \hat{\lambda}_{nj}{}^r E(\Delta_{nj}). \tag{6}$$

If the sum in (5) is close to T, the expression in (6) on the left will be close to T^r because multiplication (composition) of bounded linear operators is continuous. Hence, by (6), given $\varepsilon > 0$, there is an N such that for all $n > N$,

$$\left\| T^r - \sum_{j=1}^{n} \hat{\lambda}_{nj}{}^r E(\Delta_{nj}) \right\| < \varepsilon.$$

This proves (2) and (2*) for $p(\lambda) = \lambda^r$. From this the two formulas (2) and (2*) follow readily for an arbitrary polynomial with real coefficients. ▮

We should mention that the actual determination of the spectral family for a given bounded self-adjoint linear operator is not easy, in general. In some relatively simple cases the family may be conjectured from (1). In other cases one can use a systematic approach which is based on more advanced methods; see N. Dunford and J. T. Schwartz (1958–71), Part 2, pp. 920–921.

Let us conclude this section by listing some properties of operators

$p(T)$ which are of interest in themselves and will be helpful in the extension of the spectral theorem to general continuous functions.

9.9-2 Theorem (Properties of $p(T)$). *Let T be as in the previous theorem, and let p, p_1 and p_2 be polynomials with real coefficients. Then:*

(a) *$p(T)$ is self-adjoint.*

(b) *If $p(\lambda)=\alpha p_1(\lambda)+\beta p_2(\lambda)$, then $p(T)=\alpha p_1(T)+\beta p_2(T)$.*

(c) *If $p(\lambda)=p_1(\lambda)p_2(\lambda)$, then $p(T)=p_1(T)p_2(T)$.*

(d) *If $p(\lambda)\geqq 0$ for all $\lambda\in[m, M]$, then $p(T)\geqq 0$.*

(e) *If $p_1(\lambda)\leqq p_2(\lambda)$ for all $\lambda\in[m, M]$, then $p_1(T)\leqq p_2(T)$.*

(f) *$\|p(T)\|\leqq \max_{\lambda\in J}|p(\lambda)|$, where $J=[m, M]$.*

(g) *If a bounded linear operator commutes with T, it also commutes with $p(T)$.*

Proof. **(a)** holds since T is self-adjoint and p has real coefficients, so that $(\alpha_j T^j)^*=\alpha_j T^j$.

(b) is obvious from the definition.

(c) is obvious from the definition.

(d) Since p has real coefficients, complex zeros must occur in conjugate pairs if they occur at all. Since p changes sign if λ passes through a zero of odd multiplicity and $p(\lambda)\geqq 0$ on $[m, M]$, zeros of p in (m, M) must be of even multiplicity. Hence we can write

$$p(\lambda)=\alpha \prod_j (\lambda-\beta_j) \prod_k (\gamma_k-\lambda) \prod_l [(\lambda-\mu_l)^2+\nu_l{}^2] \tag{7}$$

where $\beta_j\leqq m$ and $\gamma_k\geqq M$ and the quadratic factors correspond to complex conjugate zeros and to real zeros in (m, M). We show that $\alpha>0$ if $p\neq 0$. For all sufficiently large λ, say, for all $\lambda\geqq\lambda_0$, we have

$$\operatorname{sgn} p(\lambda)=\operatorname{sgn}\alpha_n\lambda^n=\operatorname{sgn}\alpha_n,$$

where n is the degree of p. Hence $\alpha_n>0$ implies $p(\lambda_0)>0$ and the number of the γ_k's (each counted according to its multiplicity) must be even, to make $p(\lambda) \geqq 0$ in (m, M). Then all three products in (7) are positive at λ_0, hence we must have $\alpha > 0$ in order that $p(\lambda_0) > 0$. If $\alpha_n < 0$, then $p(\lambda_0) < 0$, the number of the γ_k's is odd, to make $p(\lambda) \geqq 0$ on (m, M). It follows that the second product in (7) is negative at λ_0, and $\alpha > 0$, as before.

We replace λ by T. Then each of the factors in (7) is a positive operator. In fact, for every $x \neq 0$, setting $v = \|x\|^{-1}x$, we have $x = \|x\| v$ and since $-\beta_j \geqq -m$,

$$\begin{aligned} \langle (T - \beta_j I)x, x \rangle &= \langle Tx, x \rangle - \beta_j \langle x, x \rangle \\ &\geqq \|x\|^2 \langle Tv, v \rangle - m\|x\|^2 \\ &\geqq \|x\|^2 \inf_{\|\tilde{v}\|=1} \langle T\tilde{v}, \tilde{v} \rangle - m\|x\|^2 = 0, \end{aligned}$$

that is, $T - \beta_j I \geqq 0$. Similarly, $\gamma_k I - T \geqq 0$. Also, $T - \mu_l I$ is self-adjoint, so that its square is positive and

$$(T - \mu_l I)^2 + \nu_l^2 I \geqq 0.$$

Since all those operators commute, their product is positive by 9.3-1, and $p(T) \geqq 0$ because $\alpha > 0$.

(e) follows immediately from (d).

(f) Let k denote the maximum of $|p(\lambda)|$ on J. Then $0 \leqq p(\lambda)^2 \leqq k^2$ for $\lambda \in J$. Hence (e) yields $p(T)^2 \leqq k^2 I$, that is, since $p(T)$ is self-adjoint, for all x,

$$\langle p(T)x, p(T)x \rangle = \langle p(T)^2 x, x \rangle \leqq k^2 \langle x, x \rangle.$$

The inequality in (f) now follows if we take square roots and then the supremum over all x of norm 1.

(g) follows immediately from the definition of $p(T)$. ∎

We shall use this theorem in the next section as an important tool for generalizing our present spectral theorem 9.9-1.

Problems

1. Verify (1) for the zero operator $T=0\colon H \longrightarrow H$.

2. Consider real numbers $\lambda_1 < \lambda_2 < \cdots < \lambda_n$ and projections $P_1, \cdots, P_n$ of a Hilbert space H onto n pairwise orthogonal subspaces of H. Assuming that $P_1 + \cdots + P_n = I$, show that

$$E_\lambda = \sum_{\lambda_k \leqq \lambda} P_k$$

defines a spectral family and list some of the properties of the corresponding operator

$$T = \int_{-\infty}^{+\infty} \lambda \, dE_\lambda.$$

3. Verify (1) if $T = I\colon H \longrightarrow H$.

4. If an operator $T\colon \mathbf{R}^3 \longrightarrow \mathbf{R}^3$ is represented, with respect to an orthonormal basis, by a matrix

$$\begin{bmatrix} 0 & 1 & 0 \\ 1 & 0 & 0 \\ 0 & 0 & 1 \end{bmatrix},$$

what is the corresponding spectral family? Using the result, verify (1) for this operator.

5. What spectral family (E_λ) corresponds to an n-rowed Hermitian matrix? Verify (1) for this case.

6. If we make the additional assumption that the self-adjoint operator T in (1) is compact, show that (1) takes the form of an infinite series or a finite sum.

7. Consider the *multiplication operator* $T\colon L^2[0, 1] \longrightarrow L^2[0, 1]$ defined by

$$y(t) = Tx(t) = tx(t).$$

Conclude from Prob. 9, Sec. 9.1, and Theorem 9.2-4 that $\sigma(T) = \sigma_c(T) = [0, 1]$. Show that the corresponding spectral family is defined by

$$E_\lambda x = \begin{cases} 0 & \text{if } \lambda < 0 \\ v_\lambda & \text{if } 0 \leqq \lambda \leqq 1 \\ x & \text{if } \lambda > 1 \end{cases}$$

where

$$v_\lambda(t) = \begin{cases} x(t) & \text{if } 0 \leqq t \leqq \lambda \\ 0 & \text{if } \lambda < t \leqq 1. \end{cases}$$

(It may be helpful to sketch the projections in the case of simple examples, such as $x(t) = t^2$ or $x(t) = \sin 2\pi t$.)

8. Find the spectral family of the operator $T: l^2 \longrightarrow l^2$ defined by $(\xi_1, \xi_2, \xi_3, \cdots) \longmapsto (\xi_1/1, \xi_2/2, \xi_3/3, \cdots)$. Find an orthonormal set of eigenvectors. What form does (1) take in this case?

9. Prove that in Prob. 8,

$$T = \sum_{j=1}^{\infty} \frac{1}{j} P_j$$

where P_j is the projection of l^2 onto the span of $e_j = (\delta_{jn})$ and the series is convergent in the norm of $B(l^2, l^2)$.

10. How could we use the idea of the proof given in the answer to Prob. 9 in the case of an arbitrary compact self-adjoint linear operator T having infinitely many different nonzero eigenvalues?

9.10 Extension of the Spectral Theorem to Continuous Functions

Theorem 9.9-1 holds for $p(T)$, where T is a bounded self-adjoint linear operator and p is a polynomial with real coefficients. We want to extend the theorem to operators $f(T)$, where T is as before and f is a continuous real-valued function. Clearly, we must first define what we mean by $f(T)$.

Let $T: H \longrightarrow H$ be a bounded self-adjoint linear operator on a complex Hilbert space H. Let f be a continuous real-valued function on $[m, M]$, where

$$(1) \qquad m = \inf_{\|x\|=1} \langle Tx, x\rangle, \qquad M = \sup_{\|x\|=1} \langle Tx, x\rangle,$$

as before. Then by the Weierstrass approximation theorem 4.11-5 there is a sequence of polynomials (p_n) with real coefficients such that

$$(2) \qquad p_n(\lambda) \longrightarrow f(\lambda)$$

uniformly on $[m, M]$. Corresponding to it we have a sequence of bounded self-adjoint linear operators $p_n(T)$. By Theorem 9.9-2(*f*),

$$\|p_n(T) - p_r(T)\| \leqq \max_{\lambda \in J} |p_n(\lambda) - p_r(\lambda)|,$$

where $J = [m, M]$. Since $p_n(\lambda) \longrightarrow f(\lambda)$, given any $\varepsilon > 0$, there is an N such that the right-hand side is smaller than ε for all $n, r > N$. Hence $(p_n(T))$ is Cauchy and has a limit in $B(H, H)$ since $B(H, H)$ is complete (cf. 2.10-2). We define $f(T)$ to be that limit; thus

$$\textbf{(3)} \qquad p_n(T) \longrightarrow f(T).$$

Of course, to justify this definition of $f(T)$, we must prove that $f(T)$ depends only on f (and T, of course) but not on the particular choice of a sequence of polynomials converging to f uniformly.

Proof. Let $(\tilde{p}_n)$ be another sequence of polynomials with real coefficients such that

$$\tilde{p}_n(\lambda) \longrightarrow f(\lambda)$$

uniformly on $[m, M]$. Then $\tilde{p}_n(T) \longrightarrow \tilde{f}(T)$ by the previous argument, and we must show that $\tilde{f}(T) = f(T)$. Clearly,

$$\tilde{p}_n(\lambda) - p_n(\lambda) \longrightarrow 0, \qquad \text{hence} \qquad \tilde{p}_n(T) - p_n(T) \longrightarrow 0,$$

again by 9.9-2(*f*). Consequently, given $\varepsilon > 0$, there is an N such that for $n > N$,

$$\|\tilde{f}(T) - \tilde{p}_n(T)\| < \frac{\varepsilon}{3}$$

$$\|\tilde{p}_n(T) - p_n(T)\| < \frac{\varepsilon}{3}$$

$$\|p_n(T) - f(T)\| < \frac{\varepsilon}{3}.$$

By the triangle inequality it follows that

$$\|\tilde{f}(T) - f(T)\| \leqq \|\tilde{f}(T) - \tilde{p}_n(T)\| + \|\tilde{p}_n(T) - p_n(T)\| + \|p_n(T) - f(T)\| < \varepsilon.$$

Since $\varepsilon > 0$ was arbitrary, $\tilde{f}(T) - f(T) = 0$ and $\tilde{f}(T) = f(T)$. ∎

With these preparations we can now extend Theorem 9.9-1 from polynomials p to general continuous real-valued functions f, as follows.

9.10-1 Spectral Theorem. *Let $T: H \longrightarrow H$ be a bounded self-adjoint linear operator on a complex Hilbert space H and f a continuous real-valued function on $[m, M]$;* cf. (1). *Then $f(T)$ has the spectral representation*[3]

(4)
$$f(T) = \int_{m-0}^{M} f(\lambda)\, dE_\lambda$$

where $\mathscr{E} = (E_\lambda)$ is the spectral family associated with T (cf. 9.8-3); *the integral is to be understood in the sense of uniform operator convergence, and for all $x, y \in H$,*

(4*)
$$\langle f(T)x, y\rangle = \int_{m-0}^{M} f(\lambda)\, dw(\lambda), \qquad w(\lambda) = \langle E_\lambda x, y\rangle$$

where the integral is an ordinary Riemann-Stieltjes integral (Sec. 4.4).

Proof. We use the same notation as in the proof of Theorem 9.9-1. For every $\varepsilon > 0$ there is a polynomial p with real coefficients

[3] The notation $m - 0$ is explained in connection with Theorem 9.9-1.

such that for all $\lambda \in [m, M]$

$$-\frac{\varepsilon}{3} \leqq f(\lambda) - p(\lambda) \leqq \frac{\varepsilon}{3}, \tag{5}$$

and then

$$\|f(T) - p(T)\| \leqq \frac{\varepsilon}{3}.$$

Furthermore, noting that $\sum E(\Delta_{nj}) = I$ and using (5), for any partition we obtain

$$-\frac{\varepsilon}{3} I \leqq \sum_{j=1}^{n} [f(\hat{\lambda}_{nj}) - p(\hat{\lambda}_{nj})] E(\Delta_{nj}) \leqq \frac{\varepsilon}{3} I.$$

It follows that

$$\left\| \sum_{j=1}^{n} [f(\hat{\lambda}_{nj}) - p(\hat{\lambda}_{nj})] E(\Delta_{nj}) \right\| \leqq \frac{\varepsilon}{3}.$$

Finally, since $p(T)$ is represented by (2), Sec. 9.9, there is an N such that for every $n > N$

$$\left\| \sum_{j=1}^{n} p(\hat{\lambda}_{nj}) E(\Delta_{nj}) - p(T) \right\| \leqq \frac{\varepsilon}{3}.$$

Using these inequalities, we can now estimate the norm of the difference between $f(T)$ and the Riemann-Stieltjes sums corresponding to the integral in (4); for $n > N$ we obtain by means of the triangle inequality

$$\begin{aligned} \left\| \sum_{j=1}^{n} f(\hat{\lambda}_{nj}) E(\Delta_{nj}) - f(T) \right\| &\leqq \left\| \sum_{j=1}^{n} [f(\hat{\lambda}_{nj}) - p(\hat{\lambda}_{nj})] E(\Delta_{nj}) \right\| \\ &\quad + \left\| \sum_{j=1}^{n} p(\hat{\lambda}_{nj}) E(\Delta_{nj}) - p(T) \right\| \\ &\quad + \left\| p(T) - f(T) \right\| \leqq \varepsilon. \end{aligned}$$

Since $\varepsilon > 0$ was arbitrary, this establishes (4) and (4*) and completes the proof. ∎

We mention the following uniqueness property.

$\mathscr{E} = (E_\lambda)$ is the only spectral family on $[m, M]$ that yields the representations (4) *and* (4*).

This becomes plausible if we observe that (4*) holds for every continuous real-valued function f on $[m, M]$ and the left-hand side of (4*) is defined in a way which does not depend on $\mathscr{E}$. A proof follows from a uniqueness theorem for Stieltjes integrals [cf. F. Riesz and B. Sz. -Nagy (1955), p. 111]; this theorem states that for any fixed x and y, the expression $w(\lambda) = \langle E_\lambda x, y \rangle$ is determined, up to an additive constant, by (4*) at its points of continuity and at $m-0$ and M. Since $E_M = I$, hence $\langle E_M x, y \rangle = \langle x, y \rangle$, and (E_λ) is continuous from the right, we conclude that $w(\lambda)$ is uniquely determined everywhere.

It is not difficult to see that the properties of $p(T)$ listed in Theorem 9.9-2 extend to $f(T)$; for later use we formulate this simple fact as

9.10-2 Theorem (Properties of $f(T)$). *Theorem* 9.9-2 *continues to hold if p, p_1, p_2 are replaced by continuous real-valued functions f, f_1, f_2 on $[m, M]$.*

9.11 Properties of the Spectral Family of a Bounded Self-Adjoint Linear Operator

It is interesting that the spectral family $\mathscr{E} = (E_\lambda)$ of a bounded self-adjoint linear operator T on a Hilbert space H reflects properties of the spectrum in a striking and simple fashion. We shall derive results of that kind from the definition of $\mathscr{E}$ (cf. Sec. 9.8) in combination with the spectral representation in Sec. 9.9.

From Sec. 9.7 we know that if H is finite dimensional, the spectral family $\mathscr{E} = (E_\lambda)$ has "points of growth" (discontinuities, jumps) precisely at the eigenvalues of T. In fact $E_{\lambda_0} - E_{\lambda_0 - 0} \neq 0$ if and only if λ_0 is an eigenvalue of T. It is remarkable, although perhaps not unexpected, that this property carries over to the infinite dimensional case:

9.11-1 Theorem (Eigenvalues). *Let $T: H \longrightarrow H$ be a bounded self-adjoint linear operator on a complex Hilbert space H and $\mathscr{E} = (E_\lambda)$ the corresponding spectral family. Then $\lambda \longmapsto E_\lambda$ has a discontinuity at any $\lambda = \lambda_0$ (that is, $E_{\lambda_0} \neq E_{\lambda_0 - 0}$) if and only if λ_0 is an eigenvalue of T. In this case, the corresponding eigenspace is*

$$\mathcal{N}(T - \lambda_0 I) = (E_{\lambda_0} - E_{\lambda_0 - 0})(H). \tag{1}$$

Proof. λ_0 is an eigenvalue of T if and only if $\mathcal{N}(T - \lambda_0 I) \neq \{0\}$, so that the first statement of the theorem follows immediately from (1). Hence it sufficies to prove (1). We write simply

$$F_0 = E_{\lambda_0} - E_{\lambda_0 - 0}$$

and prove (1) by first showing that

$$F_0(H) \subset \mathcal{N}(T - \lambda_0 I) \tag{2}$$

and then

$$F_0(H) \supset \mathcal{N}(T - \lambda_0 I). \tag{3}$$

Proof of (2):

Inequality (18) in Sec. 9.8 with $\lambda = \lambda_0 - \dfrac{1}{n}$ and $\mu = \lambda_0$ is

$$\left(\lambda_0 - \frac{1}{n}\right) E(\Delta_0) \leqq TE(\Delta_0) \leqq \lambda_0 E(\Delta_0) \tag{4}$$

where $\Delta_0 = (\lambda_0 - 1/n, \lambda_0]$. We let $n \longrightarrow \infty$. Then $E(\Delta_0) \longrightarrow F_0$, so that (4) yields

$$\lambda_0 F_0 \leqq TF_0 \leqq \lambda_0 F_0.$$

Hence $TF_0 = \lambda_0 F_0$, that is, $(T - \lambda_0 I)F_0 = 0$. This proves (2).

Proof of (3):

Let $x \in \mathcal{N}(T - \lambda_0 I)$. We show that then $x \in F_0(H)$, that is, $F_0 x = x$ since F_0 is a projection.

If $\lambda_0 \notin [m, M]$, then $\lambda_0 \in \rho(T)$ by 9.2-1. Hence in this case $\mathcal{N}(T-\lambda_0 I)=\{0\} \subset F_0(H)$ since $F_0(H)$ is a vector space. Let $\lambda_0 \in [m, M]$. By assumption, $(T-\lambda_0 I)x=0$. This implies $(T-\lambda_0 I)^2 x=0$, that is, by 9.9-1,

$$\int_a^b (\lambda-\lambda_0)^2\, dw(\lambda)=0, \qquad\qquad w(\lambda)=\langle E_\lambda x, x\rangle$$

where $a<m$ and $b>M$. Here $(\lambda-\lambda_0)^2 \geqq 0$ and $\lambda \longmapsto \langle E_\lambda x, x\rangle$ is monotone increasing by 9.7-1. Hence the integral over any subinterval of positive length must be zero. In particular, for every $\varepsilon>0$ we must have

$$0=\int_a^{\lambda_0-\varepsilon} (\lambda-\lambda_0)^2\, dw(\lambda) \geqq \varepsilon^2 \int_a^{\lambda_0-\varepsilon} dw(\lambda)=\varepsilon^2\langle E_{\lambda_0-\varepsilon}x, x\rangle$$

and

$$0=\int_{\lambda_0+\varepsilon}^{b} (\lambda-\lambda_0)^2\, dw(\lambda) \geqq \varepsilon^2 \int_{\lambda_0+\varepsilon}^{b} dw(\lambda)=\varepsilon^2\langle Ix, x\rangle-\varepsilon^2\langle E_{\lambda_0+\varepsilon}x, x\rangle.$$

Since $\varepsilon>0$, from this and 9.5-2 we obtain

$$\langle E_{\lambda_0-\varepsilon}x, x\rangle=0 \qquad \text{hence} \qquad E_{\lambda_0-\varepsilon}x=0$$

and

$$\langle x-E_{\lambda_0+\varepsilon}x, x\rangle=0 \qquad \text{hence} \qquad x-E_{\lambda_0+\varepsilon}x=0.$$

We may thus write

$$x=(E_{\lambda_0+\varepsilon}-E_{\lambda_0-\varepsilon})x.$$

If we let $\varepsilon \longmapsto 0$, we obtain $x=F_0x$ because $\lambda \longmapsto E_\lambda$ is continuous from the right. This implies (3), as was noted before. ∎

We know that the spectrum of a bounded self-adjoint linear operator T lies on the real axis of the complex plane; cf. 9.1-3. Of course, the real axis also contains points of the resolvent set $\rho(T)$. For instance, $\lambda \in \rho(T)$ if λ is real and $\lambda<m$ or $\lambda>M$; cf. 9.2-1. It is quite remarkable that *all* real $\lambda \in \rho(T)$ can be characterized by the behavior of the spectral family in a very simple fashion. This theorem will then immediately yield a characterization of points of the continuous spectrum of T and thus complete our present discussion since the residual spectrum of T is empty, by 9.2-4.

9.11-2 Theorem (Resolvent set). *Let T and $\mathscr{E}=(E_\lambda)$ be as in Theorem 9.11-1. Then a real λ_0 belongs to the resolvent set $\rho(T)$ of T if and only if there is a $\gamma>0$ such that $\mathscr{E}=(E_\lambda)$ is constant on the interval $[\lambda_0-\gamma, \lambda_0+\gamma]$.*

Proof. In part (a) we prove that the given condition is sufficient for $\lambda_0 \in \rho(T)$ and in (b) that it is necessary. In the proof we use Theorem 9.1-2 which states that $\lambda_0 \in \rho(T)$ if and only if there exists a $\gamma>0$ such that for all $x \in H$,

$$\|(T-\lambda_0 I)x\| \geqq \gamma\|x\|. \tag{5}$$

(a) Suppose that λ_0 is real and such that $\mathscr{E}$ is constant on $J=[\lambda_0-\gamma, \lambda_0+\gamma]$ for some $\gamma>0$. By Theorem 9.9-1,

$$\|(T-\lambda_0 I)x\|^2 = \langle (T-\lambda_0 I)^2 x, x\rangle = \int_{m-0}^{M} (\lambda-\lambda_0)^2 \, d\langle E_\lambda x, x\rangle. \tag{6}$$

Since $\mathscr{E}$ is constant on J, integration over J yields the value zero, and for $\lambda \notin J$ we have $(\lambda-\lambda_0)^2 \geqq \gamma^2$, so that (6) now implies

$$\|(T-\lambda_0 I)x\|^2 \geqq \gamma^2 \int_{m-0}^{M} d\langle E_\lambda x, x\rangle = \gamma^2 \langle x, x\rangle.$$

Taking square roots, we obtain (5). Hence $\lambda_0 \in \rho(T)$ by 9.1-2.

(b) Conversely, suppose that $\lambda_0 \in \rho(T)$. Then (5) with some $\gamma>0$ holds for all $x \in H$, so that by (6) and 9.9-1,

$$\int_{m-0}^{M} (\lambda-\lambda_0)^2 \, d\langle E_\lambda x, x\rangle \geqq \gamma^2 \int_{m-0}^{M} d\langle E_\lambda x, x\rangle. \tag{7}$$

We show that we obtain a contradiction if we assume that $\mathscr{E}$ is not constant on the interval $[\lambda_0-\gamma, \lambda_0+\gamma]$. In fact, then we can find a positive $\eta<\gamma$ such that $E_{\lambda_0+\eta}-E_{\lambda_0-\eta} \neq 0$ because $E_\lambda \leqq E_\mu$ for $\lambda<\mu$ (cf. 9.7-1). Hence there is a $y \in H$ such that

$$x=(E_{\lambda_0+\eta}-E_{\lambda_0-\eta})y \neq 0.$$

We use this x in (7). Then

$$E_\lambda x = E_\lambda (E_{\lambda_0+\eta} - E_{\lambda_0-\eta})y.$$

Formula (7) in Sec. 9.7 shows that this is $(E_\lambda - E_\lambda)y = 0$ when $\lambda < \lambda_0 - \eta$ and $(E_{\lambda_0+\eta} - E_{\lambda_0-\eta})y$ when $\lambda > \lambda_0 + \eta$, hence independent of λ. We may thus take $K = [\lambda_0 - \eta, \lambda_0 + \eta]$ as the interval of integration in (7). If $\lambda \in K$, then we obtain $\langle E_\lambda x, x\rangle = \langle (E_\lambda - E_{\lambda_0-\eta})y, y\rangle$ by straightforward calculation, using again (7) in Sec. 9.7. Hence (7) gives

$$\int_{\lambda_0-\eta}^{\lambda_0+\eta} (\lambda - \lambda_0)^2 \, d\langle E_\lambda y, y\rangle \geqq \gamma^2 \int_{\lambda_0-\eta}^{\lambda_0+\eta} d\langle E_\lambda y, y\rangle.$$

But this impossible because the integral on the right is positive and $(\lambda - \lambda_0)^2 \leqq \eta^2 < \gamma^2$, where $\lambda \in K$. Hence our assumption that $\mathscr{E}$ is not constant on the interval $[\lambda_0 - \gamma, \lambda_0 + \gamma]$ is false and the proof is complete. ∎

This theorem also shows that $\lambda_0 \in \sigma(T)$ if and only if $\mathscr{E}$ is not constant in any neighborhood of λ_0 on **R**. Since $\sigma_r(T) = \emptyset$ by 9.2-4 and points of $\sigma_p(T)$ correspond to discontinuities of $\mathscr{E}$ (cf. 9.11-1), we have the following theorem, which completes our discussion.

9.11-3 Theorem (Continuous spectrum). *Let T and $\mathscr{E} = (E_\lambda)$ be as in Theorem 9.11-1. Then a real λ_0 belongs to the continuous spectrum $\sigma_c(T)$ of T if and only if $\mathscr{E}$ is continuous at λ_0 (thus $E_{\lambda_0} = E_{\lambda_0-0}$) and is not constant in any neighborhood of λ_0 on* **R**.

Problems

1. What can we conclude from Theorem 9.11-1 in the case of a Hermitian matrix?

2. If T in Theorem 9.11-1 is compact and has infinitely many eigenvalues, what can we conclude about (E_λ) from Theorems 9.11-1 and 9.11-2?

3. Verify that the spectral family in Prob. 7, Sec. 9.9, satisfies the three theorems in the present section.

4. We know that if m in Theorem 9.2-1 is positive then T is positive. How does this follow from the spectral representation (1), Sec. 9.9?

5. We know that the spectrum of a bounded self-adjoint linear operator is closed. How does this follow from theorems in this section?

6. Let $T: l^2 \longrightarrow l^2$ be defined by $y = (\eta_j) = Tx$ where $x = (\xi_j)$, $\eta_j = \alpha_j \xi_j$ and (α_j) is any real sequence in a finite interval $[a, b]$. Show that the corresponding spectral family (E_λ) is defined by

$$\langle E_\lambda x, y \rangle = \sum_{\alpha_j \leqq \lambda} \xi_j \bar{\eta}_j.$$

7. (Pure point spectrum) A bounded self-adjoint linear operator $T: H \longrightarrow H$ on a Hilbert space $H \neq \{0\}$ is said to have a *pure point spectrum* or *purely discrete spectrum* if T has an orthonormal set of eigenvectors which is total in H. Illustrate with an example that this does *not* imply $\sigma_c(T) = \varnothing$ (so that this terminology, which is generally used, may confuse the beginner for a moment).

8. Give examples of compact self-adjoint linear operators $T: l^2 \longrightarrow l^2$ having a pure point spectrum such that the set of the nonzero eigenvalues (a) is a finite point set, (b) is an infinite point set and the corresponding eigenvectors form a dense set in l^2, (c) is an infinite point set and the corresponding eigenvectors span a subspace of l^2 such that the orthogonal complement of the closure of that subspace is finite dimensional, (d) as in (c) but that complement is infinite dimensional. In each case find a total orthonormal set of eigenvectors.

9. (Purely continuous spectrum) A bounded self-adjoint linear operator $T: H \longrightarrow H$ on a Hilbert space $H \neq \{0\}$ is said to have a *purely continuous spectrum* if T has no eigenvalues. If T is any bounded self-adjoint linear operator on H, show that there is a closed subspace $Y \subset H$ which reduces T (cf. Sec. 9.6, Prob. 10) and is such that $T_1 = T|_Y$ has a pure point spectrum whereas $T_2 = T|_Z$, $Z = Y^\perp$, has a purely continuous spectrum. (This reduction facilitates the investigation of T; cf. also the remark in Sec. 9.6, Prob. 10.)

10. What can we say about the spectral families $(E_{\lambda 1})$ and $(E_{\lambda 2})$ of T_1 and T_2 in Prob. 9 in terms of the spectral family (E_λ) of T?

CHAPTER 10

UNBOUNDED LINEAR OPERATORS IN HILBERT SPACE

Unbounded linear operators occur in many applications, notably in connection with differential equations and in quantum mechanics. Their theory is more complicated than that of bounded operators.

In this chapter we restrict ourselves to Hilbert spaces; this is the case of prime interest in physics. In fact, the theory of unbounded operators was stimulated by attempts in the late 1920's to put quantum mechanics on a rigorous mathematical foundation. A systematic development of the theory is due to J. von Neumann (1929–30, 1936) and M. H. Stone (1932).

The application of that theory to differential equations yields a unified approach to diverse questions and also entails substantial simplification.

The chapter is optional.

Important concepts, brief orientation about main content

For unbounded operators, considerations about domains and extension problems become of prime importance. In order that the Hilbert-adjoint operator T^* of a linear operator T exist, T must be *densely defined* in H, that is, its domain $\mathfrak{D}(T)$ must be dense in H (cf. Sec. 10.1). On the other hand, if T satisfies the relation

$$\langle Tx, y\rangle = \langle x, Ty\rangle$$

identically and is unbounded, its domain cannot be all of H (cf. 10.1-1). This relation is equivalent to $T \subset T^*$ (provided T is densely defined in H) and T is called *symmetric* (Sec. 10.2). A *self-adjoint* linear operator ($T = T^*$; cf. 10.2-5) is symmetric, but the converse is not generally true in the unbounded case.

Most unbounded linear operators occurring in practical problems are *closed* or have closed linear extensions (Sec. 10.3).

The spectrum of a self-adjoint linear operator is real, also in the unbounded case (cf. 10.4-2). A spectral representation (cf. 10.6-3) of such an operator T is obtained by means of the *Cayley transform*

$$U=(T-iI)(T+iI)^{-1}$$

of T (cf. Sec. 10.6) and the spectral theorem 10.5-4 for unitary operators.

Section 10.7 is devoted to a multiplication operator and a differentiation operator, two unbounded linear operators of particular practical importance. (These operators play a key role in Chap. 11 on quantum mechanics.)

In this chapter, the following will be convenient. We say that T is an operator **on** H if its domain is all of H, and an operator **in** H if its domain lies in H but *may* not be all of H. Furthermore, the notation

$$S\subset T$$

will mean that T is an extension of S.

10.1 Unbounded Linear Operators and their Hilbert-Adjoint Operators

Throughout this chapter we shall consider linear operators $T\colon \mathcal{D}(T)\longrightarrow H$ whose domain $\mathcal{D}(T)$ lies in a complex Hilbert space H. We admit that such an operator T may be *unbounded*, that is, T may not be bounded.

From Sec. 2.7 we remember that T is bounded if and only if there is a real number c such that for all $x\in\mathcal{D}(T)$,

$$\|Tx\|\leqq c\|x\|.$$

An important unbounded linear operator is the differentiation operator considered in Sec. 4.13.

We expect an unbounded linear operator to differ from a bounded one in various ways, and the question arises on what properties we should focus attention. A famous result (Theorem 10.1-1, below) suggests that the domain of the operator and the problem of extending the operator will play a particular role. In fact, we shall see that quite a number of properties of an operator depend on the domain and may change under extensions and restrictions.

When that theorem was discovered by E. Hellinger and O. Toeplitz (1910) it aroused both admiration and puzzlement since the theorem establishes a relation between properties of two different kinds, namely, the properties of being defined everywhere and of being bounded.

In the case of a *bounded* linear operator T on a Hilbert space H, self-adjointness of T was defined by

$$\langle Tx, y\rangle = \langle x, Ty\rangle \tag{1}$$

(cf. 3.10-1). This is a very important property. And the theorem shows that an *unbounded* linear operator T satisfying (1) cannot be defined on all of H.

10.1-1 Hellinger-Toeplitz Theorem (Boundedness). *If a linear operator T is defined on all of a complex Hilbert space H and satisfies (1) for all $x, y \in H$, then T is bounded.*

Proof. Otherwise H would contain a sequence (y_n) such that

$$\|y_n\| = 1 \qquad \text{and} \qquad \|Ty_n\| \longrightarrow \infty.$$

We consider the functional f_n defined by

$$f_n(x) = \langle Tx, y_n\rangle = \langle x, Ty_n\rangle$$

where $n = 1, 2, \cdots$, and we used (1). Each f_n is defined on all of H and is linear. For each fixed n the functional f_n is bounded since the Schwarz inequality gives

$$|f_n(x)| = |\langle x, Ty_n\rangle| \leqq \|Ty_n\| \, \|x\|.$$

Moreover, for every fixed $x \in H$, the sequence $(f_n(x))$ is bounded. Indeed, using the Schwarz inequality and $\|y_n\| = 1$, we have

$$|f_n(x)| = |\langle Tx, y_n\rangle| \leqq \|Tx\|.$$

From this and the uniform boundedness theorem 4.7-3 we conclude that $(\|f_n\|)$ is bounded, say, $\|f_n\| \leqq k$ for all n. This implies that for every $x \in H$ we have

$$|f_n(x)| \leqq \|f_n\| \, \|x\| \leqq k\|x\|$$

and, taking $x = Ty_n$, we arrive at

$$\|Ty_n\|^2 = \langle Ty_n, Ty_n \rangle = |f_n(Ty_n)| \leqq k\|Ty_n\|.$$

Hence $\|Ty_n\| \leqq k$, which contradicts our initial assumption $\|Ty_n\| \longrightarrow \infty$ and proves the theorem. ∎

Since, by this theorem, $\mathfrak{D}(T) = H$ is impossible for unbounded linear operators satisfying (1), we are confronted with the problem of determining suitable domains and with extension problems. We shall use the convenient notation

$$S \subset T;$$

by definition, this means that the operator T is an **extension** of the operator S; thus

$$\mathfrak{D}(S) \subset \mathfrak{D}(T) \qquad \text{and} \qquad S = T|_{\mathfrak{D}(S)}.$$

We call an extension T of S a *proper extension* if $\mathfrak{D}(S)$ is a proper subset of $\mathfrak{D}(T)$, that is, $\mathfrak{D}(T) - \mathfrak{D}(S) \neq \varnothing$.

In the theory of *bounded* operators, the Hilbert-adjoint operator T^* of an operator T plays a basic role. So let us first generalize this important concept to *unbounded* operators.

In the *bounded* case the operator T^* is defined by (cf. 3.9-1)

$$\langle Tx, y \rangle = \langle x, T^*y \rangle,$$

which we can write

$$\text{(2)} \qquad \text{(a)} \quad \langle Tx, y \rangle = \langle x, y^* \rangle, \qquad \text{(b)} \quad y^* = T^*y.$$

T^* exists on H and is a bounded linear operator with norm $\|T^*\| = \|T\|$; cf. 3.9-2.

In the general case we also want to use (2). Clearly, T^* will then be defined for those $y \in H$ for which there is a y^* such that (2) holds for all $x \in \mathfrak{D}(T)$.

Now comes an important point. In order that T^* be an operator (a mapping), for each y which is supposed to belong to the domain $\mathfrak{D}(T^*)$ of T^* the corresponding $y^* = T^*y$ must be unique. We claim that this holds if and only if:

T is **densely defined** in H, that is, $\mathfrak{D}(T)$ is dense in H.

Indeed, if $\mathfrak{D}(T)$ is not dense in H, then $\overline{\mathfrak{D}(T)} \neq H$, the orthogonal complement (Sec. 3.3) of $\overline{\mathfrak{D}(T)}$ in H contains a nonzero y_1, and $y_1 \perp x$ for every $x \in \mathfrak{D}(T)$, that is, $\langle x, y_1 \rangle = 0$. But then in (2) we obtain

$$\langle x, y^* \rangle = \langle x, y^* \rangle + \langle x, y_1 \rangle = \langle x, y^* + y_1 \rangle,$$

which shows non-uniqueness. On the other hand, if $\mathfrak{D}(T)$ is dense in H, then $\mathfrak{D}(T)^{\perp} = \{0\}$ by 3.3-7. Hence $\langle x, y_1 \rangle = 0$ for all $x \in \mathfrak{D}(T)$ now implies $y_1 = 0$, so that $y^* + y_1 = y^*$, which is the desired uniqueness. ∎

We agree to call T an operator **on** H if $\mathfrak{D}(T)$ is all of H, and an operator **in** H if $\mathfrak{D}(T)$ lies in H but *may* not be all of H. This is a convenient way of talking in this chapter.

Our present situation motivates the following

10.1-2 Definition (Hilbert-adjoint operator). Let $T\colon \mathfrak{D}(T) \longrightarrow H$ be a (possibly unbounded) densely defined linear operator in a complex Hilbert space H. Then the *Hilbert-adjoint operator* $T^*\colon \mathfrak{D}(T^*) \longrightarrow H$ of T is defined as follows. The domain $\mathfrak{D}(T^*)$ of T^* consists of all $y \in H$ such that there is a $y^* \in H$ satisfying

$$\langle Tx, y \rangle = \langle x, y^* \rangle \tag{2a}$$

for all $x \in \mathfrak{D}(T)$. For each such $y \in \mathfrak{D}(T^*)$ the Hilbert-adjoint operator T^* is then defined in terms of that y^* by

$$y^* = T^*y. \tag{2b}$$ ∎

In other words, an element $y \in H$ is in $\mathfrak{D}(T^*)$ if (and only if) for that y the inner product $\langle Tx, y \rangle$, considered as a function of x, can be represented in the form $\langle Tx, y \rangle = \langle x, y^* \rangle$ for all $x \in \mathfrak{D}(T)$. Also for that

y, formula (2) determines the corresponding y^* uniquely since $\mathcal{D}(T)$ is dense in H, by assumption.

The reader may show that T^* is a linear operator.

In our further work we shall need sums and products (composites) of operators. We must be very careful because different operators may have different domains, in particular in the unbounded case. Hence we must first define what we mean by sums and products in this more general situation. This can be done in a rather natural fashion as follows.

Let $S: \mathcal{D}(S) \longrightarrow H$ and $T: \mathcal{D}(T) \longrightarrow H$ be linear operators, where $\mathcal{D}(S) \subset H$ and $\mathcal{D}(T) \subset H$. Then the **sum** $S+T$ of S and T is the linear operator with domain

$$\mathcal{D}(S+T) = \mathcal{D}(S) \cap \mathcal{D}(T)$$

and for every $x \in \mathcal{D}(S+T)$ defined by

$$(S+T)x = Sx + Tx.$$

Note that $\mathcal{D}(S+T)$ is the largest set on which both S and T make sense, and $\mathcal{D}(S+T)$ is a vector space.

Note further that always $0 \in \mathcal{D}(S+T)$, so that $\mathcal{D}(S+T)$ is never empty; but it is clear that nontrivial results can be expected only if $\mathcal{D}(S+T)$ also contains other elements.

Let us define the product TS, where S and T are as before. Let M be the largest subset of $\mathcal{D}(S)$ whose image under S lies in $\mathcal{D}(T)$; thus,

$$S(M) = \mathcal{R}(S) \cap \mathcal{D}(T),$$

where $\mathcal{R}(S)$ is the range of S; cf. Fig. 65. Then the **product** TS is defined to be the operator with domain $\mathcal{D}(TS) = M$ such that for all $x \in \mathcal{D}(TS)$,

$$(TS)x = T(Sx).$$

Interchanging S and T in this definition, we see that the product ST is the operator such that for all $x \in \mathcal{D}(ST)$,

$$(ST)x = S(Tx),$$

where $\mathcal{D}(ST) = \tilde{M}$ is the largest subset of $\mathcal{D}(T)$ whose image under T

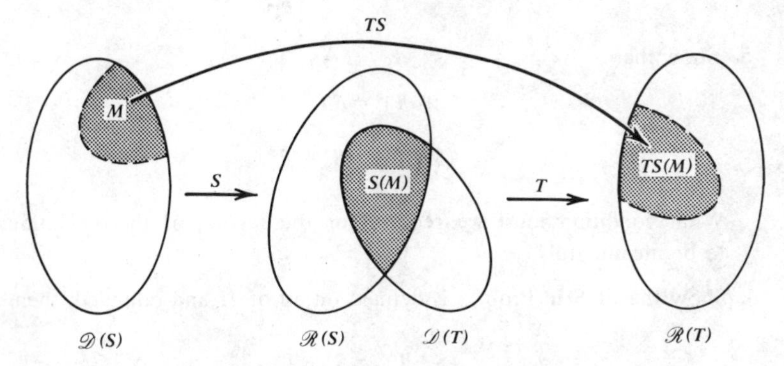

Fig. 65. Product of linear operators

lies in $\mathscr{D}(S)$; thus,

$$T(\tilde{M}) = \mathscr{R}(T) \cap \mathscr{D}(S).$$

TS and ST are linear operators. This can be readily verified; note in particular that $\mathscr{R}(S)$ is a vector space by 2.6-9, so that $S(M)$ is a vector space, which implies that M is a vector space since S is linear. And $\tilde{M}$ is a vector space by the same argument.

Problems

1. Prove that the Hilbert-adjoint operator T^* of a linear operator T is linear.

2. Show that for a *bounded* linear operator, our present definition of the Hilbert-adjoint operator yields Def. 3.9-1 with $H_1 = H_2 = H$.

3. Show that

$$(T_1T_2)T_3 = T_1(T_2T_3)$$

continues to hold for operators which may be unbounded.

4. Show that

$$(T_1 + T_2)T_3 = T_1T_3 + T_2T_3$$

$$T_1(T_2 + T_3) \supset T_1T_2 + T_1T_3.$$

Give a condition sufficient for equality in the second formula.

5. Show that

$$(\alpha T)^* = \bar{\alpha} T^*$$

$$(S+T)^* \supset S^* + T^*.$$

What condition must we require for the second of these relations to be meaningful?

6. Show that if S in Prob. 5 is defined on all of H and bounded, then

$$(S+T)^* = S^* + T^*.$$

7. Show that a bounded linear operator $T\colon \mathfrak{D}(T) \longrightarrow H$ whose domain is not dense in H always has a bounded linear extension to H whose norm equals $\|T\|$.

8. Let $T\colon \mathfrak{D}(T) \longrightarrow l^2$ be defined by

$$y = (\eta_j) = Tx, \qquad \eta_j = j\xi_j, \qquad x = (\xi_j)$$

where $\mathfrak{D}(T) \subset l^2$ consists of all $x = (\xi_j)$ with only finitely many nonzero terms ξ_j. (*a*) Show that T is unbounded. (*b*) Does T have proper linear extensions? (*c*) Can T be linearly extended to the whole space l^2?

9. If a linear operator T is defined everywhere on a complex Hilbert space H, show that its Hilbert-adjoint operator T^* is bounded.

10. Let S and T be linear operators which are defined on all of H and satisfy

$$\langle Ty, x\rangle = \langle y, Sx\rangle \qquad \text{for all } x, y \in H.$$

Show that then T is bounded and S is its Hilbert-adjoint operator.

10.2 Hilbert-Adjoint Operators, Symmetric and Self-Adjoint Linear Operators

In the following two theorems we shall state some basic properties of the Hilbert-adjoint operator. Here, by definition,

$$T^{**} = (T^*)^*.$$

10.2-1 Theorem (Hilbert-adjoint operator). *Let $S\colon \mathfrak{D}(S) \longrightarrow H$ and $T\colon \mathfrak{D}(T) \longrightarrow H$ be linear operators which are densely defined in a complex Hilbert space H. Then:*

(a) *If $S \subset T$, then $T^* \subset S^*$.*

(b) *If $\mathfrak{D}(T^*)$ is dense in H, then $T \subset T^{**}$.*

Proof. **(a)** By the definition of T^*,

$$\langle Tx, y\rangle = \langle x, T^*y\rangle \tag{1}$$

for all $x \in \mathfrak{D}(T)$ and all $y \in \mathfrak{D}(T^*)$. Since $S \subset T$, this implies

$$\langle Sx, y\rangle = \langle x, T^*y\rangle \tag{2}$$

for all $x \in \mathfrak{D}(S)$ and y as before. By the definition of S^*,

$$\langle Sx, y\rangle = \langle x, S^*y\rangle \tag{3}$$

for all $x \in \mathfrak{D}(S)$ and all $y \in \mathfrak{D}(S^*)$. From this and (2) we want to conclude that $\mathfrak{D}(T^*) \subset \mathfrak{D}(S^*)$. We explain this step in detail since similar conclusions will occur later. By the definition of the Hilbert-adjoint operator S^* the domain $\mathfrak{D}(S^*)$ includes *all* y for which one has a representation (3) of $\langle Sx, y\rangle$ with x varying throughout $\mathfrak{D}(S)$. Since (2) also represents $\langle Sx, y\rangle$ in the same form, with x varying throughout $\mathfrak{D}(S)$, the set of y's for which (2) is valid must be a (proper or improper) subset of the set of y's for which (3) holds, that is, we must have $\mathfrak{D}(T^*) \subset \mathfrak{D}(S^*)$. From (2) and (3) we now readily conclude that $S^*y = T^*y$ for all $y \in \mathfrak{D}(T^*)$, so that, by definition, $T^* \subset S^*$.

(b) Taking complex conjugates in (1), we have

$$\langle T^*y, x\rangle = \langle y, Tx\rangle \tag{4}$$

for all $y \in \mathfrak{D}(T^*)$ and all $x \in \mathfrak{D}(T)$. Since $\mathfrak{D}(T^*)$ is dense in H, the operator T^{**} exists and, by the definition,

$$\langle T^*y, x\rangle = \langle y, T^{**}x\rangle$$

for all $y \in \mathfrak{D}(T^*)$ and all $x \in \mathfrak{D}(T^{**})$. From this and (4), reasoning as in part (a), we see that an $x \in \mathscr{D}(T)$ also belongs to $\mathscr{D}(T^{**})$, and we have $T^{**}x = Tx$ for such an x. This means that $T \subset T^{**}$. ∎

Our second theorem concerns conditions under which the inverse of the adjoint equals the adjoint of the inverse. (Note that this extends Prob. 2 in Sec. 3.9 to linear operators which may be unbounded.)

10.2-2 Theorem (Inverse of the Hilbert-adjoint operator). *Let T be as in the preceding theorem. Moreover, suppose that T is injective and its range $\mathscr{R}(T)$ is dense in H. Then T^* is injective and*

$$(T^*)^{-1} = (T^{-1})^*. \tag{5}$$

Proof. T^* exists since T is densely defined in H. Also T^{-1} exists since T is injective. $(T^{-1})^*$ exists since $\mathscr{D}(T^{-1}) = \mathscr{R}(T)$ is dense in H. We must show that $(T^*)^{-1}$ exists and satisfies (5).

Let $y \in \mathscr{D}(T^*)$. Then for all $x \in \mathscr{D}(T^{-1})$ we have $T^{-1}x \in \mathscr{D}(T)$ and

$$\langle T^{-1}x, T^*y\rangle = \langle TT^{-1}x, y\rangle = \langle x, y\rangle. \tag{6}$$

On the other hand, by the definition of the Hilbert-adjoint operator of T^{-1} we have

$$\langle T^{-1}x, T^*y\rangle = \langle x, (T^{-1})^*T^*y\rangle$$

for all $x \in \mathscr{D}(T^{-1})$. This shows that $T^*y \in \mathscr{D}((T^{-1})^*)$. Furthermore, comparing this and (6), we conclude that

$$(T^{-1})^*T^*y = y \qquad\qquad y \in \mathscr{D}(T^*). \tag{7}$$

We see that $T^*y = 0$ implies $y = 0$. Hence $(T^*)^{-1}\colon \mathscr{R}(T^*) \longrightarrow \mathscr{D}(T^*)$ exists by 2.6-10. Furthermore, since $(T^*)^{-1}T^*$ is the identity operator on $\mathscr{D}(T^*)$, a comparison with (7) shows that

$$(T^*)^{-1} \subset (T^{-1})^*. \tag{8}$$

To establish (5), we merely have to show that

$$(T^*)^{-1} \supset (T^{-1})^*. \tag{9}$$

For this purpose we consider any $x \in \mathscr{D}(T)$ and $y \in \mathscr{D}((T^{-1})^*)$. Then

$Tx \in \mathcal{R}(T) = \mathcal{D}(T^{-1})$ and

$$(10) \qquad \langle Tx, (T^{-1})^* y\rangle = \langle T^{-1}Tx, y\rangle = \langle x, y\rangle.$$

On the other hand, by the definition of the Hilbert-adjoint operator of T we have

$$\langle Tx, (T^{-1})^* y\rangle = \langle x, T^*(T^{-1})^* y\rangle$$

for all $x \in \mathcal{D}(T)$. From this and (10) we conclude that $(T^{-1})^* y \in \mathcal{D}(T^*)$ and

$$(11) \qquad T^*(T^{-1})^* y = y \qquad\qquad y \in \mathcal{D}((T^{-1})^*).$$

Now, by the definition of an inverse, $T^*(T^*)^{-1}$ is the identity operator on $\mathcal{D}((T^*)^{-1}) = \mathcal{R}(T^*)$, and $(T^*)^{-1}: \mathcal{R}(T^*) \longrightarrow \mathcal{D}(T^*)$ is surjective. Hence by comparison with (11) we obtain $\mathcal{D}((T^*)^{-1}) \supset \mathcal{D}((T^{-1})^*)$, so that $(T^*)^{-1} \supset (T^{-1})^*$, which is (9). Together with (8) this gives (5). ∎

In connection with *bounded* linear operators, the Hilbert-adjoint operator was used to define self-adjointness (cf. Sec. 3.10). Since this is a very important concept in the theory as well as in applications, we wonder whether and how we can generalize it to include unbounded linear operators. For this purpose we first introduce the following concept.

10.2-3 Definition (Symmetric linear operator). Let $T: \mathcal{D}(T) \longrightarrow H$ be a linear operator which is densely defined in a complex Hilbert space H. Then T is called a *symmetric linear operator* if for all $x, y \in \mathcal{D}(T)$,

$$\langle Tx, y\rangle = \langle x, Ty\rangle.$$ ∎

It is quite remarkable that symmetry can be expressed in terms of Hilbert-adjoint operators in a simple fashion. This will be helpful in our further work. It also motivates the assumption in Def. 10.2-3 that T be densely defined.

10.2-4 Lemma (Symmetric operator). *A densely defined linear operator T in a complex Hilbert space H is symmetric if and only if*

$$(12) \qquad T \subset T^*.$$

Proof. The defining relation of T^* is

$$\langle Tx, y\rangle = \langle x, T^* y\rangle, \tag{13}$$

valid for all $x \in \mathfrak{D}(T)$ and all $y \in \mathfrak{D}(T^*)$. Assume that $T \subset T^*$. Then $T^* y = Ty$ for $y \in \mathfrak{D}(T)$, so that (13), for $x, y \in \mathfrak{D}(T)$, becomes

$$\langle Tx, y\rangle = \langle x, Ty\rangle. \tag{14}$$

Hence T is symmetric.

Conversely, suppose that (14) holds for all $x, y \in \mathfrak{D}(T)$. Then a comparison with (13) shows that $\mathfrak{D}(T) \subset \mathfrak{D}(T^*)$ and $T = T^*|_{\mathfrak{D}(T)}$. By definition this means that T^* is an extension of T. ∎

Self-adjointness may now be defined as follows.

10.2-5 Definition (Self-adjoint linear operator). Let $T\colon \mathfrak{D}(T) \longrightarrow H$ be a linear operator which is densely defined in a complex Hilbert space H. Then T is called a *self-adjoint linear operator* if

$$T = T^*. \tag{15}$$ ∎

Every self-adjoint linear operator is symmetric.

On the other hand, a symmetric linear operator need not be self-adjoint. The reason is that T^* may be a *proper* extension of T, that is, $\mathfrak{D}(T) \neq \mathfrak{D}(T^*)$. Clearly, this cannot happen if $\mathfrak{D}(T)$ is all of H. Hence we have:

For a linear operator $T\colon H \longrightarrow H$ on a complex Hilbert space H, the concepts of symmetry and self-adjointness are identical.

Note that in this case, T is bounded (cf. 10.1-1), and this explains why the concept of symmetry did not occur earlier, say, in Sec. 3.10.

Furthermore, there is an analogue of 3.10-3:

A densely defined linear operator T in a complex Hilbert space H is symmetric if and only if $\langle Tx, x\rangle$ is real for all $x \in \mathfrak{D}(T)$.

Problems

1. Show that a self-adjoint linear operator is symmetric.

2. If S and T are such that ST is densely defined in H, show that

$$(ST)^* \supset T^* S^*,$$

and if S is defined on all of H and is bounded, then

$$(ST)^* = T^*S^*.$$

3. Let H be a complex Hilbert space and $T\colon \mathfrak{D}(T) \longrightarrow H$ linear and densely defined in H. Show that T is symmetric if and only if $\langle Tx, x\rangle$ is real for all $x \in \mathfrak{D}(T)$.

4. If T is symmetric, show that T^{**} is symmetric.

5. If a linear operator T is densely defined in H and its adjoint is defined on all of H, show that T is bounded.

6. Show that $y = (\eta_j) = Tx = (\xi_j/j)$ defines a bounded self-adjoint linear operator $T\colon l^2 \longrightarrow l^2$ which has an unbounded self-adjoint inverse.

7. Let $T\colon \mathfrak{D}(T) \longrightarrow H$ be a bounded symmetric linear operator. Show that T has a bounded symmetric linear extension $\tilde{T}$ to $\overline{\mathfrak{D}(T)}$.

8. If T is symmetric and $\tilde{T}$ is a symmetric extension of T, show that $\tilde{T} \subset T^*$.

9. A **maximally symmetric** linear operator is defined to be a symmetric linear operator which has no proper symmetric extensions. Show that a self-adjoint linear operator T is maximally symmetric.

10. If a self-adjoint linear operator $T\colon \mathfrak{D}(T) \longrightarrow H$ is injective, show that (a) $\overline{\mathfrak{R}(T)} = H$ and (b) T^{-1} is self-adjoint.

10.3 Closed Linear Operators and Closures

Applications may lead to linear operators which are unbounded. But many of these operators are closed or at least have a linear extension which is closed. This explains the important role of closed linear operators in the theory of unbounded operators. In the present section we shall consider closed linear extensions and some of their properties.

We start with a review of the definition of a closed linear operator and some results from Sec. 4.13, using formulations which are convenient for Hilbert spaces.

10.3-1 Definition (Closed linear operator). Let $T\colon \mathfrak{D}(T) \longrightarrow H$ be a linear operator, where $\mathfrak{D}(T) \subset H$ and H is a complex Hilbert space.

Then T is called a *closed linear operator* if its *graph*

$$\mathscr{G}(T)=\{(x, y) \mid x \in \mathscr{D}(T), y=Tx\}$$

is closed in $H \times H$, where the norm on $H \times H$ is defined by

$$\|(x, y)\|=(\|x\|^2+\|y\|^2)^{1/2}$$

and results from the inner product defined by

$$\langle (x_1, y_1), (x_2, y_2)\rangle=\langle x_1, x_2\rangle+\langle y_1, y_2\rangle. \qquad \blacksquare$$

10.3-2 Theorem (Closed linear operator). *Let T: $\mathscr{D}(T) \longrightarrow H$ be a linear operator, where $\mathscr{D}(T) \subset H$ and H is a complex Hilbert space. Then:*

(a) *T is closed if and only if*

$$x_n \longrightarrow x \qquad [x_n \in \mathscr{D}(T)] \qquad \text{and} \qquad Tx_n \longrightarrow y$$

together imply that $x \in \mathscr{D}(T)$ and $Tx=y$. (Cf. 4.13-3.)

(b) *If T is closed and $\mathscr{D}(T)$ is closed, then T is bounded.* (Cf. 4.13-2.)

(c) *Let T be bounded. Then T is closed if and only if $\mathscr{D}(T)$ is closed.* (Cf. 4.13-5.)

Whether T is closed or not, we always have the remarkable

10.3-3 Theorem (Hilbert-adjoint operator). *The Hilbert-adjoint operator T^* defined in* 10.1-2 *is closed.*

Proof. We prove the theorem by applying Theorem 10.3-2(a) to T^*; that is, we consider any sequence (y_n) in $\mathscr{D}(T^*)$ such that

$$y_n \longrightarrow y_0 \qquad \text{and} \qquad T^* y_n \longrightarrow z_0$$

and show that $y_0 \in \mathscr{D}(T^*)$ and $z_0=T^* y_0$.

By the definition of T^* we have for every $y \in \mathscr{D}(T)$

$$\langle Ty, y_n\rangle=\langle y, T^* y_n\rangle.$$

Since the inner product is continuous, by letting $n \longrightarrow \infty$ we obtain

$$\langle Ty, y_0 \rangle = \langle y, z_0 \rangle$$

for every $y \in \mathfrak{D}(T)$. By the definition of T^* this shows that $y_0 \in \mathfrak{D}(T^*)$ and $z_0 = T^* y_0$. Applying Theorem 10.3-2(*a*) to T^*, we conclude that T^* is closed. ∎

It frequently happens that an operator is not closed but has an extension which is closed. To discuss this situation, we first formulate some relevant concepts.

10.3-4 Definition (Closable operator, closure). If a linear operator T has an extension T_1 which is a closed linear operator, then T is said to be *closable*, and T_1 is called a *closed linear extension* of T.

A closed linear extension $\bar{T}$ of a closable linear operator T is said to be *minimal* if every closed linear extension T_1 of T is a closed linear extension of $\bar{T}$. This minimal extension $\bar{T}$ of T—if it exists—is called the *closure* of T. ∎

If $\bar{T}$ exists, it is unique. (Why?)

If T is not closed, the problem arises whether T has closed extensions.

For instance, practically all unbounded linear operators in quantum mechanics are closable.

For symmetric linear operators (cf. 10.2-3) the situation is very simple, as follows.

10.3-5 Theorem (Closure). *Let $T\colon \mathfrak{D}(T) \longrightarrow H$ be a linear operator, where H is a complex Hilbert space and $\mathfrak{D}(T)$ is dense in H. Then if T is symmetric, its closure $\bar{T}$ exists and is unique.*

Proof. We define $\bar{T}$ by first defining the domain $M = \mathfrak{D}(\bar{T})$ and then $\bar{T}$ itself. Then we show that $\bar{T}$ is indeed the closure of T.

Let M be the set of all $x \in H$ for which there is a sequence (x_n) in $\mathfrak{D}(T)$ and a $y \in H$ such that

$$x_n \longrightarrow x \qquad \text{and} \qquad Tx_n \longrightarrow y. \tag{1}$$

It is not difficult to see that M is a vector space. Clearly, $\mathfrak{D}(T) \subset M$. On

M we define $\bar{T}$ by setting

$$y = \bar{T}x \qquad (x \in M) \tag{2}$$

with y given by (1). To show that $\bar{T}$ is the closure of T, we have to prove that $\bar{T}$ has all the properties by which the closure is defined.

Obviously, $\bar{T}$ has the domain $\mathfrak{D}(\bar{T}) = M$. Furthermore, we shall prove:

(*a*) To each $x \in \mathfrak{D}(\bar{T})$ there corresponds a *unique* y.
(*b*) $\bar{T}$ is a symmetric linear extension of T.
(*c*) $\bar{T}$ is closed and is the closure of T.

The details are as follows.

(a) *Uniqueness of y for every $x \in \mathfrak{D}(\bar{T})$.* In addition to (x_n) in (1) let $(\tilde{x}_n)$ be another sequence in $\mathfrak{D}(T)$ such that

$$\tilde{x}_n \longrightarrow x \qquad \text{and} \qquad T\tilde{x}_n \longrightarrow \tilde{y}.$$

Since T is linear, $Tx_n - T\tilde{x}_n = T(x_n - \tilde{x}_n)$. Since T is symmetric, we thus have for every $v \in \mathfrak{D}(T)$

$$\langle v, Tx_n - T\tilde{x}_n \rangle = \langle Tv, x_n - \tilde{x}_n \rangle.$$

We let $n \longrightarrow \infty$. Then, using the continuity of the inner product, we obtain

$$\langle v, y - \tilde{y} \rangle = \langle Tv, x - x \rangle = 0,$$

that is, $y - \tilde{y} \perp \mathfrak{D}(T)$. Since $\mathfrak{D}(T)$ is dense in H, we have $\mathfrak{D}(T)^{\perp} = \{0\}$ by 3.3-7, and $y - \tilde{y} = 0$.

(b) *Proof that $\bar{T}$ is a symmetric linear extension of T.* Since T is linear, so is $\bar{T}$ by (1) and (2), which also shows that $\bar{T}$ is an extension of T. We show that the symmetry of T implies that of $\bar{T}$. By (1) and (2), for all $x, z \in \mathfrak{D}(\bar{T})$ there are sequences (x_n) and (z_n) in $\mathfrak{D}(T)$ such that

$$x_n \longrightarrow x, \qquad Tx_n \longrightarrow \bar{T}x$$

$$z_n \longrightarrow z, \qquad Tz_n \longrightarrow \bar{T}z.$$

Since T is symmetric, $\langle z_n, Tx_n \rangle = \langle Tz_n, x_n \rangle$. We let $n \longrightarrow \infty$. Then we obtain $\langle z, \bar{T}x\rangle = \langle \bar{T}z, x\rangle$ because the inner product is continuous. Since $x, z \in \mathfrak{D}(\bar{T})$ were arbitrary, this shows that $\bar{T}$ is symmetric.

(c) *Proof that $\bar{T}$ is closed and is the closure of T.* We prove closedness of $\bar{T}$ by means of Theorem 10.3-2(a), that is, by considering any sequence (w_m) in $\mathfrak{D}(\bar{T})$ such that

$$w_m \longrightarrow x \qquad \text{and} \qquad \bar{T}w_m \longrightarrow y \tag{3}$$

and proving that $x \in \mathfrak{D}(\bar{T})$ and $\bar{T}x = y$.

Every w_m (m fixed) is in $\mathfrak{D}(\bar{T})$. By the definition of $\mathfrak{D}(\bar{T})$ there is a sequence in $\mathfrak{D}(T)$ which converges to w_m and whose image under T converges to $\bar{T}w_m$. Hence for every fixed m there is a $v_m \in \mathfrak{D}(T)$ such that

$$\|w_m - v_m\| < \frac{1}{m} \qquad \text{and} \qquad \|\bar{T}w_m - Tv_m\| < \frac{1}{m}.$$

From this and (3) we conclude that

$$v_m \longrightarrow x \qquad \text{and} \qquad Tv_m \longrightarrow y.$$

By the definitions of $\mathfrak{D}(\bar{T})$ and $\bar{T}$ this shows that $x \in \mathfrak{D}(\bar{T})$ and $y = \bar{T}x$, the relationships we wanted to prove. Hence $\bar{T}$ is closed by 10.3-2(a).

From Theorem 10.3-2(a) and our definition of $\mathfrak{D}(\bar{T})$ we see that every point of $\mathfrak{D}(\bar{T})$ must also belong to the domain of every closed linear extension of T. This shows that $\bar{T}$ is the closure of T and, moreover, it implies that the closure of T is unique. ∎

It is interesting and not difficult to see that the Hilbert-adjoint operator of the closure of a symmetric linear operator equals the Hilbert-adjoint operator of the operator itself:

10.3-6 Theorem (Hilbert-adjoint of the closure). *For a symmetric linear operator T as in the preceding theorem we have*

$$(\bar{T})^* = T^*. \tag{4}$$

Proof. Since $T \subset \bar{T}$, we have $(\bar{T})^* \subset T^*$ by Theorem 10.2-1(a). Hence $\mathfrak{D}((\bar{T})^*) \subset \mathfrak{D}(T^*)$, and all we must show is

$$y \in \mathfrak{D}(T^*) \qquad \Longrightarrow \qquad y \in \mathfrak{D}((\bar{T})^*), \tag{5}$$

because then we have $\mathfrak{D}((\bar{T})^*) = \mathfrak{D}(T^*)$, which implies (4).

Let $y \in \mathfrak{D}(T^*)$. By the definition of the Hilbert-adjoint operator, formula (5) means that we have to prove that for every $x \in \mathfrak{D}(\bar{T})$,

$$\langle \bar{T}x, y\rangle = \langle x, (\bar{T})^* y\rangle = \langle x, T^* y\rangle, \tag{6}$$

where the second equality follows from $(\bar{T})^* \subset T^*$.

By the definitions of $\mathfrak{D}(\bar{T})$ and $\bar{T}$ in the previous proof [cf. (1), (2)], for each $x \in \mathfrak{D}(\bar{T})$ there is a sequence (x_n) in $\mathfrak{D}(T)$ such that

$$x_n \longrightarrow x \qquad \text{and} \qquad Tx_n \longrightarrow y_0 = \bar{T}x.$$

Since $y \in \mathfrak{D}(T^*)$ by assumption and $x_n \in \mathfrak{D}(T)$, by the definition of the Hilbert-adjoint operator we have

$$\langle Tx_n, y\rangle = \langle x_n, T^* y\rangle.$$

If we let $n \longrightarrow \infty$ and use the continuity of the inner product, we obtain

$$\langle \bar{T}x, y\rangle = \langle x, T^* y\rangle, \qquad x \in \mathfrak{D}(\bar{T}),$$

the relationship (6) which we wanted to prove. ∎

Problems

1. Let $T: \mathfrak{D}(T) \longrightarrow l^2$, where $\mathfrak{D}(T) \subset l^2$ consists of all $x = (\xi_j)$ with only finitely many nonzero terms ξ_j and $y = (\eta_j) = Tx = (j\xi_j)$. This operator T is unbounded (cf. Prob. 8, Sec. 10.1). Show that T is not closed.

2. Clearly, the graph $\mathscr{G}(T)$ of any linear operator $T: \mathfrak{D}(T) \longrightarrow H$ has a closure $\overline{\mathscr{G}(T)} \subset H \times H$. Why does this not imply that every linear operator is closable?

3. Show that $H \times H$ with the inner product given in Def. 10.3-1 is a Hilbert space.

4. Let $T: \mathfrak{D}(T) \longrightarrow H$ be a closed linear operator. If T is injective, show that T^{-1} is closed.

5. Show that T in Prob. 1 has a closed linear extension T_1 to

$$\mathfrak{D}(T_1) = \left\{ x = (\xi_j) \in l^2 \;\middle|\; \sum_{j=1}^{\infty} j^2 \, |\xi_j|^2 < \infty \right\}$$

defined by $T_1 x = (j\xi_j)$. (Use Prob. 4.)

6. If T is a symmetric linear operator, show that T^{**} is a closed symmetric linear extension of T.

7. Show that the graph $\mathscr{G}(T^*)$ of the Hilbert-adjoint operator of a linear operator T is related to $\mathscr{G}(T)$ by

$$\mathscr{G}(T^*) = [U(\mathscr{G}(T))]^{\perp}$$

where $U: H \times H \longrightarrow H \times H$ is defined by $(x, y) \longmapsto (y, -x)$.

8. If $T: \mathscr{D}(T) \longrightarrow H$ is a densely defined closed linear operator, show that T^* is densely defined and $T^{**} = T$. (Use Prob. 7.)

9. (Closed graph theorem) Show that a closed linear operator $T: H \longrightarrow H$ on a complex Hilbert space H is bounded. (Use Prob. 8. Of course, give an independent proof, without using 4.13-2.)

10. If T is closed, show that $T_\lambda = T - \lambda I$ is closed, and if T_λ^{-1} exists, then T_λ^{-1} is closed.

10.4 Spectral Properties of Self-Adjoint Linear Operators

General properties of the spectrum of *bounded* self-adjoint linear operators were considered in Secs. 9.1 and 9.2. Several of these properties continue to hold for *unbounded* self-adjoint linear operators. In particular, the eigenvalues are real. The proof is the same as that of Theorem 9.1-1.

More generally, the whole spectrum continues to be real and closed, although it will no longer be bounded. To prove the reality of the spectrum, let us first generalize Theorem 9.1-2 which characterizes the resolvent set $\rho(T)$. The proof will be almost the same as before.

10.4-1 Theorem (Regular values). *Let $T: \mathscr{D}(T) \longrightarrow H$ be a self-adjoint linear operator which is densely defined in a complex Hilbert space H. Then a number λ belongs to the resolvent set $\rho(T)$ of T if and only if there exists a $c > 0$ such that for every $x \in \mathscr{D}(T)$,*

$$\|T_\lambda x\| \geqq c\|x\| \tag{1}$$

where $T_\lambda = T - \lambda I$.

Proof. (*a*) Let $\lambda \in \rho(T)$. Then, by Def. 7.2-1, the resolvent $R_\lambda = (T - \lambda I)^{-1}$ exists and is bounded, say, $\|R_\lambda\| = k > 0$. Consequently, since $R_\lambda T_\lambda x = x$ for $x \in \mathcal{D}(T)$, we have

$$\|x\| = \|R_\lambda T_\lambda x\| \leqq \|R_\lambda\| \, \|T_\lambda x\| = k\|T_\lambda x\|.$$

Division by k yields $\|T_\lambda x\| \geqq c\|x\|$, where $c = 1/k$.

(*b*) Conversely, suppose that (1) holds for some $c > 0$ and all $x \in \mathcal{D}(T)$. We consider the vector space

$$Y = \{y \mid y = T_\lambda x, \ x \in \mathcal{D}(T)\},$$

that is, the range of T_λ, and show that

(α) $T_\lambda\colon \mathcal{D}(T) \longrightarrow Y$ is bijective;
(β) Y is dense in H;
(γ) Y is closed.

Together this will imply that the resolvent $R_\lambda = T_\lambda^{-1}$ is defined on all of H. Boundedness of R_λ will then easily result from (1), so that $\lambda \in \rho(T)$. The details are as follows.

(α) Consider any $x_1, x_2 \in \mathcal{D}(T)$ such that $T_\lambda x_1 = T_\lambda x_2$. Since T_λ is linear, (1) yields

$$0 = \|T_\lambda x_1 - T_\lambda x_2\| = \|T_\lambda (x_1 - x_2)\| \geqq c\|x_1 - x_2\|.$$

Since $c > 0$, this implies $\|x_1 - x_2\| = 0$. Hence $x_1 = x_2$, so that the operator $T_\lambda\colon \mathcal{D}(T) \longrightarrow Y$ is bijective.

(β) We prove that $\bar{Y} = H$ by showing that $x_0 \perp Y$ implies $x_0 = 0$. Let $x_0 \perp Y$. Then for every $y = T_\lambda x \in Y$,

$$0 = \langle T_\lambda x, x_0 \rangle = \langle Tx, x_0 \rangle - \lambda \langle x, x_0 \rangle.$$

Hence for all $x \in \mathcal{D}(T)$,

$$\langle Tx, x_0 \rangle = \langle x, \bar{\lambda} x_0 \rangle.$$

By the definition of the Hilbert-adjoint operator this shows that

$x_0 \in \mathfrak{D}(T^*)$ and

$$T^* x_0 = \bar{\lambda} x_0.$$

Since T is self-adjoint, $\mathfrak{D}(T^*) = \mathfrak{D}(T)$ and $T^* = T$; thus

$$T x_0 = \bar{\lambda} x_0.$$

$x_0 \neq 0$ would imply that $\bar{\lambda}$ is an eigenvalue of T, and then $\bar{\lambda} = \lambda$ must be real. Hence $T x_0 = \lambda x_0$, that is, $T_\lambda x_0 = 0$. But now (1) yields a contradiction:

$$0 = \|T_\lambda x_0\| \geqq c\|x_0\| \qquad \Longrightarrow \qquad \|x_0\| = 0.$$

It follows that $\bar{Y}^\perp = \{0\}$, so that $\bar{Y} = H$ by 3.3-4.

(γ) We prove that Y is closed. Let $y_0 \in \bar{Y}$. Then there is a sequence (y_n) in Y such that $y_n \longrightarrow y_0$. Since $y_n \in Y$, we have $y_n = T_\lambda x_n$ for some $x_n \in \mathfrak{D}(T_\lambda) = \mathfrak{D}(T)$. By (1),

$$\|x_n - x_m\| \leqq \frac{1}{c} \|T_\lambda (x_n - x_m)\| = \frac{1}{c} \|y_n - y_m\|.$$

Since (y_n) converges, this shows that (x_n) is Cauchy. Since H is complete, (x_n) converges, say, $x_n \longrightarrow x_0$. Since T is self-adjoint, it is closed by 10.3-3. Theorem 10.3-2(*a*) thus implies that we have $x_0 \in \mathfrak{D}(T)$ and $T_\lambda x_0 = y_0$. This shows that $y_0 \in Y$. Since $y_0 \in \bar{Y}$ was arbitrary, Y is closed.

Parts (β) and (γ) imply that $Y = H$. From this and (α) we see that the resolvent R_λ exists and is defined on all of H:

$$R_\lambda = T_\lambda^{-1} \colon H \longrightarrow \mathfrak{D}(T).$$

R_λ is linear by 2.6-10. Boundedness of R_λ follows from (1), because for every $y \in H$ and corresponding $x = R_\lambda y$ we have $y = T_\lambda x$ and by (1),

$$\|R_\lambda y\| = \|x\| \leqq \frac{1}{c} \|T_\lambda x\| = \frac{1}{c} \|y\|,$$

so that $\|R_\lambda\| \leqq 1/c$. By definition this proves that $\lambda \in \rho(T)$. ∎

Generalizing Theorem 9.1-3, by the use of the theorem just proved, we can now show that the spectrum of a (possibly unbounded) self-adjoint linear operator is real:

10.4-2 Theorem (Spectrum). *The spectrum $\sigma(T)$ of a self-adjoint linear operator $T: \mathcal{D}(T) \longrightarrow H$ is real and closed; here, H is a complex Hilbert space and $\mathcal{D}(T)$ is dense in H.*

Proof. **(a)** *Reality of $\sigma(T)$.* For every $x \neq 0$ in $\mathcal{D}(T)$ we have

$$\langle T_\lambda x, x\rangle = \langle Tx, x\rangle - \lambda\langle x, x\rangle$$

and, since $\langle x, x\rangle$ and $\langle Tx, x\rangle$ are real (cf. Sec. 10.2),

$$\overline{\langle T_\lambda x, x\rangle} = \langle Tx, x\rangle - \bar{\lambda}\langle x, x\rangle.$$

We write $\lambda = \alpha + i\beta$ with real α and β. Then $\bar{\lambda} = \alpha - i\beta$, and subtraction yields

$$\overline{\langle T_\lambda x, x\rangle} - \langle T_\lambda x, x\rangle = (\lambda - \bar{\lambda})\langle x, x\rangle = 2i\beta \,\|x\|^2.$$

The left side equals $-2i \operatorname{Im} \langle T_\lambda x, x\rangle$. Since the imaginary part of a complex number cannot exceed the absolute value, we have by the Schwarz inequality

$$|\beta| \,\|x\|^2 \leqq |\langle T_\lambda x, x\rangle| \leqq \|T_\lambda x\| \,\|x\|.$$

Division by $\|x\| \neq 0$ gives $|\beta| \,\|x\| \leqq \|T_\lambda x\|$. Note that this inequality holds for all $x \in \mathcal{D}(T)$. If λ is not real, $\beta \neq 0$, so that $\lambda \in \rho(T)$ by the previous theorem. Hence $\sigma(T)$ must be real.

(b) *Closedness of $\sigma(T)$.* We show that $\sigma(T)$ is closed by proving that the resolvent set $\rho(T)$ is open. For this purpose we consider any $\lambda_0 \in \rho(T)$ and show that every λ sufficiently close to λ_0 also belongs to $\rho(T)$.

By the triangle inequality,

$$\begin{aligned}\|Tx - \lambda_0 x\| &= \|Tx - \lambda x + (\lambda - \lambda_0)x\| \\ &\leqq \|Tx - \lambda x\| + |\lambda - \lambda_0| \,\|x\|.\end{aligned}$$

This can be written

$$\|Tx-\lambda x\| \geqq \|Tx-\lambda_0 x\| - |\lambda-\lambda_0|\, \|x\|. \tag{2}$$

Since $\lambda_0 \in \rho(T)$, by Theorem 10.4-1 there is a $c>0$ such that for all $x \in \mathfrak{D}(T)$,

$$\|Tx-\lambda_0 x\| \geqq c\|x\|. \tag{3}$$

We now assume that λ is close to λ_0, say, $|\lambda-\lambda_0| \leqq c/2$. Then (2) and (3) imply that for all $x \in \mathfrak{D}(T)$,

$$\|Tx-\lambda x\| \geqq c\|x\| - \tfrac{1}{2}c\|x\| = \tfrac{1}{2}c\|x\|.$$

Hence $\lambda \in \rho(T)$ by Theorem 10.4-1. Since λ was such that $|\lambda-\lambda_0| \leqq c/2$ but otherwise arbitrary, this shows that λ_0 has a neighborhood belonging entirely to $\rho(T)$. Since $\lambda_0 \in \rho(T)$ was arbitrary, we conclude that $\rho(T)$ is open. Hence $\sigma(T) = \mathbf{C} - \rho(T)$ is closed. ∎

Problems

1. Without using Theorem 10.4-2 show that the eigenvalues of a (possibly unbounded) self-adjoint linear operator are real.

2. Show that eigenvectors corresponding to different eigenvalues of a self-adjoint linear operator are orthogonal.

3. (Approximate eigenvalues) Let $T\colon \mathfrak{D}(T) \longrightarrow H$ be a linear operator. If for a complex number λ there is a sequence (x_n) in $\mathfrak{D}(T)$ such that $\|x_n\| = 1$ and

$$(T-\lambda I)x_n \longrightarrow 0 \qquad (n \longrightarrow \infty),$$

then λ is often called an *approximate eigenvalue* of T. Show that the spectrum of a self-adjoint linear operator T consists entirely of approximate eigenvalues.

4. Let $T\colon \mathfrak{D}(T) \longrightarrow H$ be a linear operator. Characterize the fact that a λ is in $\rho(T)$, $\sigma_p(T)$, $\sigma_c(T)$ and $\sigma_r(T)$, respectively, in terms of the following properties. (A) T_λ is not injective. (B) $\mathcal{R}(T_\lambda)$ is not dense in H. (C) λ is an approximate eigenvalue (cf. Prob. 3).

5. Let $T: \mathcal{D}(T) \longrightarrow H$ be a linear operator whose Hilbert-adjoint operator T^* exists. If $\lambda \in \sigma_r(T)$, show that $\bar{\lambda} \in \sigma_p(T^*)$.

6. If $\bar{\lambda} \in \sigma_p(T^*)$ in Prob. 5, show that $\lambda \in \sigma_r(T) \cup \sigma_p(T)$.

7. (Residual spectrum) Using Prob. 5, show that the residual spectrum $\sigma_r(T)$ of a self-adjoint linear operator $T: \mathcal{D}(T) \longrightarrow H$ is empty. Note that this means that Theorem 9.2-4 continues to hold in the unbounded case.

8. If T_1 is a linear extension of a linear operator $T: \mathcal{D}(T) \longrightarrow H$, show that

$$\sigma_p(T) \subset \sigma_p(T_1),$$

$$\sigma_r(T) \supset \sigma_r(T_1),$$

$$\sigma_c(T) \subset \sigma_c(T_1) \cup \sigma_p(T_1).$$

9. Show that the point spectrum $\sigma_p(T)$ of a symmetric linear operator $T: \mathcal{D}(T) \longrightarrow H$ is real. If H is separable, show that $\sigma_p(T)$ is countable (perhaps finite or even empty).

10. If $T: \mathcal{D}(T) \longrightarrow H$ is a symmetric linear operator and λ is not real, show that the resolvent R_λ of T exists and is a bounded linear operator satisfying

$$\|R_\lambda y\| \leqq \|y\|/|\beta| \qquad (\lambda = \alpha + i\beta)$$

for every $y \in \mathcal{R}(T_\lambda)$, so that $\lambda \in \rho(T) \cup \sigma_r(T)$.

10.5 Spectral Representation of Unitary Operators

Our goal is a spectral representation of self-adjoint linear operators which may be unbounded. We shall obtain such a representation from the spectral representation of unitary operators, which are *bounded* linear operators, as we know from Sec. 3.10. In this approach we must first derive a spectral theorem for unitary operators.

We begin by showing that the spectrum of a unitary operator (cf. Def. 3.10-1) lies on the unit circle in the complex plane (circle of radius 1 with center at 0; see Fig. 66).

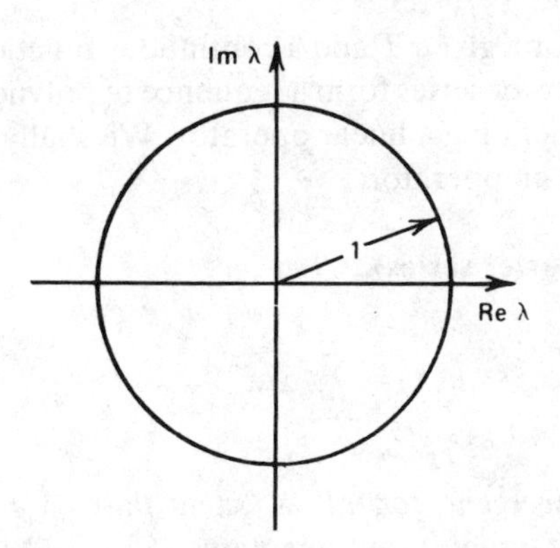

Fig. 66. Unit circle in the complex plane

10.5-1 Theorem (Spectrum). *If $U: H \longrightarrow H$ is a unitary linear operator on a complex Hilbert space $H \neq \{0\}$, then the spectrum $\sigma(U)$ is a closed subset of the unit circle; thus*

$$|\lambda| = 1 \qquad\qquad \text{for every } \lambda \in \sigma(U).$$

Proof. We have $\|U\| = 1$ by Theorem 3.10-6(*b*). Hence $|\lambda| \leqq 1$ for all $\lambda \in \sigma(U)$ by Theorem 7.3-4. Also $0 \in \rho(U)$ since for $\lambda = 0$ the resolvent operator of U is $U^{-1} = U^*$. The operator U^{-1} is unitary by Theorem 3.10-6(*c*). Hence $\|U^{-1}\| = 1$. Theorem 7.3-3 with $T = U$ and $\lambda_0 = 0$ now implies that every λ satisfying $|\lambda| < 1/\|U^{-1}\| = 1$ belongs to $\rho(U)$. Hence the spectrum of U must lie on the unit circle. It is closed by Theorem 7.3-2. ∎

There are several ways in which the spectral theorem for unitary operators U can be obtained; see, for example, J. von Neumann (1929–30), pp. 80, 119, M. Stone (1932), p. 302, K. Friedrichs (1935), and F. Riesz and B. Sz.-Nagy (1955), p. 281. We shall approach the problem by means of power series and a lemma (10.5-3, below) by F. J. Wecken (1935). This will yield a representation of unitary operators in terms of *bounded* self-adjoint linear operators. From this representation and spectral theorem 9.10-1 we shall then immediately obtain the desired spectral theorem for U. We mention that this theorem was first derived by A. Wintner (1929), p. 274.

The use of power series in connection with operators seems rather natural. We may remember the special case of the geometric series in Sec. 7.3. Furthermore, sequences of polynomials were employed in Sec. 9.10

for defining $f(T)$ for a given T and a continuous function f. Similarly, the partial sums of a power series form a sequence of polynomials and we may use the series for defining a linear operator. We shall need the following properties of such an operator.

10.5-2 Lemma (Power series). *Let*

$$h(\lambda)=\sum_{n=0}^{\infty}\alpha_n\lambda^n \qquad (\alpha_n \text{ real}) \tag{1}$$

be absolutely convergent for all λ such that $|\lambda|\leqq k$. Suppose that $S\in B(H, H)$ is self-adjoint and has norm $\|S\|\leqq k$; here H is a complex Hilbert space. Then

$$h(S)=\sum_{n=0}^{\infty}\alpha_n S^n \tag{2}$$

is a bounded self-adjoint linear operator and

$$\|h(S)\|\leqq\sum_{n=0}^{\infty}|\alpha_n|\,k^n. \tag{3}$$

If a bounded linear operator commutes with S, it also commutes with $h(S)$.

Proof. Let $h_n(\lambda)$ denote the nth partial sum of the series in (1). Since for $|\lambda|\leqq k$ that series converges absolutely (hence also uniformly), convergence of (2) follows from $\|S\|\leqq k$ and

$$\left\|\sum\alpha_n S^n\right\|\leqq\sum|\alpha_n|\,\|S\|^n\leqq\sum|\alpha_n|\,k^n,$$

because H is complete, so that absolute convergence implies convergence. We denote the sum of the series by $h(S)$. Note that this is in agreement with Sec. 9.10 because $h(\lambda)$ is continuous and $h_n(\lambda)\longrightarrow h(\lambda)$ uniformly for $|\lambda|\leqq k$. The operator $h(S)$ is self-adjoint. Indeed, the $h_n(S)$ are self-adjoint, so that $\langle h_n(S)x, x\rangle$ is real by 3.10-3; hence $\langle h(S)x, x\rangle$ is real by the continuity of the inner product, so that $h(S)$ is self-adjoint by 3.10-3 since H is complex.

We prove (3). Since $\|S\|\leqq k$, Theorem 9.2-2 gives $[m, M]\subset[-k, k]$

and Theorem 9.9-2(*f*) yields

$$\|h_n(S)\| \leqq \max_{\lambda \in J} |h_n(\lambda)| \leqq \sum_{j=0}^{n} |\alpha_j| k^j$$

where $J = [m, M]$. Letting $n \longrightarrow \infty$, we obtain (3).

The last statement follows from Theorem 9.10-2. ∎

If we have two convergent power series, we can multiply them in the usual fashion and write the resulting expression again as a power series. Similarly, if (1) converges for all λ, we can substitute for λ a convergent power series in μ, say, and write the result as a power series in μ, that is, order and arrange the result in powers of μ. It is in this sense that expressions such as $\cos^2 S$, $\sin(\operatorname{arc}\cos V)$, etc., are to be understood.

From the lemma just proved we shall now obtain our main tool, which is the following lemma by F. J. Wecken (1935). We mention that this lemma can also be used in deriving the spectral theorem for bounded self-adjoint linear operators. This was pointed out by Wecken, who formulated the lemma accordingly, as given here in its original form.

10.5-3 Wecken's Lemma. *Let W and A be bounded self-adjoint linear operators on a complex Hilbert space H. Suppose that $WA = AW$ and $W^2 = A^2$. Let P be the projection of H onto the null space $\mathcal{N}(W - A)$. Then:*

(a) *If a bounded linear operator commutes with $W - A$, it also commutes with P.*

(b) *$Wx = 0$ implies $Px = x$.*

(c) *We have $W = (2P - I)A$.*

Proof. **(a)** Suppose that B commutes with $W - A$. Since we have $Px \in \mathcal{N}(W - A)$ for every $x \in H$, we thus obtain

$$(W - A)BPx = B(W - A)Px = 0.$$

This shows that $BPx \in \mathcal{N}(W - A)$ and implies $P(BPx) = BPx$, that is,

$$PBP = BP. \tag{4}$$

We show that $PBP = PB$. Since $W - A$ is self-adjoint, (6g) in Sec. 3.9

yields

$$(W-A)B^* = [B(W-A)]^* = [(W-A)B]^* = B^*(W-A).$$

This shows that $W-A$ and B^* also commute. Hence, reasoning as before, we now obtain $PB^*P = B^*P$ as the analogue of (4). Since projections are self-adjoint (cf. 9.5-1), it follows that

$$PBP = (PB^*P)^* = (B^*P)^* = PB.$$

Together with (4) we have $BP = PB$.

(b) Let $Wx = 0$. Since A and W are self-adjoint and, moreover, $A^2 = W^2$, we obtain

$$\|Ax\|^2 = \langle Ax, Ax\rangle = \langle A^2x, x\rangle = \langle W^2x, x\rangle = \|Wx\|^2 = 0,$$

that is, $Ax = 0$. Hence $(W-A)x = 0$. This shows that $x \in \mathcal{N}(W-A)$. Consequently, $Px = x$ since P is the projection of H onto $\mathcal{N}(W-A)$.

(c) From the assumptions $W^2 = A^2$ and $WA = AW$ we have

$$(W-A)(W+A) = W^2 - A^2 = 0.$$

Hence $(W+A)x \in \mathcal{N}(W-A)$ for every $x \in H$. Since P projects H onto $\mathcal{N}(W-A)$, we thus obtain

$$P(W+A)x = (W+A)x$$

for every $x \in H$, that is,

$$P(W+A) = W+A.$$

Now $P(W-A) = (W-A)P$ by (a) and $(W-A)P = 0$ since P projects H onto $\mathcal{N}(W-A)$. Hence

$$2PA = P(W+A) - P(W-A) = W+A.$$

We see that $2PA - A = W$, which proves (c). ∎

The desired spectral theorem can now be formulated as follows.

10.5-4 Spectral Theorem for Unitary Operators. *Let $U: H \longrightarrow H$ be a unitary operator on a complex Hilbert space $H \neq \{0\}$. Then there exists a spectral family $\mathscr{E} = (E_\theta)$ on $[-\pi, \pi]$ such that*

$$U = \int_{-\pi}^{\pi} e^{i\theta} \, dE_\theta = \int_{-\pi}^{\pi} (\cos\theta + i \sin\theta) \, dE_\theta. \tag{5}$$

More generally, for every continuous function f defined on the unit circle,

$$f(U) = \int_{-\pi}^{\pi} f(e^{i\theta}) \, dE_\theta, \tag{6}$$

where the integral is to be understood in the sense of uniform operator convergence, and for all $x, y \in H$,

$$\langle f(U)x, y\rangle = \int_{-\pi}^{\pi} f(e^{i\theta}) \, dw(\theta), \qquad w(\theta) = \langle E_\theta x, y\rangle, \tag{6*}$$

where the integral is an ordinary Riemann-Stieltjes integral (cf. Sec. 4.4).

Proof. We shall prove that for a given unitary operator U there is a bounded self-adjoint linear operator S with $\sigma(S) \subset [-\pi, \pi]$ such that

$$U = e^{iS} = \cos S + i \sin S. \tag{7}$$

Once the existence of S has been proved, (5) and (6) will readily follow from spectral theorems 9.9-1 and 9.10-1. We proceed stepwise as follows.

(*a*) We prove that U in (7) is unitary, provided S exists.

(*b*) We write

$$U = V + iW \tag{8}$$

where

$$V = \frac{1}{2}(U + U^*), \qquad W = \frac{1}{2i}(U - U^*) \tag{9}$$

and prove that V and W are self-adjoint and

$$-I \leqq V \leqq I, \qquad -I \leqq W \leqq I. \tag{10}$$

(*c*) We investigate some properties of $g(V) = \text{arc}\cos V$ and $A = \sin g(V)$.

(*d*) We prove that the desired operator S is

$$S = (2P - I)(\text{arc}\cos V), \tag{11}$$

where P is the projection of H onto $\mathcal{N}(W - A)$.

The details are as follows.

(a) If S is bounded and self-adjoint, so are $\cos S$ and $\sin S$ by Lemma 10.5-2. These operators commute by the same lemma. This implies that U in (7) is unitary since, by 3.9-4,

$$\begin{aligned} UU^* &= (\cos S + i \sin S)(\cos S - i \sin S) \\ &= (\cos S)^2 + (\sin S)^2 \\ &= (\cos^2 + \sin^2)(S) = I \end{aligned}$$

and similarly, $U^*U = I$.

(b) Self-adjointness of V and W in (9) follows from 3.9-4. Since $UU^* = U^*U$ $(= I)$, we have

$$VW = WV. \tag{12}$$

Also $\|U\| = \|U^*\| = 1$ by 3.10-6, which implies by (9)

$$\|V\| \leqq 1, \qquad \|W\| \leqq 1. \tag{13}$$

Hence the Schwarz inequality yields

$$|\langle Vx, x\rangle| \leqq \|Vx\| \, \|x\| \leqq \|V\| \, \|x\|^2 \leqq \langle x, x\rangle,$$

that is, $-\langle x, x\rangle \leqq \langle Vx, x\rangle \leqq \langle x, x\rangle$. This proves the first formula in (10). The second one follows by the same argument. Furthermore, from (9)

$$V^2 + W^2 = I \tag{14}$$

by direct calculation.

(c) We consider

$$g(\lambda) = \arccos \lambda = \frac{\pi}{2} - \arcsin \lambda = \frac{\pi}{2} - \lambda - \frac{1}{6}\lambda^3 - \cdots.$$

The Maclaurin series on the right converges for $|\lambda| \leqq 1$. (Convergence at $\lambda = 1$ follows by noting that the series of $\arcsin \lambda$ has positive coefficients, hence a monotone sequence of partial sums s_n when $\lambda > 0$, which is bounded on $(0, 1)$ since $s_n(\lambda) < \arcsin \lambda < \pi/2$, so that for every fixed n we have $s_n(\lambda) \longrightarrow s_n(1) \leqq \pi/2$ as $\lambda \longrightarrow 1$. Convergence at $\lambda = -1$ follows readily from that at $\lambda = 1$.)

Since $\|V\| \leqq 1$ by (13), Lemma 10.5-2 implies that the operator

$$g(V) = \arccos V = \frac{\pi}{2} I - V - \frac{1}{6} V^3 - \cdots \tag{15}$$

exists and is self-adjoint. We now define

$$A = \sin g(V). \tag{16}$$

This is a power series in V. Lemma 10.5-2 implies that A is self-adjoint and commutes with V and, by (12), also with W. Since, by (15),

$$\cos g(V) = V, \tag{17}$$

we have

$$V^2 + A^2 = (\cos^2 + \sin^2)(g(V)) = I.$$

A comparison with (14) yields $W^2 = A^2$. Hence we can apply Wecken's lemma 10.5-3 and conclude that

$$W = (2P - I)A, \tag{18}$$

$Wx = 0$ implies $Px = x$, and P commutes with V and with $g(V)$ since these operators commute with $W - A$.

(d) We now define

$$S = (2P - I)g(V) = g(V)(2P - I). \tag{19}$$

Obviously, S is self-adjoint. We prove that S satisfies (7). We set $\kappa = \lambda^2$

and define h_1 and h_2 by

$$h_1(\kappa) = \cos\lambda = 1 - \frac{1}{2!}\lambda^2 + - \cdots$$

(20)

$$\lambda h_2(\kappa) = \sin\lambda = \lambda - \frac{1}{3!}\lambda^3 + - \cdots.$$

These functions exist for all κ. Since P is a projection, we have $(2P-I)^2 = 4P - 4P + I = I$, so that (19) yields

(21) $$S^2 = (2P-I)^2 g(V)^2 = g(V)^2.$$

Hence by (17),

$$\cos S = h_1(S^2) = h_1(g(V)^2) = \cos g(V) = V.$$

We show that $\sin S = W$. Using (20), (16) and (18), we obtain

$$\begin{aligned}\sin S &= S h_2(S^2)\\ &= (2P-I)g(V)h_2(g(V)^2)\\ &= (2P-I)\sin g(V)\\ &= (2P-I)A = W.\end{aligned}$$

We show that $\sigma(S) \subset [-\pi, \pi]$. Since $|\arccos\lambda| \leqq \pi$, we conclude from Theorem 9.10-2 that $\|S\| \leqq \pi$. Since S is self-adjoint and bounded, $\sigma(S)$ is real and Theorem 7.3-4 yields the result.

Let (E_θ) be the spectral family of S. Then (5) and (6) follow from (7) and the spectral theorem 9.10-1 for bounded self-adjoint linear operators.

Note in particular that we can take $-\pi$ (instead of $-\pi-0$) as the lower limit of integration in (5) and (6) without restricting generality. The reason is as follows. If we had a spectral family, call it $(\tilde{E}_\theta)$, such that $\tilde{E}_{-\pi} \neq 0$, we would have to take $-\pi-0$ as the lower limit of integration in those integrals. However, instead of $\tilde{E}_\theta$ we could then equally well use E_θ defined by

$$E_\theta = \begin{cases} 0 & \text{if} \quad \theta = -\pi \\ \tilde{E}_\theta - \tilde{E}_{-\pi} & \text{if} \quad -\pi < \theta < \pi \\ I & \text{if} \quad \theta = \pi \end{cases}$$

E_θ is continuous at $\theta = -\pi$, so that the lower limit of integration $-\pi$ in (5) and (6) is in order. ∎

Problems

1. If a unitary operator U has eigenvalues λ_1 and $\lambda_2 \neq \lambda_1$, show that corresponding eigenvectors x_1 and x_2 are orthogonal.

2. Show that a unitary operator is closed.

3. Show that $U\colon L^2(-\infty, +\infty) \longrightarrow L^2(-\infty, +\infty)$ defined by $Ux(t) = x(t+c)$ is unitary; here c is a given real number.

4. If λ is an eigenvalue of an isometric linear operator T, show that $|\lambda| = 1$.

5. Show that λ is an approximate eigenvalue (cf. Sec. 10.4, Prob. 3) of a linear operator $T\colon \mathcal{D}(T) \longrightarrow H$ if and only if T_λ does not have a bounded inverse.

6. Show that λ is an eigenvalue of a unitary operator $U\colon H \longrightarrow H$ if and only if $\overline{U_\lambda(H)} \neq H$.

7. Show that the *right-shift operator* $T\colon l^2 \longrightarrow l^2$ which is defined by $(\xi_1, \xi_2, \cdots) \longmapsto (0, \xi_1, \xi_2, \cdots)$ is isometric but not unitary and has no eigenvalues.

8. Show that the spectrum of the operator in Prob. 7 is the closed unit disk $M = \{\lambda \mid |\lambda| \leqq 1\}$. Conclude that Theorem 10.5-1 does *not* hold for isometric operators.

9. Show that $\lambda = 0$ is not an approximate eigenvalue of the operator in Prob. 7. (Cf. Sec. 10.4, Prob. 3.)

10. In connection with Probs. 7 to 9 it is worthwhile noting that the *left-shift operator* $T\colon l^2 \longrightarrow l^2$ defined by $y = Tx = (\xi_2, \xi_3, \cdots)$, where $x = (\xi_1, \xi_2, \cdots)$, has a spectrum which differs considerably from that of the right-shift operator. In fact, show that every λ such that $|\lambda|<1$ is an eigenvalue of the left-shift operator. What is the dimension of the corresponding eigenspace?

10.6 Spectral Representation of Self-Adjoint Linear Operators

We shall now derive a spectral representation for a self-adjoint linear operator $T\colon \mathfrak{D}(T) \longrightarrow H$ on a complex Hilbert space H, where $\mathfrak{D}(T)$ is dense in H and T may be unbounded.

For this purpose we associate with T the operator

$$U=(T-iI)(T+iI)^{-1} \tag{1}$$

U is called the **Cayley transform** of T.

The operator U is unitary, as we prove in Lemma 10.6-1 (below), and the point of the approach is that we shall be able to obtain the spectral theorem for the (possibly unbounded) T from that for the bounded operator U (cf. Theorem 10.5-4).

T has its spectrum $\sigma(T)$ on the real axis of the complex plane $\mathbf{C}$ (cf. 10.4-2), whereas the spectrum of a unitary operator lies on the unit circle of $\mathbf{C}$ (cf. 10.5-1). A mapping $\mathbf{C} \longrightarrow \mathbf{C}$ which transforms the real axis into the unit circle is[1]

$$u=\frac{t-i}{t+i} \tag{2}$$

and this suggests (1).

We shall now prove that U is unitary.

10.6-1 Lemma (Cayley transform). *The Cayley transform* (1) *of a self-adjoint linear operator $T\colon \mathfrak{D}(T) \longrightarrow H$ exists on H and is a unitary operator; here, $H \neq \{0\}$ is a complex Hilbert space.*

Proof. Since T is self-adjoint, $\sigma(T)$ is real (cf. 10.4-2). Hence i and $-i$ belong to the resolvent set $\rho(T)$. Consequently, by the definition of $\rho(T)$, the inverses $(T+iI)^{-1}$ and $(T-iI)^{-1}$ exist on a dense subset of H

[1] This is a special *fractional linear transformation* or *Möbius transformation.* These mappings are considered in most textbooks on complex analysis. See also E. Kreyszig (1972), pp. 498–506.

and are bounded operators. Theorem 10.3-3 implies that T is closed because $T = T^*$, and from Lemma 7.2-3 we now see that those inverses are defined on all of H, that is,

$$\mathcal{R}(T+iI) = H, \qquad \mathcal{R}(T-iI) = H. \tag{3}$$

We thus have, since I is defined on all of H,

$$(T+iI)^{-1}(H) = \mathcal{D}(T+iI) = \mathcal{D}(T) = \mathcal{D}(T-iI)$$

as well as

$$(T-iI)(\mathcal{D}(T)) = H.$$

This shows that U in (1) is a bijection of H onto itself. By Theorem 3.10-6(*f*), it remains to prove that U is isometric. For this purpose we take any $x \in H$, set $y = (T+iI)^{-1}x$ and use $\langle y, Ty \rangle = \langle Ty, y \rangle$. Then we obtain the desired result by straightforward calculation:

$$\begin{aligned}
\|Ux\|^2 &= \|(T-iI)y\|^2 \\
&= \langle Ty - iy, Ty - iy \rangle \\
&= \langle Ty, Ty \rangle + i\langle Ty, y \rangle - i\langle y, Ty \rangle + \langle iy, iy \rangle \\
&= \langle Ty + iy, Ty + iy \rangle \\
&= \|(T+iI)y\|^2 \\
&= \|(T+iI)(T+iI)^{-1}x\|^2 \\
&= \|x\|^2.
\end{aligned}$$

Theorem 3.10-6(*f*) now implies that U is unitary. ∎

Since the Cayley transform U of T is unitary, U has a spectral representation (cf. 10.5-4), and from it we want to obtain a spectral representation of T. For this purpose we must know how we can express T in terms of U:

10.6-2 Lemma (Cayley transform). *Let T be as in Lemma* 10.6-1 *and let U be defined by* (1). *Then*

$$T = i(I+U)(I-U)^{-1}. \tag{4}$$

Furthermore, 1 *is not an eigenvalue of U.*

Proof. Let $x \in \mathfrak{D}(T)$ and

$$y = (T + iI)x. \tag{5}$$

Then

$$Uy = (T - iI)x$$

because $(T + iI)^{-1}(T + iI) = I$. By addition and subtraction,

$$\text{(6)} \qquad \begin{aligned} &\text{(a)} \qquad (I + U)y = 2Tx \\ &\text{(b)} \qquad (I - U)y = 2ix. \end{aligned}$$

From (5) and (3) we see that $y \in \mathfrak{R}(T + iI) = H$, and (6b) now shows that $I - U$ maps H onto $\mathfrak{D}(T)$. We also see from (6b) that if $(I - U)y = 0$, then $x = 0$, so that $y = 0$ by (5). Hence $(I - U)^{-1}$ exists by Theorem 2.6-10, and is defined on the range of $I - U$, which is $\mathfrak{D}(T)$ by (6b). Hence (6b) gives

$$y = 2i(I - U)^{-1}x \qquad [x \in \mathfrak{D}(T)]. \tag{7}$$

By substitution into (6a),

$$\begin{aligned} Tx &= \tfrac{1}{2}(I + U)y \\ &= i(I + U)(I - U)^{-1}x \end{aligned}$$

for all $x \in \mathfrak{D}(T)$. This proves (4).

Furthermore, since $(I - U)^{-1}$ exists, 1 cannot be an eigenvalue of the Cayley transform U. ∎

Formula (4) represents T as a function of the unitary operator U. Hence we may apply Theorem 10.5-4. This yields the following result.

10.6-3 Spectral Theorem for Self-Adjoint Linear Operators. *Let T: $\mathfrak{D}(T) \longrightarrow H$ be a self-adjoint linear operator, where $H \neq \{0\}$ is a complex Hilbert space and $\mathfrak{D}(T)$ is dense in H. Let U be the Cayley transform* (1) *of T and (E_θ) the spectral family in the spectral representation*

(5), *Sec.* 10.5, *of* $-U$. *Then for all* $x \in \mathfrak{D}(T)$,

$$\begin{aligned} \langle Tx, x\rangle &= \int_{-\pi}^{\pi} \tan\frac{\theta}{2}\, dw(\theta) & w(\theta) = \langle E_\theta x, x\rangle \\ &= \int_{-\infty}^{+\infty} \lambda\, dv(\lambda) & v(\lambda) = \langle F_\lambda x, x\rangle \end{aligned} \tag{8}$$

where $F_\lambda = E_{2 \arctan \lambda}$.

Proof. From spectral theorem 10.5-4 we have

$$-U = \int_{-\pi}^{\pi} e^{i\theta}\, dE_\theta = \int_{-\pi}^{\pi} (\cos\theta + i \sin\theta)\, dE_\theta. \tag{9}$$

In part (*a*) of the proof we show that (E_θ) is continuous at $-\pi$ and π. This property will be needed in part (*b*), where we establish (8).

(a) (E_θ) is the spectral family of a bounded self-adjoint linear operator which we call S. Then (cf. (7) in Sec. 10.5)

$$-U = \cos S + i \sin S. \tag{10}$$

From Theorem 9.11-1 we know that a θ_0 at which (E_θ) is discontinuous is an eigenvalue of S. Then there is an $x \neq 0$ such that $Sx = \theta_0 x$. Hence for any polynomial q,

$$q(S)x = q(\theta_0)x$$

and for any continuous function g on $[-\pi, \pi]$,

$$g(S)x = g(\theta_0)x. \tag{11}$$

Since $\sigma(S) \subset [-\pi, \pi]$, we have $E_{-\pi-0} = 0$. Hence if $E_{-\pi} \neq 0$, then $-\pi$ would be an eigenvalue of S. By (10) and (11), the operator U would have the eigenvalue

$$-\cos(-\pi) - i \sin(-\pi) = 1,$$

which contradicts Lemma 10.6-2. Similarly, $E_\pi = I$, and if $E_{\pi-0} \neq I$, this would also cause an eigenvalue 1 of U.

(b) Let $x \in H$ and $y = (I - U)x$. Then $y \in \mathfrak{D}(T)$ since $I - U\colon H \longrightarrow \mathfrak{D}(T)$, as was shown in the proof of Lemma 10.6-2. From (4) it follows that

$$Ty = i(I+U)(I-U)^{-1}y = i(I+U)x.$$

Since $\|Ux\| = \|x\|$ by 3.10-6, using (9), we thus obtain

$$\begin{aligned}
\langle Ty, y\rangle &= \langle i(I+U)x, (I-U)x\rangle \\
&= i(\langle Ux, x\rangle - \langle x, Ux\rangle) \\
&= i(\langle Ux, x\rangle - \overline{\langle Ux, x\rangle}) \\
&= -2\,\mathrm{Im}\,\langle Ux, x\rangle \\
&= 2\int_{-\pi}^{\pi} \sin\theta \; d\langle E_\theta x, x\rangle.
\end{aligned}$$

Hence

$$\langle Ty, y\rangle = 4\int_{-\pi}^{\pi} \sin\frac{\theta}{2}\cos\frac{\theta}{2}\; d\langle E_\theta x, x\rangle. \tag{12}$$

From the last few lines of the proof of Theorem 10.5-4 we remember that (E_θ) is the spectral family of the bounded self-adjoint linear operator S in (10). Hence E_θ and S commute by 9.8-2, so that E_θ and U commute by 10.5-2. Using (6*), Sec. 10.5, we thus obtain

$$\begin{aligned}
\langle E_\theta y, y\rangle &= \langle E_\theta(I-U)x, (I-U)x\rangle \\
&= \langle (I-U)^*(I-U)E_\theta x, x\rangle \\
&= \int_{-\pi}^{\pi} (1+e^{-i\varphi})(1+e^{i\varphi})\; d\langle E_\varphi z, x\rangle
\end{aligned}$$

where $z = E_\theta x$. Since $E_\varphi E_\theta = E_\varphi$ when $\varphi \leqq \theta$ by (7), Sec. 9.7, and

$$(1+e^{-i\varphi})(1+e^{i\varphi}) = (e^{i\varphi/2} + e^{-i\varphi/2})^2 = 4\cos^2\frac{\varphi}{2},$$

we obtain

$$\langle E_\theta y, y\rangle = 4\int_{-\pi}^{\theta} \cos^2\frac{\varphi}{2}\, d\langle E_\varphi x, x\rangle.$$

Using this, the continuity of E_θ at $\pm\pi$ and the rule for transforming a Stieltjes integral, we finally have

$$\int_{-\pi}^{\pi} \tan\frac{\theta}{2}\, d\langle E_\theta y, y\rangle = \int_{-\pi}^{\pi} \tan\frac{\theta}{2}\left(4\cos^2\frac{\theta}{2}\right) d\langle E_\theta x, x\rangle$$

$$= 4\int_{-\pi}^{\pi} \sin\frac{\theta}{2}\,\cos\frac{\theta}{2}\, d\langle E_\theta x, x\rangle.$$

The last integral is the same as in (12). This gives the first formula in (8), except for the notation (y instead of x). The other one follows by the indicated transformation $\theta = 2\arctan\lambda$. Note that (F_λ) is indeed a spectral family; in particular, $F_\lambda \longrightarrow 0$ as $\lambda \longrightarrow -\infty$ and $F_\lambda \longrightarrow I$ as $\lambda \longrightarrow +\infty$. ∎

Problems

1. Find the inverse of (2) and compare it with (4). Comment.

2. Let U be defined by (1). Show that $1 \in \rho(U)$ if and only if the self-adjoint linear operator T is bounded.

3. (Commuting operators) A bounded linear operator $S: H \longrightarrow H$ on a Hilbert space H is said to *commute* with a linear operator $T: \mathcal{D}(T) \longrightarrow H$, where $\mathcal{D}(T) \subset H$, if $ST \subset TS$, that is, if $x \in \mathcal{D}(T)$ implies $Sx \in \mathcal{D}(T)$ as well as $STx = TSx$. (Note that if $\mathcal{D}(T) = H$, then $ST \subset TS$ is equivalent to $ST = TS$.) Show that if S commutes with T in (1), then S also commutes with U given by (1).

4. Prove that if $SU = US$ in Prob. 3, then $ST \subset TS$, that is, S also commutes with T.

5. If $T: \mathcal{D}(T) \longrightarrow H$ is a symmetric linear operator, show that its Cayley transform (1) exists and is isometric.

6. Show that if T in Prob. 5 is closed, so is the Cayley transform of T.

7. If $T\colon \mathcal{D}(T) \longrightarrow H$ is a closed symmetric linear operator, show that the domain $\mathcal{D}(U)$ and the range $\mathcal{R}(U)$ of its Cayley transform (1) are closed. Note that, in the present case, we may have $\mathcal{D}(U) \neq H$ or $\mathcal{R}(U) \neq H$ or both.

8. If the Cayley transform (1) of a symmetric linear operator $T\colon \mathcal{D}(T) \longrightarrow H$ is unitary, show that T is self-adjoint.

9. (Deficiency indices) In Prob. 7, the Hilbert dimensions (cf. Sec. 3.6) of the orthogonal complements $\mathcal{D}(U)^{\perp}$ and $\mathcal{R}(U)^{\perp}$ are called the *deficiency indices* of T. Show that these indices are both zero if and only if T is self-adjoint.

10. Show that the *right-shift operator* $U\colon l^2 \longrightarrow l^2$ defined by the formula $(\xi_1, \xi_2, \cdots) \longmapsto (0, \xi_1, \xi_2, \cdots)$ is isometric but not unitary. Verify that U is the Cayley transform of $T\colon \mathcal{D}(T) \longrightarrow l^2$ defined by $x \longmapsto y = (\eta_j)$, where

$$\eta_1 = i\xi_1, \qquad \eta_j = i(2\xi_1 + \cdots + 2\xi_{j-1} + \xi_j), \qquad j = 2, 3, \cdots,$$

and

$$\mathcal{D}(T) = \{x = (\xi_j) \mid |\xi_1|^2 + |\xi_1 + \xi_2|^2 + |\xi_1 + \xi_2 + \xi_3|^2 + \cdots < \infty\}.$$

10.7 Multiplication Operator and Differentiation Operator

In this section we shall consider some properties of two unbounded linear operators, namely, the operator of multiplication by the independent variable and a differentiation operator. We mention that these operators play a basic role in atomic physics. (Readers interested in those applications will find details in Chap. 11, in particular in Secs. 11.1 and 11.2. The present section is self-contained and independent of Chap. 11 and vice versa.)

Since we do not presuppose the theory of Lebesgue measure and integration, in the present section we shall have to present some of the facts without proof.

The first of the two operators is

$$\text{(1)} \qquad \begin{aligned} T\colon \mathcal{D}(T) &\longrightarrow L^2(-\infty, +\infty) \\ x &\longmapsto tx \end{aligned}$$

where $\mathcal{D}(T) \subset L^2(-\infty, +\infty)$.

The domain $\mathcal{D}(T)$ consists of all $x \in L^2(-\infty, +\infty)$ such that we have $Tx \in L^2(-\infty, +\infty)$, that is,

$$\int_{-\infty}^{+\infty} t^2 |x(t)|^2 \, dt < \infty. \tag{2}$$

This implies that $\mathcal{D}(T) \neq L^2(-\infty, +\infty)$. For instance, an $x \in L^2(-\infty, +\infty)$ not satisfying (2) is given by

$$x(t) = \begin{cases} 1/t & \text{if } t \geqq 1 \\ 0 & \text{if } t < 1; \end{cases}$$

hence $x \notin \mathcal{D}(T)$.

Obviously, $\mathcal{D}(T)$ contains all functions $x \in L^2(-\infty, +\infty)$ which are zero outside a compact interval. It can be shown that this set of functions is dense in $L^2(-\infty, +\infty)$. Hence $\mathcal{D}(T)$ is dense in $L^2(-\infty, +\infty)$.

10.7-1 Lemma (Multiplication operator). *The multiplication operator T defined by* (1) *is not bounded.*

Proof. We take (Fig. 67)

$$x_n(t) = \begin{cases} 1 & \text{if } n \leqq t < n+1 \\ 0 & \text{elsewhere.} \end{cases}$$

Clearly, $\|x_n\| = 1$ and

$$\|Tx_n\|^2 = \int_n^{n+1} t^2 \, dt > n^2.$$

This shows that $\|Tx_n\|/\|x_n\| > n$, where we can choose $n \in \mathbf{N}$ as large as we please. ∎

Note that the unboundedness results from the fact that we are dealing with functions on an infinite interval. For comparison, in the

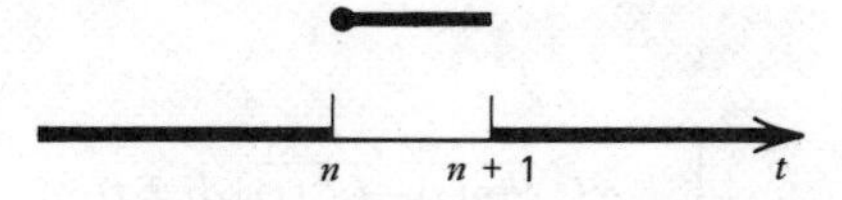

Fig. 67. Function x_n in the proof of Lemma 10.7-1

case of a *finite* interval $[a, b]$ the operator

$$\tilde{T}\colon \mathfrak{D}(\tilde{T}) \longrightarrow L^2[a, b] \tag{3}$$

$$x \longmapsto tx$$

is bounded. In fact, if $|b| \geqq |a|$, then

$$\|\tilde{T}x\|^2 = \int_a^b t^2 |x(t)|^2 \, dt \leqq b^2 \|x\|^2,$$

and if $|b| < |a|$, the proof is quite similar. Furthermore, this also shows that $x \in L^2[a, b]$ implies $\tilde{T}x \in L^2[a, b]$. Hence $\mathfrak{D}(\tilde{T}) = L^2[a, b]$, that is, the operator $\tilde{T}$ is defined on all of $L^2[a, b]$.

10.7-2 Theorem (Self-adjointness). *The multiplication operator* T *defined by* (1) *is self-adjoint.*

Proof. T is densely defined in $L^2(-\infty, +\infty)$, as was mentioned before. T is symmetric because, using $t = \bar{t}$, we have

$$\begin{aligned}\langle Tx, y\rangle &= \int_{-\infty}^{+\infty} tx(t)\overline{y(t)}\, dt \\ &= \int_{-\infty}^{+\infty} x(t)\overline{ty(t)}\, dt = \langle x, Ty\rangle.\end{aligned}$$

Hence $T \subset T^*$ by 10.2-4, and it sufficies to show that $\mathfrak{D}(T) \supset \mathfrak{D}(T^*)$. This we do by proving that $y \in \mathfrak{D}(T^*)$ implies $y \in \mathfrak{D}(T)$. Let $y \in \mathfrak{D}(T^*)$. Then for all $x \in \mathfrak{D}(T)$,

$$\langle Tx, y\rangle = \langle x, y^*\rangle \qquad\qquad y^* = T^* y$$

(cf. 10.1-2), written out

$$\int_{-\infty}^{+\infty} tx(t)\overline{y(t)}\, dt = \int_{-\infty}^{+\infty} x(t)\overline{y^*(t)}\, dt.$$

This implies

$$\int_{-\infty}^{+\infty} x(t)[\overline{ty(t)} - \overline{y^*(t)}]\, dt = 0. \tag{4}$$

In particular, this holds for every $x \in L^2(-\infty, +\infty)$ which is zero outside an arbitrary given bounded interval (a, b). Clearly, such an x is in $\mathfrak{D}(T)$. Choosing

$$x(t) = \begin{cases} ty(t) - y^*(t) & \text{if } t \in (a, b) \\ 0 & \text{elsewhere,} \end{cases}$$

we have from (4)

$$\int_a^b |ty(t) - y^*(t)|^2 \, dt = 0.$$

It follows that $ty(t) - y^*(t) = 0$ almost everywhere[2] on (a, b), that is, $ty(t) = y^*(t)$ almost everywhere on (a, b). Since (a, b) was arbitrary, this shows that $ty = y^* \in L^2(-\infty, +\infty)$, so that $y \in \mathfrak{D}(T)$. We also have $T^*y = y^* = ty = Ty$. ∎

Note that Theorem 10.3-3 now implies that T is closed, because $T = T^*$.

Important spectral properties of the operator T are as follows.

10.7-3 Theorem (Spectrum). *Let T be the multiplication operator defined by* (1) *and $\sigma(T)$ its spectrum. Then:*

(a) *T has no eigenvalues.*

(b) *$\sigma(T)$ is all of* $\mathbf{R}$.

Proof. **(a)** For any λ, let $x \in \mathfrak{D}(T)$ be such that $Tx = \lambda x$. Then $(T - \lambda I)x = 0$. Hence, by the definition of T,

$$0 = \|(T - \lambda I)x\|^2 = \int_{-\infty}^{+\infty} |t - \lambda|^2 \, |x(t)|^2 \, dt.$$

Since $|t - \lambda| > 0$ for all $t \neq \lambda$, we have $x(t) = 0$ for almost all $t \in \mathbf{R}$, that is, $x = 0$. This shows that x is not an eigenvector, and λ is not an eigenvalue of T. Since λ was arbitrary, T has no eigenvalues.

[2] That is, on (a, b), possibly except for a set of Lebesgue measure zero.

(b) We have $\sigma(T)\subset \mathbf{R}$ by 10.7-2 and 10.4-2. Let $\lambda\in\mathbf{R}$. We define (Fig. 68)

$$v_n(t)=\begin{cases}1 & \text{if } \lambda-\dfrac{1}{n}\leqq t\leqq \lambda+\dfrac{1}{n}\\ 0 & \text{elsewhere}\end{cases}$$

and consider $x_n=\|v_n\|^{-1}v_n$. Then $\|x_n\|=1$. Writing $T_\lambda=T-\lambda I$, as usual, we have from the definition of T,

$$\begin{aligned}\|T_\lambda x_n\|^2&=\int_{-\infty}^{+\infty}(t-\lambda)^2\,|x_n(t)|^2\,dt\\ &\leqq\frac{1}{n^2}\int_{-\infty}^{+\infty}|x_n(t)|^2\,dt=\frac{1}{n^2};\end{aligned}$$

here we used that $(t-\lambda)^2\leqq 1/n^2$ on the interval on which v_n is not zero. Taking square roots, we have

$$\|T_\lambda x_n\|\leqq\frac{1}{n}. \tag{5}$$

Since T has no eigenvalues, the resolvent $R_\lambda=T_\lambda^{-1}$ exists, and $T_\lambda x_n\neq 0$ because $x_n\neq 0$, by 2.6-10. The vectors

$$y_n=\frac{1}{\|T_\lambda x_n\|}\,T_\lambda x_n$$

are in the range of T_λ, which is the domain of R_λ, and have norm 1. Applying R_λ and using (5), we thus obtain

$$\|R_\lambda y_n\|=\frac{1}{\|T_\lambda x_n\|}\,\|x_n\|\geqq n.$$

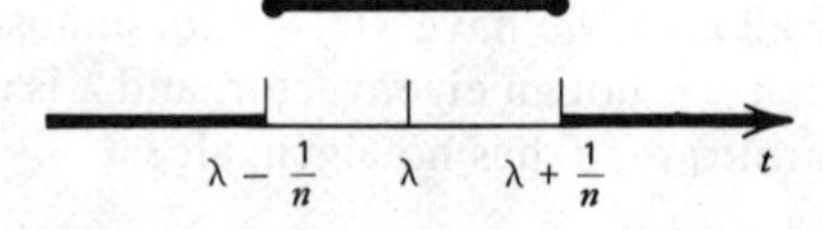

Fig. 68. Function v_n in the proof of Theorem 10.7-3

This shows that the resolvent R_λ is unbounded; hence $\lambda \in \sigma(T)$. Since $\lambda \in \mathbf{R}$ was arbitrary, $\sigma(T) = \mathbf{R}$. ∎

The spectral family of T is (E_λ), where $\lambda \in \mathbf{R}$ and

$$E_\lambda \colon L^2(-\infty, +\infty) \longrightarrow L^2(-\infty, \lambda)$$

is the projection of $L^2(-\infty, +\infty)$ onto $L^2(-\infty, \lambda)$, considered as a subspace of $L^2(-\infty, +\infty)$; thus

$$E_\lambda x(t) = \begin{cases} x(t) & \text{if } t < \lambda \\ 0 & \text{if } t \geqq \lambda. \end{cases} \tag{6}$$

∎

The other operator to be considered in this section is the **differentiation operator**

$$\begin{aligned} D\colon \mathfrak{D}(D) &\longrightarrow L^2(-\infty, +\infty) \\ x &\longmapsto ix' \end{aligned} \tag{7}$$

where $x' = dx/dt$ and i helps to make D self-adjoint, as we shall state below (in 10.7-5). By definition, the domain $\mathfrak{D}(D)$ of D consists of all $x \in L^2(-\infty, +\infty)$ which are absolutely continuous[3] on every compact interval on $\mathbf{R}$ and such that $x' \in L^2(-\infty, +\infty)$.

$\mathfrak{D}(D)$ contains the sequence (e_n) in 3.7-2 involving the Hermite polynomials, and it was stated in 3.7-2 that (e_n) is total in $L^2(-\infty, +\infty)$. Hence $\mathfrak{D}(D)$ is dense in $L^2(-\infty, +\infty)$.

10.7-4 Lemma (Differentiation operator). *The differentiation operator D defined by* (7) *is unbounded.*

Proof. D is an extension of

$$D_0 = D|_Y$$

[3] x is said to be *absolutely continuous* on $[a, b]$ if, given $\varepsilon > 0$, there is a $\delta > 0$ such that for every finite set of disjoint open subintervals $(a_1, b_1), \cdots, (a_n, b_n)$ of $[a, b]$ of total length less than δ we have

$$\sum_{j=1}^{n} |x(b_j) - x(a_j)| < \varepsilon.$$

Then x is differentiable almost everywhere on $[a, b]$, and $x' \in L[a, b]$. Cf. H. L. Royden (1968), p. 106.

where $Y = \mathfrak{D}(D) \cap L^2[0, 1]$ and $L^2[0, 1]$ is regarded as a subspace of $L^2(-\infty, +\infty)$. Hence if D_0 unbounded, so is D. We show that D_0 is unbounded.

Let (Fig. 69)

$$x_n(t) = \begin{cases} 1 - nt & \text{if } 0 \leqq t \leqq 1/n, \\ 0 & \text{if } 1/n < t \leqq 1. \end{cases}$$

The derivative is

$$x_n'(t) = \begin{cases} -n & \text{if } 0 < t < 1/n, \\ 0 & \text{if } 1/n < t < 1. \end{cases}$$

We calculate

$$\|x_n\|^2 = \int_0^1 |x_n(t)|^2 \, dt = \frac{1}{3n}$$

and

$$\|D_0 x_n\|^2 = \int_0^1 |x_n'(t)|^2 \, dt = n$$

and the quotient

$$\frac{\|D_0 x_n\|}{\|x_n\|} = n\sqrt{3} > n.$$

This shows that D_0 is unbounded. ∎

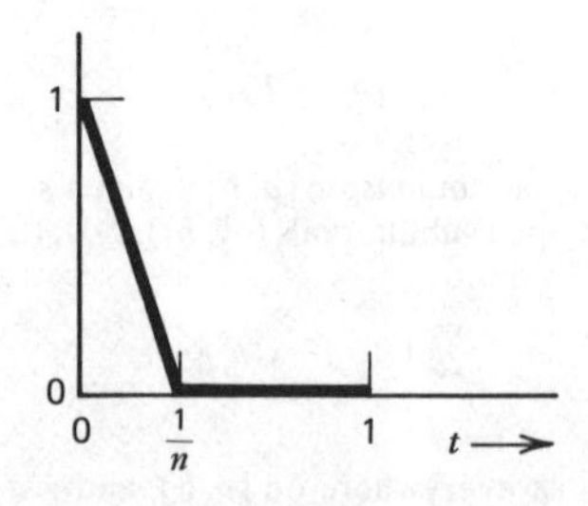

Fig. 69. Function x_n in the proof of Lemma 10.7-4

The following comparison is interesting. The multiplication operator T in (1) is unbounded because $(-\infty, +\infty)$ is an infinite interval, whereas the multiplication operator $\tilde{T}$ in (3) is bounded. In contrast to this, the differentiation operator is unbounded, even if we considered it for $L^2[a, b]$, where $[a, b]$ is a compact interval. This fact is clearly shown by the previous proof.

10.7-5 Theorem (Self-adjointness). *The differentiation operator D defined by* (7) *is self-adjoint.*

A proof of this theorem requires some tools from the theory of Lebesgue integration and can be found, e.g., in G. Helmberg (1969), p. 130.

We finally mention that D does not have eigenvalues and the spectrum $\sigma(D)$ is all of $\mathbf{R}$.

Applications of the operators (1) and (7) are included in the next chapter, where those operators play a basic role (and the notation is changed to a standard notation used in physics; cf. at the beginning of the chapter).

CHAPTER 11
UNBOUNDED LINEAR OPERATORS IN QUANTUM MECHANICS

Quantum mechanics is a part of quantum theory. The latter was initiated in 1900, when Max Planck announced his revolutionary concept of a quantum. This decisive event is usually considered to be the dividing point between classical physics and modern or quantum physics. The new period of physics was caused by many new basic discoveries—X-rays, the electron, radioactivity—and the desire to create corresponding theories.

Quantum mechanics provided the impetus for much of the Hilbert space theory, in particular in connection with unbounded self-adjoint linear operators. In the present chapter we shall explain some of the major reasons for that fact and discuss the role of unbounded linear operators in quantum mechanics.

This chapter is optional. It is kept independent of Chap. 10.

Notation

In this chapter we use a notation which is standard in physics:

	Notation in this chapter	Notation in the other chapters
Independent variable	q	t
Functions	$\psi, \varphi, \cdots$	$x, y, \cdots$

Important concepts, brief orientation about main content

We start with the physical system consisting of a single particle constrained to one dimension. In this case we have to consider the complex Hilbert space $L^2(-\infty, +\infty)$, whose elements $\psi, \varphi, \cdots$ are called *states*, and self-adjoint linear operators $T, Q, D, \cdots$, which are called *observables* and whose domains and ranges are in $L^2(-\infty, +\infty)$. This terminology is motivated in Sec. 11.1. The inner product $\langle T\psi, \psi\rangle$

is an integral which can be interpreted in terms of probability theory, where ψ helps to define a probability density. That inner product may then be called a *mean value* since it characterizes the average value of the observable T which one can expect in experiments if the physical system is in state ψ. The most important observables in this theory are the *position operator* Q defined by $\psi(q) \longmapsto q\psi(q)$ (cf. Sec. 11.1) and the *momentum operator* D defined by $\psi(q) \longmapsto (h/2\pi i)d\psi/dq$ (cf. Sec. 11.2). These operators do not commute, and this leads, via the variance of an observable, to the famous *Heisenberg uncertainty relation* 11.2-2.

In these considerations, the time t is kept constant, so that t is a parameter which does not appear explicitly. For constant t the states of a system can be obtained as solutions of the *time-independent Schrödinger equation* (Sec. 11.3). In this way one can determine various properties of physical systems, in particular the possible energy levels.

The time dependence of states is governed and described by the *time-dependent Schrödinger equation* (Sec. 11.5), which involves the *Hamilton operator* (Sec. 11.4). The latter is obtained if one replaces position and momentum in the classical Hamilton function by the position and momentum operators, respectively.

Basic physical systems and phenomena treated in the text and in the problem sets include the harmonic oscillator (cf. 11.3-1, 11.4-1), the oscillator in three dimensions (Sec. 11.3), plane waves (Sec. 11.3), potential steps and tunnel effect (Sec. 11.4), an electron in a spherically symmetric field and the hydrogen atom (Sec. 11.5).

11.1 Basic Ideas. States, Observables, Position Operator

To explain basic ideas and concepts of quantum mechanics, we consider a single particle which is constrained to one dimension (that is, **R**). This physical system is simple but fundamental and will be suitable for that purpose. More general systems will be discussed later.

We consider the system at an arbitrary fixed instant, that is, we regard time as a parameter which we keep fixed.

In *classical mechanics* the state of our system at some instant is described by specifying position and velocity of the particle. Hence, classically the instantaneous state of the system is described by a pair of numbers.

In *quantum mechanics* the state of the system is described by a function

$$\psi.$$

This notation ψ is standard in physics, so we adopt it, too (instead of our usual notation x for a function). The function ψ is complex-valued and is defined on $\mathbf{R}$; hence it is a function of a single real variable

$$q.$$

Also q is a standard notation in physics, so we adopt it (instead of our usual letter t, which we reserve for the time to be considered in the later sections of this chapter).

We assume that ψ is an element of the Hilbert space

$$L^2(-\infty, +\infty).$$

This is suggested to a large extent by the physical interpretation of ψ. The latter is as follows.

ψ is related to the probability that the particle will be found in a given subset $J \subset \mathbf{R}$; more precisely, this probability is

$$\int_J |\psi(q)|^2 \, dq. \tag{1}$$

To the whole one-dimensional space $\mathbf{R}$ there should correspond the probability 1, that is, we want the particle to be *somewhere* on the real line. This imposes the normalizing condition

$$\|\psi\|^2 = \int_{-\infty}^{+\infty} |\psi(q)|^2 \, dq = 1. \tag{2}$$

Clearly, the integral in (1) remains unchanged if we multiply ψ by a complex factor of absolute value 1.

Our consideration shows that the deterministic description of a state in classical mechanics is replaced by a probabilistic description of a state in quantum mechanics. And the situation suggests that we define a **state** (of our physical system at some instant) to be an element

$$\psi \in L^2(-\infty, +\infty), \qquad \|\psi\| = 1; \tag{3}$$

more precisely, an equivalence class of such elements where

$$\psi_1 \sim \psi_2 \quad \Longleftrightarrow \quad \psi_1 = \alpha\psi_2, \quad |\alpha| = 1.$$

For the sake of simplicity we denote these equivalence classes again by letters such as ψ, φ, etc.

Note that ψ in (3) generates a one-dimensional subspace

$$Y = \{\varphi \mid \varphi = \beta\psi, \ \beta \in \mathbf{C}\}$$

of $L^2(-\infty, +\infty)$. Hence we could equally well say that a *state* of our system is a one-dimensional subspace $Y \subset L^2(-\infty, +\infty)$ and then use a $\varphi \in Y$ of norm 1 in defining a probability according to (1).

From (1) we see that $|\psi(q)|^2$ plays the role of the density of a probability distribution[1] on $\mathbf{R}$. By definition, the corresponding *mean value* or *expected value* is

$$\mu_\psi = \int_{-\infty}^{+\infty} q \, |\psi(q)|^2 \, dq, \tag{4}$$

the *variance* of the distribution is

$$\mathrm{var}_\psi = \int_{-\infty}^{+\infty} (q - \mu_\psi)^2 \, |\psi(q)|^2 \, dq, \tag{5}$$

and the *standard deviation* is $\mathrm{sd}_\psi = \sqrt{\mathrm{var}_\psi}$ $(\geqq 0)$. Intuitively, μ_ψ measures the average value or central location and var_ψ the spread of the distribution.

Hence μ_ψ characterizes the "average position" of our particle for a given state ψ. Now comes an important point. We note that we can write (4) in the form

$$\mu_\psi(Q) = \langle Q\psi, \psi \rangle = \int_{-\infty}^{+\infty} Q\psi(q)\overline{\psi(q)} \, dq \tag{6}$$

[1] The few concepts from probability theory which we need are explained in most textbooks on probability theory or statistics. For instance, see H. Cramér (1955), E. Kreyszig (1970), S. S. Wilks (1962).

where the operator $Q\colon \mathscr{D}(Q) \longrightarrow L^2(-\infty, +\infty)$ is defined by

$$\textbf{(7)} \qquad Q\psi(q) = q\psi(q)$$

(*multiplication by the independent variable q*). Since $\mu_\psi(Q)$ characterizes the average position of the particle, Q is called the **position operator.** By definition, $\mathscr{D}(Q)$ consists of all $\psi \in L^2(-\infty, +\infty)$ such that $Q\psi \in L^2(-\infty, +\infty)$.

From Sec. 10.7 we know that Q is an unbounded self-adjoint linear operator whose domain is dense in $L^2(-\infty, +\infty)$.

We note that (5) can now be written

$$(8) \qquad \begin{aligned} \operatorname{var}_\psi(Q) &= \langle (Q-\mu I)^2\psi, \psi\rangle \\ &= \int_{-\infty}^{+\infty} (Q-\mu I)^2\psi(q)\overline{\psi(q)}\, dq. \end{aligned} \qquad \mu = \mu_\psi(Q)$$

A state ψ of a physical system contains our entire theoretical knowledge about the system, but only implicitly, and this poses the problem of how to obtain from a ψ some information about quantities that express properties of the system which we can observe experimentally. Any such quantity is called an *observable.*

Important observables are *position, momentum* and *energy.*

We have just seen that in the case of the position, for solving that problem we have available a self-adjoint linear operator, namely, the position operator Q. This suggests that in the case of other observables we proceed in a similar fashion, that is, introduce suitable self-adjoint linear operators.

In classical mechanics we ask what value an observable will assume at a given instant. In quantum mechanics we may ask for the probability that a measurement (an experiment) will produce a value of the observable that lies in a certain interval.

The situation and our discussion suggest that we define an **observable** (of our physical system at some instant) to be a self-adjoint linear operator $T\colon \mathscr{D}(T) \longrightarrow L^2(-\infty, +\infty)$, where $\mathscr{D}(T)$ is dense in the space $L^2(-\infty, +\infty)$.

Analogously to (6) and (8) we can define the *mean value* $\mu_\psi(T)$ by

$$\textbf{(9)} \qquad \mu_\psi(T) = \langle T\psi, \psi\rangle = \int_{-\infty}^{+\infty} T\psi(q)\overline{\psi(q)}\, dq,$$

the *variance* $\mathrm{var}_\psi(T)$ by

$$\begin{aligned}\mathrm{var}_\psi(T) &= \langle (T-\mu I)^2\psi, \psi\rangle \\ &= \int_{-\infty}^{+\infty} (T-\mu I)^2\psi(q)\overline{\psi(q)}\,dq\end{aligned} \qquad \mu = \mu_\psi(T) \tag{10}$$

and the *standard deviation* by

$$\mathrm{sd}_\psi(T) = \sqrt{\mathrm{var}_\psi(T)} \qquad (\geqq 0). \tag{11}$$

$\mu_\psi(T)$ characterizes the average value of the observable T which one can about expect in experiments if the system is in state ψ. The variance $\mathrm{var}_\psi(T)$ characterizes the spread (the variability of those values about the mean value).

Problems will be included at the end of the next section.

11.2 Momentum Operator. Heisenberg Uncertainty Principle

We consider the same physical system as in the previous section, where we introduced and motivated the position operator

$$\begin{aligned} Q\colon \mathfrak{D}(Q) &\longrightarrow L^2(-\infty, +\infty) \\ \psi &\longmapsto q\psi. \end{aligned} \tag{1}$$

Another very important observable is the momentum p. The corresponding **momentum operator** is[2]

$$\begin{aligned} D\colon \mathfrak{D}(D) &\longrightarrow L^2(-\infty, +\infty) \\ \psi &\longmapsto \frac{h}{2\pi i}\frac{d\psi}{dq} \end{aligned} \tag{2}$$

where h is Planck's constant and the domain $\mathfrak{D}(D) \subset L^2(-\infty, +\infty)$

[2] The usual notation in physics is P, but since we use P for projections, we write D, which suggests "differentiation." h is a universal constant of nature; $h = 6.626\,196 \cdot 10^{-27}$ erg sec (cf. *CRC Handbook of Chemistry and Physics*, 54th ed. Cleveland, Ohio: CRC Press, 1973–7· ; p. F-101). Absolute continuity is explained in Sec. 10.7, footnote 3.

consists of all functions $\psi \in L^2(-\infty, +\infty)$ which are absolutely continuous on every compact interval on $\mathbf{R}$ and such that $D\psi \in L^2(-\infty, +\infty)$. A motivation of this definition of D can be given as follows.

By Einstein's mass-energy relationship $E = mc^2$ (c the speed of light), an energy E has mass

$$m = \frac{E}{c^2}.$$

Since a photon has speed c and energy

$$E = h\nu$$

(ν the frequency), it has momentum

$$p = mc = \frac{h\nu}{c} = \frac{h}{\Lambda} = \frac{h}{2\pi} k \tag{3}$$

where $k = 2\pi/\Lambda$ and Λ is the wavelength. In 1924, L. de Broglie suggested the concept of *matter waves* satisfying relationships that hold for light waves. Hence we may use (3) also in connection with particles. Assuming the state ψ of our physical system to be such that we can apply the classical Fourier integral theorem, we have

$$\psi(q) = \frac{1}{\sqrt{h}} \int_{-\infty}^{+\infty} \varphi(p) e^{(2\pi i/h)pq} \, dp \tag{4}$$

where

$$\varphi(p) = \frac{1}{\sqrt{h}} \int_{-\infty}^{+\infty} \psi(q) e^{-(2\pi i/h)pq} \, dq. \tag{5}$$

Physically this can be interpreted as a representation of ψ in terms of functions of constant momentum p given by

$$\psi_p(q) = \varphi(p) e^{ikq} = \varphi(p) e^{(2\pi i/h)pq} \tag{6}$$

where $k = 2\pi p/h$ by (3) and $\varphi(p)$ is the amplitude. The complex conjugate $\overline{\psi_p}$ has a minus sign in the exponent, so that

$$|\psi_p(q)|^2 = \psi_p(q)\overline{\psi_p(q)} = \varphi(p)\overline{\varphi(p)} = |\varphi(p)|^2.$$

Since $|\psi_p(q)|^2$ is the probability density of the position in state ψ_p, we see that $|\varphi(p)|^2$ must be proportional to the density of the momentum, and the constant of proportionality is 1 since we have defined $\varphi(p)$ so that (4) and (5) involve the same constant $1/\sqrt{h}$. Hence, by (5), the mean value of the momentum, call it $\tilde{\mu}_\psi$, is

$$\begin{aligned}\tilde{\mu}_\psi &= \int_{-\infty}^{+\infty} p\,|\varphi(p)|^2\,dp = \int_{-\infty}^{+\infty} p\,\varphi(p)\overline{\varphi(p)}\,dp\\ &= \int_{-\infty}^{+\infty} p\,\varphi(p)\frac{1}{\sqrt{h}}\int_{-\infty}^{+\infty}\overline{\psi(q)}\,e^{(2\pi i/h)pq}\,dq\,dp.\end{aligned}$$

Assuming that we may interchange the order of integration and that in (4) we may differentiate under the integral sign, we obtain

$$\begin{aligned}\tilde{\mu}_\psi &= \int_{-\infty}^{+\infty}\overline{\psi(q)}\int_{-\infty}^{+\infty}\varphi(p)\frac{1}{\sqrt{h}}\,pe^{(2\pi i/h)pq}\,dp\,dq\\ &= \int_{-\infty}^{+\infty}\overline{\psi(q)}\frac{h}{2\pi i}\frac{d\psi(q)}{dq}\,dq.\end{aligned}$$

Using (2) and denoting $\tilde{\mu}_\psi$ by $\mu_\psi(D)$, we can write this in the form

$$\mu_\psi(D) = \langle D\psi, \psi\rangle = \int_{-\infty}^{+\infty} D\psi(q)\overline{\psi(q)}\,dq. \tag{7}$$

This motivates the definition (2) of the momentum operator. Note that $\psi \in L^2(-\infty, +\infty)$, so that for a mathematical justification of our formal operations we would need tools from measure theory, in particular an extension of the Fourier integral theorem which is known as the *Fourier-Plancherel theorem.* For details see F. Riesz and B. Sz.-Nagy (1955), pp. 291–295.

Let S and T be any self-adjoint linear operators with domains in the same complex Hilbert space. Then the operator

$$C = ST - TS$$

is called the **commutator** of S and T and is defined on

$$\mathfrak{D}(C) = \mathfrak{D}(ST) \cap \mathfrak{D}(TS).$$

In quantum mechanics, the commutator of the position and momentum operators is of basic importance. By straightforward differentiation we have

$$DQ\psi(q) = D(q\psi(q)) = \frac{h}{2\pi i}[\psi(q) + q\psi'(q)]$$

$$= \frac{h}{2\pi i}\psi(q) + QD\psi(q).$$

This gives the important **Heisenberg commutation relation**

$$\textbf{(8)} \qquad DQ - QD = \frac{h}{2\pi i}\tilde{I}$$

where $\tilde{I}$ is the identity operator on the domain

$$(9) \qquad \mathfrak{D}(DQ - QD) = \mathfrak{D}(DQ) \cap \mathfrak{D}(QD).$$

We mention without proof that this domain is dense in the space $L^2(-\infty, +\infty)$. In fact, it is not difficult to see that the domain contains the sequence (e_n) in 3.7-2 involving the Hermite polynomials, and it was stated in 3.7-2 that (e_n) is total in $L^2(-\infty, +\infty)$. (Remember that our present q is denoted by t in 3.7-2.)

To obtain the famous Heisenberg uncertainty principle, we first prove

11.2-1 Theorem (Commutator). *Let S and T be self-adjoint linear operators with domain and range in $L^2(-\infty, +\infty)$. Then $C = ST - TS$ satisfies*

$$\textbf{(10)} \qquad |\mu_\psi(C)| \leqq 2 \operatorname{sd}_\psi(S) \operatorname{sd}_\psi(T)$$

for every ψ in the domain of C.

Proof. We write $\mu_1 = \mu_\psi(S)$ and $\mu_2 = \mu_\psi(T)$ and

$$A = S - \mu_1 I, \qquad B = T - \mu_2 I.$$

Then we can readily verify by straightforward calculation that

$$C = ST - TS = AB - BA.$$

Since S and T are self-adjoint and μ_1 and μ_2 are inner products of the form (9), Sec. 11.1, these mean values are real (cf. at the end of Sec. 10.2). Hence A and B are self-adjoint. From the definition of a mean value we thus obtain

$$\begin{aligned}\mu_\psi(C) &= \langle (AB-BA)\psi, \psi\rangle \\ &= \langle AB\psi, \psi\rangle - \langle BA\psi, \psi\rangle \\ &= \langle B\psi, A\psi\rangle - \langle A\psi, B\psi\rangle.\end{aligned}$$

The last two products are equal in absolute value. Hence by the triangle and Schwarz inequalities we have

$$|\mu_\psi(C)| \leqq |\langle B\psi, A\psi\rangle| + |\langle A\psi, B\psi\rangle| \leqq 2\|B\psi\|\,\|A\psi\|.$$

This proves (10) because B is self-adjoint, so that by (10), Sec. 11.1, we obtain

$$\|B\psi\| = \langle (T-\mu_2 I)^2\psi, \psi\rangle^{1/2} = \sqrt{\operatorname{var}_\psi(T)} = \operatorname{sd}_\psi(T)$$

and similarly for $\|A\psi\|$. ∎

From (8) we see that the commutator of the position and momentum operators is $C = (h/2\pi i)\tilde{I}$. Hence $|\mu_\psi(C)| = h/2\pi$ and (10) yields

11.2-2 Theorem (Heisenberg uncertainty principle). *For the position operator Q and the momentum operator D,*

$$\operatorname{sd}_\psi(D)\operatorname{sd}_\psi(Q) \geqq \frac{h}{4\pi}. \tag{11}$$

Physically, this inequality (11) means that we cannot make a simultaneous measurement of position and momentum of a particle with an unlimited accuracy. Indeed, the standard deviations $\operatorname{sd}_\psi(D)$ and $\operatorname{sd}_\psi(Q)$ characterize the precision of the measurement of momentum and position, respectively, and (11) shows that we cannot decrease both factors on the left at the same time. h is very small (cf. footnote 2), so that in macrophysics, $h/4\pi$ is negligibly small. However, in atomic physics this is no longer the case. The whole situation becomes better understandable if we realize that any measurement of a system is a disturbance that changes the state of the system, and if the system is small (an electron, for instance), the disturbance becomes noticeable. Of course, any measurement involves an error caused by the lack of precision of the instrument. But one could imagine that such an error might be made smaller and smaller by using more and more refined methods of measurement, so that, at least in principle, in simultaneous measurements of the instantaneous position and momentum of a particle, each of the two corresponding errors can be made smaller than any preassigned positive value. Inequality (11) shows that this is not so but that precision is limited *in principle*, not merely because of the imperfection of any method of measurement.

More generally, Theorem 11.2-1 shows that any two observables S and T whose commutator is not a zero operator cannot be measured simultaneously with unlimited precison in the sense just explained, but the precision is limited in principle.

Problems

1. Determine a normalizing factor α in

$$\psi(q) = \alpha e^{-q^2/2}$$

and graph the corresponding probability density.

2. For a linear operator T and a polynomial g, define the *expectation* $E_\psi(g(T))$ of $g(T)$ by

$$E_\psi(g(T)) = \langle g(T)\psi, \psi\rangle.$$

Show that $E_\psi(T) = \mu_\psi(T)$ and

$$\mathrm{var}_\psi(T) = E_\psi(T^2) - \mu_\psi(T)^2.$$

3. Using the notations in Prob. 2, show that $E_\psi([T - cI]^2)$ is minimum if and only if $c = \mu_\psi(T)$. (Note that this is a minimum property of the variance.)

4. Show that if in (2) we replace $(-\infty, +\infty)$ by a compact interval $[a, b]$, the resulting operator $\tilde{D}$ is no longer self-adjoint (unless we restrict its domain by imposing a suitable condition at a and b).

5. In the text it was shown that the density of the momentum is proportional to $|\varphi(p)|^2$. Then it was claimed that it is equal to $|\varphi(p)|^2$. Verify this by means of (4) and (5), assuming that the interchange of the order of integration is permissible.

6. Formulas (4) and (5) have analogues in space; using Cartesian coordinates and writing $p = (p_1, p_2, p_3)$, $q = (q_1, q_2, q_3)$ and, furthermore, $p \cdot q = p_1q_1 + p_2q_2 + p_3q_3$, we have

$$\psi(q) = h^{-3/2} \int \varphi(p)\, e^{(2\pi i/h)p\cdot q}\, dp$$

where

$$\varphi(p) = h^{-3/2} \int \psi(q)\, e^{-(2\pi i/h)p\cdot q}\, dq.$$

Extend the consideration of Prob. 5 to this case.

7. In the case of a particle in space we have three Cartesian coordinates q_1, q_2, q_3 and corresponding position operators Q_1, Q_2, Q_3 as well as momentum operators D_1, D_2, D_3, where $D_j\psi = (h/2\pi i)\, \partial\psi/\partial q_j$. Show that

$$D_jQ_j - Q_jD_j = \frac{h}{2\pi i} \tilde{I}_j,$$

whereas D_j and Q_k $(j \neq k)$ commute; here, $\tilde{I}_j$ is the identity operator on $\mathfrak{D}(D_jQ_j - Q_jD_j)$.

8. In classical mechanics, the kinetic energy of a moving particle of mass m in space is

$$E_k = \frac{mv^2}{2} = \frac{(mv)^2}{2m} = \frac{1}{2m}(p_1^2 + p_2^2 + p_3^2)$$

where p_1, p_2, p_3 are the components of the momentum vector. This suggests to define the *kinetic energy operator* $\mathscr{E}_k$ by

$$\mathscr{E}_k = \frac{1}{2m}(D_1^2 + D_2^2 + D_3^2)$$

with D_j as in Prob. 7. Show that

$$\mathscr{E}_k\psi = -\frac{h^2}{8\pi^2 m}\Delta\psi$$

where the *Laplacian* $\Delta\psi$ of ψ is given by

$$\Delta\psi = \frac{\partial^2\psi}{\partial q_1^{\,2}} + \frac{\partial^2\psi}{\partial q_2^{\,2}} + \frac{\partial^2\psi}{\partial q_3^{\,2}}.$$

9. (Angular momentum) In classical mechanics, the *angular momentum* is $M = q \times p$, where $q = (q_1, q_2, q_3)$ is the position vector and $p = (p_1, p_2, p_3)$ is the (linear) momentum vector. Show that this suggests to define the *angular momentum operators* $\mathcal{M}_1, \mathcal{M}_2, \mathcal{M}_3$ by

$$\mathcal{M}_1 = Q_2D_3 - Q_3D_2$$

$$\mathcal{M}_2 = Q_3D_1 - Q_1D_3$$

$$\mathcal{M}_3 = Q_1D_2 - Q_2D_1.$$

Prove the commutation relation

$$\mathcal{M}_1\mathcal{M}_2 - \mathcal{M}_2\mathcal{M}_1 = \frac{ih}{2\pi}\mathcal{M}_3$$

and find two similar relations for $\mathcal{M}_2, \mathcal{M}_3$ and $\mathcal{M}_3, \mathcal{M}_1$.

10. Show that the operators $\mathcal{M}_1$, $\mathcal{M}_2$ and $\mathcal{M}_3$ in Prob. 9 commute with the operator

$$\mathcal{M}^2 = \mathcal{M}_1^{\,2} + \mathcal{M}_2^{\,2} + \mathcal{M}_3^{\,2}.$$

11.3 Time-Independent Schrödinger Equation

Using the analogy between light waves and de Broglie's matter waves (cf. Sec. 11.2), we shall derive the fundamental (time-independent) Schrödinger equation.

For investigating refraction, interference and other more subtle optical phenomena one uses the **wave equation**

$$\Psi_{tt} = \gamma^2 \Delta\Psi \tag{1}$$

where $\Psi_{tt} = \partial^2\Psi/\partial t^2$, the constant γ^2 is positive, and $\Delta\Psi$ is the

Laplacian of Ψ. If q_1, q_2, q_3 are Cartesian coordinates in space, then

$$\Delta\Psi = \frac{\partial^2\Psi}{\partial q_1^{\,2}} + \frac{\partial^2\Psi}{\partial q_2^{\,2}} + \frac{\partial^2\Psi}{\partial q_3^{\,2}}.$$

(In our system considered in the last section we have only one coordinate, q, and $\Delta\Psi = \partial^2\Psi/\partial q^2$.)

As usual in connection with stationary wave phenomena, we assume a simple and periodic time dependence, say, of the form

$$\Psi(q_1, q_2, q_3, t) = \psi(q_1, q_2, q_3)\, e^{-i\omega t}. \tag{2}$$

Substituting this into (1) and dropping the exponential factor, we obtain the **Helmholtz equation** (time-independent wave equation)

$$\Delta\psi + k^2\psi = 0 \tag{3}$$

where

$$k = \frac{\omega}{\gamma} = \frac{2\pi\nu}{\gamma} = \frac{2\pi}{\Lambda}$$

and ν is the frequency. For Λ we choose the *de Broglie wave length of matter waves*, that is,

$$\Lambda = \frac{h}{mv} \tag{4}$$

[cf. also (3) in Sec. 11.2, where $v = c$]. Then (3) takes the form

$$\Delta\psi + \frac{8\pi^2 m}{h^2} \cdot \frac{mv^2}{2}\,\psi = 0.$$

Let E denote the sum of the kinetic energy $mv^2/2$ and the potential energy V, that is,

$$E = \frac{mv^2}{2} + V. \qquad \text{Then} \qquad \frac{mv^2}{2} = E - V$$

and we can write

$$\Delta\psi + \frac{8\pi^2 m}{h^2}(E - V)\psi = 0. \tag{5}$$

This is the famous time-independent **Schrödinger equation,** which is fundamental in quantum mechanics.

Note that we can write (5) in the form

$$\left(-\frac{h^2}{8\pi^2 m}\Delta + V\right)\psi = E\psi. \tag{6}$$

This form suggests that the possible energy levels of the system will depend on the spectrum of the operator defined by the left-hand side of (6).

A little reflection shows that (5) is not the only conceivable differential equation that could be obtained in the present situation. However, experimental experience and Schrödinger's work have shown that (5) is particularly useful in the following sense.

Physically meaningful solutions of a differential equation should remain finite and approach zero at infinity. A potential field being given, equation (5) has such solutions only for certain values of the energy E. These values are either in agreement with the 'permissible' energy levels of Bohr's theory of the atom or, when they disagree, they are in better agreement with experimental results than values predicted by that theory. This means that (5) both "explains" and improves Bohr's theory. It also yields a theoretical foundation for a number of basic physical effects which were observed experimentally but could not be explained sufficiently well by older theories.

11.3-1 Example (Harmonic oscillator). To illustrate the Schrödinger equation (5), we consider a physical system which is basic; we mention that it is that to which Max Planck first applied his quantum postulate. Figure 70 shows the classical model, a body of mass m attached to the lower end of a spring whose upper end is fixed. In small vertical motions we may neglect damping and assume the restoring force to be aq, that is, proportional to the displacement q from the position of static equilibrium. Then the classical differential equation of motion is

$$m\ddot{q} + aq = 0 \qquad \text{or} \qquad \ddot{q} + \omega_0^2 q = 0,$$

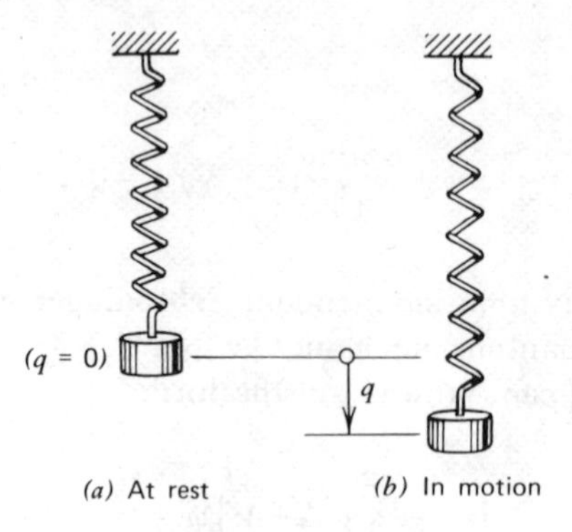

Fig. 70. Body on a spring

where $\omega_0^2 = a/m$, hence $a = m\omega_0^2$, describing harmonic motions represented by a sine or cosine function. From the restoring force aq we get the potential energy V by integration; choosing the constant of integration so that V is zero at $q = 0$, we have $V = aq^2/2 = m\omega_0^2 q^2/2$. Hence the Schrödinger equation (5) of the harmonic oscillator is

$$\psi'' + \frac{8\pi^2 m}{h^2}(E - \tfrac{1}{2}m\omega_0^2 q^2)\psi = 0. \tag{7}$$

Setting

$$\tilde{\lambda} = \frac{4\pi}{\omega_0 h} E \tag{8}$$

and multiplying (7) by $b^2 = h/2\pi m\omega_0$, we obtain

$$b^2\psi'' + \left[\tilde{\lambda} - \left(\frac{q}{b}\right)^2\right]\psi = 0.$$

Introducing $s = q/b$ as a new independent variable and writing $\psi(q) = \tilde{\psi}(s)$, we have

$$\frac{d^2\tilde{\psi}}{ds^2} + (\tilde{\lambda} - s^2)\tilde{\psi} = 0. \tag{9}$$

We determine values of the energy for which (9) has solutions in

$L^2(-\infty, +\infty)$. Substituting

$$\tilde{\psi}(s) = e^{-s^2/2} v(s)$$

into (9) and omitting the exponential factor, we obtain

$$\frac{d^2 v}{ds^2} - 2s\frac{dv}{ds} + (\tilde{\lambda} - 1)v = 0. \tag{10}$$

For

$$\tilde{\lambda} = 2n + 1, \qquad n = 0, 1, \cdots \tag{11}$$

this becomes identical with (9), Sec. 3.7, except for the notation. Hence we see that a solution is the Hermite polynomial H_n, and a total set of orthonormal eigenfunctions satisfying (9) with $\tilde{\lambda}$ given by (11) is (e_n) defined by (7), Sec. 3.7, with $s = q/b$ (instead of t) as the independent variable. The first few of these functions are shown in Fig. 71. Since the frequency is $\nu = \omega_0/2\pi$, we see from (8) that to the eigenvalues (11) there correspond the energy levels

$$E_n = \frac{\omega_0 h}{4\pi}(2n+1) = h\nu\left(n + \frac{1}{2}\right) \tag{12}$$

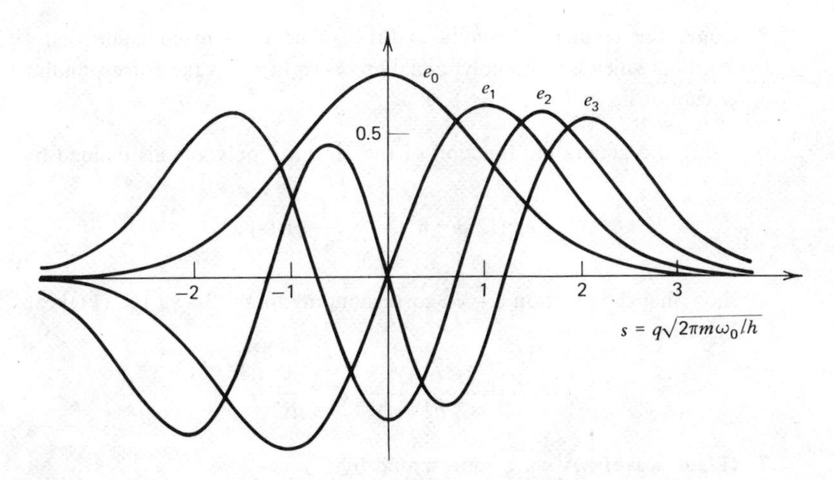

Fig. 71. First four eigenfunctions e_0, e_1, e_2, e_3 of the harmonic oscillator corresponding to the energy levels $h\nu/2$, $3h\nu/2$, $5h\nu/2$, $7h\nu/2$

where $n = 0, 1, \cdots$. These so-called "half-integral" multiples of the energy quantum $h\nu$ are characteristic of the oscillator. The "zero-point energy" (the lowest level) is $h\nu/2$, not 0 as Max Planck had assumed in his famous first investigation in 1900 which initiated the whole quantum theory. The number n which specifies the energy level is called the *principal quantum number* of the harmonic oscillator.

Problems

1. For what values of q is the expression in the parentheses in (7) equal to zero and what is the physical significance of these values in classical mechanics?

2. Can it be seen directly from (7) that (7) cannot have a nontrivial solution $\psi \in L^2(-\infty, +\infty)$ for all values of E?

3. Find a second order differential equation for

$$\psi_0(s) = e^{-s^2/2},$$

compare with (9) and comment.

4. Using the power series method for solving differential equations, show that (10) has a polynomial $v \neq 0$ as a solution if and only if $\tilde{\lambda}$ has one of the values in (11).

5. Could the recursion formula in Prob. 4 be used to conclude that a solution which is not a polynomial grows so fast that the corresponding ψ cannot be in $L^2(-\infty, +\infty)$?

6. Using the generating function of the Hermite polynomials defined by

$$\exp(2us - u^2) = \sum_{n=0}^{\infty} \frac{1}{n!} H_n(s)\, u^n$$

show that the function $\psi = \psi_n$ corresponding to $\tilde{\lambda} = 2n + 1$ [cf. (11)] can be written

$$\psi_n(s) = \frac{(-1)^n}{(2^n n! \sqrt{\pi})^{1/2}} e^{s^2/2} \frac{d^n}{ds^n}(e^{-s^2}).$$

7. (Plane wave) A wave represented by

$$\varphi(q, t) = e^{-i(\omega t - k \cdot q)}$$

is called a *plane monochromatic wave*; here $k=(k_1, k_2, k_3)$, $q=(q_1, q_2, q_3)$ and $k \cdot q$ is the dot product of k and q. Show the following. The direction of k is the direction of propagation of the wave in space. $\lambda = 2\pi/|k|$ is the wavelength, where $|k|$ is the length of k. The quantity $\nu = \omega/2\pi$ is the frequency. $v = \nu\lambda = \omega/|k|$ is the *phase velocity* (velocity of propagation of planes of equal phase). φ satisfies the wave equation (1).

8. If $a(q)$ and $b(q)$ in

$$\psi(q) = a(q)e^{ib(q)}$$

vary only slowly, an approximate solution of the Schrödinger equation $\psi'' + f(q)\psi = 0$ is obtained by substituting ψ and neglecting a''. Show that this leads to

$$b(q) = \int_0^q \sqrt{f(u)}\, du$$

and

$$a(q) = \frac{\alpha}{\sqrt[4]{f(q)}} \qquad (\alpha \text{ constant}).$$

9. (Oscillator in three dimensions) A particle of mass m is bound to the origin by a force whose component along the q_j axis is equal to $-a_jq_j$, $a_j > 0$; $j = 1, 2, 3$. Show that the Schrödinger equation of the problem is

$$\Delta\psi + \left(\lambda - \sum_{j=1}^{3} \alpha_j^2 q_j^2\right)\psi = 0$$

where

$$\lambda = \frac{8\pi^2 m}{h^2} E, \qquad \alpha_j = \frac{2\pi m}{h}\omega_j, \qquad \omega_j = \sqrt{a_j/m}.$$

Apply separation of variables, that is, substitute

$$\psi(q) = \psi_1(q_1)\psi_2(q_2)\psi_3(q_3)$$

to obtain

$$\psi_j'' + (\lambda_j - \alpha_j^2 q_j^2)\psi_j = 0 \qquad \left(\sum_{j=1}^{3} \lambda_j = \lambda\right).$$

Show that for $\lambda_j = (2n_j + 1)\alpha_j$ with integer $n_j \geqq 0$ we obtain

$$\psi_j(q_j) = c_j e^{-\alpha_j q_j^2/2} H_n(\sqrt{\alpha_j} q_j)$$

where c_j is a normalizing factor and H_n is the Hermite polynomial of order n as defined in 3.7-2.

10. An energy level is said to be *degenerate* if there is a corresponding linearly independent set consisting of more than one eigenfunction. The oscillator in Prob. 9 is said to be *isotropic* if $a_1 = a_2 = a_3 = a$. Show that in this case, the lowest energy level is $E_0 = 3h\nu/2$, where $\nu = \omega_0/2\pi = \sqrt{a/m}/2\pi$, and is nondegenerate, whereas the higher energy levels are degenerate.

11.4 Hamilton Operator

In *classical mechanics*, one can base the investigation of a conservative system of particles on the *Hamilton function* of the system; this is the total energy

$$H = E_{\text{kin}} + V \tag{1}$$

(E_{kin} = kinetic energy, V = potential energy) expressed in terms of position coordinates and momentum coordinates. Assuming that the system has n degrees of freedom, one has n position coordinates $q_1, \cdots, q_n$ and n momentum coordinates $p_1, \cdots, p_n$.

In the *quantum mechanical treatment* of the system we also determine

$$H(p_1, \cdots, p_n; q_1, \cdots, q_n).$$

This is the first step. In the second step we replace each p_j by the momentum operator [cf. (2), Sec. 11.2]

$$\textbf{(2)} \qquad \begin{aligned} D_j\colon \mathfrak{D}(D_j) &\longrightarrow L^2(\mathbf{R}^n) \\ \psi &\longmapsto \frac{h}{2\pi i}\frac{\partial \psi}{\partial q_j} \end{aligned}$$

where $\mathscr{D}(D_j) \subset L^2(\mathbf{R}^n)$. Furthermore, we replace each q_j by the position operator [cf. (7), Sec. 11.1]

$$\text{(3)} \qquad \begin{aligned} Q_j\colon \mathscr{D}(Q_j) &\longrightarrow L^2(\mathbf{R}^n) \\ \psi &\longmapsto q_j\psi \end{aligned}$$

where $\mathscr{D}(Q_j) \subset L^2(\mathbf{R}^n)$. From the above Hamilton function H we then obtain the **Hamilton operator** which we denote by $\mathscr{H}$; that is,

$$\mathscr{H}(D_1, \cdots, D_n; Q_1, \cdots, Q_n)$$

is

$$H(p_1, \cdots, p_n; q_1, \cdots, q_n)$$

with p_j replaced by D_j and q_j replaced by Q_j. By definition, $\mathscr{H}$ is assumed to be self-adjoint.

That process of replacement is called the *quantization rule.* Note that the process is not unique, since multiplication is commutative for numbers but not necessarily for operators. This is one of the weaknesses of quantum mechanics.

Equation (6) in Sec. 11.3 can now be written by the use of the Hamilton operator $\mathscr{H}$. Indeed, the kinetic energy of a particle of mass m in space is

$$\frac{m}{2}|v|^2 = \frac{m}{2}(v_1^{\,2} + v_2^{\,2} + v_3^{\,2}) = \frac{1}{2m}(p_1^{\,2} + p_2^{\,2} + p_3^{\,2}).$$

By the quantization rule the expression on the right yields

$$\frac{1}{2m}\sum_{j=1}^{3} D_j^{\,2} = \frac{1}{2m}\left(\frac{h}{2\pi i}\right)^2 \sum_{j=1}^{3} \frac{\partial^2}{\partial q_j^{\,2}} = -\frac{h^2}{8\pi^2 m}\Delta.$$

Hence (6) in Sec. 11.3 can be written

$$\text{(4)} \qquad \mathscr{H}\psi = \lambda\psi$$

where $\lambda = E$ is the energy.

If λ is in the resolvent set of $\mathscr{H}$, then the resolvent of $\mathscr{H}$ exists and (4) has only the trivial solution, considered in $L^2(\mathbf{R}^n)$. If λ is in the point spectrum $\sigma_p(\mathscr{H})$, then (4) has nontrivial solutions $\psi \in L^2(\mathbf{R}^n)$. The residual spectrum $\sigma_r(\mathscr{H})$ is empty since $\mathscr{H}$ is self-adjoint (cf. Sec. 10.4, Prob. 7). If $\lambda \in \sigma_c(\mathscr{H})$, the continuous spectrum of $\mathscr{H}$, then (4) has no solution $\psi \in L^2(\mathbf{R}^n)$, where $\psi \neq 0$. However, in this case, (4) may have nonzero solutions which are not in $L^2(\mathbf{R}^n)$ and depend on a parameter with respect to which we can perform integration to obtain a $\psi \in L^2(\mathbf{R}^n)$. In physics we say that in this process of integration we form *wave packets*. Note that in this context, $\mathscr{H}$ in (4) denotes an extension of the original operator such that the functions under consideration are in the domain of the extended operator. The process may be explained in terms of the following physical system.

We consider a free particle of mass m on $(-\infty, +\infty)$. The Hamilton function is

$$H(p, q) = \frac{1}{2m} p^2,$$

so that we obtain the Hamilton operator

$$\mathscr{H}(D, Q) = \frac{1}{2m} D^2 = -\frac{h^2}{8\pi^2 m} \frac{d^2}{dq^2}.$$

Hence (4) becomes

$$\mathscr{H}\psi = -\frac{h^2}{8\pi^2 m} \psi'' = \lambda\psi \tag{5}$$

where $\lambda = E$ is the energy. Solutions are given by

$$\eta(q) = e^{-ikq} \tag{6}$$

where the parameter k is related to the energy by

$$\lambda = E = \frac{h^2 k^2}{8\pi^2 m}.$$

These functions η can now be used to represent any $\psi \in L^2(-\infty, +\infty)$ as a *wave packet* in the form

$$\psi(q) = \frac{1}{\sqrt{2\pi}} \lim_{a \to \infty} \int_{-a}^{a} \varphi(k)\, e^{-ikq}\, dk \tag{7a}$$

where

$$\varphi(k)=\frac{1}{\sqrt{2\pi}}\lim_{b\to\infty}\int_{-b}^{b}\psi(q)\,e^{ikq}\,dq. \tag{7b}$$

The limits are in the norm of $L^2(-\infty, +\infty)$ [with respect to q in (7a) and with respect to k in (7b)]; such a limit is also called a *limit in the mean.* Formula (7), together with the underlying assumptions, is called the *Fourier-Plancherel theorem,* which was also mentioned in Sec. 11.2, where a reference to literature is given. See also N. Dunford and J. T. Schwartz (1958–71), part 2, pp. 974, 976.

The extension of this consideration to a free particle of mass m in three-dimensional space is as follows. Instead of (5) we have

$$\mathcal{H}\psi=-\frac{h^2}{8\pi^2 m}\Delta\psi=\lambda\psi \tag{8}$$

with Δ as in the previous section. Solutions are plane waves represented by

$$\eta(q)=e^{-ik\cdot q} \tag{9a}$$

where $q=(q_1, q_2, q_3)$, $k=(k_1, k_2, k_3)$ and

$$k\cdot q=k_1q_1+k_2q_2+k_3q_3,$$

and the energy is

$$\lambda=E=\frac{h^2}{8\pi^2 m}k\cdot k. \tag{9b}$$

For a $\psi\in L^2(\mathbf{R}^3)$ the Fourier-Plancherel theorem gives

$$\psi(q)=\frac{1}{(2\pi)^{3/2}}\int_{\mathbf{R}^3}\varphi(k)\,e^{-ik\cdot q}\,dk \tag{10a}$$

where

$$\varphi(k)=\frac{1}{(2\pi)^{3/2}}\int_{\mathbf{R}^3}\psi(q)\,e^{ik\cdot q}\,dq \tag{10b}$$

where the integrals are again understood as limits in the mean of corresponding integrals over finite regions in three-space.

11.4-1 Example (Harmonic oscillator). The Hamilton function of the harmonic oscillator is (cf. 11.3-1)

$$H=\frac{1}{2m}p^2+\frac{1}{2}m{\omega_0}^2q^2.$$

Hence the Hamilton operator is

$$\mathcal{H}=\frac{\omega_0}{2}\left(\frac{1}{\alpha^2}D^2+\alpha^2Q^2\right) \qquad (\alpha^2=m\omega_0). \tag{11}$$

To simplify the formulas in our further consideration, we define

$$A=\beta\left(\alpha Q+\frac{i}{\alpha}D\right) \qquad \left(\beta^2=\frac{\pi}{h}\right). \tag{12}$$

The Hilbert-adjoint operator is

$$A^*=\beta\left(\alpha Q-\frac{i}{\alpha}D\right). \tag{13}$$

By (8), Sec. 11.2,

$$\text{(14)} \quad \begin{aligned} &\text{(a)} \qquad A^*A=\frac{\pi}{h}\left(\alpha^2Q^2+\frac{1}{\alpha^2}D^2-\frac{h}{2\pi}\tilde{I}\right),\\ &\text{(b)} \qquad AA^*=\frac{\pi}{h}\left(\alpha^2Q^2+\frac{1}{\alpha^2}D^2+\frac{h}{2\pi}\tilde{I}\right). \end{aligned}$$

Hence

$$AA^*-A^*A=\tilde{I}. \tag{15}$$

From (14a) and (11),

$$\mathcal{H}=\frac{\omega_0 h}{2\pi}\left(A^*A+\frac{1}{2}\tilde{I}\right). \tag{16}$$

We show that any eigenvalue λ of $\mathcal{H}$ (if it exists) must equal one of the values given by (12), Sec. 11.3.

Let λ be an eigenvalue of $\mathcal{H}$ and ψ an eigenfunction. Then $\psi \neq 0$ and

$$\mathcal{H}\psi = \lambda\psi.$$

By (16),

$$A^*A\psi = \tilde{\lambda}\psi \qquad \text{where} \qquad \tilde{\lambda} = \frac{2\pi\lambda}{\omega_0 h} - \frac{1}{2}. \tag{17}$$

Application of A gives

$$AA^*(A\psi) = \tilde{\lambda}A\psi.$$

On the left, $AA^* = A^*A + \tilde{I}$ by (15). Hence

$$A^*A(A\psi) = (\tilde{\lambda} - 1)A\psi.$$

Similarly, by another application of A,

$$A^*A(A^2\psi) = (\tilde{\lambda} - 2)A^2\psi$$

and after j steps,

$$A^*A(A^j\psi) = (\tilde{\lambda} - j)A^j\psi. \tag{18}$$

We must have $A^j\psi = 0$ for sufficiently large j; because, otherwise, taking the inner product by $A^j\psi$ on both sides of (18), we would have for every j

$$\langle A^j\psi, A^*A(A^j\psi)\rangle = \langle A^{j+1}\psi, A^{j+1}\psi\rangle = (\tilde{\lambda} - j)\langle A^j\psi, A^j\psi\rangle,$$

that is,

$$\tilde{\lambda} - j = \frac{\|A^{j+1}\psi\|^2}{\|A^j\psi\|^2} \geqq 0 \tag{19}$$

for every j, which cannot hold since $\tilde{\lambda}$ is a certain number. Hence there is an $n \in \mathbf{N}$ such that $A^n\psi \neq 0$ but $A^j\psi = 0$ for $j > n$, in particular, $A^{n+1}\psi = 0$. For $j = n$ we thus obtain from the equality in (19)

$$\tilde{\lambda} - n = 0.$$

From this and (17), since $\omega_0 = 2\pi\nu$,

$$\lambda = \frac{\omega_0 h}{2\pi}\left(n + \frac{1}{2}\right) = h\nu\left(n + \frac{1}{2}\right),$$

in agreement with (12), Sec. 11.3.

Problems

1. Obtain (9) from (8) by the method of separation of variables.

2. If ψ_0 is a normalized eigenfunction of $\mathcal{H}$ corresponding to the smallest eigenvalue in Example 11.4-1, show by induction that

$$\psi_n = \frac{1}{\sqrt{n!}}(A^*)^n \psi_0$$

is a normalized eigenfunction of A^*A corresponding to $\tilde{\lambda} = n$.

3. Show that in Prob. 2,

$$A^*\psi_n = \sqrt{(n+1)}\,\psi_{n+1}$$

$$A\psi_n = \sqrt{n}\,\psi_{n-1}.$$

4. Calculate the mean and variance of Q for the harmonic oscillator in state ψ_0 (state of lowest energy), where $\|\psi_0\| = 1$. In what respect does the result differ from that in classical mechanics?

5. Show that the operators in Example 11.4-1 satisfy the commutation rule

$$AQ^s - Q^sA = \sqrt{\frac{h}{4\pi m\omega_0}}\, sQ^{s-1}, \qquad s = 1, 2, \cdots.$$

6. Using Prob. 5, show that the mean of Q^{2s} of the harmonic oscillator in state ψ_0 is

$$\mu_{\psi_0}(Q^{2s}) = \left(\frac{h}{4\pi m\omega_0}\right)^s (2s-1)(2s-3)\cdots 3\cdot 1.$$

7. Show that for the potential step in Fig. 72 the Schrödinger equation yields

$$\psi'' + b_1^2\psi = 0, \qquad b_1^2 = \frac{8\pi^2 m}{h^2} E \qquad (q<0),$$

$$\psi'' + b_2^2\psi = 0, \qquad b_2^2 = \frac{8\pi^2 m}{h^2}(E-U) \qquad (q \geqq 0).$$

Solve this problem for an incident wave from the left, assuming that $E>U$.

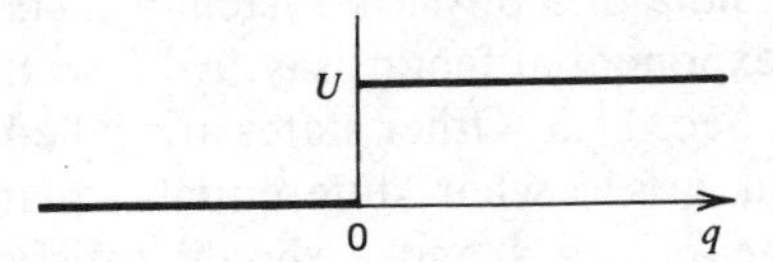

Fig. 72. Potential in Prob. 7

8. Verify that in Prob. 7 the sum of the numbers of transmitted and reflected particles equals the number of incident particles.

9. Solve Prob. 7 if $E<U$. What is the main difference between the present solution of the problem and the classical one?

10. (Tunnel effect) Show that the answer to Prob. 9 suggests that in the case of a potential barrier the wave function of a particle may look approximately as shown in Fig. 73.

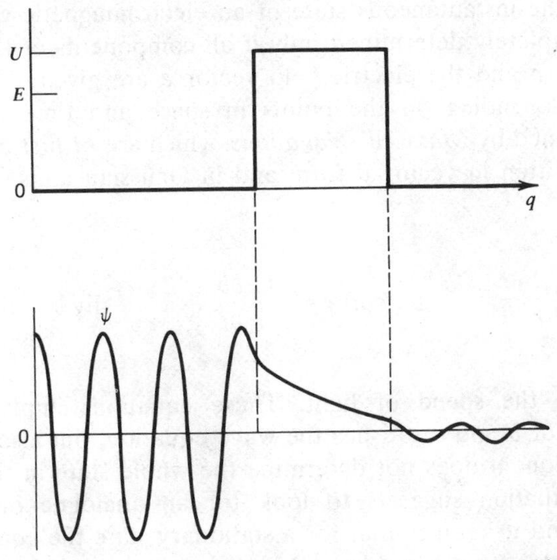

Fig. 73. Potential barrier and wave function ψ of an electron tunnelling through the barrier

11.5 Time-Dependent Schrödinger Equation

In the past four sections of this chapter we considered physical systems at some instant, that is, we always treated time as a parameter which we kept constant. In the present section we shall say a few words about time-dependence of states and observables.

A *stationary state* of a physical system is a state which depends on time only by an exponential factor, say, $e^{-i\omega t}$, so that the state is of the general form (2), Sec. 11.3. Other states are called *nonstationary states*, and the question arises what differential equation such a general function φ of the p_j's, q_j's and t should satisfy. Of course, such a fundamental equation can only be derived from experience. Since one cannot obtain direct experimental results about the form of that equation, all one can do is to consider various equations and to find out whether they are compatible with experimental results and have properties which one must require for logical reasons.

The wave equation (1), Sec. 11.3, is not suitable. One reason is as follows. We want the function φ to be determined for all t if it is given at some instant t. Since that equation involves the second derivative with respect to t, it leaves the first derivative undetermined. This fact may at first surprise the reader since that equation is used in optics. However, the instantaneous state of an electromagnetic wave in vacuum is completely determined only if all components of the magnetic field vector b and the electric field vector e are given. These are six functions depending on the points in space and time t. And they are determined by *Maxwell's equations*, which are of *first* order in t; in vacuum, written in vectorial form and in Gaussian units, these equations are

$$\operatorname{curl} b = \frac{1}{c}\frac{\partial e}{\partial t}, \qquad \operatorname{curl} e = -\frac{1}{c}\frac{\partial b}{\partial t}, \qquad \operatorname{div} b = \operatorname{div} e = 0,$$

where c is the speed of light. These equations imply that each component of b and e satisfies the wave equation, but also in optics a single component does not determine the whole state in the future.

The situation suggests to look for an analogue of Maxwell's equations and to require that for a stationary state the equation to be obtained yields the time-independent Schrödinger equation studied in Sec. 11.3. An equation of this type is the time-dependent **Schrödinger**

equation

$$\mathcal{H}\varphi = -\frac{h}{2\pi i}\frac{\partial \varphi}{\partial t} \tag{1}$$

given by Erwin Schrödinger in 1926. Since (1) involves i, a nonzero solution φ must be complex. $|\varphi|^2$ is regarded as a measure of the intensity of the wave.

A *stationary solution*, whose intensity at a point is independent of t, is obtained by setting

$$\varphi = \psi e^{-i\omega t} \tag{2}$$

where ψ does not depend on t, and $\omega = 2\pi\nu$. Substitution into (1) gives

$$\mathcal{H}\psi = -\frac{h}{2\pi i}(-2\pi i\nu)\psi$$

and, since $E = h\nu$,

$$\mathcal{H}\psi = \lambda\psi \tag{3}$$

where $\lambda = E$ is the energy of the system. This agrees with (4) in the previous section, so that our above requirement is satisfied.

Equation (1) is often called the quantum mechanical *equation of motion*, but this must be understood in the following sense.

In *classical mechanics* the (vector) differential equation of motion determines the motion of the physical system; that is, positions, velocities etc. are determined as functions of the time if initial conditions referring to some instant, say, $t = 0$, are given. In *quantum mechanics* the situation is different. The system no longer behaves in a deterministic fashion with respect to observables. It remains deterministic, however, with respect to states. In fact, if φ is given at some instant, say, $t = 0$, equation (1) determines φ for all t (provided the system is not disturbed by measurements or otherwise). This implies that the probability densities considered before are deterministic in time. Consequently, we may calculate probabilities for observables at any time in the way explained in Secs. 11.1 and 11.2.

The concluding problem set contains some further basic applications, notably in connection with spherically symmetric physical systems, such as the hydrogen atom.

Problems

1. **(Spherical waves)** Show that for ψ depending only on r, where $r^2 = q_1^{\,2} + q_2^{\,2} + q_3^{\,2}$, the Helmholtz equation (3), Sec. 11.3, becomes

$$R'' + \frac{2}{r} R' + k^2 R = 0.$$

Show that corresponding particular solutions of (1) in Sec. 11.3 are

$$\frac{1}{r} \exp\left[-i(\omega t - kr)\right] \qquad \text{and} \qquad \frac{1}{r} \exp\left[-i(\omega t + kr)\right]$$

which represent an outgoing spherical wave and an incoming spherical wave, respectively. Here, $\exp x = e^x$.

2. **(Electron in a spherically symmetric field)** If the potential V depends only on the distance r from some fixed point in space, it is advantageous to transform the Schrödinger equation[3]

$$\Delta\psi + a(E - V(r))\psi = 0, \qquad a = \frac{8\pi^2 \tilde{m}}{h^2}$$

into spherical coordinates r, θ, ϕ defined by (Fig. 74)

$$q_1 = r \sin\theta \cos\phi, \qquad q_2 = r \sin\theta \sin\phi, \qquad q_3 = r \cos\theta.$$

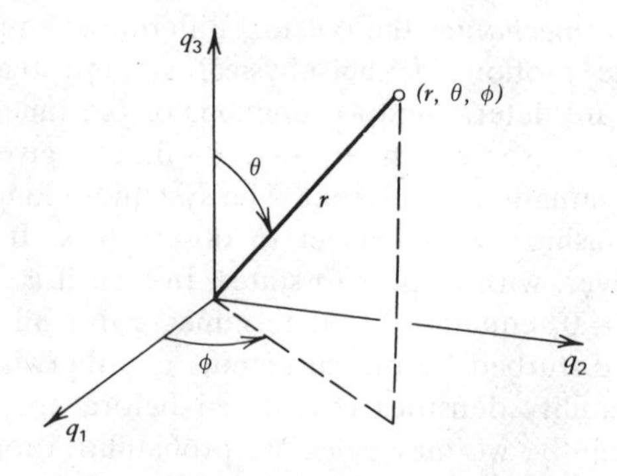

Fig. 74. Spherical coordinates in Prob. 2

[3] We denote the mass of the electron by $\tilde{m}$, to free m for the magnetic quantum number.

(Important physical systems of this type are the hydrogen atom and the singly ionized helium.) Show that

$$\Delta\psi = \frac{\partial^2\psi}{\partial r^2} + \frac{2}{r}\frac{\partial\psi}{\partial r} + \frac{1}{r^2}L\psi$$

where the "angular part" is

$$L\psi = \frac{\partial^2\psi}{\partial\theta^2} + (\cot\theta)\frac{\partial\psi}{\partial\theta} + \frac{1}{\sin^2\theta}\frac{\partial^2\psi}{\partial\phi^2}.$$

Show that we can also write

$$\Delta\psi = \frac{1}{r^2}\frac{\partial}{\partial r}\left(r^2\frac{\partial\psi}{\partial r}\right) + \frac{1}{r^2}L\psi,$$

$$L\psi = \frac{1}{\sin\theta}\frac{\partial}{\partial\theta}\left(\sin\theta\frac{\partial\psi}{\partial\theta}\right) + \frac{1}{\sin^2\theta}\frac{\partial^2\psi}{\partial\phi^2}.$$

Show that by setting

$$\psi(r, \theta, \phi) = R(r)Y(\theta, \phi)$$

and separating variables we obtain from the Schrödinger equation

$$R'' + \frac{2}{r}R' + a(E - V)R - \frac{\alpha}{r^2}R = 0,$$

where α is a separation constant, and

$$LY + \alpha Y = 0.$$

Note the remarkable fact that this equation for the angular part does not depend on the special form of $V(r)$. Setting

$$Y(\theta, \phi) = f(\theta)g(\phi)$$

and applying another separation of variables, show that

$$f'' + (\cot\theta)f' + \left(\alpha - \frac{\beta}{\sin^2\theta}\right)f = 0,$$

where β is another separation constant, and

$$g''+\beta g=0.$$

Conclude that g must be periodic of period 2π, say,

$$g(\phi)=e^{im\phi}, \qquad m=0, \pm 1, \pm 2, \cdots;$$

that is, $\beta=m^2$, where m is called the *magnetic quantum number*[4] (since it plays a role in the so-called Zeeman effect, which is a splitting of spectral lines caused by a magnetic field).

To solve the equation for f, where $\beta=m^2$, set $x=\cos\theta$ and $f(\theta)=y(x)$, and show that this leads to

$$(1-x^2)y''-2xy'+\left(\alpha-\frac{m^2}{1-x^2}\right)y=0.$$

Consider the case $m=0$. Show that for

$$\alpha=l(l+1) \qquad l=0, 1, \cdots$$

a solution is the Legendre polynomial P_l (cf. 3.7-1). [It can be shown that the infinite series obtained for other values of α do not converge at $x=\pm 1$.] l is called the *azimuthal quantum number* or the *orbital angular momentum quantum number* (since it is related to the operator $\mathcal{M}$ in Prob. 7, which is sometimes called the *angular momentum operator*).

3. (Associated Legendre functions, spherical harmonics) Consider the equation

$$(1-x^2)y''-2xy'+\left[l(l+1)-\frac{m^2}{1-x^2}\right]y=0$$

for general $m=0, 1, 2, \cdots$. Substituting

$$y(x)=(1-x^2)^{m/2}z(x),$$

show that z satisfies

$$(1-x^2)z''-2(m+1)xz'+[l(l+1)-m(m+1)]z=0.$$

[4] The letter m is standard and must not be confused with the mass (which we denote by $\tilde{m}$).

Starting from the Legendre equation for P_l and differentiating it m times, show that a solution z of the present equation is given by

$$z(x)=P_l(x)^{(m)},$$

the mth derivative of P_l. The corresponding y is given by

$$P_l^m(x)=(1-x^2)^{m/2}P_l(x)^{(m)}$$

and is called an *associated Legendre function.* Show that for negative $m=-1, -2, \cdots$ our present formulas with m replaced by $|m|$ remain valid. Show that we must require $-l \leqq m \leqq l$. The functions

$$Y_l^m(\theta, \phi)=e^{im\phi}\, P_l^m(\cos\theta)$$

are called *spherical harmonics* (or *surface harmonics*).

4. (Hydrogen atom) Consider the equation for R in Prob. 2 for the hydrogen atom, so that $V(r)=-e^2/r$, where e is the charge of an electron. Solve the equation with $\alpha=l(l+1)$ (cf. Prob. 2) and $E<0$ (the condition for a bound state of the electron). Proceed as follows. Substituting $\rho=\gamma r$, show that

$$\tilde{R}''+\frac{2}{\rho}\tilde{R}'+\left(-\frac{1}{4}+\frac{n}{\rho}-\frac{l(l+1)}{\rho^2}\right)\tilde{R}=0$$

where primes now denote derivatives with respect to ρ and $R(r)=\tilde{R}(\rho)$, and

$$\gamma^2=-4aE, \qquad n=ae^2/\gamma.$$

Substituting

$$\tilde{R}(\rho)=e^{-\rho/2}w(\rho),$$

show that

$$w''+\left(\frac{2}{\rho}-1\right)w'+\left(\frac{n-1}{\rho}-\frac{l(l+1)}{\rho^2}\right)w=0.$$

Substituting

$$w(\rho)=\rho^l u(\rho),$$

show that

$$\rho u'' + (2l+2-\rho)u' + (n-1-l)u = 0.$$

Show that a solution is

$$u(\rho) = L_{n+l}^{2l+1}(\rho) = L_{n+l}(\rho)^{(2l+1)}$$

the $(2l+1)$th derivative of the Laguerre polynomial L_{n+l} (cf. 3.7-3). The function L_{n+l}^{2l+1} is called an *associated Laguerre polynomial.* Show that, altogether, we have the result

$$R(r) = R(\rho/\gamma) = e^{-\rho/2}\rho^l L_{n+l}^{2l+1}(\rho),$$

where $\rho = \gamma r = 2r/na_0$ with the so-called *Bohr radius* given by

$$a_0 = \frac{h^2}{4\pi^2 \tilde{m} e^2} = 0.529 \cdot 10^{-8} \text{ cm}.$$

Show that we must require $l \leqq n-1$ and n must be a positive integer.

5. **(Hydrogen spectrum)** Show that the energy E in Prob. 4 depends only on n, which is called the *principal quantum number*; in fact, show that

$$E = E_n = -\frac{2\pi^2 \tilde{m} e^4}{h^2} \cdot \frac{1}{n^2} \qquad n = 1, 2, \cdots.$$

Show that to each such n there correspond n^2 different solutions

$$\psi_{nlm} = c_{nlm} R_{nl} f_{lm} g_m$$

where c_{nlm} is a normalizing constant and f and g are functions as obtained in Prob. 2. Show that, in contrast to the harmonic oscillator, the hydrogen atom has infinitely many bound states.

We mention that the transition of the electron to a state of lower energy corresponds to an emission of energy. Figure 75 shows the Lyman, Balmer and Paschen series of the hydrogen spectrum; these series correspond to the transitions

$$E_n \longrightarrow E_1, \qquad E_n \longrightarrow E_2, \qquad E_n \longrightarrow E_3,$$

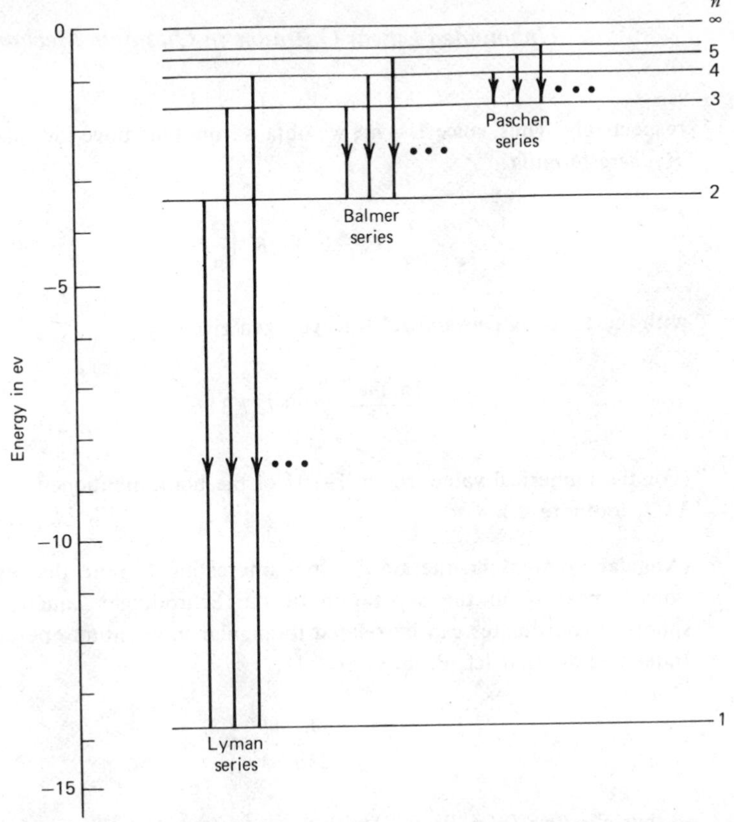

Fig. 75. Energy level diagram and spectral series of the hydrogen atom. The figure shows:

(*i*) Energy levels corresponding to principal quantum numbers $n = 1, 2, \cdots$

(*ii*) Energies in ev (where $n \longrightarrow \infty$ corresponds to 0 ev and $n = 1$ to -13.53 ev; here 13.53 ev is the ionization energy)

(*iii*) Lines of three series, with wavelengths (in Å) as follows.

Lyman series (in the ultraviolet region)

$E_2 \longrightarrow E_1$	1216 Å
$E_3 \longrightarrow E_1$	1026 Å
$E_4 \longrightarrow E_1$	973 Å

Balmer series (visible)

$E_3 \longrightarrow E_2$	6563 Å (H_α line)
$E_4 \longrightarrow E_2$	4861 Å (H_β line)
$E_5 \longrightarrow E_2$	4340 Å (H_γ line)

Paschen series (in the infrared region)

$E_4 \longrightarrow E_3$	18 751 Å
$E_5 \longrightarrow E_3$	12 818 Å
$E_6 \longrightarrow E_3$	10 938 Å

respectively. Thus, since $E = h\nu$, we obtain from the above formula the *Rydberg formula*

$$\frac{1}{\lambda} = \frac{\nu}{c} = \frac{1}{hc}(E_n - E_m) = R^*\left(\frac{1}{n^2} - \frac{1}{m^2}\right)$$

with the *Rydberg constant* R^* for hydrogen given by

$$R^* = \frac{2\pi^2 \tilde{m} e^4}{ch^3} = 109\,737.3 \text{ cm}^{-1}.$$

(For the numerical value, see p. F-104 of the book mentioned in Sec. 11.2, footnote 2.)

6. **(Angular momentum operators)** It is interesting to note that equations appearing in the separation of the Schrödinger equation in spherical coordinates can be related to angular momentum operators. Indeed, show that (cf. Prob. 9, Sec. 11.2)

$$\mathcal{M}_3\psi = \frac{h}{2\pi i}\frac{\partial\psi}{\partial\phi},$$

so that $g'' + \beta g = 0$ (cf. Prob. 2) with $\beta = m^2$, after multiplication by Rf, can be written

$$\mathcal{M}_3{}^2\psi = \frac{h^2 m^2}{4\pi^2}\psi.$$

7. **(Angular momentum operators)** Show that, in terms of spherical coordinates, the angular momentum operators in Prob. 9, Sec. 11.2, have the representations

$$\mathcal{M}_1\psi = -\frac{h}{2\pi i}\left(\sin\phi\frac{\partial\psi}{\partial\theta} + \cot\theta\cos\phi\frac{\partial\psi}{\partial\phi}\right),$$

$$\mathcal{M}_2\psi = \frac{h}{2\pi i}\left(\cos\phi\frac{\partial\psi}{\partial\theta} - \cot\theta\sin\phi\frac{\partial\psi}{\partial\phi}\right),$$

and, by Prob. 6, the operator $\mathcal{M}^2$ in Prob. 10, Sec. 11.2, has the representation

$$\mathcal{M}^2\psi = -\frac{h^2}{4\pi^2}\left[\frac{1}{\sin\theta}\frac{\partial}{\partial\theta}\left(\sin\theta\frac{\partial\psi}{\partial\theta}\right) + \frac{1}{\sin^2\theta}\frac{\partial^2\psi}{\partial\phi^2}\right].$$

Conclude from this and Prob. 3 that the Y_l^m are eigenfunctions of $\mathcal{M}^2$ corresponding to the eigenvalue

$$\frac{h^2}{4\pi^2} l(l+1) \qquad l=0, \cdots, n-1.$$

This is a celebrated result which improves the values $h^2l^2/4\pi^2$ which were predicted by Bohr's theory.

8. (Spherical Bessel functions) It is worth noting that the equation

$$R''+\frac{2}{r}R'+\left[a(E-V(r))-\frac{l(l+1)}{r^2}\right]R=0$$

obtained in Prob. 2 can also be used for other problems exhibiting spherical symmetry. For instance, let (Fig. 76)

$$V(r)=\begin{cases} -V_0<0 & \text{if } r<r_0 \\ 0 & \text{if } r\geqq r_0. \end{cases}$$

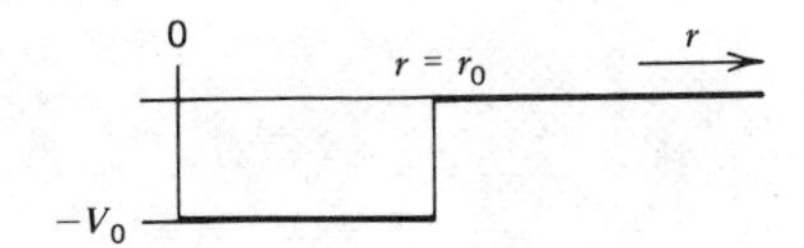

Fig. 76. Potential $V(r)$ in Prob. 8

Assume that $E<0$ and $E+V_0>0$. Find solutions for $r<r_0$ in terms of Bessel functions by showing that the equation can be transformed into the Bessel equation

$$u''+\frac{1}{\rho}u'+\left(1-\frac{(l+1/2)^2}{\rho^2}\right)u=0.$$

[For this reason the solutions R (taken with a suitable numerical factor) are called *spherical Bessel functions.*]

9. If $l=0$ in Prob. 8, show that solutions of the Bessel equation are

$$J_{1/2}(\rho)=\sqrt{\frac{2}{\pi\rho}}\sin\rho, \qquad J_{-1/2}(\rho)=\sqrt{\frac{2}{\pi\rho}}\cos\rho.$$

Using this and the *recursion relation*

$$J_{\nu-1}(\rho)+J_{\nu+1}(\rho)=\frac{2\nu}{\rho}J_{\nu}(\rho),$$

show that for all $l=1, 2, \cdots$ the solutions in Prob. 8 can be expressed by finitely many sine and cosine functions (and negative powers of ρ).

10. Solve the equation in Prob. 8 for $r>r_0$, assuming $E<0$ as before. What is the essential difference between the present solutions and those for $r<r_0$?

APPENDIX 1

SOME MATERIAL FOR REVIEW AND REFERENCE

A1.1 Sets

Sets are denoted by single capital letters $A, B, M, \cdots$ or by the use of braces, for example

$\{a, b, c\}$ denotes the set having the letters a, b, c as elements
$\{t \mid f(t)=0\}$ denotes the set of all t at which the function f is zero.

Some symbols used in set theory are

$\varnothing$	Empty set (set which has no elements)
$a \in A$	a is an element of A
$b \notin A$	b is not an element of A
$A = B$	A and B are equal (are identical, consist of the same elements)
$A \neq B$	A and B are different (not equal)
$A \subset B$	A is a subset of B (each element of A also belongs to B). This is also written $B \supset A$.
$A \subset B, A \neq B$	A is a proper subset of B (A is a subset of B and B has at least one element which is not in A)
$A \cup B$	$=\{x \mid x \in A \text{ or } x \in B\}$ Union of A and B. See Fig. 77.
$A \cap B$	$=\{x \mid x \in A \text{ and } x \in B\}$ Intersection of A and B. See Fig. 77.
$A \cap B = \varnothing$	A and B are disjoint sets (sets without common elements)
$A - B$	$=\{x \mid x \in A \text{ and } x \notin B\}$ Difference of A and B. (Here B may or may not be a subset of A.) See Fig. 78. (See also Fig. 79.)
A^C	$= X - A$ Complement of A in X (where $A \subset X$) (notation $C_X A$ if confusion concerning X seems possible). See Fig. 80.

The following formulas result directly from the definitions:

(1a) $\quad A \cup A = A \qquad A \cap A = A$

(1b) $\quad A \cup B = B \cup A \qquad A \cap B = B \cap A$

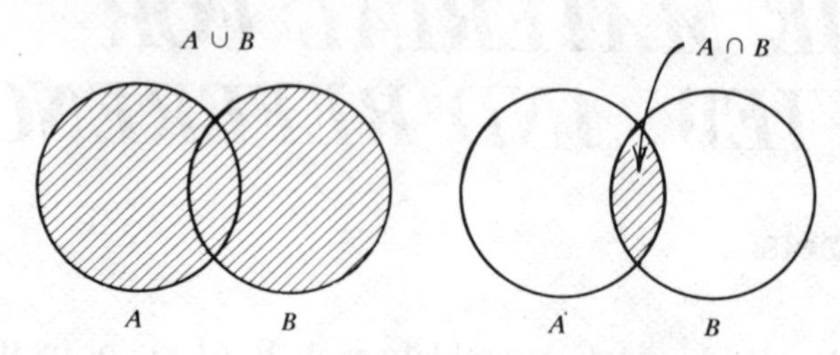

Fig. 77. Union $A \cup B$ (shaded) and intersection $A \cap B$ (shaded) of two sets A and B

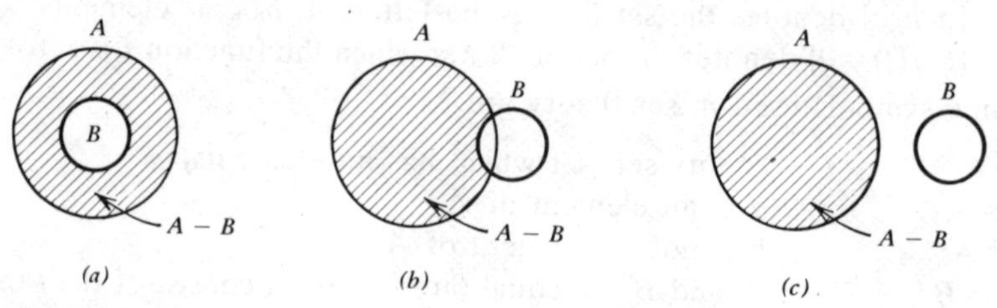

Fig. 78. Difference $A - B$ (shaded) of two sets A (large disk) and B (small disk) if (a) $B \subset A$, (b) $A \cap B \neq \varnothing$ and $B \not\subset A$, (c) $A \cap B = \varnothing$

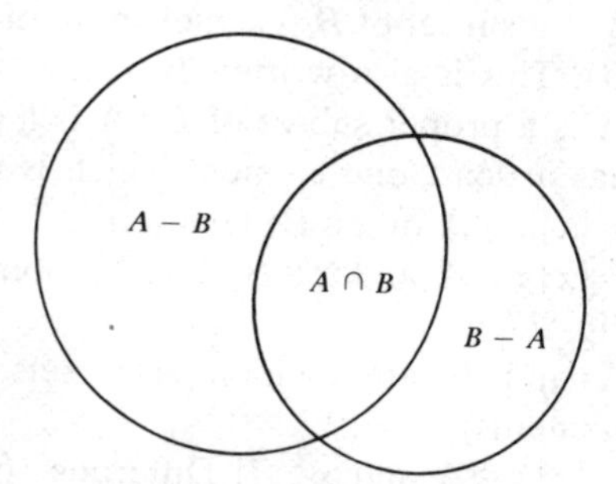

Fig. 79. Differences $A - B$ and $B - A$ and intersection $A \cap B$ of two sets A (large disk) and B (small disk)

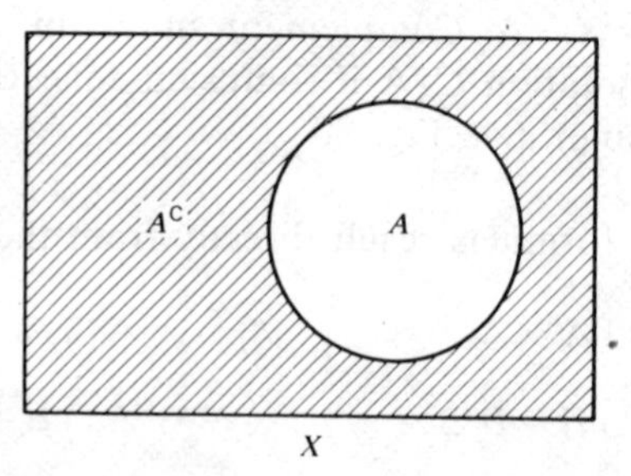

Fig. 80. Complement $A^c = X - A$ (shaded) of a subset A of a set X

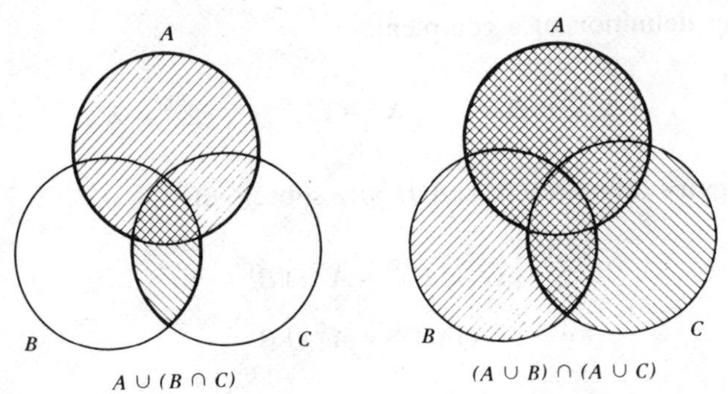

Fig. 81. Formula (1e)

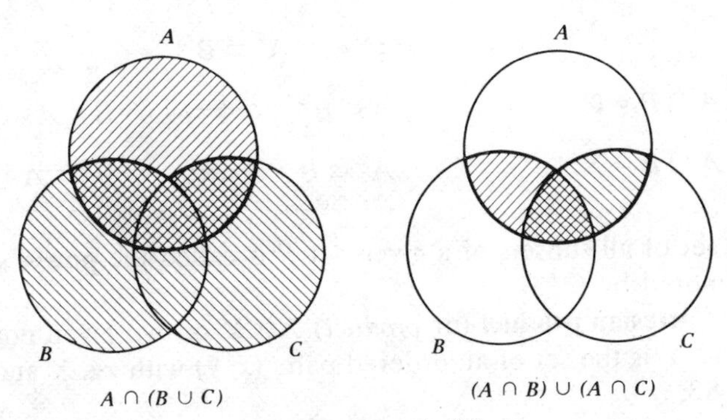

Fig. 82. Formula (1f)

(1c) $A \cup (B \cup C) = (A \cup B) \cup C,$ written $A \cup B \cup C$

(1d) $A \cap (B \cap C) = (A \cap B) \cap C$ written $A \cap B \cap C$

(1e) $$A \cup (B \cap C) = (A \cup B) \cap (A \cup C)$$

(1f) $$A \cap (B \cup C) = (A \cap B) \cup (A \cap C)$$

(1g) $A \cap B \subset A \qquad A \cap B \subset B$

(1h) $A \cup B \supset A \qquad A \cup B \supset B$

Furthermore,

$$A \subset B \iff A \cup B = B \iff A \cap B = A$$

(2) $A \subset C$ and $B \subset C \iff A \cup B \subset C$

$C \subset A$ and $C \subset B \iff C \subset A \cap B.$

From the definition of a complement,

$$(3) \qquad (A^{\mathrm{C}})^{\mathrm{C}} = A, \qquad X^{\mathrm{C}} = \varnothing, \qquad \varnothing^{\mathrm{C}} = X.$$

De Morgan's laws are (A and B any subsets of X)

$$(4) \qquad \begin{aligned} (A \cup B)^{\mathrm{C}} &= A^{\mathrm{C}} \cap B^{\mathrm{C}} \\ (A \cap B)^{\mathrm{C}} &= A^{\mathrm{C}} \cup B^{\mathrm{C}} \end{aligned}$$

Obviously,

$$(5) \qquad \begin{array}{ccccc} & & A \subset B & \Longleftrightarrow & A^{\mathrm{C}} \supset B^{\mathrm{C}} \\ A \cap B = \varnothing & \Longleftrightarrow & A \subset B^{\mathrm{C}} & \Longleftrightarrow & B \subset A^{\mathrm{C}} \\ A \cup B = X & \Longleftrightarrow & A^{\mathrm{C}} \subset B & \Longleftrightarrow & B^{\mathrm{C}} \subset A. \end{array}$$

The set of all subsets of a given set S is called the **power set** of S and is denoted by $\mathcal{P}(S)$.

The **Cartesian product** (or *product*) $X \times Y$ of two given nonempty sets X and Y is the set of all ordered pairs (x, y) with $x \in X$ and $y \in Y$. See Fig. 83.

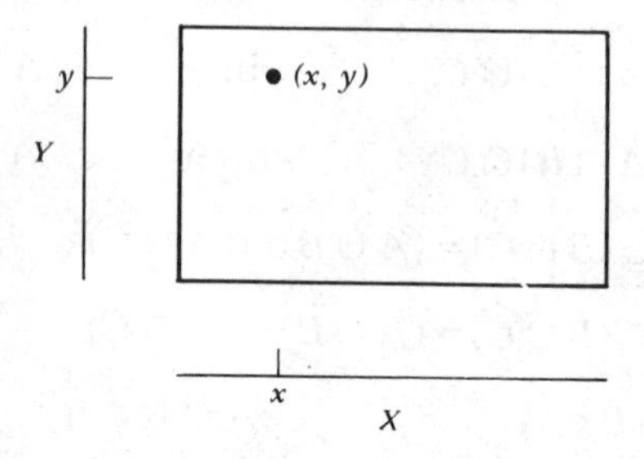

Fig. 83. A way of visualizing the Cartesian product $X \times Y$ of two sets X and Y

A set M is said to be **countable** if M is *finite* (has finitely many elements) or if we can associate positive integers with the elements of M so that to each element of M there corresponds a unique positive integer and, conversely, to each positive integer $1, 2, 3, \cdots$ there corresponds a unique element of M.

A1.2 Mappings

Let X and Y be sets and $A \subset X$ any subset. A **mapping** (or *transformation, functional relation, abstract function*) T from A into Y is obtained by associating with each $x \in A$ a single $y \in Y$, written $y = Tx$ and called the **image** *of* x *with respect to* T. The set A is called the *domain of definition* of T or, more briefly, the **domain** of T and is denoted by $\mathfrak{D}(T)$, and we write

$$T\colon\ \mathfrak{D}(T) \longrightarrow Y$$

$$x \longmapsto Tx.$$

The **range** $\mathfrak{R}(T)$ of T is the set of all images; thus

$$\mathfrak{R}(T) = \{y \in Y \mid y = Tx \text{ for some } x \in \mathfrak{D}(T)\}.$$

The *image* $T(M)$ of any subset $M \subset \mathfrak{D}(T)$ is the set of all images Tx with $x \in M$. Note that $T(\mathfrak{D}(T)) = \mathfrak{R}(T)$.

An illustration of the situation is given in Fig. 84.

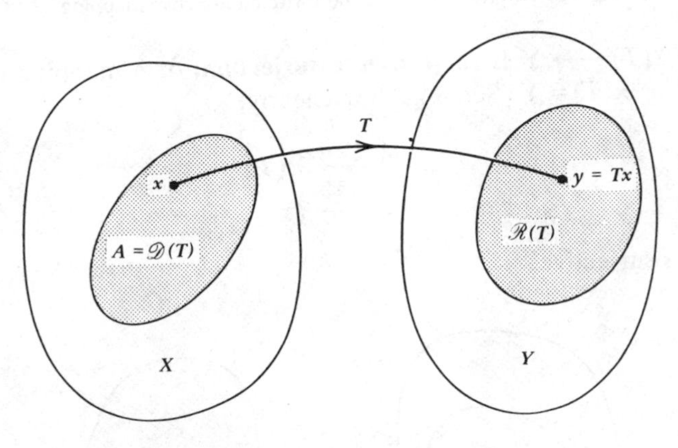

Fig. 84. Visualization of a mapping

The **inverse image** of a $y_0 \in Y$ is the set of all $x \in \mathfrak{D}(T)$ such that $Tx = y_0$. Similarly, the *inverse image* of a subset $Z \subset Y$ is the set of all $x \in \mathfrak{D}(T)$ such that $Tx \in Z$. Note that the inverse image of a $y_0 \in Y$ may be empty, a single point, or any subset of $\mathfrak{D}(T)$; this depends on y_0 and T.

A mapping T is **injective**, an **injection**, or **one-to-one** if for every $x_1, x_2 \in \mathscr{D}(T)$,

$$x_1 \neq x_2 \qquad \text{implies} \qquad Tx_1 \neq Tx_2;$$

that is, different points in $\mathfrak{D}(T)$ have different images, so that the inverse image of any point in $\mathfrak{R}(T)$ is a single point. See Fig. 85.

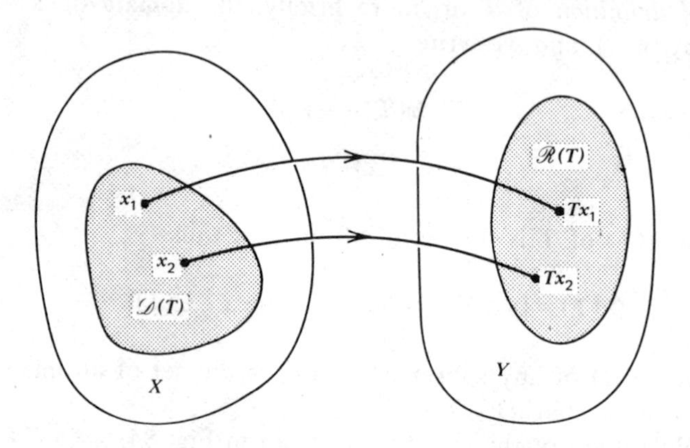

Fig. 85. Notation in connection with an injective mapping

T: $\mathfrak{D}(T) \longrightarrow Y$ is **surjective**, a **surjection**, or a mapping of $\mathfrak{D}(T)$ **onto** Y if $\mathfrak{R}(T) = Y$. See Fig. 86. Clearly,

$$\begin{aligned} \mathscr{D}(T) &\longrightarrow \mathscr{R}(T) \\ x &\longmapsto Tx \end{aligned}$$

is always surjective.

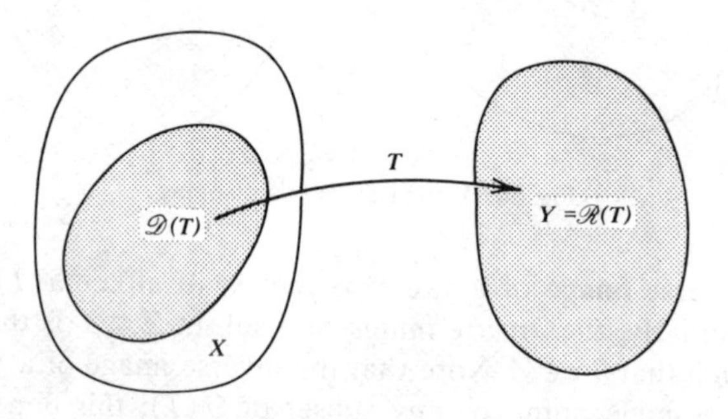

Fig. 86. Surjective mapping

T is **bijective** or a **bijection** if T is both injective and surjective. Then the **inverse mapping** T^{-1} of T: $\mathscr{D}(T) \longrightarrow Y$ is the mapping T^{-1}: $Y \longrightarrow \mathscr{D}(T)$ defined by $Tx_0 \longmapsto x_0$, that is, T^{-1} associates with each $y_0 \in Y$ that $x_0 \in \mathscr{D}(T)$ for which $Tx_0 = y_0$. See Fig. 87.

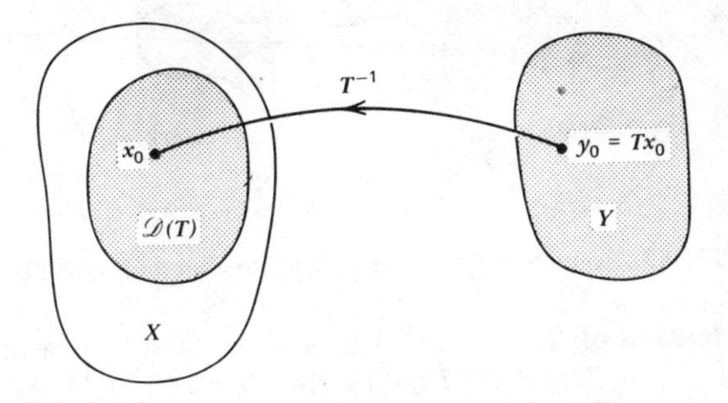

Fig. 87. Inverse $T^{-1}: Y \longrightarrow \mathscr{D}(T) \subset X$ of a bijective mapping T

For an injective mapping T: $\mathscr{D}(T) \longrightarrow Y$ the **inverse mapping** T^{-1} is defined to be the mapping $\mathscr{R}(T) \longrightarrow \mathscr{D}(T)$ such that $y_0 \in \mathscr{R}(T)$ is mapped onto that $x_0 \in \mathscr{D}(T)$ for which $Tx_0 = y_0$. See Fig. 88. Thus in this slightly more general use of the term "inverse" it is not required that T be a mapping *onto* Y; this convenient terminology employed by many authors is unlikely to cause misunderstandings in the present context.

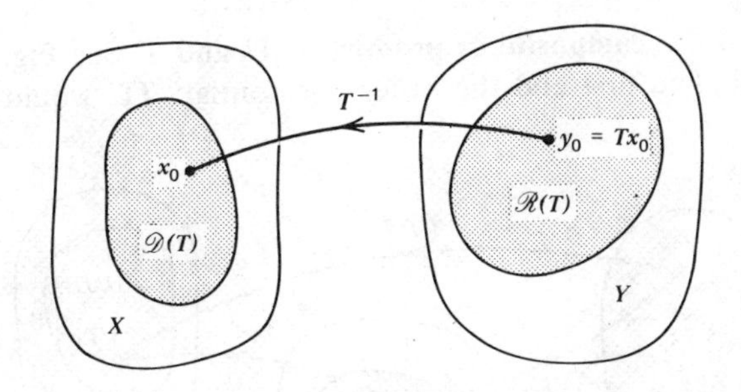

Fig. 88. Inverse $T^{-1}: \mathscr{R}(T) \longrightarrow \mathscr{D}(T)$ of an injective mapping T

Two mappings T_1 and T_2 are said to be **equal** if $\mathscr{D}(T_1) = \mathscr{D}(T_2)$ and $T_1x = T_2x$ for all $x \in \mathscr{D}(T_1) = \mathscr{D}(T_2)$.

The **restriction** $T|_B$ of a mapping T: $\mathscr{D}(T) \longrightarrow Y$ to a subset $B \subset \mathscr{D}(T)$ is the mapping $B \longrightarrow Y$ obtained from T by restricting x to

B (instead of letting it vary in the whole domain $\mathscr{D}(T)$); that is, $T|_B\colon B \longrightarrow Y$, $T|_B x = Tx$ for all $x \in B$. See Fig. 89.

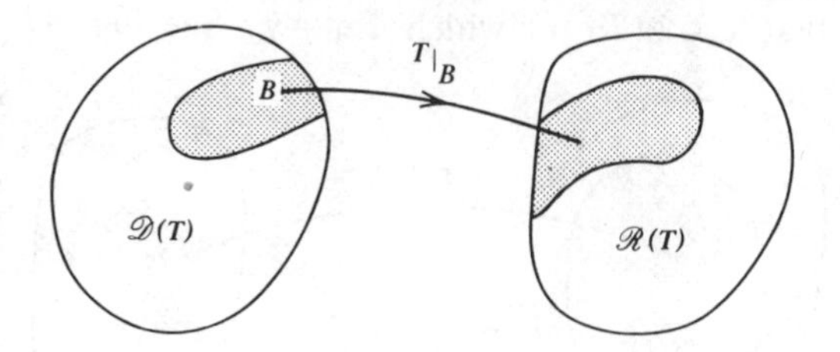

Fig. 89. Restriction $T|_B$ of a mapping T to a subset $B \subset \mathscr{D}(T)$

An **extension** of T from $\mathscr{D}(T)$ to a set $C \supset \mathscr{D}(T)$ is a mapping $\tilde{T}$ such that $\tilde{T}|_{\mathscr{D}(T)} = T$, that is, $\tilde{T}x = Tx$ for all $x \in \mathscr{D}(T)$.

An extension $\tilde{T}$ of T is said to be *proper* if $\mathscr{D}(T)$ is a proper subset of $\mathscr{D}(\tilde{T})$; thus $\mathscr{D}(\tilde{T}) - \mathscr{D}(T) \neq \varnothing$, that is, $x \in \mathscr{D}(\tilde{T})$ for some $x \notin \mathscr{D}(T)$.

Composition of mappings is defined and denoted as follows. If $T\colon X \longrightarrow Y$ and $U\colon Y \longrightarrow Z$, then

$$x \longmapsto U(Tx) \qquad (x \in X)$$

defines a mapping of X into Z which is written $U \circ T$ or simply UT, thus

$$UT\colon X \longrightarrow Z, \qquad x \longmapsto UTx \qquad (x \in X),$$

and is called the **composite** or **product** of U and T. See Fig. 90. Note that T is applied first and the order is essential: TU would not even

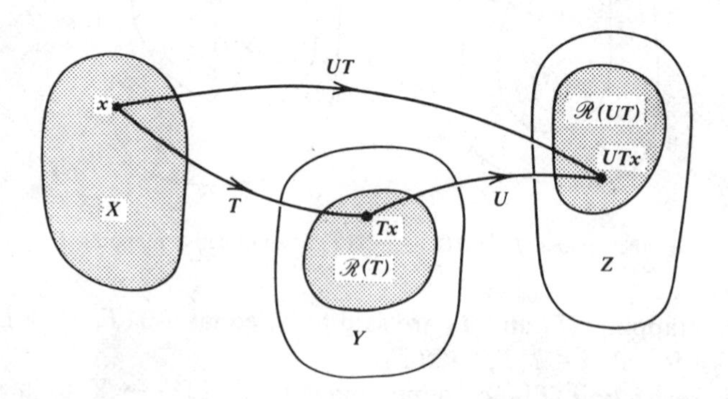

Fig. 90. Composition of two mappings

make sense, in general. If $T: X \longrightarrow Y$ and $U: Y \longrightarrow X$, both $UT: X \longrightarrow X$ and $TU: Y \longrightarrow Y$ make sense but are different if $X \neq Y$. (Even if $X = Y$, those two mappings will be different, in general.)

A1.3 Families

A **sequence** (x_n) of real or complex numbers is obtained if we associate with each positive integer n a real or complex number x_n. This process can be regarded as a mapping of $\mathbf{N} = \{1, 2, \cdots\}$ into the real or complex numbers, x_n being the image of n. The set $\mathbf{N}$ is called the **index set** of the sequence.

This process of "indexing" can be generalized. Instead of $\mathbf{N}$ we may take any nonempty set I (finite, countable or uncountable) and map I into any other given nonempty set X. This gives a **family of elements** of X, written $(x_\alpha)_{\alpha \in I}$ or simply (x_α), where $x_\alpha \in X$ is the image of $\alpha \in I$. Note that it may happen that $x_\alpha = x_\beta$ for some $\alpha \neq \beta$ in I. The set I is called the **index set** of the family. A **subfamily** of a family is obtained if we restrict the indexing mapping to a nonempty subset of the index set.

If the elements of X are subsets of a given set, we obtain a **family of subsets** $(B_\alpha)_{\alpha \in I}$ where B_α is the image of α.

The **union** $\bigcup_{\alpha \in I} B_\alpha$ of the family (B_α) is the set of elements each of which belongs to at least one B_α, and the **intersection** $\bigcap_{\alpha \in I} B_\alpha$ is the set of elements which belong to every B_α, $\alpha \in I$. If $I = \mathbf{N}$, we write

$$\bigcup_{\alpha=1}^{\infty} B_\alpha \qquad \text{and} \qquad \bigcap_{\alpha=1}^{\infty} B_\alpha;$$

and if $I = \{1, 2\}$, we write $B_1 \cup B_2$ and $B_1 \cap B_2$, respectively.

One must carefully distinguish a *family* $(x_\alpha)_{\alpha \in I}$ from the *subset* of X whose elements are the elements of the family, which is the image of the index set I under the indexing mapping.

To any nonempty subset $M \subset X$ we can always find a family of elements of X the set of whose elements is M. For instance, we may take the family defined by the *natural injection* of M into X, that is, the restriction to M of the identity mapping $x \longmapsto x$ on X.

A1.4 Equivalence Relations

Let X and Y be given nonempty sets. Any subset R of the Cartesian product $X \times Y$ (see before) is called a *(binary)* **relation**. $(x, y) \in R$ is also written $R(x, y)$.

An **equivalence relation** *on* X is a relation $R \subset X \times X$ such that

$$\begin{array}{lll} R(x, x) & \text{for all } x \in X & \textit{(Reflexivity)} \\ (1) \quad R(x, y) & \text{implies} \quad R(y, x) & \textit{(Symmetry)} \\ R(x, y) \text{ and } R(y, z) & \text{implies} \quad R(x, z) & \textit{(Transitivity)} \end{array}$$

When R is an equivalence relation on X, then $R(x, y)$ is usually written $x \sim y$, (read "*x is equivalent to y*"). In this case, (1) becomes

$$\begin{array}{ccc} & x \sim x & \\ x \sim y & \Longrightarrow & y \sim x \\ x \sim y \text{ and } y \sim z & \Longrightarrow & x \sim z. \end{array}$$

The *equivalence class* of any $x_0 \in X$ is the set of all $y \in X$ which are equivalent to x_0, and any such y is called a *representative* of the class. The equivalence classes with respect to R constitute a *partition* of X.

By definition, a **partition** of a nonempty set X is a family of nonempty subsets of X which are pairwise disjoint and whose union is X.

A1.5 Compactness

A **cover** (or *covering*) of a subset M of a set X is a family of subsets of X, say, $(B_\alpha)_{\alpha \in I}$ (I the index set), such that

$$M \subset \bigcup_{\alpha \in I} B_\alpha.$$

In particular, if (B_α) is a cover of X, then

$$\bigcup_{\alpha \in I} B_\alpha = X.$$

A cover is said to be **finite** if it consists of only finitely many sets B_α. If $X=(X, \mathcal{T})$ is a topological space (for instance, a metric space; cf. Sec. 1.3), that cover is said to be **open** if all the B_α's are open sets.

A topological space $X=(X, \mathcal{T})$ is said to be

(a) **compact** if every open cover of X contains a finite cover of X, that is, a finite subfamily which is a cover of X,

(b) **countably compact** if every countable open cover of X contains a finite cover of X,

(c) **sequentially compact** if every sequence in X contains a convergent subsequence.

A subset $M \subset (X, \mathcal{T})$ is said to be *compact* (*countably compact, sequentially compact*) if M considered as a subspace $(M, \mathcal{T}_M)$ is compact (countably compact, sequentially compact, respectively); here the *induced topology* $\mathcal{T}_M$ on M consists of all sets $M \cap A$ with $A \in \mathcal{T}$.

For a metric space, the three concepts of compactness are equivalent, that is, one implies the others.

A1.6 Supremum and Infimum

A subset E of the real line $\mathbf{R}$ is **bounded above** if E has an **upper bound**, that is, if there is a $b \in \mathbf{R}$ such that $x \leqq b$ for all $x \in E$. Then if $E \neq \varnothing$, there exists the **supremum** of E (or *least upper bound* of E), written

$$\sup E,$$

that is, the upper bound of E such that $\sup E \leqq b$ for every upper bound b of E. Also

$$\sup C \leqq \sup E$$

for every nonempty subset $C \subset E$.

Similarly, E is **bounded below** if E has a **lower bound,** that is, if there is an $a \in \mathbf{R}$ such that $x \geqq a$ for all $x \in E$. Then if $E \neq \varnothing$, there exists the **infimum** of E (or *greatest lower bound* of E), written

$$\inf E,$$

that is, the lower bound of E such that $\inf E \geqq a$ for every lower bound a of E. Also

$$\inf C \geqq \inf E$$

for every nonempty subset $C \subset E$.

E is **bounded** if E is both bounded above and bounded below. Then if $E \neq \varnothing$,

$$\inf E \leqq \sup E.$$

If for a mapping $T\colon \mathcal{D}(T) \longrightarrow \mathbf{R}$ the range $\mathcal{R}(T)$ (assumed nonempty) is bounded above, its supremum is denoted by

$$\sup_{x \in \mathcal{D}(T)} Tx,$$

and if $\mathcal{R}(T)$ is bounded below, its infimum is denoted by

$$\inf_{x \in \mathcal{D}(T)} Tx.$$

Similar notations are used in connection with subsets of $\mathcal{R}(T)$.

A1.7 Cauchy Convergence Criterion

A number a is called a *limit point* of a (real or complex) sequence of numbers (x_n) if for every given $\varepsilon > 0$ we have

$$|x_n - a| < \varepsilon \qquad \text{for infinitely many } n.$$

The *Bolzano-Weierstrass theorem* states that a *bounded* sequence (x_n) has at least one limit point. Here it is essential that a sequence has infinitely many terms, by definition.

A (real or complex) sequence (x_n) is said to be *convergent* if there is a number x such that, for every given $\varepsilon > 0$, the following condition holds:

$$|x_n - x| < \varepsilon \qquad \text{for all but finitely many } n.$$

This x is called the *limit* of the sequence (x_n).

The limit of a convergent sequence is unique. Note that it is a limit point (why?) and is the only limit point which a convergent sequence has.

We state and prove the Cauchy convergence theorem, whose importance is due to the fact that for deciding about convergence one need not know the limit.

Cauchy Convergence Theorem. *A (real or complex) sequence* (x_n) *is convergent if and only if for every* $\varepsilon > 0$ *there is an N such that*

$$|x_m - x_n| < \varepsilon \qquad \text{for all } m, n > N. \tag{1}$$

Proof. (*a*) If (x_n) converges and c is its limit, then for every given $\varepsilon > 0$ there is an N (depending on ε) such that

$$|x_n - c| < \frac{\varepsilon}{2} \qquad \text{for every } n > N,$$

so that by the triangle inequality for $m, n > N$ we obtain

$$|x_m - x_n| \leqq |x_m - c| + |c - x_n| < \frac{\varepsilon}{2} + \frac{\varepsilon}{2} = \varepsilon.$$

(*b*) Conversely, suppose that the statement involving (1) holds. Given $\varepsilon > 0$, we can choose an $n = k > N$ in (1) and see that every x_m with $m > N$ lies in the disk D of radius ε about x_k. Since there is a disk which contains D as well as the finitely many $x_n \notin D$, the sequence (x_n) is bounded. By the Bolzano-Weierstrass theorem it has a limit point a. Since (1) holds for every $\varepsilon > 0$, an $\varepsilon > 0$ being given, there is an N^* such that $|x_m - x_n| < \varepsilon/2$ for $m, n > N^*$. Choosing a fixed $n > N^*$ such that $|x_n - a| < \varepsilon/2$, by the triangle inequality we have for all $m > N^*$

$$|x_m - a| \leqq |x_m - x_n| + |x_n - a| < \frac{\varepsilon}{2} + \frac{\varepsilon}{2} = \varepsilon,$$

which shows that (x_m) is convergent with the limit a. ∎

A1.8 Groups

The definition of a group is needed only in Sec. 7.7.

A **group** $G=(G,\cdot)$ is a set G of elements $x, y, \cdots$ and a mapping

$$\begin{aligned} G\times G &\longrightarrow G \\ (x,y) &\longmapsto xy \end{aligned} \tag{1}$$

such that the following axioms are satisfied.

(G1) *Associativity.* For all $x, y, z \in G$,

$$(xy)z = x(yz).$$

(G2) *Existence of an identity* e, that is, an element e such that for all $x \in G$,

$$xe = ex = x.$$

(G3) *Existence of an inverse* x^{-1} *of* x. For every $x \in G$ there is an element of G, written x^{-1} and called the *inverse* of x, such that

$$x^{-1}x = xx^{-1} = e. \qquad \blacksquare$$

e is unique. For every $x \in G$, the inverse x^{-1} is unique. G is said to be *commutative* or *Abelian* if G also satisfies

(G4) *Commutativity.* For all $x, y \in G$,

$$xy = yx.$$

APPENDIX 2

ANSWERS TO ODD-NUMBERED PROBLEMS

Section 1.1

3. (M1) to (M3) are obvious. (M4) follows if we take square roots on both sides of

$$|x-y| \leqq |x-z|+|z-y| \leqq (|x-z|^{1/2}+|z-y|^{1/2})^2.$$

5. (*i*) $k>0$, (*ii*) $k=0$

7. The discrete metric; cf. 1.1-8.

9. (M1) to (M3) are obvious. (M4) follows from $d(x, y) \leqq 1$ and

$$d(x, z)+d(z, y) \geqq 1 \qquad (x, y, z \text{ not all equal})$$

and is trivial if $x=y=z$.

13. $d(x, z) \leqq d(x, y)+d(y, z) \implies d(x, z)-d(y, z) \leqq d(x, y)$

$d(y, z) \leqq d(y, x)+d(x, z) \implies -d(x, y) \leqq d(x, z)-d(y, z)$

15. Let $y=x$ in (M4), etc.

Section 1.2

1. The idea of the proof of the triangle inequality remains as before.

3. Take $\eta_j=1$ if $1 \leqq j \leqq n$ and $\eta_j=0$ if $j>n$, and square (11).

5. $(1/n) \notin l^1$ but $(1/n) \in l^p$ $(p>1)$ since $\sum n^{-p}<\infty$ if $p>1$.

7. $\delta(A)=\sup_{x,y \in A} d(x, y)=0 \implies d(x, y)=0 \implies x=y.$

The converse is obvious.

9. The converse does not hold.

11. (M1) to (M3) are obvious. (M4) has the form

$$\frac{d(x, y)}{1+d(x, y)} \leqq \frac{d(x, z)}{1+d(x, z)} + \frac{d(z, y)}{1+d(z, y)}$$

and follows from (M4) for d and the argument used in 1.2-1. Boundedness of X follows from $\tilde{d}(x, y) < 1$.

15. $\tilde{\tilde{d}}(x, y) = 0 \quad \Longleftrightarrow \quad d_1(x_1, y_1) = d_2(x_2, y_2) = 0 \quad \Longleftrightarrow \quad x = y.$ The triangle inequality follows from

$$\max_{k=1,2} d_k(x_k, y_k) \leqq \max_{k=1,2} [d_k(x_k, z_k) + d_k(z_k, y_k)]$$

$$\leqq \max_{i=1,2} d_i(x_i, z_i) + \max_{j=1,2} d_j(z_j, y_j).$$

Section 1.3

1. (*a*) Let $x \in B(x_0; r)$. Then $d(x, x_0) = \alpha < r$, and $B(x; (r-\alpha)/2)$ is a neighborhood of x contained in $B(x_0; r)$. (*b*) Prove that $\tilde{B}(x_0; r)^\mathrm{C}$ is open by showing that for $y \notin \tilde{B}(x_0; r)$ there is a ball about y in $\tilde{B}(x_0; r)^\mathrm{C}$.

3. $\sqrt{2}$

5. (*b*) Any subset $A \subset X$ is open since for any $a \in A$, the open ball $B(a; \frac{1}{2}) = \{a\} \subset A$. By the same argument, A^C is open, so that $(A^\mathrm{C})^\mathrm{C} = A$ is closed.

7. (*a*) The integers, (*b*) $\mathbf{R}$, (*c*) $\mathbf{C}$, (*d*) $\{z \mid |z| \leqq 1\}$

11. (*a*) $\{-1, 1\}$, (*b*) $\mathbf{R}$, (*c*) the circle $\{z \mid |z| = 1\}$

13. Let X be separable. Then X has a countable dense subset Y. Let $x \in X$ and $\varepsilon > 0$ be given. Since Y is dense in X, we have $\bar{Y} = X$ and $x \in \bar{Y}$, so that the ε-neighborhood $B(x; \varepsilon)$ of x contains a $y \in Y$, and $d(x, y) < \varepsilon$. Conversely, if X has a countable subset Y with the property given in the problem, every $x \in X$ is a point of Y or an accumulation point of Y. Hence $\bar{Y} = X$, so that X is separable.

15. $x(t) = \sin t$ defines a continuous mapping $\mathbf{R} \longrightarrow \mathbf{R}$ which maps the open set $(0, 2\pi)$ onto the closed set $[-1, 1]$.

Section 1.4

1. $d(x_n, x) < \varepsilon \quad (n > N)$ implies $d(x_{n_k}, x) < \varepsilon \quad (n_k > N)$.

9. This follows from $\tilde{\tilde{d}}(x, y) \leqq \tilde{d}(x, y) \leqq d(x, y) \leqq 2\tilde{\tilde{d}}(x, y)$.

Section 1.5

1. Cf. 1.4-7.

3. (x_n), where $x_n = (1, 1/2, 1/3, \cdots, 1/n, 0, 0, \cdots)$, is Cauchy in M because $d(x_m, x_n) = 1/(m+1)$, $m < n$, but $x_n \longrightarrow x = (1/n) \in X$, $x \notin M$.

5. X is closed in $\mathbf{R}$; use 1.4-7. *Second proof.* The terms of a Cauchy sequence (x_n) in X must be equal from some term x_n on.

7. A nonconvergent Cauchy sequence is (x_n), where $x_n = n$.

9. We show that x is continuous at any $t = t_0 \in [a, b]$. Since the convergence is uniform, for every $\varepsilon > 0$ there is an $N(\varepsilon)$ such that

$$|x(t) - x_N(t)| < \frac{\varepsilon}{3} \qquad \text{for all } t \in [a, b].$$

Since x_N is continuous at t_0, there is a $\delta > 0$ such that

$$|x_N(t) - x_N(t_0)| < \frac{\varepsilon}{3} \qquad \text{for all } t \in [a, b] \text{ such that } |t - t_0| < \delta.$$

For these t, by the triangle inequality,

$$\begin{aligned} |x(t) - x(t_0)| &\leqq |x(t) - x_N(t)| + |x_N(t) - x_N(t_0)| + |x_N(t_0) - x(t_0)| \\ &< \frac{\varepsilon}{3} + \frac{\varepsilon}{3} + \frac{\varepsilon}{3}, \end{aligned}$$

so that x is continuous at t_0.

11. Let $x_n \longrightarrow x$. Take any fixed j; then for every $\varepsilon > 0$ there is an N such that

$$\frac{1}{2^j} \frac{|\xi_j^{(n)} - \xi_j|}{1 + |\xi_j^{(n)} - \xi_j|} \leqq d(x_n, x) < \frac{\varepsilon}{2^j(1+\varepsilon)} \qquad (n > N).$$

Hence $|\xi_j^{(n)} - \xi_j| < \varepsilon \quad (n > N)$. The proof of sufficiency is immediate.

13. By direct calculation, $d(x_n, x_m) = m^{-1} - n^{-1} \quad (m < n)$.

15. For every $\varepsilon > 0$ there is an N such that for $n > m > N$,

$$d(x_n, x_m) = \sum_{j=m+1}^{n} \frac{1}{j^2} < \varepsilon.$$

But (x_n) does not converge to any $x = (\xi_j) \in X$ because $\xi_j = 0$ for j greater than some $\tilde{N}$, so that for $n > \tilde{N}$,

$$d(x_n, x) = |1 - \xi_1| + \left|\frac{1}{4} - \xi_2\right| + \cdots + \frac{1}{(\tilde{N}+1)^2} + \cdots + \frac{1}{n^2} > \frac{1}{(\tilde{N}+1)^2}$$

and $d(x_n, x) \longrightarrow 0$ is impossible since $\tilde{N}$ is fixed.

Section 1.6

3. X

5. (*b*) $\mathbf{R}$ and $(-1, 1)$ with the metric on $\mathbf{R}$; a homeomorphism is $x \longmapsto \frac{2}{\pi}$ arc tan x.

7. If $\tilde{d}(x_m, x_n) < \varepsilon < \frac{1}{2}$, then

$$d(x_m, x_n) = \frac{\tilde{d}(x_m, x_n)}{1 - \tilde{d}(x_m, x_n)} < 2\tilde{d}(x_m, x_n).$$

Hence if (x_n) is Cauchy in $(X, \tilde{d})$, it is Cauchy in (X, d), and its limit in (X, d) is its limit in $(X, \tilde{d})$.

9. $d(x_n', l) \leqq d(x_n', x_n) + d(x_n, l) \longrightarrow 0$ as $n \longrightarrow \infty$.

11. If $(x_n) \sim (y_n)$ and $(y_n) \sim (z_n)$, then $(x_n) \sim (z_n)$, as can be seen from

$$d(x_n, z_n) \leqq d(x_n, y_n) + d(y_n, z_n) \longrightarrow 0 \qquad (n \longrightarrow \infty).$$

15. Open "vertical strips" of width 2.

Section 2.1

3. The plane $\xi_1 = \xi_2$.

7. $\{e_1, \cdots, e_n, ie_1, \cdots, ie_n\}$, n, $2n$

9. $\{e_0, \cdots, e_n\}$, where $e_j(t) = t^j$, $t \in [a, b]$. No.

15. The set of all lines parallel to the ξ_1-axis, $\{0\}$, X.

Section 2.2

3. By the triangle inequality and (N3),

$$\|y\| = \|y - x + x\| \leqq \|y - x\| + \|x\|,$$
$$\|x\| = \|x - y + y\| \leqq \|y - x\| + \|y\|.$$

From this,

$$\|y\| - \|x\| \leqq \|y - x\|,$$
$$\|y\| - \|x\| \geqq -\|y - x\|.$$

5. (N1) to (N3) are readily verified, and (N4) follows from the Minkowski inequality (12), Sec. 1.2, with $p = 2$ (summations from 1 to n only).

7. (N1) to (N3) are obvious and (N4) follows from the Minkowski inequality (Sec. 1.2).

11. $\|z\| = \|\alpha x + (1 - \alpha)y\| \leqq \alpha \|x\| + (1 - \alpha)\|y\| \leqq \alpha + (1 - \alpha) = 1$

13. It does not satisfy (9b).

15. Let M be bounded, say, $\delta(M) = \sup_{x,y \in M} \|x-y\| = b < \infty$. Consider any $x \in M$. Take a fixed $x_0 \in M$ and set $c = b + \|x_0\|$. Then

$$\|x\| = \|x - x_0 + x_0\| \leqq \|x - x_0\| + \|x_0\| \leqq b + \|x_0\| = c.$$

Conversely, let $\|x\| \leqq c$ for every $x \in M$. Then for all $x, y \in M$,

$$\|x-y\| \leqq \|x\| + \|y\| \leqq 2c, \qquad \text{and} \qquad \delta(M) \leqq 2c.$$

Section 2.3

3. For instance, $x = (\xi_n) = (1/n) \in \bar{Y}$ but $x \notin Y$.

5. This follows immediately from Prob. 4.

7. $\sum_{n=1}^{\infty} \|y_n\| = \sum_{n=1}^{\infty} 1/n^2$ converges, but

$$\sum_{j=1}^{n} y_j = s_n = (1, 1/4, 1/9, \cdots, 1/n^2, 0, 0, \cdots) \qquad \longrightarrow \qquad s \notin Y.$$

9. The sequence (s_n) of the partial sums is Cauchy since for $m < n$,

$$\begin{aligned} \|s_n - s_m\| = \|x_{m+1} + \cdots + x_n\| &\leqq \|x_{m+1}\| + \cdots + \|x_n\| \\ &\leqq \|x_{m+1}\| + \|x_{m+2}\| + \cdots . \end{aligned}$$

13. If $p(x) = p(y) = 0$, then $p(\alpha x + \beta y) = 0$ by (N4), (N3) and (N1). $\|\hat{x}\|_0$ is unique since for any $v \in N$ and $x \in X$ we have $p(v) = 0$ and by (N4),

$$p(x) = p(x + v - v) \leqq p(x+v) + 0 \leqq p(x).$$

(N2) holds since $p(0) = 0$ and $\|\hat{x}\|_0 = 0$ implies $p(x) = 0$, hence $x \in N$, which is the zero element of X/N.

15. $\|x\| = 0 \quad \Longleftrightarrow \quad \|x_1\|_1 = \|x_2\|_2 = 0 \quad \Longleftrightarrow \quad x = (0, 0) = 0.$

Let $x=(x_1, x_2)$, $y=(y_1, y_2)$. Then

$$\begin{aligned}\|x+y\| &= \max (\|x_1+y_1\|_1, \|x_2+y_2\|_2)\\ &\leqq \max (\|x_1\|_1+\|y_1\|_1, \|x_2\|_2+\|y_2\|_2)\\ &\leqq \max (\|x_1\|_1, \|x_2\|_2)+\max (\|y_1\|_1, \|y_2\|_2)\\ &= \|x\|+\|y\|.\end{aligned}$$

Section 2.4

7. Let $e_1=(1, 0, \cdots, 0)$, $e_2=(0, 1, 0, \cdots, 0)$, etc. By the Cauchy-Schwarz inequality (11), Sec. 1.2,

$$\|x\| \leqq \sum |\xi_j| \, \|e_j\| \leqq b\|x\|_2 \qquad \text{where} \qquad b^2=\sum \|e_j\|^2.$$

Section 2.5

7. Let $\{b_1, \cdots, b_n\}$ be a basis for Y. Let $y_k=\sum \alpha_{kl} b_l \in Y$ and $\|y_k - v\| \longrightarrow a$. Then the α_{kl}'s constitute a bounded set (cf. Lemma 2.4-1) and (y_k) has a subsequence (y_{k_j}) such that $\alpha_{k_j l} \longrightarrow \alpha_l$ for each $l=1, \cdots, n$, and we have

$$\tilde{y}=\sum \alpha_l b_l \in Y, \qquad \|v-\tilde{y}\| \leqq \|v-y_{k_j}\| + \sum |\alpha_{k_j l}-\alpha_l| \, \|b_l\|$$

which implies $\|v-\tilde{y}\|=a$. We now repeat the argument of the proof of the lemma with (1) replaced by the equality $\|v-\tilde{y}\|=a$. We then conclude that $\tilde{z}=\|v-\tilde{y}\|^{-1}(v-\tilde{y})$ satisfies $\|\tilde{z}-y\| \geqq 1$ for every $y \in Y$.

9. Since X is compact, any sequence (x_n) in M has a subsequence (x_{n_k}) which converges in X, say $x_{n_k} \longrightarrow x \in X$, and $x \in \bar{M}$ by 1.4-6(a), thus $x \in M$ since M is closed. Hence M is compact.

Section 2.6

3. The domain is $\mathbf{R}^2$. The ranges are the ξ_1-axis, the ξ_2-axis, $\mathbf{R}^2$. The null spaces are the ξ_2-axis, the ξ_1-axis, the origin.

5. Let Tx_1, $Tx_2 \in T(V)$. Then x_1, $x_2 \in V$, $\alpha x_1 + \beta x_2 \in V$. Hence $T(\alpha x_1 + \beta x_2) = \alpha Tx_1 + \beta Tx_2 \in T(V)$.

Let x_1, x_2 be in that inverse image. Then Tx_1, $Tx_2 \in W$, $\alpha Tx_1 + \beta Tx_2 \in W$, $\alpha Tx_1 + \beta Tx_2 = T(\alpha x_1 + \beta x_2)$, so that $\alpha x_1 + \beta x_2$ is an element of that inverse image.

7. No, which also is geometrically obvious.

11. b nonsingular ($\det b \neq 0$)

13. Otherwise $\alpha_1 Tx_1 + \cdots + \alpha_n Tx_n = 0$ with some $\alpha_j \neq 0$ and, since T^{-1} exists and is linear,

$$T^{-1}(\alpha_1 Tx_1 + \cdots + \alpha_n Tx_n) = \alpha_1 x_1 + \cdots + \alpha_n x_n = 0,$$

which shows linear dependence of $\{x_1, \cdots, x_n\}$, a contradiction.

15. $\mathcal{R}(T) = X$ since for every $y \in X$ we have $y = Tx$, where

$$x(t) = \int_0^t y(\tau)\, d\tau.$$

But T^{-1} does not exist since $Tx = 0$ for every constant function. This shows that finite dimensionality is essential in Prob. 14.

Section 2.7

1. We have

$$\|T_1 T_2\| = \sup_{\|x\|=1} \|T_1 T_2 x\| \leqq \sup_{\|x\|=1} \|T_1\| \, \|T_2 x\|$$

$$= \|T_1\| \sup_{\|x\|=1} \|T_2 x\| = \|T_1\| \, \|T_2\|.$$

3. $\|x\| = \gamma < 1$ by assumption and $\|Tx\| \leqq \|T\| \gamma < \|T\|$ by (3).

5. $\|T\|=1$

7. Let $Tx=0$. Then $0=\|Tx\|\geqq b\|x\|$, $\|x\|=0$, $x=0$, so that T^{-1} exists by 2.6-10(*a*), and $T^{-1}\colon Y\longrightarrow X$ since $\mathscr{R}(T)=Y$. Let $y=Tx$. Then $T^{-1}y=x$ and boundedness of T^{-1} follows from

$$\|T^{-1}y\|=\|x\|\leqq\frac{1}{b}\|Tx\|=\frac{1}{b}\|y\|.$$

9. The subspace of all continuously differentiable functions y on $[0,1]$ such that $y(0)=0$. $T^{-1}y=y'$; T^{-1} is linear, but is unbounded since $|(t^n)'|=n\,|t^{n-1}|$ implies $\|T^{-1}\|\geqq n$; cf. also 2.7-5.

11. Yes. Yes.

13. The first statement follows from the last formula in 2.7-7. To prove the second statement, consider the unit matrix.

15. Let $\|\cdot\|_0$ be the natural norm. By Prob. 14,

$$\|A\|_0=\sup_{\|x\|_1=1}\|Ax\|_2\leqq\|A\|.$$

The case $\|A\|=0$ is trivial. If $\|A\|>0$, there is a $k=s$ such that

$$\|A\|=\max_k\sum_{j=1}^{n}|\alpha_{jk}|=\sum_{j=1}^{n}|\alpha_{js}|.$$

We choose $x=(\xi_j)$ with $\xi_s=1$ and $\xi_j=0$ $(j\neq s)$. Then $\|x\|_1=1$, and $\|Ax\|_2=\sum|\alpha_{js}|=\|A\|$, hence $\|A\|_0=\|A\|$.

Section 2.8

3. 2

5. Yes, $\|f\|=1$.

7. $g=\bar{f}$ is bounded but not linear since $g(\alpha x)=\overline{f(\alpha x)}=\bar{\alpha}g(x)$.

9. Let $\alpha=f(x)/f(x_0)$ and $y=x-\alpha x_0$. Then we have $x=\alpha x_0+y$

and $f(y)=f(x)-\alpha f(x_0)=0$, so that $y\in\mathcal{N}(f)$. *Uniqueness.* Let

$$y+\alpha x_0=\tilde{y}+\tilde{\alpha}x_0. \qquad \text{Then} \qquad y-\tilde{y}=(\tilde{\alpha}-\alpha)x_0.$$

Hence $\tilde{\alpha}=\alpha$ since otherwise

$$x_0=(\tilde{\alpha}-\alpha)^{-1}(y-\tilde{y})\in\mathcal{N}(f),$$

a contradiction. Hence also $y=\tilde{y}$.

11. $x=y+[f_1(x)/f_1(x_0)]x_0$ by Prob. 9. Since $y\in\mathcal{N}(f_1)=\mathcal{N}(f_2)$, so that $f_2(y)=0$, this gives the proportionality $f_2(x)=f_1(x)f_2(x_0)/f_1(x_0)$.

13. The assumption $f(y_0)=\gamma\neq 0$ for a $y_0\in Y$ yields the contradiction that any

$$\alpha=\frac{\alpha}{\gamma}f(y_0)=f\left(\frac{\alpha}{\gamma}y_0\right)\in f(Y).$$

15. If $\|x\|\leqq 1$, then $f(x)\leqq|f(x)|\leqq\|f\|\,\|x\|\leqq\|f\|$, but $\|f\|=\sup_{\|x\|\leqq 1}|f(x)|$ shows that for any $\varepsilon>0$ there is an x with $\|x\|\leqq 1$ such that $f(x)>\|f\|-\varepsilon$.

Section 2.9

1. $\{\alpha x_0 \mid \alpha\in\mathbf{R},\ x_0=(2,4,-7)\}$

3. $f_1=(1,0,0)$, $f_2=(0,1,0)$, $f_3=(0,0,1)$

5. n or $n-1$

7. $(\alpha_2,-\alpha_1,0)$, $(\alpha_3,0,-\alpha_1)$

11. Otherwise, $f(x)-f(y)=f(x-y)=0$ for all $f\in X^*$, and $x-y=0$ by 2.9-2, a contradiction.

13. Let $\{e_1,\cdots,e_n\}$ be a basis for X such that $\{e_1,\cdots,e_p\}$, $p<n$, is a basis for Z, and let $\{f_1,\cdots,f_n\}$ be the dual basis. Let

$$\tilde{f}=\sum_{j=1}^{p} f(e_j)f_j.$$

Then $\tilde{f}(e_k)=f(e_k)$, $k=1,\cdots,p$, hence $\tilde{f}|_Z=f$.

15. $\tilde{f}(x)=\frac{1}{2}\xi_1+k\xi_2-\frac{1}{2}\xi_3$. Yes.

Section 2.10

1. The zero operator $0\colon X \longrightarrow \{0\} \subset Y$. The operator $-T$.

3. $\mathfrak{D}(\alpha T_1 + \beta T_2) = \mathfrak{D}(T_1) \cap \mathfrak{D}(T_2)$; the two ranges must lie in the same space.

7. On X with norm defined by $\|x\|_1 = |\xi_1| + \cdots + |\xi_n|$, a linear functional f represented by $f(x) = \alpha_1 \xi_1 + \cdots + \alpha_n \xi_n$ has norm $\|f\| = \max |\alpha_j|$.

11. Use Prob. 10 with $Y = \mathbf{R}$ or $\mathbf{C}$.

13. If $f \in \overline{M^a}$, there is a sequence (f_n) in M^a such that $f_n \longrightarrow f$. For any $x \in M$ we have $f_n(x) = 0$, and $f(x) = 0$, so that $f \in M^a$, and M^a is closed. $\{0\}$, X'.

15. $(1, 1, 1)$

Section 3.1

1. We obtain

$$\begin{aligned} \|x+y\|^2 + \|x-y\|^2 &= \langle x+y, x+y\rangle + \langle x-y, x-y\rangle \\ &= \langle x, x\rangle + \langle x, y\rangle + \langle y, x\rangle + \langle y, y\rangle \\ &\qquad + \langle x, x\rangle - \langle x, y\rangle - \langle y, x\rangle + \langle y, y\rangle \\ &= 2\langle x, x\rangle + 2\langle y, y\rangle = 2\,\|x\|^2 + 2\,\|y\|^2. \end{aligned}$$

3. By assumption,

$$\begin{aligned} 0 &= \langle x+y, x+y\rangle - \|x\|^2 - \|y\|^2 \\ &= \langle x, y\rangle + \langle y, x\rangle = \langle x, y\rangle + \overline{\langle x, y\rangle} = 2 \operatorname{Re} \langle x, y\rangle. \end{aligned}$$

7. $\langle x, u-v\rangle = 0$; take $x = u - v$.

9. This follows by direct calculation.

11. No; cf. 3.1-7.

15. No,

$$\gamma_{jk} = \langle e_j, e_k \rangle = \overline{\langle e_k, e_j \rangle} = \bar{\gamma}_{kj}.$$

Section 3.2

1. For vectors $x \neq 0$ and $y \neq 0$ the dot product is

$$x \cdot y = |x|\,|y| \cos \theta, \qquad \text{hence} \qquad |x \cdot y| \leqq |x|\,|y|.$$

3. Cf. 3.2-4(*b*). Yes. No.

5. We have

$$\begin{aligned} \|x_n - x\|^2 &= \langle x_n - x, x_n - x \rangle \\ &= \|x_n\|^2 - \langle x_n, x \rangle - \langle x, x_n \rangle + \|x\|^2 \\ &\longrightarrow \quad 2\,\|x\|^2 - 2\langle x, x \rangle = 0. \end{aligned}$$

7. From

$$\langle x \pm \alpha y, x \pm \alpha y \rangle = \|x\|^2 \pm \bar{\alpha} \langle x, y \rangle \pm \alpha \langle y, x \rangle + |\alpha|^2 \|y\|^2$$

we see that orthogonality implies the given condition. Conversely, that condition implies

$$\bar{\alpha} \langle x, y \rangle + \alpha \langle y, x \rangle = 0.$$

Taking $\alpha = 1$ if the space is real and $\alpha = 1$, $\alpha = i$ if it is complex, we see that $\langle x, y \rangle = 0$.

9. Use Theorem 1.4-8 and

$$\|x\|_2{}^2 = \int_a^b |x(t)|^2 \, dt \leqq (b-a) \|x\|_\infty{}^2.$$

Section 3.3

1. (x_n) is Cauchy, since from the assumption and the parallelogram equality (4), Sec. 3.1, we obtain

$$\begin{aligned}\|x_n - x_m\|^2 &= 2\,\|x_m\|^2 + 2\,\|x_n\|^2 - \|x_n + x_m\|^2 \\ &\leqq 2\,\|x_m\|^2 + 2\,\|x_n\|^2 - 4d^2.\end{aligned}$$

5. (*a*) $\{z \mid z = \alpha(\xi_2, -\xi_1),\ \alpha \in \mathbf{R}\}$, (*b*) $\{0\}$

7. (*a*) $x \in A \quad \Rightarrow \quad x \perp A^{\perp} \quad \Rightarrow \quad x \in A^{\perp\perp} \quad \Rightarrow \quad A \subset A^{\perp\perp}$, as in the text.

(*b*) $x \in B^{\perp} \quad \Rightarrow \quad x \perp B \supset A \quad \Rightarrow \quad x \in A^{\perp} \quad \Rightarrow \quad B^{\perp} \subset A^{\perp}$.

(*c*) $A^{\perp\perp\perp} = (A^{\perp})^{\perp\perp} \supset A^{\perp}$ by (*a*); and by (*b*),

$$A \subset A^{\perp\perp} \quad \Rightarrow \quad A^{\perp} \supset (A^{\perp\perp})^{\perp}.$$

9. Let $Y = Y^{\perp\perp} = (Y^{\perp})^{\perp}$. Then Y is closed by Prob. 8. The converse is stated in Lemma 3.3-6.

Section 3.4

1. This is an immediate consequence of the Gram-Schmidt process.

3. For any x and $y \neq 0$, setting $e = \|y\|^{-1} y$, we have from (12*) with $n = 1$

$$|\langle x, e\rangle|^2 \leqq \|x\|^2$$

and multiplication by $\|y\|^2$ gives $|\langle x, y\rangle|^2 \leqq \|x\|^2 \|y\|^2$.

5. $y \in Y_n$, $x = y + (x - y)$, and $x - y \perp e_m$ since

$$\langle x - y, e_m\rangle = \langle x - \textstyle\sum \alpha_k e_k, e_m\rangle = \langle x, e_m\rangle - \alpha_m = 0.$$

7. From the Cauchy-Schwarz inequality (Sec. 1.2) and (12),

$$\sum |\langle x, e_k\rangle\langle y, e_k\rangle| \leqq \left[\sum |\langle x, e_k\rangle|^2\right]^{1/2} \left[\sum |\langle x, e_k\rangle|^2\right]^{1/2} \leqq \|x\|\,\|y\|.$$

9. $1/\sqrt{2}$, $(3/2)^{1/2} t$, $(5/8)^{1/2}(3t^2 - 1)$

Section 3.5

1. Using the orthonormality and the notations in the proof of 3.5-2, we have

$$\|s_n\|^2 = \|\alpha_1 e_1 + \cdots + \alpha_n e_n\|^2 = |\alpha_1|^2 + \cdots + |\alpha_n|^2 = \sigma_n$$

and $s_n \longrightarrow x$ implies $\|s_n\|^2 = \langle s_n, s_n \rangle \longrightarrow \langle x, x \rangle$, by 3.2-2.

3. The sum may differ from x by a function $z \perp (e_k)$. Take, for instance, $x = (1, 1, 1) \in \mathbf{R}^3$ and (e_1, e_2) in $\mathbf{R}^3$, where $e_1 = (1, 0, 0)$, $e_2 = (0, 1, 0)$.

5. (s_n), where $s_n = x_1 + \cdots + x_n$ is Cauchy since

$$\|s_n - s_m\| \leqq \sum_{j=m+1}^{n} \|x_j\| \leqq \sum_{j=m+1}^{\infty} \|x_j\| \longrightarrow 0 \quad (m \longrightarrow \infty)$$

and convergence of (s_n) follows from the completeness of H. Cf. also Probs. 7 to 9, Sec. 2.3.

7. The series converges and defines y by 3.5-2(c), and $\langle x, e_k \rangle = \langle y, e_k \rangle$ by 3.5-2(b), so that $x - y \perp e_k$ follows from

$$\langle x - y, e_k \rangle = \langle x, e_k \rangle - \langle y, e_k \rangle = 0.$$

9. Prob. 8 shows that $e_n \in \bar{M}_2$ if and only if (a) holds, and $\tilde{e}_n \in \bar{M}_1$ if and only if (b) holds. Then (a) implies that (e_n) lies in $\bar{M}_2$ and (b) implies that $(\tilde{e}_n)$ lies in $\bar{M}_1$; hence $\bar{M}_1 = \bar{M}_2$.

Section 3.6

1. No

3. Pythagorean theorem

5. This follows from Theorem 3.6-3 and the fact that the relation in Prob. 4 implies (3) and conversely.

7. In this case, one can use the Gram-Schmidt process, as proved in Prob. 6.

9. $\langle v-w, x\rangle = 0$ for all $x \in M$ implies $v - w \perp M$, hence $v-w=0$ by 3.6-2(*a*).

Section 3.7

1. We obtain

$$\int_{-1}^{1} P_m[(1-t^2)P_n']'\,dt - \int_{-1}^{1} P_n[(1-t^2)P_m']'\,dt$$
$$= (m-n)(m+n+1)\int_{-1}^{1} P_n P_m\,dt.$$

Integration by parts on the left shows that the left-hand side is zero. Hence the integral on the right must be zero when $m-n \neq 0$.

3. Develop $(1-q)^{-1/2}$ by the binomial theorem. Then substitute $q = 2tw - w^2$. Develop the powers of q by the binomial theorem. Show that in the resulting development, the power w^n has the coefficient $P_n(t)$ as given by (2c).

9. $y(t) = e^{-t^2/2}H_n(t)$

11. We obtain

$$\begin{aligned}\sum_{n=0}^{\infty} L_n(t)w^n &= \sum_{n=0}^{\infty}\sum_{m=0}^{n}(-1)^m\binom{n}{m}\frac{t^m w^n}{m!}\\ &= \sum_{m=0}^{\infty}\frac{(-1)^m t^m}{m!}\sum_{n=m}^{\infty}\binom{n}{m}w^n\\ &= \sum_{m=0}^{\infty}\frac{(-1)^m t^m}{m!}\,\frac{w^m}{(1-w)^{m+1}} = \frac{e^{-wt/(1-w)}}{1-w}.\end{aligned}$$

13. Differentiate (a) in Prob. 12. In the result express L'_{n+1} and L'_{n-1} by the use of (b) in Prob. 12. This gives (c). From (c) and (b),

$$\text{(d)} \qquad nL'_{n-1} = nL_n + (n-t)L_n'.$$

Differentiate (c), substitute (d) and simplify to get (11).

15. Consider

$$\int_0^\infty e^{-t} L_m L_n \, dt \qquad (m<n).$$

Conclude that it suffices to show that

$$\int_0^\infty e^{-t} t^k L_n \, dt = 0 \qquad (k<n).$$

Prove this by repeated integration by parts.

Section 3.8

1. In $\mathbf{R}^3$, every linear functional is bounded, and the inner product in (1) is the dot product.

3. We obtain

$$|f(x)| = |\langle x, z\rangle| \leqq \|x\| \, \|z\|, \qquad |f(x)|/\|x\| \leqq \|z\| \qquad (x \neq 0).$$

Hence $\|f\| \leqq \|z\|$. Also $\|f\| = \|z\|$ if $z = 0$. Let $z \neq 0$. Then

$$\|f\| \, \|z\| \geqq |f(z)| = \langle z, z\rangle = \|z\|^2, \qquad \|f\| \geqq \|z\|.$$

5. An isomorphism of $l^{2\prime}$ onto l^2 is $f \longmapsto z_f$, where z_f is defined by (cf. 3.8-1)

$$f(x) = \langle x, z_f\rangle.$$

(Note that for the complex space l^2, that mapping is conjugate linear since $\alpha f \longmapsto \bar{\alpha} z_f$.)

9. We have

$$M^a = \{f \mid f(x) = \langle x, z_f\rangle = 0 \text{ for all } x \in M\},$$

hence $f \in M^a \iff z_f \in M^\perp$.

11. The first statement is rather obvious and the second follows from

$$f_2(\alpha y_1 + \beta y_2) = \overline{h(x_0, \alpha y_1 + \beta y_2)} = \alpha \overline{h(x_0, y_1)} + \beta \overline{h(x_0, y_2)}.$$

13. $h(x, y) = h(y, x)$; then h is called a *symmetric bilinear form.* The condition of *positive definiteness,* that is, $h(x, x) \geqq 0$ for all $x \in X$ and $h(x, x) > 0$ if $x \neq 0$.

15. The Schwarz inequality in Prob. 14 gives the triangle inequality in a way similar to that in Sec. 3.2.

Section 3.9

3. $\|T_n^* - T^*\| = \|(T_n - T)^*\| = \|T_n - T\| \longrightarrow 0$

5. Let $T(M_1) \subset M_2$. Then $M_1^\perp \supset T^*(M_2^\perp)$ by Prob. 4. Conversely, let $M_1^\perp \supset T^*(M_2^\perp)$. Then $T^{**}(M_1^{\perp\perp}) \subset M_2^{\perp\perp}$ by Prob. 4, where $T^{**} = T$ by 3.9-4 and $M_1^{\perp\perp} = M_1$, $M_2^{\perp\perp} = M_2$ by 3.3-6.

7. Use 3.9-3(b).

9. Let $\{b_1, \cdots, b_n\}$ be an orthonormal basis for $T(H) = \mathcal{R}(T)$. Let $x \in H$ and $Tx = \sum \alpha_j(x) b_j$. Then $\langle Tx, b_k \rangle = \alpha_k(x) = \langle x, T^* b_k \rangle$ and

$$Tx = \sum_{j=1}^{n} \langle x, v_j \rangle w_j \qquad \text{where} \qquad v_j = T^* b_j, \ w_j = b_j.$$

Section 3.10

1. Use 3.9-4.

3. Use 3.10-4.

5. $T^* x = (\xi_1 + \xi_2, -i\xi_1 + i\xi_2)$, hence

$$T_1 x = \left(\xi_1 + \frac{1+i}{2}\xi_2, \frac{1-i}{2}\xi_1\right),$$

$$T_2 x = \left(\frac{1+i}{2}\xi_2, \frac{1-i}{2}\xi_1 - \xi_2\right).$$

7. This follows from $\bar{U}^{\mathsf{T}} U = U^{-1} U = I$.

9. $\mathcal{R}(T)$ is a subspace $Y \subset H$ by 2.6-9. For $y \in \bar{Y}$ there is a sequence (y_n) in Y such that $y_n \longrightarrow y$. Let $y_n = Tx_n$. Then (x_n) is Cauchy (by isometry), $x_n \longrightarrow x$ since H is complete, $y = Tx \in Y$ by 1.4-8, so that Y is closed. If $Y = H$, then T would be unitary.

11. $S^* = (UTU^*)^* = UT^*U^* = UTU^* = S$; cf. 3.9-4.

13. $\|TT^* - T^*T\| \leqq \|TT^* - T_nT_n^*\| + \|T_nT_n^* - T_n^*T_n\| + \|T_n^*T_n - T^*T\|$. The second term on the right is zero. $T_n \longrightarrow T$ implies that $T_n^* \longrightarrow T^*$ (Prob. 3, Sec. 3.9), so that each of the two other terms on the right approaches zero as $n \longrightarrow \infty$.

15. We have, using 3.9-3(*b*), for all x

$$\begin{aligned} \|T^*x\|^2 = \|Tx\|^2 \quad &\Longleftrightarrow \quad \langle T^*x, T^*x\rangle = \langle Tx, Tx\rangle \\ &\Longleftrightarrow \quad \langle TT^*x, x\rangle = \langle T^*Tx, x\rangle \\ &\Longleftrightarrow \quad \langle [TT^* - T^*T]x, x\rangle = 0 \\ &\Longleftrightarrow \quad TT^* - T^*T = 0. \end{aligned}$$

From $\|T^*x\| = \|Tx\|$ with $x = Tz$ we have $\|T^*Tz\| = \|T^2z\|$ and, by (6e) in Sec. 3.9,

$$\|T^2\| = \sup_{\|z\|=1} \|T^2z\| = \sup_{\|z\|=1} \|T^*Tz\| = \|T^*T\| = \|T\|^2.$$

Section 4.1

5. Use induction with respect to the number of elements of A.

7. 12, 24, 36, $\cdots$ (all $x \in \mathbf{N}$ divisible by 4 and 6). 1, 2.

Section 4.2

5. $p(x) \leqq \gamma$, $p(y) \leqq \gamma$ and $\alpha \in [0, 1]$ implies $1 - \alpha \geqq 0$ and

$$p(\alpha x + (1-\alpha)y) \leqq \alpha p(x) + (1-\alpha)p(y) \leqq \alpha\gamma + (1-\alpha)\gamma = \gamma.$$

9. If $\alpha > 0$, then $f(x) = p(\alpha x_0) = p(x)$. If $\alpha < 0$, then by Prob. 4,

$$f(x) = \alpha p(x_0) \leqq -\alpha p(-x_0) = p(\alpha x_0) = p(x).$$

Section 4.3

1. $p(0)=p(0x)=0p(x)=0$; this implies

$$0=p(0)=p(x+(-x))\leqq p(x)+p((-1)x)=2p(x).$$

7. $\tilde{f}(x)=\langle x, x_0\rangle/\|x_0\|$ by Riesz's theorem 3.8-1.

9. Extend f to the space $Z_1=\operatorname{span}(Z\cup\{y_1\})$, $y_1\in X-Z$, by setting $g_1(z+\alpha y_1)=f(z)+\alpha c$, determine c as in part (c) of the proof of Theorem 4.2-1 with p as in (9), this section. By countably many such steps we obtain an extension of f to a set which is dense in X, and Theorem 2.7-11 yields the result.

11. $f(x)-f(y)=f(x-y)=0$. Apply 4.3-4.

13. $\hat{f}=\|x_0\|^{-1}\tilde{f}$

15. $\|x_0\|>c$ would imply the existence of an $\tilde{f}\in X'$ such that $\|\tilde{f}\|=1$ and $\tilde{f}(x_0)=\|x_0\|>c$, by Theorem 4.3-3.

Section 4.5

3. $((S+T)^\times g)(x)=g((S+T)x)=g(Sx)+g(Tx)=(S^\times g)(x)+(T^\times g)(x)$

5. $((ST)^\times g)(x)=g(STx)=(S^\times g)(Tx)=(T^\times(S^\times g))(x)=(T^\times S^\times g)(x)$

7. $(AB)^\mathsf{T}=B^\mathsf{T}A^\mathsf{T}$

9.
$$\begin{aligned} g\in M^a \quad &\Longleftrightarrow \quad 0=g(Tx)=(T^\times g)(x) \text{ for all } x\in X\\ &\Longleftrightarrow \quad T^\times g=0\\ &\Longleftrightarrow \quad g\in \mathcal{N}(T^\times)\end{aligned}$$

Section 4.6

1. $x=(\xi_1,\cdots,\ \xi_n)$, $f(x)=\alpha_1\xi_1+\cdots+\alpha_n\xi_n$, $g_x(f)=\alpha_1\xi_1+\cdots+\alpha_n\xi_n$ (ξ_j fixed)

3. Let $h\in X'''$. For every $g\in X''$ there is an $x\in X$ such that $g=Cx$ since X is reflexive. Hence $h(g)=h(Cx)=f(x)$ defines a bounded linear functional f on X and $C_1 f = h$, where $C_1\colon X' \longrightarrow X'''$ is the canonical mapping. Hence C_1 is surjective, so that X' is reflexive.

5. $h = \delta^{-1}\tilde{f}$

7. If $Y \neq X$, there is an $x_0 \in X - Y$, and $\delta = \inf_{y \in Y} \|y - x_0\| > 0$ since Y is closed. By Lemma 4.6-7 there is an $\tilde{f} \in X'$ which is zero on Y but not zero at x_0, which contradicts our assumption.

9. If M is not total, $Y = \overline{\text{span } M} \neq X$ and Lemma 4.6-7 shows that there is an $\tilde{f} \in X'$ which is zero everywhere on Y, hence on M, but not zero at an $x_0 \in X - Y$. If M is total, then $Y = X$ and the condition in the problem is satisfied.

Section 4.7

1. (*a*) First, (*b*) first.

3. $\varnothing$, because every subset of X is open.

5. The closure of $(\bar{M})^C$ is all of X if and only if $\bar{M}$ has no interior points, so that every $x \in \bar{M}$ is a point of accumulation of $(\bar{M})^C$.

7. Immediate consequence of Theorem 4.7-3.

9. $\|x\|$; $\|T_n x\|^2 = |\xi_{2n+1}|^2 + |\xi_{2n+2}|^2 + \cdots \longrightarrow 0 \quad (n \longrightarrow \infty)$; 1; 1.

11. Use the fact that a Cauchy sequence is bounded (cf. Sec. 1.4) and apply Theorem 4.7-3.

13. Let us write $f(x_n) = g_n(f)$. Then $(g_n(f))$ is bounded for every f, so that $(\|g_n\|)$ is bounded by 4.7-3, and $\|x_n\| = \|g_n\|$ by 4.6-1.

15. $\dfrac{1}{2} + \dfrac{2}{\pi}\left(\sin t + \dfrac{1}{3}\sin 3t + \dfrac{1}{5}\sin 5t + \cdots\right)$

Section 4.8

1. A bounded linear functional on $C[a, b]$ is δ_{t_0} defined by $\delta_{t_0}(x) = x(t_0)$, where $t_0 \in [a, b]$, and $\delta_{t_0}(x_n) \longrightarrow \delta_{t_0}(x)$ means that $x_n(t_0) \longrightarrow x(t_0)$.

3. This follows from the linearity of the functionals on X.

5. Otherwise, the distance δ from x_0 to $\bar{Y}$ is positive. By 4.6-7 there is an $\tilde{f} \in X'$ such that $\tilde{f}(x_0) = \delta$ and $\tilde{f}(x) = 0$ for all $x \in \bar{Y}$. Hence $\tilde{f}(x_n) = 0$, so that $(\tilde{f}(x_n))$ does not converge to $\tilde{f}(x_0)$. But this contradicts $x_n \xrightarrow{w} x_0$.

7. Use Prob. 6.

9. Otherwise A would contain an unbounded sequence (x_n) such that $\lim \|x_n\| = \infty$. Then $\lim \|x_{n_j}\| = \infty$ for every subsequence (x_{n_j}) of (x_n), so that (x_n) has no weak Cauchy subsequences, by Prob. 8. This contradicts the assumptions.

Section 4.9

1. $\|T_n x - Tx\| = \|(T_n - T)x\| \leqq \|T_n - T\| \, \|x\| \longrightarrow 0$ as $n \longrightarrow \infty$.

3. This follows immediately from Theorem 4.8-4(a) applied to $y_n = T_n x$ and $y = Tx$ in place of x_n and x.

5. Since $x \in l^1$, the series $\sum |\xi_n|$ converges. Hence for every $x \in l^1$ we have $\xi_n = f_n(x) \longrightarrow 0$ as $n \longrightarrow \infty$. But $\|f_n\| = 1$.

7. By assumption, $(T_n x)$ converges for every $x \in X$. Hence $(\|T_n x\|)$ is bounded by 1.4-2, and $(\|T_n\|)$ is bounded; cf. 4.7-3.

9. $(\|T_n\|)$ is bounded. Since $\|T_n x\| \leqq \|T_n\| \, \|x\|$ and the norm is continuous,

$$\|Tx\| = \lim_{n \to \infty} \|T_n x\| \leqq \underline{\lim}_{n \to \infty} \|T_n\| \, \|x\|.$$

Section 4.10

1.

$$A = \begin{bmatrix} 1 & 0 & 0 & 0 & \cdots \\ \frac{1}{2} & \frac{1}{2} & 0 & 0 & \cdots \\ \frac{1}{3} & \frac{1}{3} & \frac{1}{3} & 0 & \cdots \\ \cdot & \cdot & \cdot & \cdot & \cdots \end{bmatrix}$$

3. $\xi_1 = \eta_1,\ \xi_n = n\eta_n - (n-1)\eta_{n-1}$; $(1, 0, 0, \cdots)$

5. H_1 gives $(1, -1, 1, -1, \cdots)$; H_2 gives $(1, 0, \frac{1}{3}, 0, \frac{1}{5}, 0, \cdots)$; hence the sequence is H_2-summable but not H_1-summable.

7. Use induction with respect to k.

Section 4.11

9. Formula (16) does not contain $x'''(0)$.

Section 4.12

1. T maps open balls onto open intervals, so that the statement follows from Sec. 1.3, Prob. 4.—No.

3. $\{\alpha, 2\alpha, 3\alpha, 4\alpha\}$, $\{1+w, 2+w, 3+w, 4+w\}$, $\{2, 3, \cdots, 8\}$

5. $\|T\| = 1$; $1 = \|x\| = \|T^{-1}y\| = k\,\|y\|$, where $x = (\delta_{kj})$ has 1 as the kth term and all other terms zero. Thus $\|T^{-1}\| \geqq k$. No, since X is not complete.

7. This follows from the bounded inverse theorem.

9. T: $X_2 \longrightarrow X_1$ defined by $x \longmapsto x$ is bijective and continuous since $\|x\|_1/\|x\|_2 \leqq c$, and T^{-1} is continuous by 4.12-2.

Section 4.13

5. T^{-1} is linear by 2.6-10. The graph of T^{-1} can be written $\mathscr{G}(T^{-1}) = \{(Tx, x) \mid x \in \mathscr{D}(T)\} \subset Y \times X$ and is closed since $\mathscr{G}(T) \subset X \times Y$ is closed and the mapping $X \times Y \longrightarrow Y \times X$ defined by $(x, y) \longmapsto (y, x)$ is isometric.

7. T: $X \longrightarrow Y$ is bounded and linear, and $\mathscr{D}(T) = X$ is closed. Hence T is closed by 4.13-5(*a*). Since T^{-1}: $Y \longrightarrow X$ exists by assumption, T^{-1} is closed (proof in the answer to Prob. 5) and is continuous by the closed graph theorem because $\mathscr{D}(T^{-1}) = Y$ is closed.

9. Any closed subset $K \subset Y$ is compact (Prob. 9, Sec. 2.5) and its inverse image is closed (Prob. 8). Hence T is continuous (Prob. 14, Sec. 1.3) and bounded by 2.7-9.

11. Use Theorem 4.13-3.

13. T^{-1} is closed (Prob. 5), hence $\mathcal{R}(T) = \mathcal{D}(T^{-1})$ is closed by 4.13-5(*b*).

15. (*a*) Since a linear operator maps 0 onto 0, the condition is necessary.
(*b*) Since $\mathcal{G}(T)$ is a vector space, so is $\overline{\mathcal{G}(T)}$. Suppose that $(x, y_1), (x, y_2) \in \overline{\mathcal{G}(T)}$. Then

$$(x, y_1) - (x, y_2) = (0, y_1 - y_2) \in \overline{\mathcal{G}(T)},$$

and $y_1 - y_2 = 0$ by that condition, so that $\tilde{T}$ is a mapping. Since $\overline{\mathcal{G}(T)}$ is a vector space, $\tilde{T}$ is linear. Since $\overline{\mathcal{G}(T)}$ is closed, $\tilde{T}$ is a closed linear operator.

Section 5.1

1. (*a*) Uniform dilatation, (*b*) reflection of the plane in a straight line, rotation about a fixed axis in space, projection of the plane onto any straight line, identity mapping.

5. The existence of two fixed points x and $y \neq x$ would imply the contradiction

$$d(x, y) = d(Tx, Ty) < d(x, y).$$

7. $d(Tx_1, Tx_2) < \varepsilon$ for $d(x_1, x_2) < \delta = \varepsilon/\alpha$.

11. By the mean value theorem of differential calculus,

$$|g(x) - g(y)| = |x - y|\,|g'(\xi)| \leqq \alpha\,|x - y|,$$

where ξ lies between x and y. Apply 5.1-4, use Prob. 9.

13. (*a*) $x_1 = 0.500$, $x_2 = 0.800$, $x_3 = 0.610$. Yes.
(*b*) $|g'(x)| \leqq 3\sqrt{3}/8 < 0.65 = \alpha$ (from $g''(x) = 0$ which gives $x = 1/\sqrt{3}$, where $|g'|$ has a maximum); this α yields the error bounds 0.93, 0.60, 0.39 (errors 0.18, 0.12, 0.07, respectively).

(*c*) The derivative of $1-x^3$ has absolute value greater than 1 near the root (0.682 328), so that we cannot expect convergence.

15. Since $f(\hat{x})=0$, the mean value theorem gives

$$|f(x)|=|f(x)-f(\hat{x})|=|f'(\xi)|\,|x-\hat{x}|\leqq k_1|x-\hat{x}| \qquad (k_1>0).$$

Since $\hat{x}$ is simple, $f'(x)\neq 0$ on a closed neighborhood N of $\hat{x}$, $N\subset[a, b]$, f'' is bounded on N, and for any $x\in N$,

$$|g'(x)|=\frac{|f(x)f''(x)|}{f'(x)^2}\leqq k_2|f(x)|\leqq k_1k_2|x-\hat{x}|<\tfrac{1}{2}$$

if $|x-\hat{x}|<1/2k_1k_2$.

17. True when $m=1$. Assuming the formula to hold for any $m\geqq 1$, we have

$$\begin{aligned} d(T^{m+1}x, S^{m+1}x) &\leqq d(TT^mx, TS^mx)+d(TS^mx, SS^mx)\\ &\leqq \alpha d(T^mx, S^mx)+\eta \leqq \alpha\eta\frac{1-\alpha^m}{1-\alpha}+\eta. \end{aligned}$$

19. The formula is an error estimate for y_m and follows from

$$\begin{aligned} d(x, y_m) &\leqq d(x, T^my_0)+d(T^my_0, S^my_0)\\ &\leqq \frac{\alpha^m}{1-\alpha}\,d(y_0, Ty_0)+\eta\frac{1-\alpha^m}{1-\alpha}\\ &\leqq \frac{\alpha^m}{1-\alpha}[d(y_0, Sy_0)+\eta]+\eta\frac{1-\alpha^m}{1-\alpha}. \end{aligned}$$

Section 5.2

3. (*a*) $\begin{bmatrix}1.00\\1.00\\0.75\\0.75\end{bmatrix}, \begin{bmatrix}0.9375\\0.9375\\0.6875\\0.6875\end{bmatrix}, \begin{bmatrix}0.90625\\0.90625\\0.65625\\0.65625\end{bmatrix}, \cdots$

(*b*) $\begin{bmatrix} 1.0000 \\ 1.0000 \\ 0.7500 \\ 0.6875 \end{bmatrix}, \quad \begin{bmatrix} 0.9375 \\ 0.9063 \\ 0.6563 \\ 0.6407 \end{bmatrix}, \quad \cdots$

5. To the two methods there correspond the two sequences

$$\begin{bmatrix} 0 \\ 0 \\ 0 \end{bmatrix}, \quad \begin{bmatrix} 2 \\ 2 \\ 2 \end{bmatrix}, \quad \begin{bmatrix} 0 \\ 0 \\ 0 \end{bmatrix}, \quad \begin{bmatrix} 2 \\ 2 \\ 2 \end{bmatrix}, \quad \cdots$$

and

$$\begin{bmatrix} 0 \\ 0 \\ 0 \end{bmatrix}, \quad \begin{bmatrix} 2.0 \\ 1.0 \\ 0.5 \end{bmatrix}, \quad \begin{bmatrix} 1.2500 \\ 1.1250 \\ 0.8125 \end{bmatrix}, \quad \begin{bmatrix} 1.031\ 2500 \\ 1.078\ 1250 \\ 0.945\ 3125 \end{bmatrix}, \quad \cdots$$

7.
$$\begin{aligned} d_1(Tx, Tz) &= \sum_{j=1}^{n} \left| \sum_{k=1}^{n} c_{jk}(\xi_k - \zeta_k) \right| \\ &\leqq \sum_{j=1}^{n} \sum_{k=1}^{n} |c_{jk}|\,|\xi_k - \zeta_k| \\ &\leqq \left(\max_{\kappa} \sum_{j=1}^{n} |c_{j\kappa}| \right) \sum_{k=1}^{n} |\xi_k - \zeta_k|. \end{aligned}$$

9. In (10) we have $D^{-1} = \operatorname{diag}(1/a_{jj})$.

Section 5.3

1. Apply the mean value theorem of differential calculus.

3. Not in regions which include points of the t-axis ($x = 0$).

5. By (2), the solution curve must lie between the two straight lines which pass through (t_0, x_0) and have slopes $-c$ and c; and for any $t \in [t_0 - a, t_0 + a]$ such that $|t - t_0| < b/c$ the curve cannot leave R. Furthermore, $\beta k = \alpha < 1$ implies that T is a contraction.

7. The proof in the text shows that for the new choice, T remains a contraction on $\tilde{C}$ into itself.

9. Not in a region containing $x=0$.

Section 5.4

1. We obtain

$$x_n(t)=v(t)+\mu k_0 e^t(1+\mu+\cdots+\mu^{n-1}),$$

$$x(t)=v(t)+\frac{\mu}{1-\mu}k_0 e^t, \quad k_0=\int_0^1 e^{-\tau}v(\tau)\,d\tau.$$

3. (*a*) A nonlinear Volterra equation

$$x(t)=x_0+\int_{t_0}^t f(\tau, x(\tau))\,d\tau.$$

(*b*) By two differentiations one can verify that the equation is

$$x(t)=\int_{t_0}^t (t-\tau)f(\tau, x(\tau))\,d\tau+(t-t_0)x_1+x_0.$$

5. (*a*) $x(t)=1+\mu+\mu^2+\cdots=1/(1-\mu)$.
(*b*) The integral is an unknown constant c. Hence $x(t)-\mu c=1$, $x(t)=1+\mu c$. Substitute this under the integral sign to get the value $c=1/(1-\mu)$.

9. $k_{(2)}=0$, $k_{(3)}=0, \cdots$, $x(t)=v(t)+\mu\int_0^{2\pi} k(t,\tau)v(\tau)\,d\tau$.

Section 6.2

1. If and only if $x\in Y$ (by 2.4-3).

3. This follows from the triangle inequality; in fact, letting $\beta=(\beta_1, \cdots, \beta_n)$ and using (2) in Sec. 2.2, we obtain

$$|f(\alpha)-f(\beta)|\leqq\left\|\sum(\beta_j-\alpha_j)e_j\right\|\leqq\max_k|\beta_k-\alpha_k|\sum\|e_j\|.$$

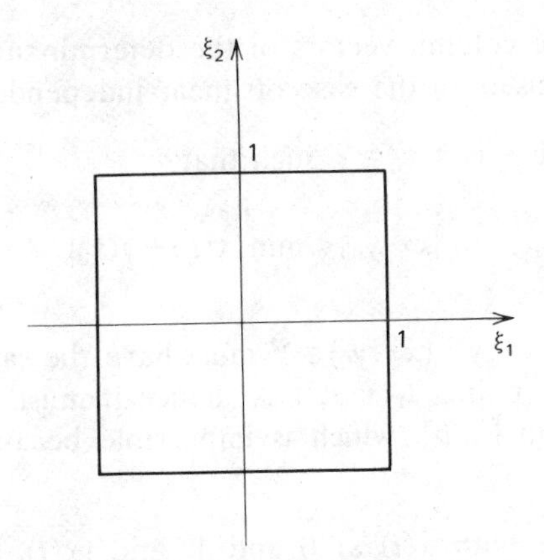

Fig. 91. Unit sphere in Prob. 7

5. $x=(1,0)$, $y=(0,1)$ give $\|x+y\|_1=2$.

7. $\|(1,0)+(1,1)\|=2=\|(1,0)\|+\|(1,1)\|$

9. (*a*) $(0,0)$, (*b*) the segment $\xi_1=0$, $-1\leqq\xi_2\leqq 1$, (*c*) $(0,0)$

11. This follows immediately from Lemma 6.2-1.

13. We set

$$x_1=\frac{1}{\|x\|}\,x, \qquad y_1=\frac{1}{\|y\|}\,y, \qquad \alpha=\frac{\|x\|}{\|x\|+\|y\|}.$$

Then $\|x_1\|=\|y_1\|=1$, and the given equality yields

$$1=\frac{\|x+y\|}{\|x\|+\|y\|}=\left\|\frac{x}{\|x\|+\|y\|}+\frac{y}{\|x\|+\|y\|}\right\|=\|\alpha x_1+(1-\alpha)y_1\|.$$

Since X is strictly convex, $x_1=y_1$ by Prob. 12, hence $x=cy$ where $c=\|x\|/\|y\|>0$.

15. This follows immediately from the fact that in the case of strict convexity the unit sphere does not contain a segment of a straight line.

Section 6.3

1. This follows from the statement involving (1), which expresses the linear independence of the n row vectors in (1).

3. These are the column vectors of the determinant in (1), which is not zero precisely in the case of linear independence.

5. Otherwise there is a $y_0 \in Y$ such that

$$\|x-y_0\| < \min_j |x(t_j)-y(t_j)|.$$

Then $y_0 - y = x - y - (x - y_0) \in Y$ must have the same sign as $x-y$ at those $n+1$ points $t_1, \cdots, t_{n+1}$; hence it must be zero at n or more points in $[a, b]$, which is impossible because of the Haar condition.

7. $\tilde{y}(t) = t$ agrees with $x(t)$ at 0 and 1, and $(x(t)-\tilde{y}(t))' = 0$ yields $\cos(\pi t/2) = 2/\pi$, $t = t_0 = (2/\pi)\operatorname{arc\,cos}(2/\pi) = 0.56$. Furthermore, $x(t_0) - \tilde{y}(t_0) = 0.211$, $y(t) = \tilde{y}(t) + 0.211/2$.

9. Regarding $\beta_1, \cdots, \beta_r$ as r values of a function x at $t_1, \cdots, t_r$ and $\gamma_{1k}, \cdots, \gamma_{rk}$ (k fixed) as values of a y_k at $t_1, \cdots, t_r$, we see that $\zeta_1, \cdots, \zeta_n$ correspond to $\alpha_1, \cdots, \alpha_n$ in $y = \sum \alpha_k y_k$ and (1) becomes

$$\begin{vmatrix} \gamma_{j_1 1} & \gamma_{j_2 1} & \cdots & \gamma_{j_n 1} \\ \gamma_{j_1 2} & \gamma_{j_2 2} & \cdots & \gamma_{j_n 2} \\ \cdot & \cdot & \cdots & \cdot \\ \gamma_{j_1 n} & \gamma_{j_2 n} & \cdots & \gamma_{j_n n} \end{vmatrix} \neq 0$$

where $\{j_1, \cdots, j_n\}$ is an n-tuple taken from $\{1, 2, \cdots, r\}$.

Section 6.4

1. $T_6(t) = 32t^6 - 48t^4 + 18t^2 - 1$

3. $\cos n\theta = 0$ in $[0, \pi]$ at $\theta_j = (2j-1)\pi/2n$, $j = 1, \cdots, n$ and $t = \cos\theta$, so that the zeros are $t = \cos[(2j-1)\pi/2n]$.

5. Otherwise T_{n-2} would be zero at the same point, by (10), and by repeating this conclusion we would arrive at $T_0(t) = 0$ for some t, which is impossible since $T_0(t) = 1$.

7. Let $v(\theta)=\cos n\theta$. Then $v''+n^2v=0$. Set $t=\cos\theta$.

9. Setting $t=\cos\theta$, we see that the integral becomes

$$\int_{\pi}^{0}\frac{1}{\sin\theta}\cos n\theta\cos m\theta\,(-\sin\theta)\,d\theta.$$

Section 6.5

3. Use Theorem 6.5-1 and the fact that a subset of a linearly independent set is linearly independent. Similarly,

$$G(y_1,\cdots,y_n,y_{n+1},\cdots,y_p)=0.$$

5. The inequality holds when $n=1$. Assuming that it holds for any n and using (5) with $x=y_{n+1}$, we obtain

$$G(y_1,\cdots,y_{n+1})=G(y_{n+1},y_1,\cdots,y_n)=\|z\|^2G(y_1,\cdots,y_n)\geqq 0.$$

The second statement now follows immediately from Theorem 6.5-1.

7. Use (5) and the following inequalities which are obvious because of the interpretation of z in Theorem 6.5-2:

$$\min_{\alpha}\|y_k-\alpha_{k+1}y_{k+1}-\cdots-\alpha_ny_n\|$$

$$\leqq\min_{\beta}\|y_k-\beta_{k+1}y_{k+1}-\cdots-\beta_my_m\|$$

and

$$\min_{\alpha}\|y_m-\alpha_{m+1}y_{m+1}-\cdots-\alpha_ny_n\|\leqq\|y_m\|.$$

9. The first statement follows immediately from Prob. 8. To get the second, take $y_j=(\alpha_{j1},\cdots,\alpha_{jn})$. Then $(\det A)^2=G(y_1,\cdots,y_n)$ follows from the familiar formula for representing the product of two determinants as a determinant. Also $\langle y_j,y_j\rangle=a_j$.

Section 6.6

1. $n+3$

3. We find

$$y(t)=\begin{cases} -2t^3-t^2 & \text{if } -1 \leqq t < 0 \\ 2t^3-t^2 & \text{if } \ 0 \leqq t \leqq 1. \end{cases}$$

5. 1/8 versus 1/16, but this does not contradict the minimum property of the Chebyshev polynomials since the spline function is not a polynomial.

7. $y(t)=-4t^3/\pi^3+3t/\pi$

9. The orthogonality $\langle y, x-y\rangle_2=0$ implies

$$p(x-y)^2=p(x)^2-p(y)^2 \geqq 0,$$

so that $p(x)^2 \geqq p(y)^2$, which is (6).

Section 7.1

1. 3, $\begin{bmatrix}1\\1\end{bmatrix}$, 9, $\begin{bmatrix}1\\4\end{bmatrix}$; $a+ib$, $\begin{bmatrix}1\\i\end{bmatrix}$, $a-ib$, $\begin{bmatrix}1\\-i\end{bmatrix}$

3. We have $Ax=\lambda x$ $(x\neq 0)$, $\bar{x}^{\mathsf{T}}Ax=\bar{x}^{\mathsf{T}}\lambda x=\lambda \bar{x}^{\mathsf{T}}x$; hence

$$\lambda=\frac{\bar{x}^{\mathsf{T}}Ax}{\bar{x}^{\mathsf{T}}x}$$

$\bar{x}^{\mathsf{T}}x$ is real. The numerator $N=\bar{x}^{\mathsf{T}}Ax$ is pure imaginary or zero since

$$\bar{N}=\bar{N}^{\mathsf{T}}=\overline{(\bar{x}^{\mathsf{T}}Ax)}^{\mathsf{T}}=(x^{\mathsf{T}}\bar{A}\bar{x})^{\mathsf{T}}=\bar{x}^{\mathsf{T}}\bar{A}^{\mathsf{T}}x=-\bar{x}^{\mathsf{T}}Ax=-N.$$

5. This follows from Probs. 2 and 4. See also 3.10-2.

7. A^{-1} exists if and only if $\det A\neq 0$, and $\det A$ is the product of the n eigenvalues of A since $\det A$ is the constant term of the characteristic polynomial, whose leading coefficient is $(-1)^n$. To get the second statement, premultiply $Ax_j = \lambda_j x_j$ by A^{-1}

9. Use induction and premultiply $A^{m-1}x_j = \lambda_j^{m-1}x_j$ by A to get

$$A^m x_j = \lambda_j^{m-1} A x_j = \lambda_j^{m-1} \lambda_j x_j.$$

11. Since $x_j = Cy_j$, we obtain

$$C^{-1}ACy_j = C^{-1}Ax_j = C^{-1}\lambda_j x_j = \lambda_j C^{-1} x_j = \lambda_j y_j.$$

13. $\lambda = 1$, algebraic multiplicity n, geometric multiplicity 1, eigenvector $(1 \quad 0 \quad 0 \quad \cdots \quad 0)^{\mathsf{T}}$.

15. $Tx = x' = \lambda x$, $\lambda = 0$, $x(t) = 1$ since $x(t) = e^{\lambda t}$ with $\lambda \neq 0$ does not define a polynomial; algebraic multiplicity n, geometric multiplicity 1.

Section 7.2

1. $\sigma(I) = \{1\} = \sigma_p(I)$, the eigenspace corresponding to 1 is X, and $R_\lambda(I) = (1-\lambda)^{-1}I$ is bounded for all $\lambda \neq 1$.

5. $Y_n = \operatorname{span}\{e_n, e_{n+1}, \cdots\}$.

7. Let $\lambda \in \sigma_r(T_1)$. Then $T_{1\lambda}^{-1}$ exists and its domain is not dense in X. Now $\mathfrak{D}(T_1) \supset \mathfrak{D}(T)$ implies $\mathfrak{D}(T_{1\lambda}) \supset \mathfrak{D}(T_\lambda)$ and $\mathfrak{R}(T_{1\lambda}) \supset \mathfrak{R}(T_\lambda)$, so th . $\mathfrak{R}(T_\lambda)$ cannot be dense in X, and $\lambda \in \sigma_r(T)$.

9. L.. $\lambda \in \rho(T_1)$. Then $T_{1\lambda}^{-1}$ exists and is bounded and $\mathfrak{R}(T_{1\lambda})$ is dense in X. Hence T_λ^{-1} exists and is bounded and its domain $\mathfrak{R}(T_\lambda) \subset \mathfrak{R}(T_{1\lambda})$ may be dense in X [then $\lambda \in \rho(T)$] or not [then $\lambda \in \sigma_r(T)$].

Section 7.3

1. $\sigma(T)$ is the range of v, which is a closed interval since v is continuous and has a maximum and a minimum on the compact set $[0, 1]$.

3. $\{\lambda\}$

5. $T_\lambda(l^2)$ is dense in l^2, hence $\lambda \notin \sigma_r(T)$, so that $\lambda \in \sigma_c(T)$.

7. Let $|\lambda| > \|T\|$ and $y = T_\lambda x$. Then

$$\|y\| = \|\lambda x - Tx\| \geqq |\lambda| \, \|x\| - \|Tx\| \geqq (|\lambda| - \|T\|) \, \|x\|;$$

hence

$$\|R_\lambda(T)\| = \sup_{y \neq 0} (\|x\|/\|y\|) \leqq 1/(|\lambda| - \|T\|).$$

9. (*a*) $\|T\| = 1$; use 7.3-4. (*b*) $T_\lambda x = (\xi_2 - \lambda\xi_1, \ \xi_3 - \lambda\xi_2, \cdots) = 0$, $Y = \{x \in X \mid x = (\alpha, \alpha\lambda, \alpha\lambda^2, \cdots), \alpha \in \mathbf{C}\}$.

Section 7.4

1. Take inverses on $(T - \lambda I)(T - \mu I) = (T - \mu I)(T - \lambda I)$.

3. Since $R_\lambda(S)S_\lambda = I$ and $T_\lambda R_\lambda(T) = I$, we obtain

$$\begin{aligned} R_\lambda(S)(T - S)R_\lambda(T) &= R_\lambda(S)(T_\lambda - S_\lambda)R_\lambda(T) \\ &= (R_\lambda(S)T_\lambda - I)R_\lambda(T) \\ &= R_\lambda(S) - R_\lambda(T). \end{aligned}$$

7. Use 7.4-2.

9. (*a*) For $|\lambda| > 1$ we obtain

$$R_\lambda(T) = -\lambda^{-1}[I - T + T(1 + \lambda^{-1} + \lambda^{-2} + \cdots)]$$

so that

$$R_\lambda(T) = -\lambda^{-1}(I - T) - (\lambda - 1)^{-1}T,$$

which also holds for any $\lambda \neq 0, 1$.

(*b*) $p(T) = T^2 - T = 0$, $p(\lambda) = \lambda^2 - \lambda = 0$.

Section 7.5

1. $\sigma(T) = \{0\}$ by (10).

3. $(1 - \lambda^2)^{-1}(A + \lambda I)$

5. From (10) we obtain

$$\begin{aligned} r_\sigma(ST) &= \lim \|(ST)^n\|^{1/n} = \lim \|S^n T^n\|^{1/n} \\ &\leqq \lim \|S^n\|^{1/n} \lim \|T^n\|^{1/n} = r_\sigma(S) r_\sigma(T). \end{aligned}$$

7. $\|T\|=2$, $\|T^2\|^{1/2}=\sqrt{2}$, $\|T^3\|^{1/3}=\sqrt[3]{4}$, etc.

9. Use (10) and Prob. 15 in Sec. 3.10.

Section 7.6

1. $\dim X<\infty$; cf. 2.4-2.

3. Use $\|x\|=\max(|\xi_1|, \cdots, |\xi_n|)$ and define multiplication by

$$(\xi_1, \cdots, \xi_n)(\eta_1, \cdots, \eta_n)=(\xi_1\eta_1, \cdots, \xi_n\eta_n).$$

9. $x^{-1}y=x^{-1}y(xx^{-1})=x^{-1}xyx^{-1}=yx^{-1}$

Section 7.7

1. Trivial consequence of Theorem 7.7-1.

3. Immediate consequence of Theorem 7.7-1.

7. Otherwise A would contain an x such that $x=\lambda e$ for no $\lambda \in \mathbf{C}$, so that $x-\lambda e \neq 0$ for all $\lambda \in \mathbf{C}$, and $\sigma(x)=\varnothing$, which contradicts Theorem 7.7-4.

9. Consider any $x \neq 0$. By assumption, $vx=e$ for some $v \in A$. Then $v \neq 0$ since otherwise $0=vx=e$. Set $xv=w$. Then $w \neq 0$ since otherwise

$$v=ev=vxv=vw=0.$$

By assumption $yw=e$ for some $y \in A$, that is, $yxv=e$. Hence v has a left inverse yx and a right inverse x. The two are equal (cf. Sec. 7.6, Prob. 8), $yx=x$. Since $yxv=e$ (see before), we have $xv=e$. Together with $vx=e$ this shows that the arbitrary $x \neq 0$ has an inverse.

Section 8.1

3. Consider any $T \in \overline{C(X, Y)}$. By 1.4-6(*a*) there is a sequence (T_n) in $C(X, Y)$ which converges to T in the norm on $B(X, Y)$. Hence T is compact by 8.1-5, that is, $T \in C(X, Y)$.

7. This follows from Theorem 8.1-3.

9. Cf. 8.1-4(*a*).

11. This follows from 8.1-4(*a*).

15. $\bar{A}$ is compact. $T(\bar{A})$ is compact (by 2.5-6) and closed (by 2.5-2). Hence $T(A) \subset T(\bar{A})$ implies $\overline{T(A)} \subset \overline{T(\bar{A})} = T(\bar{A})$, so that $\overline{T(A)}$ is compact (by Prob. 9, Sec. 2.5) and $T(A)$ is relatively compact.

Section 8.2

1. For a given $\varepsilon > 0$ the space X has an $\varepsilon/2$-net $M = \{x_1, \cdots, x_s\}$. Hence Y lies in the union of the s balls $B(x_1; \varepsilon/2), \cdots, B(x_s; \varepsilon/2)$. Since Y is infinite, one of the balls must contain an infinite subset Z of Y.

5. Since X is compact, it is totally bounded.

7. $Tx = (\eta_j) = (\xi_j/\sqrt{j})$ defines a compact linear operator, but $\sum\sum |\alpha_{jk}|^2 = \sum n^{-1}$ diverges.

9. Yes

Section 8.3

1. Since $S = T^p$ is compact, the statement holds for S. Now apply the spectral mapping theorem 7.4-2.

3. Use 8.3-1 and 8.3-3.

5. Let (x_n) be bounded, say, $\|x_n\| \leqq c$ for all n. Then (Sx_n) is bounded since

$$\|Sx_n\| \leqq \|S\| \, \|x_n\| \leqq \|S\| \, c.$$

Hence (Sx_n) contains a subsequence (Sx_{n_k}) such that (TSx_{n_k}) converges. This shows that TS is compact.

7. T^* is linear and bounded (cf. 3.9-2), TT^* is compact by Lemma 8.3-2, $TT^* = (T^*)^*T^*$, and T^* is compact by Prob. 6.

9. We write $\mathcal{N} = \mathcal{N}(T_\lambda)$. We assume that $\dim \mathcal{N} = \infty$. Then $\mathcal{N}$ has an infinite linearly independent subset, say, (x_n). Consider $K_m = \operatorname{span}\{x_1, \cdots, x_m\}$. Then $K_1 \subset K_2 \subset \cdots$ are closed subspaces of $\mathcal{N}$ and all these inclusions are proper. Let $y_1 = \|x_1\|^{-1}x_1$. By 2.5-4 (with $\theta = 1/2$) there is a $y_2 \in K_2$ such that $\|y_2\| = 1$, $\|y_2 - y_1\| \geqq 1/2$, and a $y_3 \in K_3$ such that $\|y_3\| = 1$ and $\|y_3 - y_2\| \geqq 1/2$ as well as $\|y_3 - y_1\| \geqq 1/2$, etc. This gives an infinite sequence (y_m) such that $\|y_m\| = 1$, $\|y_m - y_q\| \geqq 1/2$ if $m \neq q$. Hence

$$\text{(A)} \qquad \|\lambda y_m - \lambda y_q\| \geqq |\lambda|/2 \qquad (m \neq q).$$

Since $y_m \in \mathcal{N}$, we have $0 = T_\lambda y_m = (T - \lambda I)y_m$. Hence

$$\text{(B)} \qquad Ty_m = \lambda y_m.$$

Since $\lambda \neq 0$, relations (A) and (B) show that (Ty_m) has no convergent subsequence, but this is a contradiction since (y_m) is bounded and T is compact. Hence $\dim \mathcal{N} = \infty$ is impossible.

11. $T = I$ is not compact if $\dim X = \infty$ [cf. 8.1-2(*b*)] and we have $\dim \mathcal{N}(T_\lambda) = \dim X = \infty$ if $\lambda = 1$. The operator $T = 0$ is compact, but if $\dim X = \infty$ then for $\lambda = 0$ we have $\dim \mathcal{N}(T_\lambda) = \dim X = \infty$.

13. As in the answer to Prob. 9, if we assume that $\dim \mathcal{N}(T_\lambda{}^n) = \infty$, we now have from 2.5-4 a sequence (y_m), $\|y_m\| = 1$, $y_m \in \mathcal{N}(T_\lambda{}^n)$; also $\|y_m - y_q\| \geqq 1/2$ for $m \neq q$. Hence (see near the end of the proof of 8.3-4 in the text)

$$0 = T_\lambda{}^n y_m = (W - \mu I)y_m.$$

Since T^p is compact and S^p is bounded, it follows that the operator $W^p = (TS)^p = (ST)^p = S^pT^p$ is compact. Hence $(W^p y_m) = (\mu^p y_m)$ should have a convergent subsequence, but this is impossible since $\lambda \neq 0$, hence $\mu \neq 0$ and

$$\|\mu^p y_m - \mu^p y_q\| \geqq |\mu^p|/2 \qquad (m \neq q).$$

15. $\{x \mid \xi_{2k} = 0\}$ if $\lambda = 0$, $\{x \mid \xi_{2k-1} = 0\}$ if $\lambda = 1$, $\{0\}$ if $\lambda \neq 0, 1$. No.

Section 8.4

1. Apply T to (3) to get, for $n > m$,

$$T^2 y_n - T^2 y_m = \lambda^2 (y_n - x_2) \qquad x_2 \in \mathcal{N}_{n-1}$$

$$\cdots\cdots\cdots\cdots\cdots$$

$$T^p y_n - T^p y_m = \lambda^p (y_n - x_p) \qquad x_p \in \mathcal{N}_{n-1}$$

$$\|T^p y_n - T^p y_m\| \geqq |\lambda^p|/2,$$

etc.

3. $\lambda \in \rho(\tilde{T})$ implies $\lambda \in \rho(T) \cup \sigma_r(T)$; cf. Prob. 9, Sec. 7.2.

5. $Tx = \lambda x$, $0 = \lambda \xi_1$, $\xi_{n-1}/(n-1) = \lambda \xi_n$ $(n = 2, 3, \cdots)$, $x = 0$. Every $\lambda \neq 0$ is in $\rho(T)$. If $\lambda = 0$, then $\eta_1 = 0$, where $Tx = (\eta_j)$, $\overline{\mathcal{R}(T)} \neq l^2$, $0 \notin \sigma_c(T)$, hence $0 \in \sigma_r(T)$ since $\sigma_p(T) = \varnothing$.

7. Every α_j is an eigenvalue of T. Apply 8.3-1.

9. If $\lambda \notin [0, 1]$, then $T_\lambda^{-1} x(t) = x(t)/(t - \lambda)$; $\sigma(T) = [0, 1]$. Use 8.3-1 and 8.4-4.

Section 8.6

3. We have $\sum_k \alpha_{jk} \xi_k = \eta_j$. Let $f = (\varphi_1, \cdots, \varphi_n)$ be such that

$$\sum_k \alpha_{kj} \varphi_k = 0.$$

Multiplying the first formula by φ_j and summing, we get

$$\sum \sum \alpha_{jk} \xi_k \varphi_j = \sum \sum \alpha_{kj} \varphi_k \xi_j = \sum \varphi_j \eta_j = f(y) = 0.$$

5. Let $A = T - \lambda I$. Then (1) becomes $Ax = y$, which has a solution x if and only if any $w = (\omega_j)$ satisfying $\sum_j \alpha_{jk} \omega_j = 0$, $k = 1, \cdots, n$, also satisfies $\sum_j \eta_j \omega_j = 0$. Using dot products and the column vectors

$a_1, \cdots, a_n$ of A, we see that the condition becomes

$$w \cdot a_k = 0 \quad (k = 1, \cdots, n) \qquad \Longrightarrow \qquad w \cdot y = 0;$$

that is, any vector w which is orthogonal to all column vectors of A is also orthogonal to all column vectors of the augmented matrix, so that the two matrices have the same rank.

11. $z_1, z_2, \cdots; y_1, y_2, \cdots; \langle z_k, y_j \rangle = \delta_{kj}$, as a consequence of Riesz's theorem 3.8-1.

13. The two systems

$$\sum_k \alpha_{jk}\xi_k = 0 \qquad (j = 1, \cdots, n)$$

and

$$\sum_j \alpha_{jk}\eta_j = 0 \qquad (k = 1, \cdots, n)$$

have the same number of linearly independent solutions (namely, only the trivial solution if $r = \text{rank } A = n$ and $n - r$ linearly independent solutions if $r < n$).

15. The given series converges uniformly, and by termwise integration we obtain a Fourier series representation for Tx, and $Tx = 0$ if and only if $x = 0$ since $Tx \in C[0, \pi]$. But $Tx = y$ is not solvable if $y(0) \neq 0$ since each term of that series is zero at $s = 0$.

Section 8.7

1. In this case, T in (2) is an n-rowed square matrix and x and y are column vectors. Either the nonhomogeneous system has a unique solution for every given vector y on the right or the corresponding homogeneous system has at least one nontrivial solution. In the first case the same holds for the transposed system. In the second case the homogeneous system has the same number $(n - r)$ linearly independent solutions as its transposed system, where r is the rank of the coefficient matrix.

3. For instance, $k(s, t) = 1$ if $s < 1/2$ and 0 if $s \geqq 1/2$, $s, t \in [0, 1]$. Hence T is not a mapping into $C[0, 1]$.

5. Note that the integral is an unknown constant c. Hence $x(s)=1+\mu c$. Substitution in the given equation gives $c=1/(1-\mu)$, $\mu\neq 1$, and $x(s)=1/(1-\mu)$, $\mu\neq 1$. The Neumann series is the geometric series

$$x(s)=1+\mu+\mu^2+\cdots=1/(1-\mu), \qquad (|\mu|<1).$$

For the homogeneous equation we obtain $x(s)=0$ if $\mu\neq 1$, $x(s)=c$ (arbitrary) if $\mu=1$. This agrees with 8.7-3.

9. $k_{(2)}=0,\ k_{(3)}=0, \cdots,\ x(s)=\tilde{y}(s)+\mu\int_0^{2\pi} k(s,t)\tilde{y}(t)\,dt.$

11. $\lambda=1/\mu=e^2-1$; eigenfunction e^s.

13. (x_n) on $[-1,1]$, where $x_n(t)=|t|^{1/n}$; the convergence cannot be uniform since the limit function is not continuous. Another example is (x_n) on $[0,1]$, where $x_n(t)=t^n$.

15. (*a*) We obtain

$$\left(1-\frac{1}{2}\mu\right)c_1-\mu c_2=y_1, \qquad -\frac{1}{3}\mu c_1+\left(1-\frac{1}{2}\mu\right)c_2=y_2,$$

$$x(s)=\tilde{y}(s)+\mu\int_0^1\frac{6(\mu-2)(s+t)-12\mu st-4\mu}{\mu^2+12\mu-12}\tilde{y}(t)\,dt.$$

(*b*) The eigenvalues and eigenfunctions are

$$\mu_1=-6+4\sqrt{3}, \qquad \mu_2=-6-4\sqrt{3}, \qquad \lambda_1=1/\mu_1, \qquad \lambda_2=1/\mu_2,$$

$$x(s)=\frac{2\mu}{2-\mu}s+1 \qquad (\mu=\mu_1,\mu_2).$$

Section 9.1

1. If A is an n-rowed Hermitian matrix, $\bar{x}^{\mathsf{T}}Ax$ has a real value for every $x\in\mathbf{C}^n$, cf. 3.10-2. A Hermitian matrix has real eigenvalues, and to different eigenvalues there correspond orthogonal eigenvectors.

3. Writing $T_\lambda x = y$, we have $\|x\| = \|R_\lambda y\| \leqq c^{-1}\|y\|$.

5. $\langle W^*TWx, y\rangle = \langle TWx, Wy\rangle = \langle Wx, TWy\rangle = \langle x, W^*TWy\rangle$

7. $Tx = \lambda_j x$, where $x = (\xi_n)$, $\xi_n = \delta_{nj}$. $\sigma(T) \supset [a, b]$ if (λ_j) is dense on $[a, b]$; here we use that $\sigma(T)$ is closed; cf. 7.3-2.

9. Self-adjointness of $T|_X$ follows by noting that t is real; and for T on $L^2[0, 1]$, self-adjointness also follows from the integral representation of the inner product, where the integral now is a Lebesgue integral. $R_\lambda(T)x(t) = (t-\lambda)^{-1}x(t)$ shows that we have $\sigma(T) = [0, 1]$, and for $\lambda \in [0, 1]$ we see that

$$T_\lambda x(t) = (t-\lambda)x(t) = 0$$

implies $x(t) = 0$ for all $t \neq \lambda$, that is, $x = 0$ (the zero element in $L^2[0, 1]$), so that λ cannot be an eigenvalue of T.

Section 9.2

3. $m = 0$, $M = 1$

5. This follows immediately from Theorem 9.2-3.

7. Eigenvalues $1, 1/2, 1/3, \cdots$, and $\sigma(T) = \sigma_p(T) \cup \{0\}$. Since $Tx = 0$ implies $x = 0$, we have $0 \notin \sigma_p(T)$. Since T is self-adjoint, we see that $\sigma_c(T) = \{0\}$ follows from Theorem 9.2-4.

9. The first statement follows from Theorems 9.2-1 and 9.2-3, and implies the second statement since A maps Y_j into itself.

Section 9.3

1. $0 \leqq \langle (T-S)x, x\rangle$, $0 \leqq \langle (S-T)x, x\rangle$, hence $\langle (T-S)x, x\rangle = 0$ for all x, and $T - S = 0$ by Lemma 3.9-3(*b*).

3. $S = B - A \geqq 0$, $ST = TS$, and Theorem 9.3-1 implies $ST \geqq 0$; this yields the result.

5. This follows from Theorems 9.2-1 and 9.2-3.—Let A be an n-rowed Hermitian matrix (cf. Sec. 3.10). Then $\bar{x}^\mathsf{T} Ax \geqq 0$ for all $x \in \mathbf{C}^n$ if and only if all eigenvalues of A are nonnegative.

7. Self-adjointness is obvious from

$$\langle T_1^2 T_2 x, y\rangle = \langle x, T_2 T_1^2 y\rangle = \langle x, T_1^2 T_2 y\rangle.$$

Writing $y = T_1 x$, we obtain

$$\langle T_1^2 T_2 x, x\rangle = \langle T_2 T_1 x, T_1 x\rangle = \langle T_2 y, y\rangle \geqq 0.$$

(Note that the result also follows from Prob. 6.)

9. Let $(I+T)x = 0$. Then $-x = Tx$ and, since $T \geqq 0$,

$$0 \leqq \langle Tx, x\rangle = -\langle x, x\rangle = -\|x\|^2 \leqq 0$$

which implies $x = 0$, so that $(I+T)^{-1}$ exists; cf. 2.6-10.

13. This follows from Prob. 12 and Theorem 9.2-1.

15. $\langle Tx, Tx\rangle \geqq c^2\langle x, x\rangle$, $T^*T \geqq c^2 I$, T^*T is not compact (by 8.1-2(*b*) and Prob. 14), and T is not compact (cf. Prob. 6, Sec. 8.3).

Section 9.4

1. For instance, the operators represented by the following matrices where a_{12} and a_{21} are arbitrary. $I^{1/2} = I$.

$$\begin{bmatrix} 1 & 0 \\ 0 & 1 \end{bmatrix}, \quad \begin{bmatrix} -1 & 0 \\ 0 & -1 \end{bmatrix}, \quad \begin{bmatrix} 1 & 0 \\ a_{21} & -1 \end{bmatrix}, \quad \begin{bmatrix} -1 & a_{12} \\ 0 & 1 \end{bmatrix}.$$

3. Yes. Yes. Yes. $Ax = (0, \quad 0, \quad \xi_3, \quad \xi_4, \quad \cdots)$.

5. Since $T = T^{1/2}T^{1/2}$ and $T^{1/2}$ is self-adjoint,

$$\begin{aligned} |\langle Tx, y\rangle| &= |\langle T^{1/2}x, T^{1/2}y\rangle| \leqq \|T^{1/2}x\|\,\|T^{1/2}y\| \\ &= \langle T^{1/2}x, T^{1/2}x\rangle^{1/2}\langle T^{1/2}y, T^{1/2}y\rangle^{1/2} \\ &= \langle Tx, x\rangle^{1/2}\langle Ty, y\rangle^{1/2}. \end{aligned}$$

7. If $Tx = 0$, that inequality holds. Let $Tx \neq 0$. Writing $y = Tx$, we obtain

$$\|Tx\|^2 \leqq \langle Tx, x\rangle^{1/2}\langle T^2x, Tx\rangle^{1/2}.$$

Since

$$\langle T^2x, Tx\rangle \leqq \|T^2x\| \, \|Tx\| \leqq \|T\| \, \|Tx\|^2,$$

we have

$$\|Tx\|^2 \leqq \langle Tx, x\rangle^{1/2} \|T\|^{1/2} \|Tx\|$$

and division by $\|Tx\|$ yields the result.

9. $DD^\mathsf{T} = D^\mathsf{T}D = I$

Section 9.5

1. Use Theorem 9.5-2. Clearly, $P = 0$ if P projects onto $\{0\}$, and $P = I$ if P projects onto H.

3. For instance, T represented by the following matrix, where a_{21} is arbitrary, not zero.

$$\begin{bmatrix} 1 & 0 \\ a_{21} & 0 \end{bmatrix}$$

5. If the spaces $Y_j = P_j(H)$, $j = 1, \cdots, m$, are orthogonal in pairs, P is a projection, as follows by induction. Conversely, if P is a projection, then

$$\|Px\|^2 = \langle P^2x, x\rangle = \langle Px, x\rangle, \qquad \|P_kx\|^2 = \langle P_kx, x\rangle,$$

hence for all x,

$$\|P_1x\|^2 + \|P_2x\|^2 \leqq \sum_{k=1}^{m} \langle P_kx, x\rangle = \langle Px, x\rangle = \|Px\|^2 \leqq \|x\|^2.$$

For every y and $x = P_1y$ we thus have $P_1x = P_1{}^2y = P_1y$ and

$$\|P_1y\|^2 + \|P_2P_1y\|^2 \leqq \|x\|^2 = \|P_1y\|^2$$

so that $P_2P_1y = 0$, that is, $P_2P_1 = 0$, and $Y_1 \perp Y_2$ by 9.5-3. Similarly, $Y_j \perp Y_k$ for all j and $k \neq j$.

9. Let (e_k) be an orthonormal sequence in an inner product space X. Then P_k defined by $P_k x = \langle x, e_k \rangle e_k$ is the projection onto the space $Y_k = P_k(X)$. From Theorem 9.5-4 we see that $P_1 + \cdots + P_n$ is a projection. Since

$$\|P_k x\|^2 = |\langle x, e_k \rangle|^2 \|e_k\|^2 = |\langle x, e_k \rangle|^2,$$

Prob. 8 yields (12*), Sec. 3.4, which implies (12) in 3.4-6.

Section 9.6

3. $(P_2 - P_1)x = ([\xi_1 - \xi_2]/2, [\xi_2 - \xi_1]/2, 0)$. No (cf. 9.5-4).

5. For instance, let P_n be the projection of l^2 onto the subspace consisting of all sequences $x = (\xi_j)$ such that $\xi_j = 0$ for all $j > n$.

7. $P(H) = \bigcap_{n=1}^{\infty} P_n(H)$

9. $T^*(Y^\perp) \subset Y^\perp$ if and only if $Y^{\perp\perp} \supset (T^*)^*(Y^{\perp\perp})$ by Prob. 5, Sec. 3.9, $Y^{\perp\perp} = Y$ by (8), Sec. 3.3, and $(T^*)^* = T$ by 3.9-4.

11. If $\dim Y = r$, $\dim Y^\perp = n - r$ and $Y = \operatorname{span}\{e_1, \cdots, e_r\}$, where $(e_1, \cdots, e_n)$ is a basis for H, then the matrix has all zeros in the intersection of the first r rows and the last $n - r$ columns as well as in the intersection of the last $n - r$ rows and first r columns.

13. We obtain

$$TP_2 = T(I - P_1) = T - TP_1 = T - P_1 T = (I - P_1)T = P_2 T.$$

15. Let $y \in Y$ and $z \in Y^\perp$. Then $Ty \in Y$ by assumption, and $Tz \in Y^\perp$ follows from the self-adjointness since

$$\langle Tz, y \rangle = \langle z, Ty \rangle = 0.$$

Section 9.8

1. $F_\lambda = E_{\lambda - 0}$

5. (*a*) Replace all negative elements by zeros.

(*b*) Replace all positive elements by zeros and omit the minus signs of the negative elements.
(*c*) Omit the minus signs of the negative elements.

7. Diagonal matrix with principal diagonal elements (*a*) $t_{jj}-\lambda$ (t_{jj} the principal diagonal elements of $\tilde{T}$), (*b*) $\max(t_{jj}-\lambda, 0)$, (*c*) $\max(-t_{jj}+\lambda, 0)$, (*d*) $|t_{jj}-\lambda|$.

9. $E_\lambda = 0$ if $\lambda < 0$, and $E_\lambda = I$ if $\lambda \geqq 0$.

Section 9.9

1. $E_\lambda = 0$ if $\lambda < 0$, $E_\lambda = I$ if $\lambda \geqq 0$,

$$T = \int_{0-0}^{0} \lambda \, dE_\lambda = 0(E_0 - E_{0-0}) = 0(I-0) = 0.$$

3. $E_\lambda = 0$ if $\lambda < 1$, $E_\lambda = I$ if $\lambda \geqq 1$. Hence

$$T = \int_{1-0}^{1} \lambda \, dE_\lambda = 1(E_1 - E_{1-0}) = 1(I-0) = I.$$

5. E_λ is the projection onto the sum of the eigenspaces of all those eigenvalues of the matrix which do not exceed λ.

9. Let $x = (\xi_j) \in l^2$. Then

$$\left\| \left(T - \sum_{j=1}^{m} \frac{1}{j} P_j \right) x \right\|^2 = \left\| \sum_{j=m+1}^{\infty} \frac{1}{j} \xi_j e_j \right\|^2 = \sum_{j=m+1}^{\infty} \frac{1}{j^2} |\xi_j|^2$$

$$\leqq \frac{1}{(m+1)^2} \sum_{j=m+1}^{\infty} |\xi_j|^2 \leqq \frac{1}{(m+1)^2} \|x\|^2,$$

so that

$$\left\| T - \sum_{j=1}^{m} \frac{1}{j} P_j \right\| \leqq \frac{1}{m+1} \longrightarrow 0 \qquad (m \longrightarrow \infty).$$

Section 9.11

1. $E_\lambda = 0$ for $\lambda < \lambda_1$ (the smallest eigenvalue), E_λ has "jumps" precisely at the eigenvalues and reaches I when $\lambda = \lambda_n$ (the largest eigenvalue). Of course, this merely confirms our consideration at the beginning of Sec. 9.7.

3. $\lambda \longmapsto E_\lambda$ is continuous ($\sigma_p(T)=\varnothing$), is constant for $\lambda<0$ and $\lambda \geqq 1$ and nonconstant on $[0,1]=\sigma(T)=\sigma_c(T)$.

5. If a real λ_0 is in $\rho(T)$, Theorem 9.11-2 implies that $\lambda \in \rho(T)$ for all $\lambda \in \mathbf{R}$ and sufficiently close to λ_0. Hence $\rho(T) \cap \mathbf{R}$ is an open subset of $\mathbf{R}$ and its complement $\sigma(T)$ on $\mathbf{R}$ is closed.

7. T in Prob. 7, Sec. 9.2, is an example.

9. Let Y be the closure of the span of all eigenvectors of T. Then $T_1 = T|_Y$ has a pure point spectrum and $T_1(Y) \subset Y$. Also T_1 is self-adjoint on Y. Similarly, $T_2 = T|_Z$ is self-adjoint on $Z = Y^\perp$ and has a purely continuous spectrum, as follows from the construction of Y.

Section 10.1

5. $\mathfrak{D}(S+T)$ dense in H.

7. Extend T to $\overline{\mathfrak{D}(T)}$ by Theorem 2.7-11. Extend the resulting operator $\tilde{T}$ to H, for instance, by setting $\hat{T}x=0$ for $x \in \overline{\mathfrak{D}(T)}^\perp$.

9. Use the idea of the proof of the Hellinger-Toeplitz theorem.

Section 10.2

3. Use the idea of the proof of 3.10-3.

5. $T \subset T^{**}$ by 10.2-1(b), and T^{**} is bounded, by Prob. 9, Sec. 10.1. Hence T is bounded.

7. Use 2.7-11. Symmetry of $\tilde{T}$ follows from that of T and the continuity of the inner product (cf. 3.2-2).

9. Let S be a symmetric extension of T. Then

$$T \subset S \subset S^* \subset T^* = T$$

(cf. 10.2-1(a)); hence $T = S$.

Section 10.3

1. This follows from Theorem 10.3-2(*a*) since, for instance,

$$x_n = \left(1, \frac{1}{4}, \frac{1}{9}, \cdots, \frac{1}{n^2}, 0, 0, \cdots\right) \longrightarrow x = (1/j^2) \notin \mathfrak{D}(T),$$

$$Tx_n = \left(1, \frac{1}{2}, \frac{1}{3}, \cdots, \frac{1}{n}, 0, 0, \cdots\right) \longrightarrow y = (1/j) \in l^2.$$

3. Let (w_n) be Cauchy in $H \times H$, where $w_n = (x_n, y_n)$. Then

$$\|w_n - w_m\|^2 = \|x_n - x_m\|^2 + \|y_n - y_m\|^2$$

shows that (x_n) and (y_n) are Cauchy in H, hence $x_n \longrightarrow x$ and $y_n \longrightarrow y$. Then $w_n \longrightarrow w = (x, y)$.

5. We have $T_1 = S^{-1}$, where $S\colon l^2 \longrightarrow l^2$ is defined by $Sx = (\xi_j/j)$. Clearly, S is bounded. Since $\mathfrak{D}(S) = l^2$ is closed, S is closed by 10.3-2(*c*) and $S^{-1} = T_1$ is closed by Prob. 4.

7. Let $(x_0, y_0) \in [U(\mathscr{G}(T))]^\perp$. Then for all $x \in \mathfrak{D}(T)$,

$$0 = \langle (x_0, y_0), (Tx, -x)\rangle = \langle x_0, Tx\rangle + \langle y_0, -x\rangle,$$

that is, $\langle Tx, x_0\rangle = \langle x, y_0\rangle$. Hence $x_0 \in \mathfrak{D}(T^*)$ and $y_0 = T^* x_0$, so that $(x_0, y_0) \in \mathscr{G}(T^*)$. Conversely, starting from $(x_0, y_0) \in \mathscr{G}(T^*)$ and going backward, we arrive at $(x_0, y_0) \in [U(\mathscr{G}(T))]^\perp$.

9. T^* is bounded by Prob. 9, Sec. 10.1, and closed by 10.3-3. Hence $\mathfrak{D}(T^*)$ is closed by 10.3-2(*c*) and dense by Prob. 8, so that $\mathfrak{D}(T^*) = H$, and T^{**} is bounded by Prob. 9, Sec. 10.1, applied to T^* instead of T. And $T^{**} = T$ by Prob. 8.

Section 10.4

1. The proof is literally the same as that of Theorem 9.1-1(*a*).

3. This follows from 10.4-1.

5. T_λ^{-1} exists and $\overline{\mathfrak{D}(T_\lambda^{-1})} \neq H$. Hence there is a $y \neq 0$ such that for all $x \in \mathfrak{D}(T_\lambda) = \mathfrak{D}(T)$

$$0 = \langle T_\lambda x, y\rangle = \langle Tx, y\rangle - \langle x, \bar{\lambda} y\rangle,$$

which shows that $T^* y = \bar{\lambda} y$.

7. Let $\lambda \in \sigma_r(T)$. Then $\bar{\lambda} \in \sigma_p(T^*)$ by Prob. 5, which, by 10.4-2 and $T = T^*$, implies that $\lambda \in \sigma_p(T)$, a contradiction.

9. $Tx = \lambda x$ with $x \neq 0$ implies

$$\lambda\langle x, x\rangle = \langle \lambda x, x\rangle = \langle Tx, x\rangle = \langle x, Tx\rangle = \bar{\lambda}\langle x, x\rangle,$$

hence $\lambda = \bar{\lambda}$. Eigenvectors corresponding to different eigenvalues are orthogonal; this follows as in Theorem 9.1-1(*b*). Countability of $\sigma_p(T)$ now results from Theorem 3.6-4(*a*).

Section 10.5

1. $Ux_1 = \lambda_1 x_1$, $Ux_2 = \lambda_2 x_2$, and

$$\langle x_1, x_2\rangle = \langle Ux_1, Ux_2\rangle = \lambda_1 \bar{\lambda}_2 \langle x_1, x_2\rangle,$$

hence $\langle x_1, x_2\rangle = 0$ since $\lambda_1 \bar{\lambda}_2 \neq 1$.

5. If λ is an eigenvalue, T_λ has no inverse at all and vice versa. Let λ be an approximate eigenvalue and $\|T_\lambda x_n\| \longrightarrow 0$, $\|x_n\| = 1$, and suppose that T_λ^{-1} exists. Then

$$y_n = \|T_\lambda x_n\|^{-1} T_\lambda x_n \in \mathcal{R}(T_\lambda) = \mathcal{D}(T_\lambda^{-1}),$$

$\|y_n\| = 1$ and

$$\|T_\lambda^{-1} y_n\| = \|T_\lambda x_n\|^{-1} \|x_n\| = \|T_\lambda x_n\|^{-1} \longrightarrow \infty,$$

which shows that T_λ^{-1} is unbounded.

Conversely, if T_λ^{-1} is unbounded, there is a sequence (y_n) in $\mathcal{D}(T_\lambda^{-1})$ such that $\|y_n\| = 1$ and $\|T_\lambda^{-1} y_n\| \longrightarrow \infty$. Taking $x_n = \|T_\lambda^{-1} y_n\|^{-1} T_\lambda^{-1} y_n$, we have $\|x_n\| = 1$ and

$$\|T_\lambda x_n\| = \|T_\lambda^{-1} y_n\|^{-1} \|y_n\| = \|T_\lambda^{-1} y_n\|^{-1} \longrightarrow 0.$$

7. $Tx - \lambda x = (-\lambda \xi_1, \xi_1 - \lambda \xi_2, \xi_2 - \lambda \xi_3, \cdots) = 0$ implies $x = 0$.

9. This follows directly from the definition.

Section 10.6

1. $t = i(1+u)/(1-u)$

3. For every $x \in H$ we have $(T+iI)^{-1}x \in \mathcal{D}(T)$, hence

$$S(T+iI)^{-1}x \in \mathcal{D}(T)$$

and

$$(T+iI)S(T+iI)^{-1}x = S(T+iI)(T+iI)^{-1}x = Sx$$
$$S(T+iI)^{-1}x = (T+iI)^{-1}Sx$$
$$SUx = (T-iI)(T+iI)^{-1}Sx = USx.$$

5. For every $x \in \mathcal{D}(T)$, since T is symmetric,

$$\begin{aligned}\|(T \pm iI)x\|^2 &= \|Tx\|^2 \pm \langle Tx, ix\rangle \pm \langle ix, Tx\rangle + \|x\|^2 \\ &= \|Tx\|^2 + \|x\|^2 \geqq \|x\|^2.\end{aligned} \tag{A}$$

Hence $(T+iI)x = 0$ implies $x = 0$ and $(T+iI)^{-1}$ exists by 2.6-10 and is bounded by (A). We set $y = (T+iI)x$ and use (1) and (A) to get

$$\begin{aligned}\|Uy\|^2 = \|(T-iI)x\|^2 &= \|Tx\|^2 + \|x\|^2 \\ &= \|(T+iI)x\|^2 = \|y\|^2.\end{aligned}$$

7. $\mathcal{D}(U)$ is closed by Prob. 6 and 4.13-5(*b*), so that $\mathcal{R}(U)$ is closed since U is isometric by Prob. 5.

Section 11.2

1. $\pi^{-1/4}$

3. $T - cI = T - \mu I + (\mu - c)I$ implies

$$\begin{aligned}E_\psi([T-cI]^2) &= \mathrm{var}_\psi(T) + 2(\mu - c)E_\psi(T-\mu I) + (\mu - c)^2 \\ &\geqq \mathrm{var}_\psi(T)\end{aligned}$$

since $E_\psi(T - \mu I) = 0$; here $\mu = \mu_\psi(T)$.

5. By (2), Sec. 11.1, and (4), this section,

$$1=\int \psi\bar{\psi}\,dq=\int \bar{\psi}\int \frac{1}{\sqrt{h}}\,\varphi e^{(2\pi i/h)pq}\,dp\,dq$$

$$=\int \varphi\int \frac{1}{\sqrt{h}}\,\bar{\psi} e^{(2\pi i/h)pq}\,dq\,dp.$$

The integral over q is $\overline{\varphi(p)}$, by (5).

7. For instance,

$$(D_1Q_2-Q_2D_1)\psi=\frac{\partial}{\partial q_1}(q_2\psi)-q_2\frac{\partial}{\partial q_1}\psi=0.$$

9. The $\mathcal{M}_j$'s are suggested by the form of the components of the vector product. The two other relations are

$$\mathcal{M}_2\mathcal{M}_3-\mathcal{M}_3\mathcal{M}_2=\frac{ih}{2\pi}\mathcal{M}_1$$

$$\mathcal{M}_3\mathcal{M}_1-\mathcal{M}_1\mathcal{M}_3=\frac{ih}{2\pi}\mathcal{M}_2.$$

They follow from Prob. 7 by straightforward calculation, or by cyclic permutation of the subscripts in the first commutation relation.

Section 11.3

1. $q=\pm\sqrt{2E/m\omega_0{}^2}$ are the points of maximum displacement, where $E=V$; hence $E_{\text{kin}}=0$.

3. $\psi_0''+(1-s^2)\psi_0=0$; ψ_0 corresponds to $\tilde{\lambda}=1$ in (11); this is compatible with the fact that $H_0(s)=1$.

5. From the recursion formula,

$$\frac{\alpha_{m+2}}{\alpha_m}\sim\frac{2}{m}.$$

Writing

$$e^{s^2} = 1 + \beta_2 s^2 + \cdots + \beta_m s^m + \beta_{m+2} s^{m+2} + \cdots$$

we have for even m

$$\frac{\beta_{m+2}}{\beta_m} = \frac{1\Big/\left(\frac{m}{2}+1\right)!}{1\Big/\left(\frac{m}{2}\right)!} = \frac{2}{m+2}$$

which shows that if the series does not terminate, the corresponding solution grows about as fast as $\exp(s^2)$ if $|s|$ is large, so that ψ grows about as fast as $\exp(s^2/2)$.

Section 11.4

1. We have

$$\Delta\eta + \frac{8\pi^2 m}{h^2} E\eta = 0,$$

$$\eta(q) = \eta_1(q_1)\eta_2(q_2)\eta_3(q_3), \qquad E = A_1 + A_2 + A_3,$$

$$\frac{\eta_1''}{\eta_1} + \frac{\eta_2''}{\eta_2} + \frac{\eta_3''}{\eta_3} + \frac{8\pi^2 m}{h^2}(A_1 + A_2 + A_3) = 0,$$

$$\eta_j'' + \frac{8\pi^2 m}{h^2} A_j \eta_j = 0,$$

etc.

3. By Prob. 2,

$$\psi_{n+1} = \frac{1}{\sqrt{(n+1)!}} A^*(A^{*n}\psi_0) = \frac{1}{\sqrt{(n+1)!}} A^* \sqrt{n!}\,\psi_n.$$

Also, by (15),

$$A\psi_n = \frac{1}{\sqrt{n}} AA^* \psi_{n-1}$$

$$= \frac{1}{\sqrt{n}} (A^*A + \tilde{I})\psi_{n-1} = \frac{1}{\sqrt{n}} (n - 1 + 1)\psi_{n-1}.$$

5. True for $s=1$ since, by (13) to (15),

$$AQ-QA=\frac{1}{2\alpha\beta}[A(A+A^*)-(A+A^*)A]$$

$$=\frac{1}{2\alpha\beta}[AA^*-A^*A]=\sqrt{\frac{h}{4\pi m\omega_0}}\,\tilde{I}.$$

We make the induction hypothesis that the formula holds for any fixed s. Application of Q from the left and from the right gives

$$QAQ^s-Q^{s+1}A=\sqrt{\frac{h}{4\pi m\omega_0}}\,sQ^s$$

$$AQ^{s+1}-Q^sAQ=\sqrt{\frac{h}{4\pi m\omega_0}}\,sQ^s.$$

By addition,

$$AQ^{s+1}-Q^{s+1}A+Q(AQ^{s-1}-Q^{s-1}A)Q=\sqrt{\frac{h}{4\pi m\omega_0}}\,2sQ^s$$

where

$$AQ^{s-1}-Q^{s-1}A=\sqrt{\frac{h}{4\pi m\omega_0}}\,(s-1)Q^{s-2}.$$

7. By (2), Sec. 11.3, we may write $\Psi(q,t)=\psi(q)e^{-i\omega t}$, where

$$\psi(q)=\psi_1(q)=A_1e^{ib_1q}+B_1e^{-ib_1q} \qquad (q<0),$$

$$\psi(q)=\psi_2(q)=A_2e^{ib_2q}+B_2e^{-ib_2q} \qquad (q\geqq 0).$$

The first term in ψ_1 represents the incident wave and we may choose $A_1=1$. The second term in ψ_1 represents a reflected wave. The first term in ψ_2 represents a transmitted wave, and $B_2=0$ since there is no wave incident from the right, by assumption. The Schrödinger equation shows that the discontinuity in the potential causes ψ'' to be discontinuous at $q=0$, and the continuity of ψ and

ψ' at 0 gives the two conditions

$$1+B_1=A_2$$
$$ib_1(1-B_1)=ib_2A_2.$$

Hence

$$B_1=\frac{b_1-b_2}{b_1+b_2}, \qquad A_2=\frac{2b_1}{b_1+b_2}.$$

Note that $b_2^2>0$ so that b_2 is real and the transmitted wave is sinusoidal.

9. $b_2^2<0$, $b_2=i\beta_2$ (β_2 real positive). ψ_1 remains sinusoidal but

$$\psi_2(q)=A_2e^{-\beta_2 q}$$

(exponential decay). The wave penetrates into the region $q>0$ where the classical particle cannot go.

Section 11.5

3. $|m|\leqq l$ since P_l is of degree l and z must not be identically zero.

9. $u(\rho)=\rho^{-1/2}v(\rho)$ gives $v''+v=0$, which yields $J_{1/2}$ and $J_{-1/2}$. Solutions which are finite at 0 are

$$(l=0) \qquad J_{1/2}(\rho)=\sqrt{\frac{2}{\pi\rho}}\sin\rho$$

$$(l=1) \qquad J_{3/2}(\rho)=\sqrt{\frac{2}{\pi\rho}}\left(\frac{\sin\rho}{\rho}-\cos\rho\right)$$

$$(l=2) \qquad J_{5/2}(\rho)=\sqrt{\frac{2}{\pi\rho}}\left(3\frac{\sin\rho}{\rho^2}-3\frac{\cos\rho}{\rho}-\sin\rho\right),$$

etc.

APPENDIX 3
REFERENCES

Banach, S. (1922), Sur les opérations dans les ensembles abstraits et leur application aux équations intégrales. *Fundamenta Math.* **3,** 133–181

Banach, S. (1929), Sur les fonctionnelles linéaires II. *Studia Math.* **1,** 223–239

Banach, S. (1932), *Théorie des opérations linéaires.* New York: Chelsea

Banach, S., et H. Steinhaus (1927), Sur le principe de la condensation de singularités. *Fundamenta Math.* **9,** 50–61

Berberian, S. (1961), *Introduction to Hilbert Space.* New York: Oxford University Press

Bernstein, S. N. (1912), Démonstration du théorème de Weierstrass fondée sur le calcul des probabilités. *Comm. Soc. Math. Kharkow* **13,** 1–2

Bielicki, A. (1956), Une remarque sur la méthode de Banach-Cacciopoli-Tikhonov. *Bull. Acad. Polon. Sci.* **4,** 261–268

Birkhoff, G. (1967), *Lattice Theory.* 3rd ed. Amer. Math. Soc. Coll. Publ. **25.** Providence, R. I.: American Mathematical Society

Birkhoff, G., and S. Mac Lane (1965), *A Survey of Modern Algebra.* 3rd. ed. New York: Macmillan

Bohnenblust, H. F., and A. Sobczyk (1938), Extensions of functionals on complex linear spaces. *Bull. Amer. Math. Soc.* **44,** 91–93

Bourbaki, N. (1955), *Éléments de mathématique,* livre V. *Espaces vectoriels topologiques.* Chap. III à V. Paris: Hermann

Bourbaki, N. (1970), *Éleménts de mathématique, Algèbre.* Chap. 1 à 3. Paris: Hermann

Cheney, E. W. (1966), *Introduction to Approximation Theory.* New York: McGraw-Hill

Churchill, R. V. (1963), *Fourier Series and Boundary Value Problems.* 2nd ed. New York: McGraw-Hill

Courant, R., and D. Hilbert (1953–62), *Methods of Mathematical Physics.* 2 vols. New York: Interscience/Wiley

Cramér, H. (1955), *The Elements of Probability Theory and Some of its Applications.* New York: Wiley

Day, M. M. (1973), *Normed Linear Spaces.* 3rd ed. New York: Springer

Dieudonné, J. (1960), *Foundations of Modern Analysis.* New York: Academic Press

Dixmier, J. (1953), Sur les bases orthonormales dans les espaces préhilbertiens. *Acta Math. Szeged* **15,** 29–30

Dunford, N., and J. T. Schwartz (1958–71), *Linear Operators.* 3 parts. New York: Interscience/Wiley

Edwards, R. E. (1965), *Functional Analysis.* New York: Holt, Rinehart and Winston

Enflo, P. (1973), A counterexample to the approximation property. *Acta Math.* **130,** 309–317

Erdélyi, A., W. Magnus, F. Oberhettinger and F. G. Tricomi (1953–55), *Higher Transcendental Functions.* 3 vols. New York: McGraw-Hill

Fejér, L. (1910), Beispiele stetiger Funktionen mit divergenter Fourierreihe. *Journal Reine Angew. Math.* **137,** 1–5

Fréchet, M. (1906), Sur quelques points du calcul fonctionnel. *Rend. Circ. Mat. Palermo* **22,** 1–74

Fredholm, I. (1903), Sur une classe d'équations fonctionnelles. *Acta Math.* **27,** 365–390

Friedrichs, K. (1935), Beiträge zur Theorie der Spektralschar. *Math. Annalen* **110,** 54–62

Gantmacher, F. R. (1960), *The Theory of Matrices.* 2 vols. New York: Chelsea

Gelfand, I. (1941), Normierte Ringe. *Mat. Sbornik* (*Recueil mathématique*) N. S. **9,** (51), 3–24

Gram, J. P. (1883), Ueber die Entwickelung reeller Functionen in Reihen mittelst der Methode der kleinsten Quadrate. *Journal Reine Angew. Math.* **94,** 41–73

Haar, A. (1918), Die Minkowskische Geometrie und die Annäherung an stetige Funktionen. *Math. Annalen* **78,** 294–311

Hahn, H. (1922), Über Folgen linearer Operationen. *Monatshefte Math. Phys.* **32,** 3–88

Hahn, H. (1927), Über lineare Gleichungssysteme in linearen Räumen. *Journal Reine Angew. Math.* **157,** 214–229

Halmos, P. R. (1958), *Finite-Dimensional Vector Spaces.* 2nd ed. New York: Van Nostrand Reinhold

Hamming, R. W. (1950), Error detecting and error correcting codes. *Bell System Tech. Journal* **29,** 147–160

Hellinger, E., und O. Toeplitz (1910), Grundlagen für eine Theorie der unendlichen Matrizen. *Math. Annalen* **69,** 289–330

Helmberg, G. (1969), *Introduction to Spectral Theory in Hilbert Space.* New York: American Elsevier

Hewitt, E., and K. Stromberg (1969), *Real and Abstract Analysis.* Berlin: Springer

Hilbert, D. (1912), *Grundzüge einer allgemeinen Theorie der linearen Integralgleichungen.* Repr. 1953. New York: Chelsea

Hille, E. (1973), *Analytic Function Theory.* Vol. I. 2nd ed. New York: Chelsea

Hille, E., and R. S. Phillips (1957), *Functional Analysis and Semi-Groups.* Amer. Math. Soc. Coll. Publ. **31.** Rev. ed. Providence, R. I.: American Mathematical Society

Hölder, O. (1889), Über einen Mittelwertsatz. *Nachr. Akad. Wiss. Göttingen. Math.-Phys. Kl.,* 38–47

Ince, E. L. (1956), *Ordinary Differential Equations.* New York: Dover

James, R. C. (1950), Bases and reflexivity of Banach spaces. *Annals of Math.* (2) **52,** 518–527

James, R. C. (1951), A non-reflexive Banach space isometric with its second conjugate space. *Proc. Nat. Acad. Sci. U.S.A.* **37,** 174–177

Kelley, J. L. (1955), *General Topology.* New York: Van Nostrand

Kelley, J. L., and I. Namioka (1963), *Linear Topological Spaces.* New York: Van Nostrand

Kreyszig, E. (1970), *Introductory Mathematical Statistics.* New York: Wiley

Kreyszig, E. (1972), *Advanced Engineering Mathematics.* 3rd ed. New York: Wiley

Lebesgue, H. (1909), Sur les intégrales singulières, *Ann. de Toulouse* (3) **1,** 25–117

Lorch, E. R. (1939), On a calculus of operators in reflexive vector spaces. *Trans. Amer. Math. Soc.* **45,** 217–234

Lorch, E. R. (1962), *Spectral Theory.* New York: Oxford University Press

Löwig, H. (1934), Komplexe euklidische Räume von beliebiger endlicher oder transfiniter Dimensionszahl. *Acta Sci. Math. Szeged* **7,** 1–33

McShane, E. J. (1944), *Integration.* Princeton, N. J.: Princeton University Press

Merzbacher, E. (1970), *Quantum Mechanics.* 2nd ed. New York: Wiley

Minkowski, H. (1896), *Geometrie der Zahlen.* Leipzig: Teubner

Murray, F. J. (1937), On complementary manifolds and projections in spaces L_p and l_p. *Trans. Amer. Math. Soc.* **41,** 138–152

Naimark, M. A. (1972), *Normed Algebras.* 2nd ed. Groningen: Wolters-Noordhoff

Neumann, J. von (1927), Mathematische Begründung der Quantenmechanik. *Nachr. Ges. Wiss. Göttingen. Math.-Phys. Kl.*, 1–57

Neumann, J. von (1929–30), Allgemeine Eigenwerttheorie Hermitescher Funktionaloperatoren. *Math. Annalen* **102,** 49–131

Neumann, J. von (1929–30b), Zur Algebra der Funktionaloperationen und Theorie der normalen Operatoren. *Math. Annalen* **102,** 370–427

Neumann, J. von (1936), Über adjungierte Funktionaloperatoren. *Annals of Math.* (2) **33,** 294–310

Poincaré, H. (1896), La méthode de Neumann et le problème de Dirichlet. *Acta Math.* **20,** 59–142

Pólya, G. (1933), Über die Konvergenz von Quadraturverfahren. *Math. Zeitschr.* **37,** 264–286

Rellich, F. (1934), Spektraltheorie in nichtseparablen Räumen. *Math. Annalen* **110,** 342–356

Riesz, F. (1909), Sur les opérations fonctionnelles linéaires. *Comptes Rendus Acad. Sci. Paris* **149,** 974–977

Riesz, F. (1918), Über lineare Funktionalgleichungen. *Acta Math.* **41,** 71–98

Riesz, F. (1934), Zur Theorie des Hilbertschen Raumes. *Acta Sci. Math. Szeged* **7,** 34–38

Riesz, F., and B. Sz.-Nagy (1955), *Functional Analysis.* New York; Ungar

Rogosinski, W. (1959), *Fourier Series.* 2nd ed. New York: Chelsea

Royden, H. L. (1968), *Real Analysis.* 2nd ed. New York: Macmillan

Sard, A., and S. Weintraub (1971), *A Book of Splines.* New York: Wiley

Schauder, J. (1930), Über lineare, vollstetige Funktionaloperationen. *Studia Math.* **2,** 1–6

Schiff, L. I. (1968), *Quantum Mechanics.* 3rd ed. New York: McGraw-Hill

Schmidt, E. (1907), Entwicklung willkürlicher Funktionen nach Systemen vorgeschriebener. *Math. Annalen* **63,** 433–476

Schmidt, E. (1908), Über die Auflösung linearer Gleichungen mit unendlich vielen Unbekannten. *Rend. Circ. Mat. Palermo* **25,** 53–77

Schur, I. (1921), Über lineare Transformationen in der Theorie der unendlichen Reihen. *Journal Reine Angew. Math.* **151,** 79–111

Sobczyk, A. (1941), Projections in Minkowski and Banach spaces. *Duke Math. Journal* **8,** 78–106

Stone, M. H. (1932), *Linear Transformations in Hilbert Space and their Applications to Analysis.* Amer. Math. Soc. Coll. Publ. **15.** New York: American Mathematical Society

Szegö, G. (1967), *Orthogonal Polynomials.* 3rd ed. Amer. Math. Soc. Coll. Publ. **23.** Providence, R. I.: American Mathematical Society

Taylor, A. E. (1958), *Introduction to Functional Analysis.* New York: Wiley

Todd, J. (1962), *Survey of Numerical Analysis.* New York: McGraw-Hill

Wecken, F. J. (1935), Zur Theorie linearer Operatoren. *Math. Annalen* **110,** 722–725

Weierstrass, K. (1885), Über die analytische Darstellbarkeit sogenannter willkürlicher Functionen reeller Argumente. *Sitzungsber. Kgl. Preuss. Akad. Wiss. Berlin,* 633–639, 789–805

Wiener, N. (1922), Limit in terms of continuous transformation. *Bull. Soc. Math. France* (2) **50,** 119–134

Wilks, S. S. (1962), *Mathematical Statistics.* New York: Wiley

Wintner, A. (1929), Zur Theorie der beschränkten Bilinearformen. *Math. Zeitschr.* **30,** 228–282

Yosida, K. (1971), *Functional Analysis.* 3rd ed. Berlin: Springer

Zaanen, A. C. (1964), *Linear Analysis.* Amsterdam: North-Holland Publ.

Zakon, E. (1973), *Mathematical Analysis.* Part II. Lecture Notes. Department of Mathematics, University of Windsor, Windsor, Ont.

INDEX

Notations are listed at the beginning of the book, after the table of contents.

D